Communications in Computer and Information Science   1875

Rationale

The CCIS series is devoted to the publication of proceedings of computer science conferences. Its aim is to efficiently disseminate original research results in informatics in printed and electronic form. While the focus is on publication of peer-reviewed full papers presenting mature work, inclusion of reviewed short papers reporting on work in progress is welcome, too. Besides globally relevant meetings with internationally representative program committees guaranteeing a strict peer-reviewing and paper selection process, conferences run by societies or of high regional or national relevance are also considered for publication.

Topics

The topical scope of CCIS spans the entire spectrum of informatics ranging from foundational topics in the theory of computing to information and communications science and technology and a broad variety of interdisciplinary application fields.

Information for Volume Editors and Authors

Publication in CCIS is free of charge. No royalties are paid, however, we offer registered conference participants temporary free access to the online version of the conference proceedings on SpringerLink (http://link.springer.com) by means of an http referrer from the conference website and/or a number of complimentary printed copies, as specified in the official acceptance email of the event.

CCIS proceedings can be published in time for distribution at conferences or as postproceedings, and delivered in the form of printed books and/or electronically as USBs and/or e-content licenses for accessing proceedings at SpringerLink. Furthermore, CCIS proceedings are included in the CCIS electronic book series hosted in the SpringerLink digital library at http://link.springer.com/bookseries/7899. Conferences publishing in CCIS are allowed to use Online Conference Service (OCS) for managing the whole proceedings lifecycle (from submission and reviewing to preparing for publication) free of charge.

Publication process

The language of publication is exclusively English. Authors publishing in CCIS have to sign the Springer CCIS copyright transfer form, however, they are free to use their material published in CCIS for substantially changed, more elaborate subsequent publications elsewhere. For the preparation of the camera-ready papers/files, authors have to strictly adhere to the Springer CCIS Authors' Instructions and are strongly encouraged to use the CCIS LaTeX style files or templates.

Abstracting/Indexing

CCIS is abstracted/indexed in DBLP, Google Scholar, EI-Compendex, Mathematical Reviews, SCImago, Scopus. CCIS volumes are also submitted for the inclusion in ISI Proceedings.

How to start

To start the evaluation of your proposal for inclusion in the CCIS series, please send an e-mail to ccis@springer.com.

Donatello Conte · Ana Fred · Oleg Gusikhin ·
Carlo Sansone
Editors

Deep Learning Theory and Applications

4th International Conference, DeLTA 2023
Rome, Italy, July 13–14, 2023
Proceedings

 Springer

Editors
Donatello Conte
Université de Tours
Tours, France

Oleg Gusikhin
Ford Motor Company
Commerce Township, MI, USA

Ana Fred
Instituto de Telecomunicações and University
of Lisbon
Lisbon, Portugal

Carlo Sansone
University of Naples Federico II
Naples, Italy

ISSN 1865-0929 ISSN 1865-0937 (electronic)
Communications in Computer and Information Science
ISBN 978-3-031-39058-6 ISBN 978-3-031-39059-3 (eBook)
https://doi.org/10.1007/978-3-031-39059-3

This Springer imprint is published by the registered company Springer Nature Switzerland AG
The registered company address is: Gewerbestrasse 11, 6330 Cham, Switzerland

Preface

This book contains the proceedings of the 4th International Conference on Deep Learning Theory and Applications (DeLTA 2023), held in Rome, Italy from 13 to 14 July 2023. This conference was sponsored by the Institute for Systems and Technologies of Information, Control and Communication (INSTICC), endorsed by the International Association for Pattern Recognition (IAPR), and held in cooperation with the ACM Special Interest Group on Artificial Intelligence (ACM SIGAI), the International Neural Network Society (INNS), the Società Italiana di Reti Neuroniche (SIREN), and the European Society for Fuzzy Logic and Technology (EUSFLAT).

The Conference Program included oral presentations (full papers and short papers) and posters, organized in five simultaneous tracks: "Models and Algorithms", "Machine Learning", " Big Data Analytics", " Computer Vision Applications" and "Natural Language Understanding". We are proud to announce that the program also included two plenary keynote lectures, given by internationally distinguished researchers, namely: Davide Bacciu (University of Pisa, Italy) and Luís Paulo Reis (University of Porto, Portugal).

DeLTA received 42 paper submissions from 20 countries. To evaluate each submission, a double-blind paper review was performed by the Program Committee, whose members are highly qualified researchers in DeLTA topic areas. Based on the classifications provided, 9 papers were selected as full papers, which led to a full paper acceptance ratio of 21%. These strict acceptance ratios show the intention to preserve a high-quality forum, which we expect to develop further next year.

The organization distributed three paper awards at the conference closing session. A "Best Paper Award", a "Best Student Paper Award", and a "Best Poster Award" were conferred on the author(s) of the best papers presented at the conference, selected by the Program/Conference Chairs based on the best combined marks from the paper reviews, assessed by the Program Committee, and paper presentation quality, assessed by session chairs at the conference venue.

We would like to express our thanks to all participants. First of all, to the authors, whose quality work is the essence of this conference; secondly to all members of the Program Committee and auxiliary reviewers, who helped us with their expertise and valuable time. We would also like to deeply thank the invited speakers for their excellent contribution in sharing their knowledge and vision. Finally, words of appreciation for the hard work of the secretariat: organizing a conference of this level is a task that can only be achieved by the collaborative effort of a dedicated and highly capable team.

July 2023

Donatello Conte
Ana Fred
Oleg Gusikhin
Carlo Sansone

Organization

Conference Co-chairs

Oleg Gusikhin — Ford Motor Company, USA
Carlo Sansone — University of Naples Federico II, Italy

Program Co-chairs

Donatello Conte — Université de Tours, France
Ana Fred — Instituto de Telecomunicações and University of Lisbon, Portugal

Program Committee

Marco Aceves-Fernández — Autonomous University of Queretaro, Mexico
Abdul Bais — University of Regina, Canada
Marco Buzzelli — University of Milano-Bicocca, Italy
Claudio Cusano — University of Pavia, Italy
Arturo de la Escalera — Universidad Carlos III De Madrid, Spain
Gilles Guillot — CSL Behring / Swiss Institute for Translational and Entrepreneurial Medicine, Switzerland
Rakesh Kumar — SRI International, USA
Chih-Chin Lai — National University of Kaohsiung, Taiwan, Republic of China
Chang-Hsing Lee — Chung Hua University, Taiwan, Republic of China
Marco Leo — National Research Council of Italy, Italy
Fuhai Li — Washington University Saint Louis, USA
Yung-Hui Li — National Central University, Taiwan, Republic of China
Marc Masana Castrillo — Computer Vision Center, Spain
Perry Moerland — Amsterdam UMC, University of Amsterdam, The Netherlands
Parma Nand — Auckland University of Technology, New Zealand
Aline Paes — Universidade Federal Fluminense, Brazil
Stefania Perri — University of Calabria, Italy
Mircea-Bogdan Radac — Politehnica University of Timisoara, Romania

Sivaramakrishnan Rajaraman	National Library of Medicine, USA
Jitae Shin	Sungkyunkwan University, South Korea
Sunghwan Sohn	Mayo Clinic, USA
Minghe Sun	University of Texas at San Antonio, USA
Ron Sun	Rensselaer Polytechnic Institute, USA
Ryszard Tadeusiewicz	AGH University of Science and Technology, Poland
F. Boray Tek	Istanbul Technical University, Turkey
Jayaraman Valadi	Shiv Nadar University, India
Theodore Willke	Intel Corporation, USA
Jianhua Xuan	Virginia Tech, USA

Invited Speakers

| Davide Bacciu | University of Pisa, Italy |
| Luís Paulo Reis | University of Porto, Portugal |

Invited Talks

Pervasive AI: (Deep) Learning into the Wild

Davide Bacciu

University of Pisa, Italy

Abstract. The deployment of intelligent applications in real-world settings poses significant challenges which span from the AI methodology itself to the computing, communication and orchestration support needed to execute it. Such challenges can inspire compelling research questions which can drive and foster novel developments and research directions within the deep learning field. The talk will explore some current research directions in the context of pervasive deep learning, touching upon efficient neural networks, non-dissipative neural propagation, neural computing on dynamical systems and continual learning.

Brief Biography

Davide Bacciu is Associate Professor at the Department of Computer Science, University of Pisa, where he is the founder and head of the Pervasive Artificial Intelligence Laboratory. He holds a Ph.D. in Computer Science and Engineering from the IMT Lucca Institute for Advanced Studies, for which he has been awarded the 2009 E.R. Caianiello prize for the best Italian Ph.D. thesis on neural networks. He has co-authored over 160 research works on (deep) neural networks, generative learning, Bayesian models, learning for graphs, continual learning, and distributed and embedded learning systems. He is the coordinator of two EC-funded projects on embedded, distributed and lifelong learning AI, and of several national/industrial projects. He is the chair of the IEEE CIS Neural Network Technical Committee and a Vice President of the Italian Association for AI

Deep Reinforcement Learning to Improve Traditional Supervised Learning Methodologies

Luís Paulo Reis

Faculty of Engineering/LIACC, University of Porto, Porto, Portugal

Abstract. This talk focuses on the intersection of Deep Reinforcement Learning (DRL) and traditional Supervised Learning (SL) methodologies, exploring how DRL can enhance performance and overcome challenges in tasks typically approached via SL. Despite the success of SL in various domains, its limitations, including the inability to handle sequential decision-making and non-stationary environments, are obvious, making DRL a potentially useful tool. The talk will outline the fundamental principles of DRL, including its distinguishing features, such as learning from delayed rewards, handling the exploration-exploitation trade-off, and operating in complex, dynamic environments. It will also focus on the integration of DRL into traditionally SL-dominated areas, providing real-world examples from several fields. The talk will discuss how DRL can automate and optimise processes within the machine learning pipeline that have traditionally been manual and heuristic, such as hyperparameter tuning and feature engineering. By using DRL, the talk will showcase how these processes can be transformed into learnable tasks, improving the efficiency and performance of the supervised learning system. The talk will also present the latest research and techniques on the incorporation of DRL into traditionally SL-focused domains and feature interesting examples from several projects developed at the University of Porto on these areas of DRL and DRL for SL, such as the DRL methodologies included in our RoboCup world champion team in the humanoid 3D Simulation League

Brief Biography

Luis Paulo Reis is an Associate Professor at the University of Porto in Portugal and Director of LIACC – Artificial Intelligence and Computer Science Laboratory. He is an IEEE Senior Member, and he is the President of APPIA - Portuguese Association for Artificial Intelligence. He is also Co-Director of LIACD - First Degree in Artificial Intelligence and Data Science. During the last 25 years, he has lectured courses, at the University, on Artificial Intelligence, Intelligent Robotics, Multi-Agent Systems, Simulation and Modelling, Games and Interaction, Educational/Serious Games and Computer Programming. He was the principal investigator of more than 30 research projects in those areas. He won more than 60 scientific awards including winning more than 15 RoboCup international competitions and best papers at conferences such as ICEIS, Robotica, IEEE ICARSC and ICAART. He supervised 22 PhD and 150 MSc theses to completion and is supervising 12 PhD theses. He was a plenary speaker at several international conferences, organised more than 60 international scientific events and belonged to the Program Committee of more than 250 scientific events. He is the author of more than 400 publications in international conferences and journals (indexed at SCOPUS or ISI Web of Knowledge).

Contents

Synthetic Network Traffic Data Generation and Classification of Advanced Persistent Threat Samples: A Case Study with GANs and XGBoost

T. J. Anande and M. S. Leeson$^{(\boxtimes)}$ (iD)

School of Engineering, University of Warwick, Coventry CV4 7AL, UK
{Tertsegha-Joseph.Anande,mark.leeson}@warwick.ac.uk

Abstract. The need to develop more efficient network traffic data generation techniques that can reproduce the intricate features of traffic flows forms a central element in secured monitoring systems for networks and cybersecurity. This study investigates selected Generative Adversarial Network (GAN) architectures to generate realistic network traffic samples. It incorporates Extreme Gradient Boosting (XGBoost), an Ensemble Machine Learning algorithm effectively used for the classification and detection of observed and unobserved Advanced Persistent Threat (APT) attack samples in the synthetic and new data distributions. Results show that the Wasserstein GAN architectures achieve optimal generation with a sustained Earth Mover distance estimation of 10^{-3} between the Critic loss and the Generator loss compared to the vanilla GAN architecture. Performance statistics using XGBoost and other evaluation metrics indicate successful generation and detection with an accuracy of 99.97% a recall rate of 99.94%, and 100% precision. Further results show a 99.97% f_1 score for detecting APT samples in the synthetic data, and a Receiver Operator Characteristic Area Under the Curve (ROC_AUC) value of 1.0, indicating optimum behavior, surpassing previous state-of-the-art methods. When evaluated on unseen data, the proposed approach maintains optimal detection performance with 100% recall, 100% Area Under the Curve (AUC) and precision above 90%.

Keywords: Advanced persistent threats (APTs) · Generative adversarial networks (GANs) · Binary cross-entropy (BCE) · Wasserstein loss · eXtreme gradient boosting (XGBoost)

1 Introduction

The modern data-driven world has increasing needs for security given data volume and sensitivity. The evolution of networking and cloud computing has enabled many systems to migrate to online platforms. However, this has increased complexity and vulnerability in the face of continually evolving cyber-attacks. Such attacks include Advanced Persistent Threats (APTs) that stealthily gain unauthorized access to networks and cyber

Supported by the University of Warwick.

systems, remaining undetected for extended periods and possess significant sophistication and resources [4]. They are specifically designed to target hosts and include system security architecture to evade detection to stealthily exploit various vulnerabilities over long periods, using a wide range of deployment methods and dynamic attack patterns [2]. The fact that they mimic the legitimate behavior of network users and gather information via stealthy means, with very sophisticated methods, makes them even more difficult to avoid or exterminate [12]. Growth in the number of these attacks and their increasing complexity necessitates techniques to expose and combat them across all phases of the APT attack lifecycle [18]. However, it is difficult to obtain access to realistic data sets since these are often commercially sensitive or contain restricted information and this has limited efforts in the area.

Deep Learning (DL) has been extensively utilized for generative tasks that augment insufficient real data with new synthetic data [38]. Generative Adversarial Networks (GANs) have led this field, offering the capability to learn and reproduce intricate data distributions from noise inputs [5]. Thus, prospects for DL in security applications are excellent for, inter alia, the developing Internet of Things (IoT), Distributed and Cyber-Physical systems and Intrusion Detection Systems (IDSs) [47]. Thus, we investigate several GAN architectures to generate realistic synthetic network traffic data and apply the Extreme Gradient Boosting (XGBoost) classifier [10] for evaluation, detection, and APT prediction of observed and unobserved data samples.

The remainder of the paper is divided as follows. Section 2 discusses various Machine Learning (ML) and DL defense methods implemented against APTs and APT-related attacks. In Sect. 3, the methodology employed is described, including the GAN architectures used for network traffic data generation and the principles of XGBoost. Section 4 details the simulation process including data preparation and model training. In Sect. 5, the results are presented, evaluated, and analyzed. The paper draws its conclusions in Sect. 6, where prospects and recommendations for possible improvements are also presented.

2   Background and Related Work

APT defense methods have been categorized into three types by [4], namely Monitoring, Detection and Mitigation methods. The first surveil system memory, disk, Command and Control Center (C&C) communication, and system codes and logs. The second comprise anomaly detection methods that learn and adapt to the attackers' techniques by collecting data and learning from it to make predictions that enable the detection of potential attacks, and pattern matching methods that observe network traffic patterns and behavior to achieve the detection of behavior that matches signatures registered as malicious patterns. The third may be subdivided into reactive techniques that identify system vulnerabilities and perform analysis to identify potential attack paths and proactive methods that are deceptive techniques used to lure attackers away from their targets. The most effective APT defense methods should thus operate against the various APT attack stages and optimization is critical. Nath and Mehtre compared various ML techniques for analyzing malware generally but focused their analysis on static malware, whereas current APTs evade detection by traditional or non-dynamic detection methods [41]. Furthermore, a decision tree to classify accuracy, latency, and throughput of Internet traffic flow has been employed [34]. DL was also applied to nonlinear spatiotemporal flow effects for predicting road traffic flows during special or unexpected events [16].

DL models have delivered excellent results for Network Intrusion and malware detection. These include enhanced intrusion detection of both known and unknown attacks [36], implementing adversarial synthesis and defenses on distributed systems [17], autonomously learning high-quality password guesses as a measure to combating malicious leaks [22], and incorporating hybrid models for more robust, stable and resilient defense systems [30]. Particular success has been achieved with rule-based and signature-based controls, such as Self-Taught Learning (STL) [25]. This was a rule-based DL System that combined Classification and a Sparse Autoencoder (AE), implemented on Network Intrusion Detection Systems (NIDS) against unknown attacks, showing better detection rates compared with previous non-DL methods but requiring other DL methods to enhance system performance, ruling out real-time operation.

Malware is increasingly either polymorphic (where the code changes via a variable encryption key) or metamorphic (where the malicious code randomly mutates as it decrypts and installs itself) [28]. Convolutional Neural Networks (CNNs) have been shown to provide superior detection of malware employing these techniques than traditional signature-based methods [19]. Nevertheless, further performance improvements have been achieved using Deep Neural Networks (DNNs) for detecting and tracking zero-day flash malware [26]. Moreover, a DNN [32], long short-term memory (LSTM) [49] and a Recurrent Neural Network (RNN) [48] were separately implemented for Bot detection but the good performance delivered applied only to Botnets. This was also true of Bot-GAN, a GAN-based DL platform that was more effective in detecting unknown botnets than existing methods [53]. DL has also delivered excellent results in anomaly detection applications. An RNN-based Classifier, using LSTM and Gated Recurrent Units (GRUs) was implemented for keystroke dynamics in verifying human beings [29]. Although this showed improved performance over approaches previously employed, its training time was inefficient vis-à-vis the other methods, preventing optimal performance.

The onion router or Tor network [15] has also been subject to analysis by a Deep Feed Forward Neural Network [24] since its multi-layered encapsulation of messages makes it ideal for attackers. [54] implemented a CNN-based system that detected web attacks by analyzing Hyper Text Transfer Protocol (HTTP) request packets. Artificial Neural Networks (ANNs) and DL methods operating as a Stacked AE were applied by [52] to network traffic identification for feature learning, protocol classification, anomalous protocol detection and unknown protocol identification. A range of other ML techniques have been successfully employed in controlled scenarios but have often met with limited success in real environments [42]. To enhance classification and detection of APT attacks in Cloud Systems, [1] implemented an AE neural network with an incorporated Softmax regression layer to achieve a One-Time Password (OTP) based two-factor authentication mechanism for boosting security against APT attacks. This approach achieved 98.32% detection accuracy even though implementation was specific to the cloud environment.

Deep Unsupervised Learning has continued to show high accuracy for anomaly detection [43] but GANs have been at the forefront of current research due to their versatility, scalability, dynamic applicability, and adversarial training capability. They have become established in a range of image-related tasks, including text-to-image processing, speech synthesis, and digit labeling and classification [3] but in addition, recent progress has been made in malware classification and detection, as well as traffic generation and synthesis to combat adversarial and malicious attacks [37].

A GAN-based Intrusion Detection System (GIDS) was implemented for In-Vehicle networks to detect unknown adversaries [46], demonstrating a detection rate of over 98% using only four unknown attack categories. [36] tested the robustness and strength of an existing IDS with IDSGAN, which modified original malicious traffic to adversarial traffic that evaded IDS detection. Another GAN-based IDS was introduced into a black-box system and trained using seen and unseen adversarial attacks [50]. Ferdowsi and Saad's distributed GAN-based IDS for detecting adversarial attacks over the Internet of Things (IoT) outperformed the existing centralized version [17].

[13] implemented XGBoost on the Network Socket Layer-Knowledge Discovery in Databases (NSL-KDD) dataset and compared their classification accuracy with other methods. This established XGBoost as a strong classification method since it outperformed other state-of-the-art methods on that dataset, achieving 98.70% accuracy. APTs cannot be dealt with using traditional, signature-based methods but require learning methods that are able to automatically spot anomalous patterns [8]. Most defense models proposed and implemented thus far (including those based on GANs) are aimed at general malware or specific and non-APT attacks making them largely inadequate for combating trending APTs. There is thus an urgent need to harness existing and potential GAN models to develop enhanced defense models to deal with APT strategies, providing direction for this research.

3 Methodology

We now describe the elements employed in the simulations, namely GANs, XGBoost and the evaluation metrics.

3.1 GANs

GANs are designed with a hybrid structure to incorporate and train two competing models, the Generator (G) and Discriminator (D) [5]. Both models learn intricate data distributions using the backpropagation of signals with G taking in random noise (z) as input and generating a data instance ($G(z)$) that is used as D's input together with real data samples, $P_{data}(x)$. D attempts to distinguish $G(z)$ from $P_{data}(x)$ while updating its weight via backpropagation where G receives losses as feedback from D, trains to improve $G(z)$ and feeds the improved sample back to D to re-classify it as real or fake. This process continues to alternate while $G(z)$ improves until it succeeds in sufficiently confusing D to pass $G(z)$ as real data. GANs use loss functions to determine the distance between $G(z)$ and $P_{data}(x)$ during the learning process, as they aim to reproduce realistic synthetic data instances that are passed as real samples [51]. We first outline the initial architecture (Vanilla GAN) proposed by [20] before discussing its limitations and later GAN architectures developed to address these.

3.2   The Vanilla GAN Architecture

This architecture implements a standard loss function, the minimax loss, that is based on the Binary Cross-Entropy (BCE) loss function [23], where G strives to maximize the probability of the fake data being accepted as real samples by minimizing the Generator loss function. This loss is determined from D's classification of $G(z)$, and is represented as minimizing (1) by means of parameters θ_g

$$\nabla_{\theta_g} \frac{1}{m} \sum_{i=1}^{m} \log[1 - D(G(z^i))] \tag{1}$$

This approach may produce G saturation, where learning stops, and is commonly replaced by the maximization of (2) [20]

$$\nabla_{\theta_g} \frac{1}{m} \sum_{i=1}^{m} \log D(G(z^i)) \tag{2}$$

On the other hand, D aims to correctly classify the fake versus the real data samples by maximizing its loss function via parameters θ_d

$$\nabla_{\theta_d} \frac{1}{m} \sum_{i=1}^{m} [\log D(x^i) + \log(1 - D(G(z^i)))] \tag{3}$$

The minimax objective function measures the difference between the fake and real probability distributions and is represented thus:

$$\min_G \max_D V(D, G) = E_{x \sim P_{data}(x)} [\log D(x)] + E_{z \sim P_z(z)} [\log(1 - D(G(z)))] \tag{4}$$

where $D(x)$ is the probability estimate of a real data instance, $E_{x \sim P_{data}(x)}$ is the expected value given all $P_{data}(x)$, $D(G(z))$ is the probability estimate that $G(z)$ is real, and $E_{z \sim P_z(z)}$ is the expected value given all $G(z)$. D penalizes itself by maximizing (4) and rewards G for wrongly classifying $P_{data}(x)$ or $D(G(z))$ but penalizes G by minimizing (4) when it correctly classifies $P_{data}(x)$ or $G(z)$ as real or fake, respectively.

Although the Vanilla GAN was seminal it possesses several weaknesses that have led to its further modification [33]. First, it suffers from *mode collapse* [6], where G is unable to learn all data distribution modes and so fails to improve generation, resulting in the constant production of just one mode. Second, training makes use of gradient descent methods and these may suffer from numerical issues such as *vanishing gradients*, where the generator updates become vanishingly small, preventing convergence. Finally, *divergent* or *oscillatory behavior* of G and D is also common, when further training does not produce stability, particularly when G surpasses D's ability to correctly classify data as false.

3.3   The Conditional GAN Architecture

This GAN [39] is an extension to the Vanilla architecture inputs to condition them on auxiliary/extra information (y), employing class labels to condition the model learning process and enable multimodal generation of associated attributes of $P_{data}(x)$. The standard GAN loss function is modified to include conditional probability for both G and D as represented in the modified two-player minimax objective function

$$\min_G \max_D V(D,G) = E_{x \sim P_{data}(x)}[\log D(x|y)] + E_{z \sim P_z(z)}[\log(1 - D(G(z|y)))] \quad (5)$$

While this modification improved generation and training stability, the architecture continued to suffer from mode collapse and vanishing gradients.

3.4   The Wasserstein GAN with Gradient Penalty (WGAN-GP) Architecture

The Wasserstein architecture, based on the Wasserstein loss (W-Loss) that estimates the Earth Mover (EM) distance [9], which measures the distance between the probability distributions of $P_{data}(x)$ and $G(z)$ and the amount moved. The architecture replaces D with a Critic (C) that utilizes a linear output activation function (rather than the *sigmoid* function) to produce a score rather than the probability of the data distribution. [21] added a penalty term to produce an objective function:

$$E_{x \sim P_{data}(x)}[C(x)] - E_{\tilde{x} \sim P_g}[C(\tilde{x})] + \lambda E_{\hat{x} \sim P_{\hat{x}}}[(||\nabla_{\hat{x}} C(\hat{x})||_2 - 1)^2] \quad (6)$$

where $\tilde{x}$ are data from P_g, the generated distribution, and $P_{\hat{x}}$ are uniformly sampled data points ($\hat{x}$) between $P_{data}(x)$ and P_g. The gradient penalty coefficient parameter ensures that convergence failure is avoided and using W-Loss addresses the mode collapse and vanishing gradient problems.

3.5   The Condition Wasserstein GAN with Gradient Penalty (WCGAN-GP) Architecture

Conditioning further improves the performance of WGAN-GP to produce the objective function [55]:

$$E_{x \sim P_{data}(x)}[C(x|y)] - E_{\tilde{x} \sim P_g}[C(\tilde{x}|y)] + \lambda E_{\hat{x} \sim P_{\hat{x}}}[(||\nabla_{\hat{x}} C(\hat{x}|y)||_2 - 1)^2] \quad (7)$$

This retains the advantages of WGAN-GP but also uses the auxiliary conditional information to increase the quality of the synthetic data generated.

3.6   XGBoost

A powerful way to enhance the performance of ML algorithms is provided by the ensemble gradient-boosted decision tree algorithm XGBoost [10]. This uses the Similarity Score and Gain as shown in (8) and (9) to locate optimum data splitting points

$$\text{SimilarityScore} = \frac{(\sum_{i=1}^{n} R_i)^2}{\sum_{i=1}^{n}[P_i \times (1 - P_i)] + \lambda} \quad (8)$$

where R_i is the residual formed from the (observed value - the predicted value), P_i is the probability of an event calculated in the previous step and λ is a regularization parameter.

$$\text{Gain} = \text{LeftLeaf}_{\text{Similarity}} + \text{RightLeaf}_{\text{Similarity}} - \text{Root}_{\text{Similarity}} \qquad (9)$$

The Similarity Score is used in computing (9) is used in conjunction with the split point (node split) with the highest Gain to select the optimal split for the tree.

XGBoost has been severally applied to handle regression, classification, prediction, and ranking problems, and is popular for its scalability, speed, and performance [7]. To achieve system optimization, the sequential tree-building process is parallelized by interchanging the order of loops used to build base learners, thereby improving system computation run time. Trees are pruned starting backward using a depth-first approach, with regularization to prevent overfitting, automatic learning of best missing values and model fitness evaluation using a built-in cross-validation technique during iterations. This research applies XGBoost to cross-validate the ratio between normal and malicious features in the dataset, as well as classify, detect, and predict APT samples in both real and synthetic data.

3.7 Evaluation Metrics

This study implements various evaluation metrics to determine the generation quality, detection, and predictive capability against observed and unobserved APT attacks. These enable the quality of prediction to be ascertained and are given as follows [44]. The ratio of the correct predictions given the total evaluated instances is determined by

$$\text{accuracy} = \frac{TP + TN}{TP + TN + FP + FN} \qquad (10)$$

where TP, TN, FP and FN represent the number of True Positive, True Negative, False Positive and False Negative predictions made, respectively. The actual rate of positive predictions given the total positive predicted instances is determined by

$$\text{precision} = \frac{TP}{TP + FP} \qquad (11)$$

The rate of actual positive instances correctly predicted determines the model sensitivity

$$\text{recall} = \frac{TP}{TP + FN} \qquad (12)$$

The f_1 score ensures a harmonic balance (mean) between precision and recall thus

$$f_1 = 2 \times \frac{\text{precision} \times \text{recall}}{\text{precision} + \text{recall}} \qquad (13)$$

The Receiver Operator Characteristic Area Under the Curve (ROC_AUC) is the summary plot of the probability of True Positive Rate (TPR) against False Positive Rate

(FPR) measured at different thresholds and the measured classification capability of the model. The TPR is the recall, while the FPR is determined by

$$\text{FPR} = \frac{FP}{FP + TN} \tag{14}$$

The positive predictive value (PPV) and negative predictive value (NPV) are calculated from

$$\text{PPV} = \frac{TP}{TP + FP} \tag{15}$$

and

$$\text{NPV} = \frac{TN}{TN + FN} \tag{16}$$

4   Simulation Experiments

4.1   Data Collection and Pre-processing

The UNSW_NB15 dataset used in this work was a captured record of real network traffic that includes normal and malicious flows with comprehensive details such as network addresses, transmission sizes and so on [40]. The dataset contained 2540047 tuples and 48 features (fields), including 304930 malicious flows divided into nine attack types. Of these, we excluded Denial of Service (DoS) flows, since these would not serve as a clandestine route for system compromise. Feature names were added to the raw dataset and the *label* feature deleted as it was a duplication of the attack category feature. The dataset had 9 categorical features and 41 numeric features. All null values were replaced with *normal* for the attack category feature (as there were no normal flows in the raw set) and 0 in flows with command in File Transfer Protocol (FTP) session feature, while all non-binary values in FTP sessions accessed by user and password features were converted to binary, with all other null values being removed. Values in the protocol and state/dependent protocol features were mapped to numeric values, while attack labels were mapped to 0 (for *normal*), 1 (for all other attacks hereon called APTs) and 2 (for DoS to be removed).

4.2   Data Extraction and Processing

A matrix of Pearson's correlation coefficients [45] was used to find the correlation of the various features to remove those with values greater than 0.9 to enhance training and prediction accuracy. Address-related features were also removed and the Python MinMaxScaler (part of the Scikit-Learn library [31]) used to address imbalance and skew in the dataset. This function rescaled all features to the range [0,1] which also compressed inliers and enhanced training performance. Two flow classes were created (0: normal; 1: APT related) and 70% of the set used for training, with 30% for testing (both employing XGBoost).

Table 1. Training parameters for the GAN models used

GAN Parameter	Vanilla GAN	CGAN	WGAN-GP	WCGAN-GP
Learning Rate	2×10^{-4}	2×10^{-4}	2×10^{-4}	2×10^{-4}
Optimization	Adam	Adam	Adam	Adam
Loss	Minimax	Minimax	Wasserstein	Wasserstein
Training Epochs	5000	5000	5000	5000
Batch Size	128	128	128	128
Layer Activation	Leaky ReLU	Leaky ReLU	Leaky ReLU	Leaky ReLU
G Output Activation	tanh	tanh	tanh	tanh
D Output Activation	sigmoid	sigmoid	–	–
Dropout	0.2	0.2	0.2	0.2
Critic Update Steps	–	–	5	5

Table 2. Training parameters for XGBoost

XGBoost Parameter	
Learning Rate	0.1
Gamma	0.1
Regularization Lambda	10
Objective	Binary Log
Iterations	100
Max Depth	4
Min Child Weight	1
Scale POS Weight	3
Subsample	0.9
ColSample by Tree	0.5
Random State by Tree	0
Evaluation Metric	Recall, AUCPR

4.3 Model Architecture and Training

The simulation experiment was implemented on the Anaconda JupyterLab platform in Python 3.9 with the Keras library and TensorFlow backend. All models were trained for 5000 epochs, with a learning rate of 2×10^{-4}, optimized with the Adam optimizer [27]. Training Vanilla GAN and CGAN took almost one hour 40 min, while for WGAN-GP and WCGAN it lasted approximately three hours, using a 3.00 GHz Core i7 CPU (8 Cores) with 32GB RAM. G was built with five layers including three hidden layers and Rectified Linear Unit (*ReLU*) activation for each layer, while D had an input layer, five hidden layers, *ReLU* activation for each layer except the output layer which had a *sigmoid* activation function. C, however, did not use the sigmoid activation function at the output layer. Common activation functions are discussed by [14]. The piecewise linear

ReLU activation function is popularly used for neural network training as it overcomes the vanishing gradient problem that prevents the backpropagation of gradient information through the network. The *sigmoid* or *Logistic* activation function is a non-linear activation function used in models when predicting output probabilities, bringing values into the range [0, 1]. For ease of reference, we summarize the GAN parameters used in Table 1 and the XGBoost parameters in Table 2.

5   Results and Analysis

5.1   Pre-training with XGBoost

After training with XGBoost on the testing set, the algorithm was able to detect malicious samples with an accuracy of 98.76%, 99.81% ROC_AUC, 98.76% recall, and 90.75% precision indicating that fewer than 2% of the predicted malicious samples were normal data. The Feature Score (f-score) [11] of each feature was also used to determine the 10 most important features in the dataset for identifying malicious samples as shown in Fig. 1. Subsequently, a sorted list of the top eight of these features relevant for detecting attacks was utilized for visualizing training results.

5.2   Synthetic Data Classification and Generation

The k-means clustering algorithm [35] was used to classify all malicious data into two classes. Vanilla GAN and WGAN-GP did not utilize auxiliary (class) information during training and therefore generated output as one class while CGAN and WCGAN-GP generated results by class. Vanilla GAN started learning the original data but was not able to converge or sustain the learning process and eventually experienced mode collapse as the shape and range of the learned distributions no longer improved and the losses increased. Figure 2 shows that CGAN performed better with convergence as it learned each class's distributions to a limited extent but reached a limit in its performance after approximately 3200 training steps.

Unlike Vanilla GAN and CGAN, both WGAN-GP and WCGAN-GP did not experience mode collapse, vanishing gradients or convergence failure. Figure 3 shows both architectures steadily and quickly learned data distributions with improved generation as the loss slope approached zero as desired for EM Distance Estimation. The difference in generation between both WGAN-GP and WCGAN-GP was relatively insignificant when used to determine their performance, despite WCGAN-GP performing slightly better than WCGAN-GP.

While WGAN-GP learned and produced non-normal distribution outputs without class information, WCGAN-GP did the same but with class information, thus generating the individual data classes.

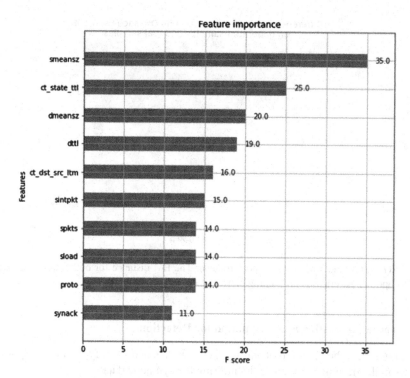

Fig. 1. Plot showing 10 most important features for identifying malicious samples in the test set with each feature's *f-score*.

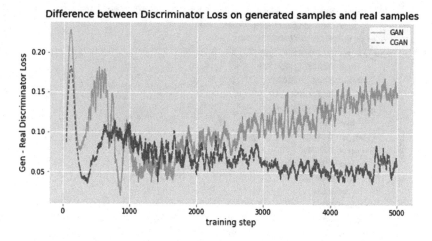

Fig. 2. Vanilla GAN and CGAN performance: the former experiences mode collapse, then fails to converge; auxiliary labeling in the latter produces better convergence but performance plateaus.

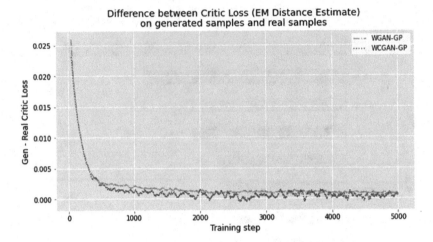

Fig. 3. WGAN-GP and WCGAN-GP performance: The EM distance for both approaches zero, showing optimal generation and training convergence.

5.3   Generation Testing and Evaluation for Detection

As seen in Fig. 4, the *f-score* plot was used to visualize and identify the most important features in the training set used for identifying the synthetic data.

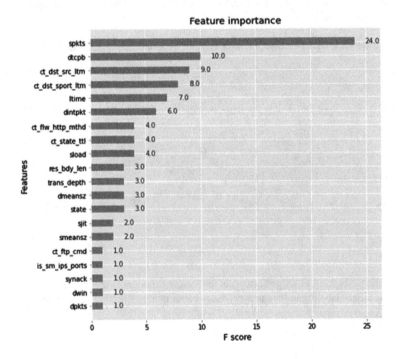

Fig. 4. Plot of feature importance and *f-score* used for identifying generated data.

XGBoost was used to test and evaluate the generation accuracy and detection rate. Figure 5 shows that the algorithm was able to detect synthetic samples with 99.97% accuracy. It can also be seen that the Vanilla GAN architecture displays deteriorating accuracy results after training step 2800 despite reaching 100% at that point.

Figure 6 indicates 100% ROC_AUC and further results also show 99.94% recall, 100% precision and an f_1 score of 99.97%, indicating optimal detection of APT samples in the synthetic data.

Further evaluation revealed that the predictions generally agreed with the target labels indicating that the model correctly predicted normal traffic 99.94% of the time using the NPV and accurately predicted APTs 100% of the time using the PPV.

5.4 Detection Using Observed and Unobserved Data Samples

To evaluate the detection ability against unseen attacks, 70% of the normal data and 20% of malicious data formed a training set while 30% of normal and attack data were used as a test set for training with the WCGAN-GP architecture since it performed best among the GAN models. Figure 6 shows that training with additional synthetic attack data did not decrease recall, indicating the ability of the XGBoost classifier to retain information used in detecting attacks from the real data (20% used for training) while maintaining the same detection rate when more synthetic data are added. However, training with supplementary unobserved attack data revealed a slight increase in recall as seen in Fig. 7.

Similarly, precision did not increase or decrease with additional synthetic data samples. However, introducing unobserved data samples showed a precision rate reduction to 91.19% as seen in Fig. 8.

Finally, Fig. 9 shows that the AUC remained unchanged with additional synthetic data samples and the introduction of the unobserved attack samples. This indicated the optimum detection capability of the classification model.

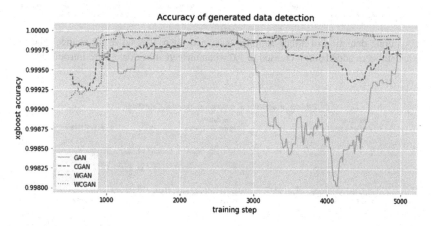

Fig. 5. Plot showing XGBoost's classification accuracy of the synthetic samples for the GAN architectures considered.

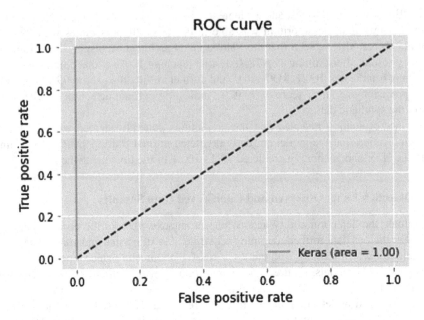

Fig. 6. The ROC Curve shows 100% prediction rate (1.0)indicating optimal prediction/detection performance.

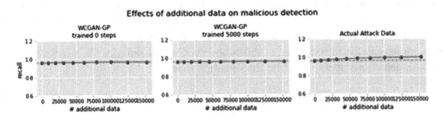

Fig. 7. Accurate detection of attack samples was seen; there was no recall change when more synthetic data were added and increased to 1.0 when unseen attack data were introduced.

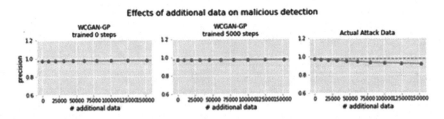

Fig. 8. Precision did not change when more synthetic data were added but decreased to 0.91 when actual attack data were used.

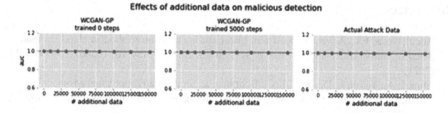

Fig. 9. The unchanging AUC as more synthetic data were added or unseen attack data samples introduced.

6   Conclusions

In building an enhanced, robust, and sustainable IDS against Cyber and Cyber related attacks, increased research has been geared toward developing models that can match the sophistication and dynamism associated with today's attack patterns. To achieve this, careful attention is given to the detailed capabilities of such methods to determine the effectiveness of their performance when implemented on synthetic and real data. This study has highlighted the significant contribution that may be made by GANs in enhancing synthetic data generation. By using GANs it is possible to overcome the limitations of insufficient access to real network traffic data that is suitable for analyzing and designing defense systems. Four GAN architectures have been implemented for the generation task with the aim of achieving optimal data generation that satisfies the requirement for the classification and prediction problem. The Vanilla architectures (GAN and CGAN) were not able to attain optimal generation due to mode collapse, vanishing gradients and convergence failure. However, the Wasserstein architectures (WGAN-GP and WCGAN-GP) overcome the inherent limitations of the Vanilla architectures and attain optimum generation with 99.97% accuracy. This is important when designing defense and detection systems as the synthetic data is required to be as real as possible for consistent performance results. Known for its high performance in classification tasks, XGBoost is implemented in this study to cross-validate the ratio between the normal and malicious flows in the pre-trained data, and further for classification, prediction, and detection of APT samples in both real and synthetic data. We tested and evaluated the model performance using detailed metrics, particularly for the detection and prediction of attack samples. All metrics evaluated indicated optimum performance, surpassing previous state-of-the-art methods with 100% accurate prediction of APTs. The study further assessed the detection capability of the model on unobserved data samples, results showed a consistent detection rate of 91.19% and a False Alarm Rate (FAR) of 1.5×10^{-4}. While these results demonstrate excellent performance, the implementation was only for the UNSW_NB15 dataset. In the future, this technique can be implemented on other cyber datasets to further determine its validity on different data distribution patterns. The method also needs to be evaluated using unseen data samples to ascertain its performance on both known and unknown attacks.

Acknowledgements. This research is supported by the University of Warwick School of Engineering and Cyber Security Global Research Priorities - Early Career Fellowships (Cyber Security GRP).

References

1. Abdullayeva, F.J.: Advanced persistent threat attack detection method in cloud computing based on autoencoder and softmax regression algorithm. Array **10**, 100067-1–100067-11 (2021)
2. Ahmad, A., Webb, J., Desouza, K.C., Boorman, J.: Strategically-motivated advanced persistent threat: definition, process, tactics and a disinformation model of counterattack. Comput. Secur. **86**, 402–418 (2019)
3. Alqahtani, S.H., Thorne, M.K., Kumar, G.: Applications of generative adversarial networks (GANs): an updated review. Arch. Comput. Methods Eng. **28**(2), 525–552 (2021)
4. Alshamrani, A., Myneni, S., Chowdhary, A., Huang, D.: A survey on advanced persistent threats: techniques, solutions, challenges, and research opportunities. IEEE Commun. Surv. Tutor. **21**(2), 1851–1877 (2019)
5. Anande, T.J., Leeson, M.S.: Generative adversarial networks (GANs): a survey on network traffic generation. Int. J. Mach. Learn. Comput. **12**(6), 333–343 (2022)
6. Anande, T.J., Al-Saadi, S., Leeson, M.S.: Generative adversarial networks for network traffic feature generation. Int. J. Comput. Appl. 1–9 (2023). https://doi.org/10.1080/1206212X.2023.2191072
7. Bentéjac, C., Csörgő, A., Martínez-Muñoz, G.: A comparative analysis of gradient boosting algorithms. Artif. Intell. Rev. **54**(3), 1937–1967 (2021)
8. Biggio, B., Šrndić, N.: Machine learning for computer security. In: Joseph, A.D., Laskov, P., Roli, F., Tygar, J.D., Nelson, B. (eds.) Machine Learning Methods for Computer Security, vol. 3, pp. 5–10. Dagstuhl Manifestos, Dagstuhl (2012)
9. Chan, T.N., Yiu, M.L., U, L.H.: The power of bounds: answering approximate earth mover's distance with parametric bounds. IEEE Trans. Knowl. Data Eng. **33**(2), 768–781 (2021)
10. Chen, T., Guestrin, C.: XGBoost: a scalable tree boosting systems. In: 2nd ACM SIGKDD International Conference on Knowledge Discovery and Data Mining (KDD 2016), pp. 785–794. ACM, New York (2016)
11. Chen, Y.W., Lin, C.J.: Combining SVMs with various feature selection strategies. In: Guyon, I., Nikravesh, M., Gunn, S., Zadeh, L.A. (eds.) Feature Extraction. Studies in Fuzziness and Soft Computing, vol. 207, pp. 314–324. Springer, Berlin (2006). https://doi.org/10.1007/978-3-540-35488-8_13
12. Chen, P., Desmet, L., Huygens, C.: A study on advanced persistent threats. In: De Decker, B., Zúquete, A. (eds.) CMS 2014. LNCS, vol. 8735, pp. 63–72. Springer, Heidelberg (2014). https://doi.org/10.1007/978-3-662-44885-4_5
13. Dhaliwal, S., Nahid, A., Abbas, R.: Effective intrusion detection system using XGBoost. Information **9**(7), 149-1–149-24 (2018)
14. Ding, B., Qian, H., Zhou, J.: Activation functions and their characteristics in deep neural networks. In: Chinese Control and Decision Conference (CCDC), pp. 1836–1841. IEEE, Piscataway (2018)
15. Dingledine, R., Mathewson, N., Syverson, P.: Tor: the second-generation onion router. In: 13th USENIX Security Symposium, pp. 303–320. USENIX Association (2004)
16. Dixon, M.F., Polson, N.G., Sokolov, V.O.: Deep learning for spatio-temporal modeling: dynamic traffic flows and high frequency trading. Appl. Stoch. Model. Bus. Ind. **35**(3), 788–807 (2019)
17. Ferdowsi, A., Saad, W.: Generative adversarial networks for distributed intrusion detection in the internet of things. In: IEEE Global Communications Conference (GLOBECOM), pp. 1–6. IEEE, Piscataway (2019)
18. Ghafir, I., Prenosil, V.: Proposed approach for targeted attacks detection. In: Sulaiman, H.A., Othman, M.A., Othman, M.F.I., Rahim, Y.A., Pee, N.C. (eds.) Advanced Computer

and Communication Engineering Technology. LNEE, vol. 362, pp. 73–80. Springer, Cham (2016). https://doi.org/10.1007/978-3-319-24584-3_7

19. Gibert, D., Mateu, C., Planes, J., Vicens, R.: Using convolutional neural networks for classification of malware represented as images. J. Comput. Virol. Hacking Tech. **15**(1), 15–28 (2019)

20. Goodfellow, I.J., et al.: Generative adversarial nets. In: Ghahramani, Z., Welling, M., Cortes, C., Lawrence, N., Weinberger, K. (eds.) Proceedings of Advances in Neural Information Processing Systems (NIPS 2014), vol. 27, pp. 2672–2680. Curran Associates Inc., Red Hook (2014)

21. Gulrajani, I., Ahmed, F., Arjovsky, M., Dumoulin, V., Courville, A.C.: Improved training of Wasserstein GANs. In: 30th Conference on Advances in Neural Information Processing Systems (NIPS 2017), pp. 5767–5777. Curran Associates Inc., Red Hook (2017)

22. Hitaj, B., Gasti, P., Ateniese, G., Perez-Cruz, F.: PassGAN: a deep learning approach for password guessing. In: NeurIPS 2018 Workshop on Security in Machine Learning (2018)

23. Hurtik, P., Tomasiello, S., Hula, J., Hynar, D.: Binary cross-entropy with dynamical clipping. Neural Comput. Appl. (1), 1–13 (2022). https://doi.org/10.1007/s00521-022-07091-x

24. Ishitaki, T., Obukata, R., Oda, T., Barolli, L.: Application of deep recurrent neural networks for prediction of user behavior in tor networks. In: Barolli, L., Takizawa, M., Enokido, T., Hsu, H.H., Lin, C.Y. (eds.) 31st International Conference on Advanced Information Networking and Applications Workshops (WAINA), vol. 56, pp. 238–243. IEEE, Piscataway (2017)

25. Javaid, A., Niyaz, Q., Sun, W., Alam, M.: Deep learning for spatio-temporal modeling: dynamic traffic flows and high frequency trading. In: 9th EAI International Conference on Bio-Inspired Information and Communications Technologies, vol. 3, pp. 21–26. Institute for Computer Sciences, Social-Informatics and Telecommunications Engineering (2016)

26. Jung, W., Kim, S., Choi, S.: Deep learning for zero-day flash malware detection. In: 36th IEEE Symposium on Security and Privacy. IEEE (2015, poster)

27. Kingma, D.P., Ba, L.J.: Adam: a method for stochastic optimization. In: International Conference on Learning Representations (ICLR). ICLR (2015, poster)

28. Kiperberg, M., Resh, A., Zaidenberg, N.: Malware analysis. In: Lehto, M., Neittaanmäki, P. (eds.) Cyber Security. CMAS, vol. 56, pp. 475–484. Springer, Cham (2022). https://doi.org/10.1007/978-3-030-91293-2_21

29. Kobojek, P., Saeed, K.: Application of recurrent neural networks for user verification based on keystroke dynamics. J. Telecommun. Inf. Technol. **3**, 80–90 (2016)

30. Kos, J., Fischer, I., Song, D.: Adversarial examples for generative model. In: IEEE Security and Privacy Workshops (SPW), pp. 36–42. IEEE, Piscataway (2018)

31. Kramer, O.: Machine Learning for Evolution Strategies. Springer, Cham (2016)

32. Kudugunta, S., Ferrara, E.: Deep neural networks for bot detection. Inf. Sci. **467**, 312–322 (2018)

33. Li, A.J., Madry, A., Peebles, J., Schmidt, L.: On the limitations of first-order approximation in GAN dynamics. Proc. Mach. Learn. Res. **80**, 3005–3013 (2018)

34. Li, W., Moore, A.: A machine learning approach for efficient traffic classification. In: 15th International Symposium on Modelling, Analysis, and Simulation of Computer and Telecommunication Systems (MASCOTS), pp. 310–317. IEEE, Piscataway (2007)

35. Li, Y., Wu, H.: A clustering method based on k-means algorithm. Phys. Procedia **25**, 1104–1109 (2012)

36. Lin, Z., Shi, Y., Xue, Z.: IDSGAN: generative adversarial networks for attack generation against intrusion detection. In: Gama, J., Li, T., Yu, Y., Chen, E., Zheng, Y., Teng, F. (eds.) PAKDD 2022. LNAI, vol. 13282, pp. 79–91. Springer, Cham (2022). https://doi.org/10.1007/978-3-031-05981-0_7

37. Liu, Z., Li, S., Zhang, Y., Yun, X., Cheng, Z.: Efficient malware originated traffic classification by using generative adversarial networks. In: IEEE Symposium on Computers and Communications (ISCC), pp. 1–7. IEEE, Piscataway (2020)
38. de Melo, C.M., Torralba, A., Guibas, L., DiCarlo, J., Chellappa, R., Hodgins, J.: Next-generation deep learning based on simulators and synthetic data. Trends Cogn. Sci. **26**(2), 174–187 (2022)
39. Mirza, M., Osindero, S.: Conditional generative adversarial nets. arXiv e-prints (2014)
40. Moustafa, N., Slay, J.: UNSW-NB15: a comprehensive data set for network intrusion detection systems (UNSW-NB15 network data set). In: Military Communications and Information Systems Conference (MilCIS), pp. 1–6. IEEE, Piscataway (2015)
41. Nath, H.V., Mehtre, B.M.: Static malware analysis using machine learning methods. In: Martínez Pérez, G., Thampi, S.M., Ko, R., Shu, L. (eds.) SNDS 2014. CCIS, vol. 420, pp. 440–450. Springer, Heidelberg (2014). https://doi.org/10.1007/978-3-642-54525-2_39
42. Nikos, V., Oscar, S., Luc, D.: Big data analytics for sophisticated attack detection. ISASCA J. **3**, 1–8 (2014)
43. Pang, G., Shen, C., Cao, L., Hengel, A.V.D.: Deep learning for anomaly detection: a review. ACM Comput. Surv. **54**(2), 1–38 (2022)
44. Powers, D.M.W.: Evaluation: from precision, recall and f-measure to roc, informedness, markedness and correlation. Int. J. Mach. Learn. Technol. **2**(1), 37–63 (2011)
45. Sedgwick, P.: Pearson's correlation coefficient. Br. Med. J. **345**, e4483-1–e4483-2 (2012)
46. Seo, E., Song, H.M., Kim, H.K.: GIDS: GAN based intrusion detection system for in-vehicle network. In: 16th Annual Conference on Privacy, Security and Trust (PST), pp. 1–6. IEEE, Piscataway (2018)
47. Thakkar, A., Lohiyan, R.: A review on machine learning and deep learning perspectives of ids for IoT: recent updates, security issues, and challenges. Arch. Comput. Methods Eng. **28**(4), 3211–3243 (2021)
48. Torres, P., Catania, C., Garcia, S., Garino, C.G.: An analysis of recurrent neural networks for botnet detection behavior. In: Biennial Congress of Argentina (ARGENCON), pp. 1–6. IEEE (2016)
49. Tran, D., Mac, H., Tong, V., Tran, H.A., Nguyen, L.G.: A LSTM based framework for handling multiclass imbalance in DGA botnet detection. Neurocomputing **275**, 2401–2413 (2018)
50. Usama, M., Asim, M., Latif, S., Qadir, J., Ala-Al-Fuqaha: Generative adversarial networks for launching and thwarting adversarial attacks on network intrusion detection systems. In: 15th International Wireless Communications & Mobile Computing Conference (IWCMC), pp. 78–83. IEEE, Piscataway (2019)
51. Wang, K., Gou, C., Duan, Y., Lin, Y., Zheng, X., Wang, F.Y.: Generative adversarial networks: introduction and outlook. IEEE/CAA J. Automatica Sinica **4**(4), 588–598 (2017)
52. Wang, Z.: The applications of deep learning on traffic identification (2015). https://www.blackhat.com/docs/us-15/materials/us-15-Wang-The-Applications-Of-Deep-Learning-On-Traffic-Identification-wp.pdf. Accessed 9 Nov 2022
53. Yin, C., Zhu, Y., Liu, S., Fei, J., Zhang, H.: An enhancing framework for botnet detection using generative adversarial networks. In: International Conference on Artificial Intelligence and Big Data, pp. 228–234. IEEE (2018)
54. Zhang, M., Xu, B., Bai, S., Lu, S., Lin, Z.: A deep learning method to detect web attacks using a specially designed CNN. In: Liu, D., Xie, S., Li, Y., Zhao, D., El-Alfy, E.-S.M. (eds.) ICONIP 2017. LNCS, vol. 10638, pp. 828–836. Springer, Cham (2017). https://doi.org/10.1007/978-3-319-70139-4_84
55. Zheng, M., et al.: Conditional Wasserstein generative adversarial network-gradient penalty-based approach to alleviating imbalanced data classification. Inf. Sci. **512**, 1009–1023 (2020)

Improving Primate Sounds Classification Using Binary Presorting for Deep Learning

Michael Kölle[✉], Steffen Illium, Maximilian Zorn, Jonas Nüßlein,
Patrick Suchostawski, and Claudia Linnhoff-Popien

Institute of Informatics, LMU Munich, Oettingenstraße 67, Munich, Germany
{michael.koelle,steffen.illium,jonas.nuesslein,
linnhoff}@ifi.lmu.de

Abstract. In the field of wildlife observation and conservation, approaches involving machine learning on audio recordings are becoming increasingly popular. Unfortunately, available datasets from this field of research are often not optimal learning material; Samples can be weakly labeled, of different lengths or come with a poor signal-to-noise ratio. In this work, we introduce a generalized approach that first relabels subsegments of MEL spectrogram representations, to achieve higher performances on the actual multi-class classification tasks. For both the binary pre-sorting and the classification, we make use of convolutional neural networks (CNN) and various data-augmentation techniques. We showcase the results of this approach on the challenging *ComparE 2021* dataset, with the task of classifying between different primate species sounds, and report significantly higher Accuracy and UAR scores in contrast to comparatively equipped model baselines.

Keywords: Audio classification · Binary presorting · Relabelling · Background noise · CNN

1 Introduction

Along the worldwide loss of biodiversity, measures to preserve wildlife are becoming increasingly more important. Life on earth is largely determined by its diversity, but also by the complex interaction of fauna and flora (cf. [36]). However, to take the necessary protective measures, thorough research is needed to gain knowledge about our environment. One crucial part of this process often involves long-term observation, e.g., of the behavior and development of a wide variety of wild animals. By continuous documentation, possibly harmful changes can be detected at an early stage and the necessary protective countermeasures can be taken.

Among other data samples like movement patterns, population densities, social structures, the collection of acoustic data is often used to detect, recognize, locate and track animals of interest [2]. Wildlife observation through audio recordings bear several advantages, in contrast to, e.g., camera traps [40]: **1.** they do not have to be precisely aimed or accurately placed and do not need to be moved as often; **2.** group behavior of animals can be studied, even in inaccessible habitats, like underwater or in dark spaces,

D. Conte et al. (Eds.): DeLTA 2023, CCIS 1875, pp. 19–34, 2023.
https://doi.org/10.1007/978-3-031-39059-3_2

while introducing no intrusion; **3.** fewer amounts of data are generated, which reduces complications in storing, securing, distributing, or processing the data [14, 15, 39].

Recent advanced in machine learning (ML) fueled the involvement of evaluating such wildlife audio recordings. However, for ML models to learn efficiently and consistently, clean data of adequate size is important. Audio-based datasets, specifically those recording high-noise environments (primate habitats) in the jungle, are made all the more challenging due to misleading noises (e.g., birds that mimic their acoustic surrounding), a poor signal-to-noise ratio (SNR, S/N) from strong background noise or electronically introduced inferences [17]. General problems like irregular class distribution (underrepresentation), varying sample length or 'weak labeling' further increase the analysis difficulty. While it is still possible to reduce noise in recordings and classify specimen correctly [11], for such methods to work robust, sufficient pre-labeled training datasets are needed. The processes to generate these labels (human hearing, crowed-sourcing) can unfortunately be error-prone on their own, which often results in partially mislabeled datasets [27].

In this work, we address these data problems and present an approach which aims to achieve more accurate ml-model predictions for noisy and weakly labeled audio datasets. Our approach is then evaluated on the 'Central African Primate Vocalization dataset' [46] (as part of 'ComparE 2021' Challenge [35]). The goal here is, to develop a multi-class classification model that can distinguish between the sounds of different primates specimen and pure jungle noise ('background'). For this purpose, we combine several methods, such as data-augmentation, pre-sorting by binary pre-sorting and thresholding on the scope of smaller segments of the original files.

The work at hand is structured as follows: We start by describing the dataset, it's challenging aspects and the concepts behind them in Sect. 3. Details on the methods we use are specified in Sect. 4. Our results are presented in Sect. 6, which we then differentiate to comparable and related work in Sect. 2, before concluding in Sect. 7.

2   Related Work

Before the introduction of the used methods, we utilize this chapter to briefly compare ours - Binary pre-sorting and the CNN architecture for audio classification - to established work in the literature.

2.1   Pre-Sorting

Lin et al., were first to show the usefulness of pre-sorting when the underlying dataset has a high number of features and, consequently, distributional inconsistencies in the samples [27]. We therefore consider the challenging ComparE data a natural fit for this technique. The idea of adjusting data labels (pre-sorting) during the training process can also be found in the work of [44]. For this purpose, they use a combination of different methods, consisting of a modified, generalized Rectified Linear Unit (ReLU) activation function – called MaxOut – in the CNN, three different networks for training, and (later) 'semantic bootstrapping', cf. [43]. Semantic bootstrapping describes a procedure in which certain patterns are extracted from already learned information to gain more and more detailed information. Semantic bootstrapping on the data features is performed iteratively on increasingly more general patterns with a pre-trained deep network, so that later during the pre-sorting inconsistent labels can be detected and subsequently changed (or individual, unhelpful samples removed completely). In our work, we also employ a pre-trained model for the pre-sorting, but we substitute the notion of semantic bootstrapping with the binary classification task as our preferred way to learn a generalized feature understanding. A similar approach to this is also taken by [10]. Here, a CNN classifier was trained on a weakly labeled audio dataset. Shorter audio segments were used for training and pseudo-strong labels were assigned to them. Then, the probabilities of the predictions are compared with the original and pseudo-labels to determine the new label. This corresponds to the approach in this work, i.e., the majority-vote of multiple binary-classifiers. For the general multi-class training, Dinkel et al. did not use any form of augmentation, however. [24] were also able to show that pre-sorting data can have a significant positive effect on training. They trained an auxiliary classifier on a small data set and used it to adjust labels. It was shown that only a small amount of data is needed to train a meaningful auxiliary classifier. However, unlike the dataset in this work, their samples have labels with a ground truth, so the auxiliary classifier could learn much more confidently than in our case.

2.2 Audio Classification With CNNs

To extract relevant features from audio files, it is possible to use the raw audio signal [26, 45] or to extract 2D spectrograms in image form from audio files [6,32,42] to generate audio features from the image representation.

The application of CNNs for audio ML tasks has consistently shown promising results, from audio classification, e.g., [19,29,30], speech recognition or audio recognition in general [21,33]. Contrasting the pure CNN approach is work utilizing recurrent neural networks (RNNs) [8,13,28], and, more recently, a combination of both; Convolutional Recurrent Neural Networks (CRNNs) [1,4,7,42] also show promising suitability to audio recognition tasks and might be an interesting consideration for future work extensions on this topic.

Since the choice of the ML model architecture has a major impact on the resulting outcome, [33] have compared different pre-trained models for training. For example, Inception [41], ResNet [16], and DenseNet [20]. In each case, five independently trained models were merged for predictions using an Ensemble. In doing so, they were able to demonstrate that CNN models can predict bounding values of energy distribution in spectrograms (in general) to classify them. Ensembling made it possible to increase the performance of the predictions. [9], for instance, also uses a ensemble voting system to determine the final predictions of the different models. [31] use such an ensemble approach for the classification of audio data with a focus on surveying different, popular CNN architectures. Kong et al. [25] describe sound event detection on weakly labeled data using CNN and CRNN with automatic boundary optimization. Here, an audio signal is also divided into segments and the CNN model is trained on them. This determines whether certain audio signals, based on their segments, contain usable information or not. The segments are given all the properties of the original data, such as the label. Subsequent classification checks whether the original label matches the prediction. MEL spectrograms were used as input to the CNN. We have modeled our binary classifier on sample segments with a similar intent.

3   Primate Dataset (ComparE)

The 'Central African Primate Vocalization dataset' [46] was collected in a primate sanctuary in Cameroon. The samples contain recordings of the five classes; Chimpanzees, mandrills, collared mangabeys (Redcapped Mangabeys), guenons, as well as forest-noise ('background' or 'BKGD'). The rate of primates samples to background sound samples is 1:1 (10,000 each, in total 20,000 samples). As shown by Fig. 1a, unfortunately the data classes are highly imbalanced, which makes the classification more difficult.

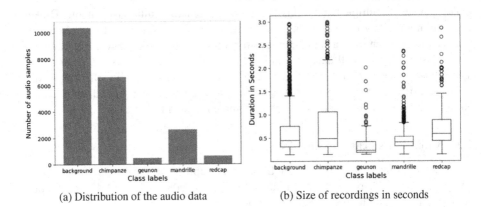

    (a) Distribution of the audio data   (b) Size of recordings in seconds

Fig. 1. In 1a the distribution of the audio data of the individual classes in the entire data set is depicted. In total there are 20,000 data samples. In 1b the size of recordings in seconds for audio samples of the respective classes in the entire data set is shown. The duration of the individual data is between 0.145 and 3 s.

Another problem lies within the distribution of the audio durations per sample (cf. Fig. 1b); Data points appear with lengths between 0.145(!) and 3 s, with samples from guenons and mandrills having particularly short audio durations. Considering both the imbalanced distribution of the classes and the highly varying audio durations, in addition to the high amount of noisy samples, we have to consider this dataset of advanced difficulty. Correctly classifying the lesser represented classes (guenons and mandrills), will be a major challenge, even for deep learning models.

In addition – as described by [46] – the so-called *signal-to-noise* ratio for two of the classes (mandrills and collar mangabees) is a particularly significant problem, as both classes are more difficult to distinguish those from general noise in the background. Signal-to-noise ratio (often abbreviated as *SNR* or *S/N*) is a measure that indicates the ratio of a signal to background noise defined by

$$SNR_{\mathrm{dB}} = 10 \, log_{10} \left(\frac{\mathrm{signal}}{\mathrm{noise}} \right) \tag{1}$$

For our purpose, this ratio of energy from the signal in contrast to the noise energy is calculated and then converted to decibels (dB) [18]. A low SRN value therefore indicates a high amount of noise in the audio signal relative to the informative signal share and vice versa. Consequently, a high value is desirable most of the time.

Noise in this concept is often regarded as a some kind of static introduced by, e.g., the recording equipment. Unfortunately, this is not the kind of noise this dataset is influenced by, rather than natural background noise such as 'forest noise', human talking etc. Background noise, as such, is often mixed with the pure informative signal itself, but also present in samples, that consist of many in-consecutive informative parts (such as complex multipart primate screams). In addition to the regular classification task, a valid automated approach would have to distinguish between background noise and the actual signal first. This fact, coupled with the class and length distribution issues discussed above, renders the automated detection and labeling of the CompareE data samples a considerably tricky problem.

4 Method

As is, this data set represents a collection of quite undesirable, challenging properties. To address these issues, we utilize a selection of preprocessing methods, like segmentation, thresholding, data augmentation and, most importantly, a pre-sorting approach on the equal length segments of every data sample. As is common practice with audio-related ML classification, we first convert the recordings into MEL spectrogram images, by converting the signal to the frequency domain using a (Fast) Fourier Transform (FFT) [3]. Spectrograms are particularly useful for audio analysis because they contain a lot of information about the audio file visually and can be used directly as input for our convolutional neural network (CNN) classifier. The formula for converting a FFT spectrogram to the MEL representation is given by:

$$MEL(f) = 2595 \, log_{10} \left(1 + \frac{f}{700} \right) \tag{2}$$

where f is the spectrogram frequency in Hertz. Using image data instead of pure audio also offers the advantage of many easy and fast to use data augmentation methods, fast training, easy low-level feature (LLF) extraction (through the MEL spectrogram) and small model sizes. A schematic overview of our complete training pipeline is shown in Fig. 2.

4.1 Segmentation

Since the lengths of the individual audio data differs greatly, we segment the data samples. This not only increases our net-amount of data, but also ensures that training data passed to the model is of uniform length. The samples are divided into sections of equal lengths, directly on the MEL spectrograms image of the audio data. (Segment length is a variable hyperparameter.) The individual segments inherit the label of the original audio file. Certain windows can thereby contain no primate sounds but are labeled otherwise and vice versa. We remedy this problem afterwards by pre-sorting, which is

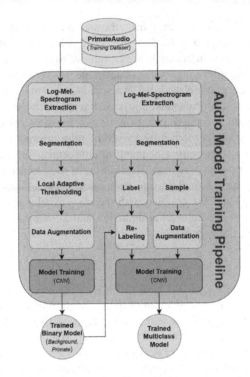

Fig. 2. Pipeline for training a multi-class classifier with binary pre-sorting by a binary classifier trained with thresholding. On the left, the pipeline for training a binary model is shown. Here, pre-sorting is used to adjust the label of a segment in the process of multi-class training.

described in more detail in Sect. 4.4. Note that the predetermined segment length for binary training must be identical to that of multi-class training, otherwise the size of the network for predictions will not match that of multi-class training. In this process, the augmentation of the data is performed on the individual segments of the audio data.

4.2 Thresholding

Thresholding is a technique where we omit sections of a sample, for which the MEL signal falls below an assigned threshold, to ensure that learning can be done on data that contains the most relevant information possible. Since the audio for the dataset was recorded in the wild, there are many noises of varying intensity present in the recordings. Because of the high SNR variance, even between the primate classes, it would not make sense to use a global threshold. To counteract this, *local adaptive thresholding* [37] is applied to all sample segments by iterating over the segment with a separate thresholding window (size determined as hyperparameter) and summing the MEL frequency of the respective window. This adds noise and other interfering sounds to the relevant sounds, so that irrelevant sounds are set relative to the actual information content. Subsequently, the threshold (also a hyperparameter) is set relative to the normalized

sum of the individual windows of the MEL spectrogram, e.g., the same threshold value
is lower for more quiet samples. Segment windows below a calculated threshold are
blackened out accordingly so that background noise is removed as far as possible (e.g.,
Fig. 3). The threshold is only intended to differentiate between background noise or a
primate cry in a segment. Since this is not relevant for the primary task of multi-class
classification, we only use the thresholding for the binary training.

4.3   Data Augmentation

If a model is trained with an unbalanced data set, it will also be biased in its predictions so that the evaluation classification accuracy differs significantly from the training
accuracy. This pitfall is commonly known as overfitting. To further counteract the class
distribution imbalance already shown in Fig. 1a, there are two common approaches;
Under sampling instances of the most frequently occurring class - not ideal considering we already do not have that many samples to train on. Nonetheless, to ensure a
balanced training process, we use a *Weighted Random Sampler* that draws samples proportional to their class size. *Oversampling*, on the other hand, generates synthetic data
of underrepresented classes to match the number of the most frequently occurring class.
Thus, several instances of data augmentation are undertaken in this work, in addition
to Weighted Random Sampling. In general, data augmentation takes advantage of the
assumption that more information can be extracted from the original data to enlarge a
data set. For *data warping* methods, existing data is transformed to inflate the size of
a data set. On the other hand, *methods of oversampling* synthesize entirely new data to
add to a data set [38].

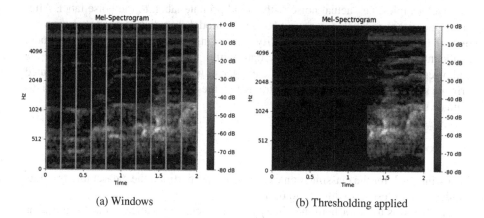

(a) Windows   (b) Thresholding applied

Fig. 3. MEL spectrogram thresholding demonstration. We iterate over the samples with a separate
thresholding window of a predefined size and sum the MEL frequency of the respective window.
The signal frequency is represented by color. The brighter the color in the MEL spectrogram, the
stronger is also the expression of the intensity of the signal in this frequency range. The color
distribution is indexed in decibels (cf. color scale on the right).

In this work, augmentations such as noise injection, shifting, pitching, and frequency and time masking are used. Their effectiveness in providing solid performance on this very dataset has been shown in [22,23]. A simple *Loudness* adjustment minimally amplifies the signal across all samples. By adding random values (Gaussian / random-normal) to an audio signal (*Noise Injection*) stochasticity is introduced without altering the overall character of the sample. This is done on the background noise samples as well as on the regions of the signal relevant for training. *Shifting* shifts the signal left or right by a certain value on the time axis. *Pitching* increases the overall mean of the signal, increasing or decreasing the frequency of the sound. *Frequency masking* sets random frequencies on a MEL spectrogram to 0. The same applies to *time masking*, where time segments that are close to each other on the time axis in the MEL spectrogram have their signal are reduced to 0. Afterwards, all segments are brought to the same length (also called *padding*). Data augmentation is performed on every individual sample-segment of the audio data.

4.4   Pre-sorting With Binary Prediction

With all data samples split into multiple segments, some with reduced signal from the thresholding, padding or other modification from the data augmentation, we have gained considerably more data to work with. However, the problem remains that the split has introduced a significant amount of false positive samples, i.e., purely noisy segments that have inherited a primate class label of the whole data point. To correctly re-label all samples, we first pretrain the multi-class predictor for a binary classification task, where we reduce the prediction of the five given classes to a binary problem ('Background' and 'Primates'). We then use this pretrained model to re-label all the (possibly false positive) samples, i.e., actual noise with an old primate label, to the noise label. After all segments are re-labeled, we can change the model's prediction head and continue on with the multi-class training. To clarify, pre-sorting only takes place when the model confidently predicts 'Background' for input with a primate label. By pre-sorting the audio segments, the multi-class model can later train more confidently on augmented primate sounds and, crucially, gets pretrained on the general difference of noise versus primate sounds.

5   Model Architecture

Following [46], for the classification experiments, we partition the dataset into training, validation, and test sets with a ratio of 3:1:1.

A CNN was used for both the binary and multi-class classification tasks, the same one, in fact. The Adam Optimizer was chosen as the optimizer, with a learning scheduler with a step size of 100 epochs and a reduction of the learning rate by 0.05. To calculate the loss, the binary cross entropy-loss (BCE-Loss) was used to train the binary classifier. Furthermore, we used focal-loss with a value of 2 for γ for the multi-class approach, a dynamically scaled cross entropy loss which is suited for datasets with high-class imbalance. Batch normalization, as well as dropout, were not used for training the binary model. For (pre-)training, various hyperparameter configurations were examined and compared using a mix of Unweighted Average Recall (UAR), as well as

Table 1. Hyperparameters and their set values for binary, as well as multi-class training. The sample rate is given in Hertz and the segment length, as well as the window size of the threshold (Threshold Gap Size) in seconds. Randomness indicates the probability with which the respective augmentations, such as loudness, shift, noise and masking, are applied. A value of 50% is used for each of these augmentations. Consequently, it is also possible that, for example, no loudness is used, but shift and noise are. The hyperparameter for the time or frequency masking is specified here as mask. This value determines how many adjacent samples on each of the two axes should be converted to zero. Note that the binary model was only trained for 150 epochs.

Parameter	Val.	Parameter	Val.
Epochs	200	Batch size	32
Learning Rate	1e-4	Sample Rate	16 K
n FFT's	1024	Hop size	128
n MEL Bands	128	Segment length	0.7 s
Threshold	0.3	Gap Size	0.4 s
Randomness	0.5	Loudness	0.3
Shift	0.3	Noise	0.3
Mask	5	Dropout	0.2

Accuracy and F1-Score. The F1 score [5] was used to determine the most informative model to be used as a binary classifier. As mentioned above, for the pretrained generalization to be transferable to the multi-class task, it is crucial that the model is trained with the identical segment length as the binary model used for training and pre-sorting. A collection of the (final) parameters we used for training is shown in Table 1.

The CNN model used is composed of five hidden layers. Each of the hidden layers consists of a convolutional layer, batch normalization, the activation function ReLU, max-pooling and dropout. The convolutional layer has a window or kernel size of 3×3, stride of size 1 and padding with the value 2. The following batch normalization corresponds to the size of the output of the previous convolutional layer. Thereupon, a max-pooling layer with a window size of 2×2 is applied, with a subsequent dropout. The max-pooling makes it possible for the model to look more closely at contrasts in the MEL spectrogram and thus better detect relevant information in audio data. Finally, in the final dense layer, all values from the last hidden layer are taken and reduced to one dimension so that they can be passed to a suitable activation function. For the multi-class model, a typical Softmax activation function was used for multi-class classification, which maps the output to a probability distribution of the individual labels. For the binary classifier, a sigmoid activation function was applied instead.

6 Experiments

Our experiments involve the following models: A multi-class model, to which neither thresholding nor pre-sorting was applied as a baseline (only audio data into MEL spec. conversion, segmentation & data augmentation); two binary models (w/wo thresholding); and two multi-class models involving the binary models for pre-sorting. Finally, we investigate the impact of the hyperparameters segment length, hop length, as well as the number of FFT's.

6.1 Evaluation

The results of each training run are measured using the expected and actual prediction of a model. Based on these results, the metrics Unweighted Average Recall (UAR), as well as Accuracy and F1-Score are calculated. Accuracy and UAR are defined as

$$\text{Accuracy} = \frac{TP + TN}{TP + TN + FP + FN} \tag{3}$$

and

$$\text{UAR} = \frac{1}{N} \sum_{n=1}^{N} \frac{TP_n}{TP_n + FN_n} \tag{4}$$

where TP, TN, FP, FN, P, N are *True Positives, True Negatives, False Positives* and *False Negatives, Positives (Total), Negatives (Total),* and number of classes (n) respectively. These metrics indicate how precisely a model has been trained under certain conditions. For binary training, additional mismatch matrices are used to be able to determine exactly which classes have been incorrectly labeled. Thus, a binary model with an insufficient mismatch matrix could be discarded. Similarly, confusion matrices are used to observe exactly which classes have been incorrectly assigned in multi-class training. To serve as comparison, the baseline model was trained with 200 epochs and achieved an Accuracy of 95.23% after 125 epochs, and a UAR of 82.86% on the test set. Guenons were incorrectly labeled as background in 18.9%, chimpanzees in 17.99%, and collar mangabees in 19.05% of cases.

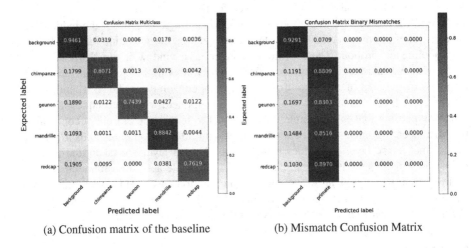

(a) Confusion matrix of the baseline   (b) Mismatch Confusion Matrix

Fig. 4. Confusion matrix of the baseline used for this work (4a). The X-axis describes the predicted labels of the classifier, whereas the Y-axis describes the actual true labels. The diagonal represents the correctly predicted labels of the individual classes. Mismatch Confusion Matrix (4b) for the trained binary model without thresholding. The second column indicates which primates were correctly classified. were correctly classified. The first column represents which audio data was detected as noise or background.

6.2 Binary Pre-sorting

We trained two binary models, one with and one without thresholding. The goal was to compare the two approaches under the same initial conditions. The binary model without thresholding achieved an F1 score of 90.2% on the validation dataset. Other metrics scored between 90% and 91%. Figure 4b shows the resulting mismatch confusion matrix, which clearly indicates which classes have been mislabeled, which classes have been mislabeled. Thereby, guenons with 16.97% and mandrills with 14.84% were incorrectly labeled as background.

A binary model was then trained with thresholding. Different window sizes for thresholding were investigated to counteract the SNR problem. For this, two binary models were trained with a threshold window size of 0.2 s and 0.4 s, and a threshold of 30% for each. The remaining hyperparameters for the training can be taken from the Table 1.

The binary model with thresholding and a window size of 0.2 s achieved a UAR of 85.94%, while the second model with a size of 0.4 s was able to produce a much better UAR of 89.15%. The miss-classifications with a small window size are very high for chimpanzees with 21.49%, guenons with 30.3% and collar mangabees of 38.21%. The mismatch matrix for thresholding with a window size of 0.4 s is significantly lower in values than the matrix without thresholding (cf. Fig. 4b) Consequently, the model with the window size of 0.4 s was used for the pre-sorting of the multi-class model.

6.3 Multi-Class Classifier

Finally, two multi-class models were trained with binary pre-sorting. For this step, one binary model with thresholding and one without were examined (cf. Fig. 5). Both models, in fact, produce quite competitive results, both in terms of accuracy and UAR. We observe, however, that the model without thresholding performs notably better than the one with this technique. The high SNR in many of the class samples is the likely culprit, as thresholding (even the locally adaptive variant), has difficulties setting clean threshold values. We have collected an overview of the test scores of works attempting the ComparE dataset challenge (including our models) in Table 2.

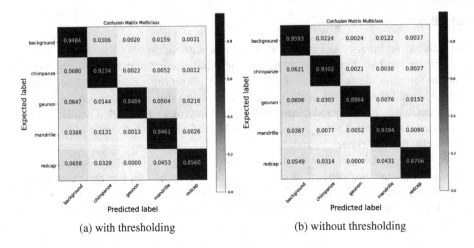

(a) with thresholding   (b) without thresholding

Fig. 5. (Re-labeled) Multi-class confusion matrices pretrained with and without thresholding

Table 2. Results of the different multi-class models on the validation and test data set. The baseline, a multi-class model with pre-sorting but without thresholding, and a multi-class model with pre-sorting and thresholding were compared. Thresholding was used only for training the binary relabeler.

multi-class-Model	Acc. Val	Acc. Test	UAR Val	UAR Test
Baseline (our)	95.69	95.23	85.48	82.86
Re-Label (our) w/o Threshold	**97.83**	**97.81**	**92.84**	**91.72**
Re-Label w/ Threshold (our)	97.54	97.48	92.29	90.46
ComParE 21 Baseline (Fusion)[1]	–	–	–	87.5
CNN10-16k[2] [34]	–	–	<u>93.6</u>	<u>92.5</u>
ComParE + FV (FBANK) + auDeep[3]	–	–	88.2	89.8
ComParE + FV (FBANK)[3]	–	–	87.5	88.8
DeepBiLSTM[4]	–	–	89.4	89.3
ResNet22-16k[2]	–	–	92.6	–
Vision Transformer[5]	–	–	91.5	88.3

Note: Additional scores are cited as reported by the authors:
[1]Schuller et al., 2021,
[2]Pellegrini, 2021,
[3]Egas-López et al., 2021,
[4]Müller et al., 2021,
[5]Illium et al., 2021.

7   Conclusion

Let's summarize: In this work, we have presented and evaluated a dataset pre-sorting approach as a pre-processing step for training of CNN-based classifiers. We have tested our training pipeline - including segmenting, local adaptive thresholding, data-augmentation and pre-sorting - on the challenging CompareE dataset of primate vocalizations and observed competitive results to other participants of the challenge. We managed to show that the binary-pre-sorting approach significantly outperformed the baseline without this step, indicating the value of splitting and pre-sorting difficult, i.e., high-variance, imbalanced data, resulting in a more confident final multi-class classification. Compared to our basic baseline, the UAR could be increased from 82.86% to 91.72%. As a bonus, this binary differentiation task between background noise and any primate sound, serves a beneficial pre-training for the actual prediction task, as the model can start the training on a general notion 'primate or noise' and then fine-tune its weights on the multiple-primate task.

We have also observed that thresholding, as a pre-processing step for training of the binary pre-sorting model, has had no significant positive effect on the resulting test accuracy metrics (on the contrary). For the final model (with binary-pre-sorting) we observed an UAR of 89.15%, which is better than the baseline but worse than the model without thresholding. The reason for this may lie in the data itself, the weakly labeled samples, or the highly imbalanced (class and length) distribution. The low duration of some audio samples, in particular, additionally prevents the thresholding from finding beneficial application. It is important to mention that especially the low SNR of mandrills and collar mangabees in the audio data set are responsible for the fact that the affected classes are difficult to distinguish from background noise. While thresholding may certainly offer promising application on other data sets, for high SNR ratio data like the CompareE dataset, we find the suitability rather pool.

For future work, the effects of certain hyperparameters and their dependencies could be investigated in more detail, e.g., like we have shown that the number of FFT's to hop length significantly affected the performance of binary training. Additionally, it is unclear how promising the binary presort works on other datasets. Therefore, it would be possible to investigate and compare this approach on other data sets. to compare. Furthermore, it would be conceivable to train an additional model on the noisy segments. on the noisy segments. This would make it possible to use another model for pre-sorting to provide more meaningful data and labels to the multi-class model. Especially for current emerging technologies like quantum machine learning, binary presorting could help to decrease the amount of input data that is needed for classification tasks. Finally, it might also make for interesting future work to examine the transfer-ability of pre-trained models on normal images to spectrogram imagery and related audio tasks.

Acknowledgements. This paper was partially funded by the German Federal Ministry of Education and Research through the funding program "quantum technologies—from basic research to market" (contract number: 13N16196).

References

1. Adavanne, S., Pertilä, P., Virtanen, T.: Sound event detection using spatial features and convolutional recurrent neural network. In: 2017 IEEE International Conference on Acoustics, Speech and Signal Processing (ICASSP), pp. 771–775. IEEE (2017)
2. Blumstein, D.T., et al.: Acoustic monitoring in terrestrial environments using microphone arrays: applications, technological considerations and prospectus. J. Appl. Ecol. **48**(3), 758–767 (2011)

3. Bracewell, R.N., Bracewell, R.N.: The Fourier Transform and Its Applications, vol. 31999. McGraw-Hill, New York (1986)
4. Cakır, E., Parascandolo, G., Heittola, T., Huttunen, H., Virtanen, T.: Convolutional recurrent neural networks for polyphonic sound event detection. IEEE/ACM Trans. Audio Speech Lang. Process. **25**(6), 1291–1303 (2017)
5. Chinchor, N.: Muc-4 evaluation metrics. In: Proceedings of the 4th Conference on Message Understanding, pp. 22–29. MUC4 '92, Association for Computational Linguistics, USA (1992). https://doi.org/10.3115/1072064.1072067, https://doi.org/10.3115/1072064.1072067
6. Choi, K., Fazekas, G., Sandler, M.: Automatic tagging using deep convolutional neural networks. arXiv preprint arXiv:1606.00298 (2016)
7. Choi, K., Fazekas, G., Sandler, M., Cho, K.: Convolutional recurrent neural networks for music classification. In: 2017 IEEE International Conference on Acoustics, Speech and Signal Processing (ICASSP), pp. 2392–2396. IEEE (2017)
8. Dai, J., Liang, S., Xue, W., Ni, C., Liu, W.: Long short-term memory recurrent neural network based segment features for music genre classification. In: 2016 10th International Symposium on Chinese Spoken Language Processing (ISCSLP), pp. 1–5. IEEE (2016)
9. Dietterich, T.G.: Ensemble methods in machine learning. In: Multiple Classifier Systems. MCS 2000. LNCS, vol. 1857, pp. 1–15. Springer, Berlin, Heidelberg (2000). https://doi.org/10.1007/3-540-45014-9_1
10. Dinkel, H., Yan, Z., Wang, Y., Zhang, J., Wang, Y.: Pseudo strong labels for large scale weakly supervised audio tagging. In: ICASSP 2022–2022 IEEE International Conference on Acoustics, Speech and Signal Processing (ICASSP), pp. 336–340. IEEE (2022)
11. Dogra, M., Borwankar, S., Domala, J.: Noise removal from audio using CNN and denoiser. In: Biswas, A., Wennekes, E., Hong, TP., Wieczorkowska, A. (eds.) Advances in Speech and Music Technology. Advances in Intelligent Systems and Computing, vol. 1320, pp. 37–48. Springer, Singapore (2021). https://doi.org/10.1007/978-981-33-6881-1_4
12. Egas-López, J.V., Vetráb, M., Tóth, L., Gosztolya, G.: Identifying conflict escalation and primates by using ensemble x-vectors and fisher vector features. In: Proceedings of the Interspeech 2021, pp. 476–480 (2021). https://doi.org/10.21437/Interspeech.2021-1173
13. Gimeno, P., Viñals, I., Ortega, A., Miguel, A., Lleida, E.: Multiclass audio segmentation based on recurrent neural networks for broadcast domain data. EURASIP J. Audio Speech Music Process. **2020**(1), 1–19 (2020). https://doi.org/10.1186/s13636-020-00172-6
14. Hamel, S., et al.: Towards good practice guidance in using camera-traps in ecology: influence of sampling design on validity of ecological inferences. Methods Ecol. Evol. **4**(2), 105–113 (2013)
15. Harris, G., Thompson, R., Childs, J.L., Sanderson, J.G.: Automatic storage and analysis of camera trap data. Bull. Ecol. Soc. Am. **91**(3), 352–360 (2010)
16. He, K., Zhang, X., Ren, S., Sun, J.: Deep residual learning for image recognition. In: Proceedings of the IEEE Conference on Computer Vision and Pattern Recognition, pp. 770–778 (2016)
17. Heinicke, S., Kalan, A.K., Wagner, O.J., Mundry, R., Lukashevich, H., Kühl, H.S.: Assessing the performance of a semi-automated acoustic monitoring system for primates. Methods Ecol. Evol. **6**(7), 753–763 (2015)
18. Hemalatha, S., Acharya, U.D., Renuka, A.: Wavelet transform based steganography technique to hide audio signals in image. Procedia Comput. Sci. **47**, 272–281 (2015)
19. Hershey, S., et al.: CNN architectures for large-scale audio classification. In: 2017 IEEE International Conference on Acoustics, Speech and Signal Processing (ICASSP), pp. 131–135. IEEE (2017)

20. Huang, G., Liu, Z., Van Der Maaten, L., Weinberger, K.Q.: Densely connected convolutional networks. In: Proceedings of the IEEE Conference on Computer Vision and Pattern Recognition, pp. 4700–4708 (2017)
21. Huang, J.J., Leanos, J. J. A.: Aclnet: efficient end-to-end audio classification CNN. arXiv preprint arXiv:1811.06669 (2018)
22. Illium, S., Müller, R., Sedlmeier, A., Linnhoff-Popien, C.: Surgical mask detection with convolutional neural networks and data augmentations on spectrograms. arXiv preprint arXiv:2008.04590 (2020)
23. Illium, S., Müller, R., Sedlmeier, A., Popien, C.L.: Visual transformers for primates classification and COVID detection. In: 22nd Annual Conference of the International Speech Communication Association, INTERSPEECH 2021, pp. 4341–4345 (2021)
24. Iqbal, T., Cao, Y., Kong, Q., Plumbley, M.D., Wang, W.: Learning with out-of-distribution data for audio classification. In: ICASSP 2020–2020 IEEE International Conference on Acoustics, Speech and Signal Processing (ICASSP), pp. 636–640. IEEE (2020)
25. Kong, Q., Xu, Y., Wang, W., Plumbley, M.D.: Sound event detection of weakly labelled data with CNN-transformer and automatic threshold optimization. IEEE/ACM Trans. Audio Speech Lang. Process. **28**, 2450–2460 (2020)
26. Lee, J., Park, J., Kim, K.L., Nam, J.: Sample-level deep convolutional neural networks for music auto-tagging using raw waveforms. arXiv preprint arXiv:1703.01789 (2017)
27. Lin, C.H., Weld, D.S., et al.: To re (label), or not to re (label). In: Second AAAI Conference on Human Computation and Crowdsourcing (2014)
28. Müller, R., Illium, S., Linnhoff-Popien, C.: A deep and recurrent architecture for primate vocalization classification. In: Interspeech, pp. 461–465 (2021)
29. Müller, R., et al.: Acoustic leak detection in water networks. arXiv preprint arXiv:2012.06280 (2020)
30. Müller, R., Ritz, F., Illium, S., Linnhoff-Popien, C.: Acoustic anomaly detection for machine sounds based on image transfer learning. arXiv preprint arXiv:2006.03429 (2020)
31. Nanni, L., Maguolo, G., Brahnam, S., Paci, M.: An ensemble of convolutional neural networks for audio classification. Appl. Sci. **11**(13), 5796 (2021)
32. Nasrullah, Z., Zhao, Y.: Music artist classification with convolutional recurrent neural networks. In: 2019 International Joint Conference on Neural Networks (IJCNN), pp. 1–8. IEEE (2019)
33. Palanisamy, K., Singhania, D., Yao, A.: Rethinking CNN models for audio classification. arXiv preprint arXiv:2007.11154 (2020)
34. Pellegrini, T.: Deep-learning-based central African primate species classification with MixUp and SpecAugment. In: Proceedings of the Interspeech 2021, pp. 456–460 (2021). https://doi.org/10.21437/Interspeech.2021-1911
35. Schuller, B.W., et al.: The interspeech 2021 computational paralinguistics challenge: COVID-19 cough, COVID-19 speech, escalation & primates. arXiv preprint arXiv:2102.13468 (2021)
36. Schulze, E.D., Mooney, H.A.: Biodiversity and Ecosystem Function. Springer Science & Business Media, Berlin, Heidelberg (2012). https://doi.org/10.1007/978-3-642-58001-7
37. Shafait, F., Keysers, D., Breuel, T.M.: Efficient implementation of local adaptive thresholding techniques using integral images. In: Yanikoglu, B.A., Berkner, K. (eds.) Document Recognition and Retrieval XV, vol. 6815, p. 681510. International Society for Optics and Photonics, SPIE (2008). https://doi.org/10.1117/12.767755, https://doi.org/10.1117/12.767755
38. Shorten, C., Khoshgoftaar, T.M.: A survey on image data augmentation for deep learning. J. Big Data **6**(1), 1–48 (2019). https://doi.org/10.1186/s40537-019-0197-0, https://journalofbigdata.springeropen.com/articles/10.1186/s40537-019-0197-0

39. Sundaresan, S.R., Riginos, C., Abelson, E.S.: Management and analysis of camera trap data: alternative approaches (response to Harris et al. 2010). Bull. Ecol. Soc. Am. **92**(2), 188–195 (2011)
40. Swann, D.E., Perkins, N.: Camera trapping for animal monitoring and management: a review of applications. Camera Trapp. Wildl. Manag. Res. 3–11 (2014)
41. Szegedy, C., et al.: Going deeper with convolutions. In: Proceedings of the IEEE Conference on Computer Vision and Pattern Recognition, pp. 1–9 (2015)
42. Wang, Z., Muknahallipatna, S., Fan, M., Okray, A., Lan, C.: Music classification using an improved CRNN with multi-directional spatial dependencies in both time and frequency dimensions. In: 2019 International Joint Conference on Neural Networks (IJCNN), pp. 1–8. IEEE (2019)
43. Wu, W., Li, H., Wang, H., Zhu, K.Q.: Semantic bootstrapping: a theoretical perspective. IEEE Trans. Knowl. Data Eng. **29**(2), 446–457 (2016)
44. Wu, X., He, R., Sun, Z., Tan, T.: A light CNN for deep face representation with noisy labels. IEEE Trans. Inf. Forensics Secur. **13**(11), 2884–2896 (2018)
45. Zhu, Z., Engel, J.H., Hannun, A.: Learning multiscale features directly from waveforms. arXiv preprint arXiv:1603.09509 (2016)
46. Zwerts, J.A., Treep, J., Kaandorp, C.S., Meewis, F., Koot, A.C., Kaya, H.: Introducing a central African primate vocalisation dataset for automated species classification. arXiv preprint arXiv:2101.10390 (2021)

Towards Exploring Adversarial Learning for Anomaly Detection in Complex Driving Scenes

Nour Habib[✉][iD], Yunsu Cho[✉][iD], Abhishek Buragohain[✉][iD],
and Andreas Rausch[✉][iD]

Institute for Software and Systems Engineering, Technische Universität Clausthal,
Arnold-Sommerfeld-Straße 1, 38678 Clausthal-Zellerfeld, Germany
{nour.habib,yunsu.cho,abhishek.buragohain,
andreas.rausch}@tu-clausthal.de
https://www.isse.tu-clausthal.de/

Abstract. One of the many Autonomous Systems (ASs), such as autonomous driving cars, performs various safety-critical functions. Many of these autonomous systems take advantage of Artificial Intelligence (AI) techniques to perceive their environment. But these perceiving components could not be formally verified, since, the accuracy of such AI-based components has a high dependency on the quality of training data. So Machine learning (ML) based anomaly detection, a technique to identify data that does not belong to the training data could be used as a safety measuring indicator during the development and operational time of such AI-based components. Adversarial learning, a subfield of machine learning has proven its ability to detect anomalies in images and videos with impressive results on simple data sets. Therefore, in this work, we investigate and provide insight into the performance of such techniques on a highly complex driving scenes dataset called Berkeley DeepDrive.

Keywords: Adversarial learning · Artificial intelligence · Anomaly detection · Berkeley DeepDrive (BDD)

1 Introduction

Autonomous systems have achieved tremendous success in various domains, such as autonomous cars, smart office systems, smart security systems, and surveillance systems. With such advancements, nowadays, autonomous systems have become very common in our daily life, where we use such systems regularly even in various safety-critical application domains such as financial analysis. All these current developments in various autonomous systems are due to increased performance in the field of machine learning techniques. Many of these autonomous systems have been developed as hybrid systems combining classically engineered subsystems with various Artificial (AI) techniques. One such example is autonomous driving Vehicles. In autonomous vehicles, the trajectory planning subsystem is usually designed in classic engineered ways, whereas the perception part of such vehicles to understand the surrounding environment is based on AI techniques. Both these parts are combined, so they could perform as a single system to execute various safety-critical functions [22].

© The Author(s), under exclusive license to Springer Nature Switzerland AG 2023
D. Conte et al. (Eds.): DeLTA 2023, CCIS 1875, pp. 35–55, 2023.
https://doi.org/10.1007/978-3-031-39059-3_3

During the design and development phase of the perception subsystems in autonomous vehicles, first perceptions engineers label the training data. Then they use these labeled data and machine learning frameworks to train an interpreted function. To explain this concept of training an interpreted function, we can take the example of a perception task, where the trained interpreted function can perform the classification of traffic signal images into its correct traffic sign, as illustrated in Fig. 1. In general training data is the mapping of a finite set of input data to its corresponding output information. In this case, input data can be compared to images of traffic signs and output information can be compared to its correct label which is traffic sign class. So once this machine-learned interpreted function is trained using this traffic sign training data, during the test time, it could process any image to one of the specified traffic sign classes in the training data's output information. However if we consider such machine learned interpreted function to be a part of a real Autonomous vehicle's perception sub-system stack, one important question arises, that is to what extent, the output $if_{ml}(x)$ of such interpreted function if_{ml} is reliable or safe enough, so that other subsystems in the AVs can rely on [15].

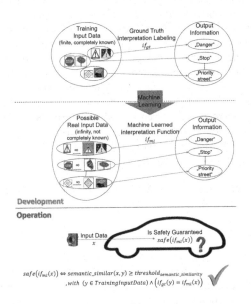

Fig. 1. Operational time check of dependability requirements [14].

The process behind the development of classical engineered systems vastly differs from AI-based systems [16]. During the development process of a classical engineered system, first, semi-formal requirements specifications are created. Such requirement specifications are later used during the testing and verification process of the developed engineered system. But in the case of the development phase of AI-based systems, the creation of such requirement specifications is never done. And instead, it is replaced by a collection of training data. But such training data always does not contain all the information required and is sometimes incomplete [15]. In other cases, due to inaccurate labeling, some amount of these training data consist of errors.

The manufacturers of AI-based systems especially, in a domain like ASs, need to meet very high standards, so that they can satisfy the requirements of the regulations imposed on them. But the current engineering methods struggle to guarantee the dependability requirements from the perspective of safety security, and privacy in a cost-effective manner. Because of such limitations, the development engineers struggle to completely guarantees all the requirements of such AI-based systems during the development phase. Aniculaesei et al. [4] introduced the concept of a dependability cage, which can be used to test a system's dependability requirements both during the development and operational phase.

In the concept of a dependability cage as is shown in the Fig. 2, the Quantitative monitor plays an important role to validate the output of the machine-learned interpreted functions. The concept of the Quantitative Monitor is such that, it tries to check if the sensor data currently being configured by the system as the input data for the ML-interpreted function is semantically similar enough to the ground truth of the training data, used during the development time of the machine learned interpreted function. In this way, the Quantitative monitor tries to verify if the output information of the machine-learned interpreted function is correct and safe. If the current input data to the interpreted function is not semantically similar enough to the training data, then, it is an indication that the output information of this function is not safe or reliable to be used for making safety-critical tasks. Anomaly detection is a promising method, which is used to identify if the input test data (x), is similar or not similar to the ground truth(y) of training data based on some semantic similarity metric $(semantic\_similar(x, y) \geq threshold_{semantic\_similarity})$(cf. Fig. 1). With the evolution of the Artificial intelligence field, many promising methods for anomaly detection such as Generative Adversarial Networks, Autoencoder, and Variational Autoencoder have been experimented to provide the measurement of semantic similarity between the test data and the training data.

The fundamental principle for using auto-encoder for the task of anomaly detection was introduced by Japkowicz et al. [8]. Here the auto-encoder is first trained to minimize the distance error between the input image to the auto-encoder and its output, that is the reconstructed image. And once trained, during the test time, the auto-encoder takes a new image as input and tries to produce a reconstructed image closer to the original input image. Now if the distance error between the input and the reconstructed output image is higher than a certain threshold, then the input image is considered to be an anomaly, or else, it is classified as not an anomaly. Amini et al. [2] also demonstrated the same concept but with training data containing daylight driving scenes and anomaly images consisting of nighttime driving scenes.

But unlike Autoencoder approaches, many of Generative adversarial networks (GANs) based anomaly detection follows a slightly different underlying principle for discriminating between anomaly and not anomaly data. Generative Adversarial Networks (GANs) have some advantages over Autoencoders and Variational Autoencoders. One of the main advantages of GANs for anomaly detection is their ability to generate realistic data samples [19]. In a GAN, a generator network learns to create new samples that are similar to the training data, while a discriminator network learns to distinguish between real and fake samples [19]. This means that a GAN can learn to generate highly realistic samples of normal data, which can be useful for detecting anomalous samples that differ significantly from the learned distribution [19]. Another advantage of GANs is that they can be trained to detect anomalies directly, without the need for a separate anomaly detection algorithm. This is achieved by training the discriminator network to not only distinguish between real and fake samples but also between normal and anomalous samples. This approach is known as adversarial anomaly detection and has been shown to be effective in detecting anomalies in a variety of domains [19]. The anomaly detection using GANs-based techniques will further be discussed in Sect. 3 of the paper.

However, most of these papers have evaluated their techniques on simple data sets such as MNIST [10], CIFAR [9], UCSD [11]. Images in such data sets have very simple features, for instance, in MNIST, only one number is present per image. Similar is the case in CIFAR, where one object class is present per image. Another issue with these data sets is that they are of low resolution. Because in the real world, the driving scene images contain various class of objects, taken in various lighting conditions and weather such as raining day, night, day, etc. So the application of such anomaly detection techniques on complex driving scenes is still needed to be evaluated.

So to evaluate the performance of these GAN-based anomaly detection techniques on real-world driving scenes, we will first reproduce their work using their settings. Once we are able to reproduce their work, then we will evaluate their technique on a complex real-world driving scenes dataset of our choice. For the purpose of this evaluation, we have selected the BDD dataset [23] as our driving scenes dataset. The reason behind using this dataset is, it has high variability in terms of class of objects, number of objects, weather conditions, and environment.

The rest of the paper is organized as follows: In Sect. 2, we provide a brief introduction to dependability cage monitoring architecture for autonomous systems; In Sect. 3, we provide an overview of GAN-based novelty detection works; in Sect. 4, the research questions for our work are introduced; In Sect. 5, we present the description, concept, and dataset for the selected GAN technique; In Sect. 6, we present the evaluation of the selected technique; in Sect. 7, we provide a short summary of the contribution of our work and future work in the direction of GAN based technique for anomaly detection.

2   Dependability Cage - A Brief Overview

To overcome the challenges posed by engineering-dependent autonomous systems, the concept of a dependability cage has been proposed in [4,6,7,12,13]. These dependability cages are derived by the engineers from existing development artifacts. The fundamental concept behind these dependability cages is a continuous monitoring framework for various subsystems of an autonomous system as shown in Fig. 2.

Firstly a high-level functional architecture of an Autonomous Vehicle(AV) has been established. It consists of three parts 1. environment self-perception, 2. situation comprehension, and action decision. 3. trajectory planning and vehicle control [6,7,12]. The continuous monitoring framework resolves two issues that could arise. 1. shows the system the correct behavior based on the dependability requirements. the component handling this issue is called the qualitative monitor. 2. Makes sure the system operates in a situation or environment that has been considered and tested during the development phase of the autonomous system. The component handling this issue is called the quantitative monitor as shown in Fig. 2.

For the monitors to operate reliably, both of them require consistent and abstract data access from the system under consideration. This is handled by the input abstraction and output abstraction components. In Fig. 2, these components are shown as the interfaces between the autonomous systems and the two monitors, as monitoring interface. Both these input and output abstraction components convert the autonomous systems' data into an abstract representation based on certain user-defined interfaces. The type and data values of this representation are decided based on the requirements specification and the dependability criteria derived by the engineers, during the development phase of the autonomous systems [15].

The quantitative monitor observes the abstract representation of the environment, it receives from the autonomous system's input and output abstraction components. For every situation, the monitors evaluate the abstract situation, in real-time, if it is known or tested during the development time of the autonomous system. The information about these tested situations is provided by a knowledge base to the quantitative monitor.

Since the study in this work is carried out taking quantitative monitoring into consideration, we will talk about quantitative monitoring in detail. For a better understanding of the qualitative monitor, refer to the work of Rausch et al. [15]. If one of the above-mentioned monitors detects any unsafe and incorrect behavior in the system, an appropriate safety decision has to be taken to uphold the dependability requirements. For this purpose, the continuous monitoring framework has a fail operation reaction component which receives the information from both the monitors and must bring the corrupted systems to a safe state. One such fail operational reaction could be a graceful degradation of the autonomous system as stated in [5]. As part of the safety decision, the system's data will be recorded automatically. These recorded data then can then be transferred back to the system development so that, they could be analyzed and be used to resolve any unknown system's faulty behaviors.

Finally to realize such a Quantitative Monitor, an efficient solution to classify between known and unknown sensor data is required. As mentioned in the previous section, for this we will first, re-evaluate some of the state-of-the-art generative adversarial network (GAN) based anomaly detection techniques on data sets published in their work. Once we successfully complete the first step, we will evaluate their performance on a driving scenes dataset quite similar to scenes encountered by an autonomous vehicle in the real world. A study of these anomaly detection techniques and the selection of some of these techniques for the purpose of the evaluation will be described in the following sections.

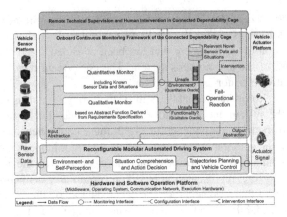

Fig. 2. Continuous Monitoring framework in the dependability cage [15].

3 Previous Works on GAN-Based Anomaly Detection

Anomaly detection has become a genuine concern in various fields, such as security, communication, and data analysis, which makes it a significant interest to researchers. In this paper, we wanted to explore its ability to detect unknown and known driving scenes from the perspective of an autonomous vehicle. Unknown driving scenes can be considered as anomalies that were not considered during the development time of the autonomous vehicle. Whereas known driving scenes are detected as not anomaly, which was considered during the development time of the autonomous vehicle. Various theoretical and applied research have been published in the scope of detecting anomalies in the data. In the following subsections, we will review some of the research papers from which, we selected the approaches. These selected approaches will be our reference for further evaluation in this paper later on.

3.1 Generative Adversarial Network-Based Novelty Detection Using Minimized Reconstruction Error

This paper provides an investigation of the traditional semi-supervised approach of deep convolutional generative adversarial networks (DC-GAN) for detecting the novelty in both the MNIST digit database and the Tennessee Eastman (TE) Dataset [21]. Figure 3 presents the structure of the network of DC-GAN that has been used in the paper. The generator used a random vector Z (Latent space) as input for generating samples G(Z), and the discriminator discriminates whether those samples belong to the Dataset distribution, i.e., indicating them as not-novel images, or they don't belong to the Dataset distribution and they are indicated as novel images [21]. Only data with normal classes were used during the training, so the discriminator learns the normal features of the dataset to discriminate the outlier in the data during the evaluation. The loss function of GAN that is used in the paper is minimax as presented in Eq. 1 [21].

$$\min_G \max_D V(D, G) = E_{x \sim P_{\text{data}(x)}} [log D(x)] + E_{x \sim P_z(z)} [log(1 - D(G(z)))] \quad (1)$$

The generator tries to minimize the error of this loss function and reconstruct better images G(z), while the discriminator D(X) tries to maximize the error of this loss function and indicate the generated samples as fake images [21]. Figure 3 presents the structure of the network of DC-GAN that has been used in the paper.

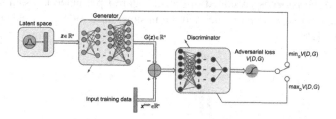

Fig. 3. The structure of DC-GAN [21]

The discriminator classifies whether those samples belong to the dataset distribution, i.e., indicating them as not-novel images, or they don't belong to the dataset distribution and they are indicated as novel images [21]. Data with normal classes were used during the training so the discriminator learns the normal features of the dataset to discriminate the outlier in the data during the evaluation [21]. The loss function of GAN that is used in the paper is the minimax loss function as presented in Eq. 1. The evaluation metrics that are used for model evaluation of the MNIST dataset are the novelty score fg(x) or called the G-score, and the D-score fd(x) [21]. The G-score is calculated as presented in Eq. 2 to minimize the reconstruction error between the generated sample G(z) and the reference actual image x. While the D-score is the discriminator output(decision), and it is varied between 0 (Fake - Novel) to 1 (real - not novel) [21] as presented in Eqs. 2 and 3 [21].

$$f_g(x) = \min_{z \in \mathbb{R}^s} \| x - G(z) \|^2 \tag{2}$$

$$f_d(x) = -D(x) \tag{3}$$

The paper applied another approach using Principal Component Analysis (PCA)-based novelty detection methods on the data for benchmarking. And it used Hotelling's T2 and squared prediction error (SPE) statistics for comparison [21]. The applied approaches in this paper were able to detect the anomaly successfully and with high accuracy.

3.2 Unsupervised and Semi-supervised Novelty Detection Using Variational Autoencoders in Opportunistic Science Missions

In the context of ensuring the safety of the robots, which are sent on scientific exploratory missions to other planets, and ensuring that the robots reach their goals and carry out the desired investigation operations without performing any unplanned actions or deviating from the desired goals, this paper provided an unsupervised and semi-supervised approach in the anomaly detection area [20]. The approach is based on Variational Autoencoders (VAE) model and focuses on detecting the anomaly in the camera data using the data from previous scientific missions, besides providing a study about the VAE-based loss functions for generating the best reconstruction errors in detecting the anomaly features [20]. The structure of the novelty detection approach that is used in the paper is presented in Fig. 4. Figure 4 presented the samples generator, both the semi-supervised model and the unsupervised model, for calculating the anomaly score [20].

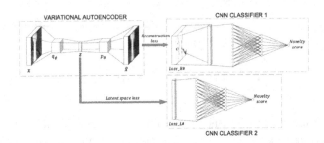

Fig. 4. The structure of GAN for novelty detection approach [20].

The paper used a variety of different losses for best performance and was able to provide comparable and state-of-the-art results using the novelty score, which is the output of the applied neural networks.

3.3   Adversarially Learned One-Class Classifier for Novelty Detection

This paper provides a one-class classification model in terms of separating the outlier samples in the data from inlier ones. The approach in the paper is based on end-to-end Generative adversarial networks in a semi-supervised manner with slight modifications in the generator [17]. The approach in the paper is meant to encode the typical normal images to their latent space before decoding them and reconstructing them again. The reconstructed images are sent as input to the Discriminator network for learning the normal features in those images [17]. Based on that, anomaly scores are given to the corresponding images to help detect whether those images are considered novel or normal, as presented in Fig. 5.

3.4   GANomaly: Semi-supervised Anomaly Detection via Adversarial Training

This paper follows up on many other approaches [3] to investigate, how inverse mapping the reconstructed image to latent space is more efficient and objective than the reconstruction error between the original image and reconstructed image for anomaly detection purpose [1]. The generative characteristics of the variational autoencoder give the ability to analyze the data to determine the anomaly's cause. This approach takes into account the distribution of variables [1]. [19] hypothesizes that the latent space

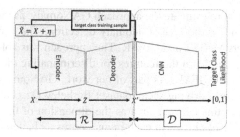

Fig. 5. The Generative adversarial Network Structure [17].

of generative adversarial networks represents the accurate distribution of the data [1]. The approach proposed remapping the GAN based on the latent space. This approach [24] provided statistically and computationally ideal results by simultaneously mapping from image space to latent space [1]. This paper proposes a genetic anomaly detection architecture comprising an adversarial training framework based on the previous three approaches. The approach used normal images (simple images in terms of size and the number of classes that contain) for the training (MNIST dataset and CIFAR dataset), providing excellent results.

For the purpose of our further evaluation and investigations, in this work, we will only consider GANomaly from Subsect. 3.4. We selected the approach of GANomaly since it showed very promising results in terms of accuracy on the data sets used in their work. Another interesting factor for considering their work is that they used the distance error between the latent space of both the original image and its corresponding reconstructed output as a factor to find the anomaly images.

4 Research Questions in This Work

An anomaly in the data is considered a risk that leads to unexpected behavior in the system. It may lead to incorrect results; in some cases, it can be catastrophic and threatening. An anomaly is a deviation from the dataset distribution, an unexpected item, event, or technical glitch in the dataset that deviates from normal behavior or the standard pattern of the data. This deviation in the data may lead to the erratic behavior of the system and abnormal procedures outside the scope that the system was trained to perform. Therefore, the discovery of anomalies has become an interest to many researchers due to its significant role in solving real-world problems, such as detecting abnormal patterns in the data and giving the possibility to take prior measures to prevent wrong actions from the Camera. Inspired by the work of Rausch et al. [15], where research was done with the motivation to detect anomalies in the driving scene for the safety of the autonomous driving system. Their research was successfully validated and proved with the image dataset MNIST. The approach was able to reconstruct the input images using a fully-connected autoencoder network; the autoencoder network was able to learn the features of not novel images to be used later as a factor to discriminate the novelty images from the not novelty images.

In this work, we will re-evaluate the approach GANomaly, as mentioned at the end of the previous section, for the task of anomaly detection. The approach was applied to some simple datasets MNIST and CIFAR. The approach was able to reconstruct the input images correctly and learn the features for discriminating the anomalies and not anomalies on MNIST and CIFAR. As part of our work, GANomlay will be applied to a more complex real-world driving scene dataset Berkeley DeepDrive [23]. The complexity of the images in this dataset, such as the dimension of the image, RGB color channel, and the number of classes in each image, poses a challenge for the approach. We have formulated the contribution of our work in the following research questions (RQ) below.

- **RQ.1:** *Can we reproduce the work of GANomaly on one of their used datasets?*
- **RQ.2:** *Does such GAN approach have the ability to reconstruct high dimensional RGB driving scene images?*
- **RQ.3:** *Can such GAN approach be applied on highly complex driving scenes data set for task of anomaly detection?*

5 Evaluation Process for This Work

In this section, we will explain the GANomaly architecture, the data sets, the training process, and the evaluation metric considered, as part of the evaluation process.

5.1 GANnomaly Architecture:

GANomaly is an unsupervised approach that is derived from GAN, and it was developed for anomaly detection purposes. The approach structure is a follow-up to the approach implemented in the paper [1]. The approach consists of three networks. The overall structure of GANomaly is illustrated in Fig. 6. The generator network G, which consists of encoder model G_E and decoder model G_D, is responsible for learning the normal data distribution which is free of any outlier classes, and generating realistic samples. The Encoder network E maps the reconstructed image $\hat{X}$ to the latent space $\hat{Z}$ and finds the feature representation of the image. The discriminator network D classifies the image, whether it is real or fake.

Generator G is an encoder-decoder model. The encoder is responsible for compressing the input sample and reducing the dimensions to vector Z (Latent space), which represents the most important features of the input sample. The decoder, on the other side, decompresses the latent space and reconstructs the input sample as realistically as possible.

The training flow of the Generator is as follows: The Generator Encoder G_E reads the input data X and maps it to its latent space Z, the bottleneck of the autoencoder, using three sequence groups of layers (convolutional layer, batch normalization and finally, LeakyRelu layer), downscaling the data to the smallest dimensions that should contain the best representation of the data, having the important features of the data. The Generator Decoder G_D decodes the latent space Z and reconstructs the image again as $\hat{X}$. G_D acts like the architecture of DCGAN Generator, using three groups of layers

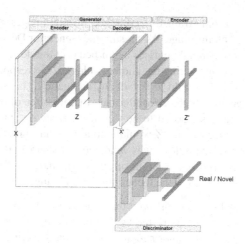

Fig. 6. The structure of GANomaly [1]

(Deconvolutional layer, batch normalization, and finally, Relu layer), followed by the final layer, Tanh layer, so the values normalized between $[-1, 1]$, upscales the vector Z and reconstructs the input image as $\hat{X}$. The Generator reconstructs the image $\hat{X} = G_D(Z)$ based on Latent space $Z = G_E(X)$.

The Encoder E acts exactly like the Generator Encoder G_E. However, E downscales the reconstructed image $\hat{X}$ to its latent space $\hat{Z}$ with a different parametrization than G_E. E learns to compress $\hat{X}$ to the smallest dimension, which should have the best representation of $\hat{X}$ but with its parametrization. The dimension of $\hat{Z} = E(\hat{X})$ is exactly like the dimension of $Z = G_E(X)$. The Encoder is essential for the testing stage as it is part of calculating the anomaly score of the images.

The Discriminator D, which follows the architecture of DCGAN Discriminator and it is responsible for classifying the images between Fake and Normal, using five groups of layers (convolutional layer, batch normalization, and finally, LeakyRelu layer), followed at the end with sigmoid layer so the results would be normalized between [0, 1]. However, in the GANamly approach, anomaly detection does not rely on the Discriminator classification results. The main use of the Discriminator in this approach is for feature matching using the values in the last hidden layers before the sigmoid layer. It reduces the distance between the extracted features of the input image X and the reconstructed image $\hat{X}$ and feeds the results back to the Generator to improve its reconstruction performance. The Discriminator is trained using binary cross entropy loss with target class 1 for input images X and with target class 0 for the reconstructed images $\hat{X}$. The loss function of the discriminator is as the following Eq. 4 [1].

$$Loss_D = \nabla_\theta \frac{1}{m} \sum_{i=1}^{m} [log(D(x^i)) + log(1 - D(G(x^i)))] \qquad (4)$$

The GANomaly approach hypothesizes that after compressing the abnormal image to its latent space $Z = G_E(X)$, the latent space would be free of any anomalies features.

That is because the G_E is only trained to compress normal images, which contain normal classes, during the training stage. As a result, the Generator Decoder G_D would not be able to reconstruct the anomaly classes again because the developed parameterization is not suitable for reconstructing the anomaly classes. Correspondingly, the reconstructed images $\hat{X} = G_D(Z)$ would be free of any anomalies. The Encoder compresses the reconstructed image $\hat{X}$, which it hypothesized that it is free of anomaly classes, to its latent space $\hat{Z}$, which is supposed to be free from anomaly features as well. As a result, the difference between Z and $\hat{Z}$ would increase, declaring an anomaly detection in the image. The previous hypothesis was validated using three different types of loss functions; each of them optimizes different sub-network, and, as a final action, the total results of them would be passed to the Generator for updating its weights.

Adversarial Loss: This Loss has followed the approach that is proposed by Salimans et al. [18]; feature matching helps to reduce the instability in GANs training. Using the values of the features in the intermediate layer in the Discriminator to reduce the distance between the features representation of the input image X, that follows the data distribution θ, and the reconstructed image $\hat{X}$, respectively. This Loss is also meant to fool the Discriminator that the reconstructed image is real. Let f represent the function of the output of the intermediate layer of the Discriminator. The adversarial Loss L_{adv} is calculated as illustrated in the Eq. 5 [1].

$$Loss_{adv} = \mathbb{E}_{X \sim \theta} \left\| f(X) - \mathbb{E}_{X \sim \theta} f(\hat{X}) \right\|_2 \tag{5}$$

Contextual Loss: This Loss is meant to improve the quality of the reconstructed image by penalizing the Generator by calculating the distance between the input image X and the reconstructed image $\hat{X}$ as the Eq. 6 [1]:

$$Loss_{con} = \mathbb{E}_{X \sim \theta} \left\| X - \hat{X} \right\|_1 \tag{6}$$

Encoder Loss: This Loss is for reducing the distance between the latent space Z that is mapped from the original image X using the Generator Encoder Z= $G_E(X)$ and the latent space $\hat{Z}$ that is mapped from the reconstructed image $\hat{X}$ using the Encoder $\hat{Z} = E(\hat{X})$ as the Eq. 7 [1]:

$$Loss_{enc} = \mathbb{E}_{X \sim \theta} \left\| G_E(X) - E(\hat{X}) \right\|_2 \tag{7}$$

The Generator learns to reconstruct images that are free of anomaly classes by both learning the Generator Encoder and the Encoder to compress normal features of the images, and they would fail to compress the abnormal features. As a result, the distance between the features of the normal image and the reconstructed image would increase and declares the anomaly in the image. The total Loss that the Generator would depend on it for updating its weights is calculated as the Eq. 8 [1].

$$Loss_G = w_{adv} Loss_{adv} + w_{con} Loss_{con} + w_{enc} Loss_{enc} \tag{8}$$

w_{adv}, w_{con}, and w_{enc} are the weights parameters of the overall loss function of the Generator, which updates the Generator. The initial weights used in this approach are $w_{adv} = 1$, $w_{con} = 20$ and $w_{enc} = 1$

The model is optimized using Adam optimization with a learning rate of 0.0002 with a Momentum of 0.5.

5.2 Datasets

Berkeley DeepDrive (BDD) dataset includes high-resolution images (1280 px × 720 px) of real-live driving scenes. These images were taken in varying locations (cities streets, highways, gas stations, tunnels, residential, parking places, and villages), at three different times of the day (daytime, night, dawn), and in six different weather conditions (rainy, foggy, cloudy, snowy, clear and overcast) [23]. The dataset includes two packages. The first package contains 100k images. Those images include several sequences of driving scenes besides videos of those tours. The second package contains 10K images, and it is not a subset of the 100k images, but there are many overlaps between the two packages [23].

BDD's usage covers many topics, including Lane detection, Road Object Detection, Semantic Segmentation, Panoptic Segmentation, and tracking. In this thesis, a 10k package is used, and this package has two components called Things and Stuff. The Things include countable objects such as people, flowers, birds, and animals. The Stuff includes repeating patterns such as roads, sky, buildings, and grass). Mainly 10k package is labeled under 19 different classes of objects (road, sidewalk, building, wall, fence, pole, traffic light, traffic sign, vegetation, terrain, sky, person, rider, car, truck, bus, train, motorcycle, bicycle) Fig. 7 illustrates some samples of the BDD 10k dataset.

Fig. 7. Samples of BDD dataset [23]

Four of the 19 labels were considered novel objects for our novelty detection purpose, so the dataset is separated into two parts. The first part, the Novel dataset, contains images that have one of the following objects listed in their labels (rider, train, motorcycle, and bicycle) The second part, the Normal dataset, contains images that do not have any of the four novel labels mentioned lately.

6   Evaluation

In this section, we evaluate the performance of GANomaly based on the research questions, we formulated previously in Sect. 4 and the evaluation process from the previous section.

6.1   Evaluation for RQ.1

We replicated the results that are illustrated in the GANomaly reference paper [1], to ensure that our architecture and training method are efficient in detecting anomalies using the same dataset as the reference paper. The GANomaly setup was trained using MNIST Dataset with several anomaly parameterizations. Each time the GANomaly setup was trained by considering one digit as a novel (abnormal) while the other digits were normal. And to provide comparable results, a quantitative evaluation was applied to the results by calculating the area under the curve (AUC) for the result of each applied neural network to detect the abnormal digit. In the GANnomaly approach that is applied in this paper, two types of anomaly scores were applied that indicate the anomalies in the reconstructed images were calculated. The original method is calculated as presented in Eq. 9 which uses the Generator Encoder to map the input image X to the latent space Z and the Encoder which maps the reconstructed image $\hat{X}$ to the latent space $\hat{Z}$ and calculate the difference. The blue line in Fig. 8 indicates the original method. The second method presented in Eq. 10, gets an advantage from the Generator Encoder to map both the original image X and the reconstructed image $\hat{X}$ to their latent space Z and $\hat{Z}$ respectively. The red line in Fig. 8 indicates the second method. Both methods are explained in detail in the Sect. 6.3.

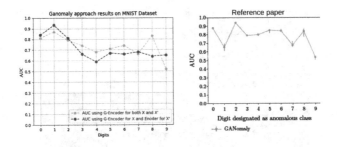

Fig. 8. Left: reproduced results of the GANomaly; right: original results in reference paper of GANomaly [1].

As shown in the Fig. 8, we were able to approximate the results obtained in the reference research paper with slight differences due to modifications in parameters and hyper-parameters to get the best quality in reconstructing the complex Berkeley DeepDrive dataset, to which the approach was applied. Figure 8 shows that the model achieved an excellent anomaly detection result due to the high AUC (Area under the curve) values for the digits 0, 1, 2 and 8. The results were better than the reference paper for digit 1,8, and they were a little less than the reference paper results for other digits 3, 4, 5, 6, and 7 and equal for some digits like 8, 9. So in reference to RQ.1, we can conclude, we were able to successfully reproduce the paper's results.

6.2   Evaluation for RQ.2

During the training, only normal images were used for training with size 6239 images. For evaluation, the test sub-dataset was used, which includes 902 normal images, which are free from outlier classes, and 957 abnormal images, which contain outliers classes. The same training method mentioned on reference page [1] was followed and replicated using the MNIST dataset. With some modifications to the architecture of GANomlay, the architecture was able to reconstruct the images with high resolution with minimum reconstruction error. Figure 9 illustrates the performance of the GANomaly setup in reconstructing the images. As it is illustrated in Fig. 9, the top sample contains a motorcycle (abnormal class) in the bottom corner of the image. The Generator of GANomlay was still able to reconstruct the abnormal classes as efficiently as the normal classes. The region of the abnormal class was reconstructed properly without any distortion.

Fig. 9. GANomaly performance in reconstructing the images (top with anomaly class

The unsatisfied results regarding detecting anomalies in Berkeley DeepDrive dataset can be referred to as the high complexity and the unsuitability of selected abnormal objects in the dataset.

Analyzing the Berkeley DeepDrive dataset images show many challenges and drawback. Some of the images are labeled with some of the classes defined as abnormal in our approach. Which indicate existing abnormal classes in the images, however, the abnormal classes are not visible or recognizable like in Fig. 10a or not fully or clearly visible like in Fig. 10b. In addition, some of the classes that are defined as abnormal have high similarity in terms of features with classes that are defined as normal as Fig. 10c.

The GANomaly approach didn't succeed in detecting the abnormal classes in the reconstructed images. The Generator, despite the training process using only normal classes, was still able to reconstruct the abnormal classes during testing. This is due to the similarity in features matching between the normal and abnormal classes as mentioned previously.

(a) The image has a "Train" label but the train is not visible.

(b) Abnormal classes are barely visible or small and not clear.

(c) The train has high similarity in terms of features with a building block.

Fig. 10. Challenges and Drawback with BDD

So in reference to RQ.2, we could conclude that the GANomaly technique could successfully reconstruct driving scenes of the Berkeley DeepDrive dataset.

6.3   Evaluation for RQ.3

GANomaly is one of the newly developed approaches of GAN, and it aims to detect anomalies in the dataset rather than learning to generate samples that belong to the original data distribution. Moreover, GANomlay consists of multiple models, and it does not depend on the discriminator for discriminating the samples and classifying them into the novel and normal ones.

As mentioned in the GANomaly reference paper [1], it hypothesizes that the generator should not be able to reconstruct the outliers in the abnormal images. As a result, the Generator encoder G_E can map the input image X to its latent space Z without the abnormal features. On the other hand, the encoder should be able to extract the full features of the image (normal and abnormal) and map it to its latent space $\hat{Z}$. So it is expected that the difference between Z and $\hat{Z}$ will increase with increasing the outliers in the image. To evaluate this approach, the encoder loss L_{enc} is applied to the test data D_{test} by assigning an anomaly score A(x) or S_x to the given samples x to have a set of anomaly scores as illustrated in Eq. 9 [1].

$$S = \left\{ s_i : A(x_i), x_i \in D_{test} \right\}$$

Equation 9: illustrates the anomaly score function [1].

$$A(x_i) = \left\| G_E(x_i) - E(G(x_i)) \right\| = \left\| G_E(x_i) - E(\hat{x_i})) \right\| \qquad (9)$$

Another approach was applied in calculating the anomaly score. During the Generator testing, the Generator was able to reconstruct the input images successfully with its outliers but with slight distortion. So the Generator Encoder was applied on both the input images and the reconstructed samples, mapping them to their latent space $\hat{Z}$; as a result, we expected the abnormal features would be more extractable in the reconstructed images. The anomaly score function is transformed to the form, illustrated in Eq. 10 [1].

$$A(x_i) = \left\| G_E(x_i) - G_E(G(x_i)) \right\| = \left\| G_E(x_i) - G_E(\hat{x_i})) \right\| \qquad (10)$$

Finally, the anomaly scores in both approaches were scaled to be between 0 and 1 as the Eq. 11 [1].

$$\hat{s}_i = \frac{s_i - min(S)}{max(S) - min(S)} \tag{11}$$

The anomaly score was calculated for each image and then scaled to be between [0, 1] depending on the max and min anomaly score between all the images. The threshold was selected by calculating the anomaly scores of the training images and the anomaly scores of 98 abnormal images, which are a subset of the abnormal images of the test images. The threshold that makes the best separation between the normal and abnormal images was selected. The evaluation was applied using a different threshold in the range [0.4, 0.6], and the threshold 0.5 made the best separation. Figure 11a illustrates the scatter diagram of the first approach after separation depending on threshold 0.5.

Figure 11b illustrates the scatter diagram of the second approach after separation depending on threshold 0.5.

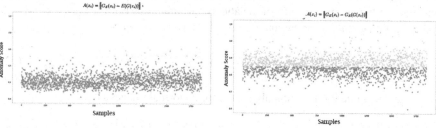

(a) Scatter diagram of GANomaly scores depending on the Generator Encoder and the Encoder.

(b) Scatter diagram of GANomaly scores Depending on the Encoder of the Generator only.

Fig. 11. Comparing results between the two approaches.

The confusion matrix for both GANomaly score approaches was provided to provide a comparable result with the GANomaly reference paper and the derived metrics were calculated. Table 1a presents the confusion matrix and derived metrics results using the Generator encoder for both the input image and the reconstructed image (we called from now GANomlay score 1) and Table 1b presents the confusion matrix and derived metrics results using the Generator encoder for the input image and the encoder for reconstructed image (we called from now GANomlay score 2). So in reference to RQ.3, we could conclude that the GANomaly technique which was successful in detecting anomalies on the MNIST dataset, was not successful when we applied this technique for anomaly detection on the driving scenes dataset we consider in our work.

6.4   Evaluation Supporting RQ.3

Another scaling approach was applied for scaling the anomaly scores for the samples. This approach depends on manipulating the threshold that separates the novel from

Table 1. Quantitative results of the two approaches.

(a) A(X) = Ge(X) - E(G(X))

A(X) = Ge(X) - E(G(X)) Epoch 190		True Values	
		Normal	Novel
Predicted values	Normal	886	954
	Novel	16	3
f1-score: 0.64	ACC: 0.47	P: 0.48	Sn:0.98

(b) A(X) = Ge(X) - Ge(G(X))

A(X) = Ge(X) - Ge(G(X)) Epoch 190		True Values	
		Normal	Novel
Predicted values	Normal	460	448
	Novel	442	509
f1-score: 0.50	ACC: 0.52	P: 0.50	Sn:0.50

normal images. In addition, manipulating the scaling range [Max and Min] that we depend on with scaling the anomaly scores between [0, 1]. This approach depends on using both the train and test data sets in figuring out the scaling ranges. The anomaly scores for all the normal samples were scaled depending on the max and min values between the values of the anomaly scores of the normal samples as Eq. 12 [1]. The max and min anomaly scores for the abnormal samples were calculated as well. The anomaly scores for all the abnormal samples were scaled depending on the max and min values of the anomaly scores of the abnormal samples as Eq. 13 [1].

The scaling equation for anomaly scores for normal images is:

$$s_{\hat{NromalImages}} = \frac{s_i - min(S_{nromal})}{max(S_{nromal}) - min(S_{nromal})} \tag{12}$$

The scaling equation for anomaly scores for abnormal images is:

$$s_{\hat{AbnromalImages}} = \frac{s_i - min(S_{abnromal})}{max(S_{abnromal}) - min(S_{abnromal})} \tag{13}$$

The threshold was tested in the range between [0.4 to 0.55]. The threshold, that gives the best separation of the novel and normal samples, is selected.

That approach provided excellent accuracy in classifying the normal image from the novel images. Table 2 presents the confusion matrix and the accuracy of the approach using threshold 0.47 which gives the best separation of the normal and abnormal images. Figure 12 illustrates how the anomaly scores are scattered.

Table 2. Results using different scaling ranges for anomaly

Threshold = 0.47 acc = 96.3%		True Values	
		Normal	Novel
Predicted values	Normal	878	44
	Novel	24	913

As stated before, the fundamental idea behind anomaly detection is to detect unknown occurrences of input data not used during the training of the ML method (in our case its GANomaly), without any prior knowledge about these unknown data.

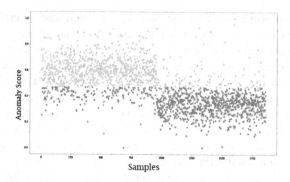

Fig. 12. Scatter diagram for the scaled results using different scaling ranges. Blue points are declared as normal while pink is declared as fake images. (Color figure online)

In the case of GANomaly, when we used the unknown set of images to find a threshold value for classifying the input images into anomaly and not anomaly, it was able to classify most of the known images as not anomaly, and most of the unknown images as not anomaly. So based on the fundamental idea of anomaly detection, GANomaly was not able to classify the unknown driving scenes as an anomaly.

7 Summary and Outlook

The GANomaly approach that is considered in this work originally demonstrated its performance of anomaly detection on image datasets MNIST and CIFAR. These image datasets contain simple images in terms of size, quality, and number of object classes in the images compared to Berkeley DeepDrive. The method was applied to RGB 32 × 32 px images of CIFAR, as well as on the MNIST dataset, which contains gray-scale 28 × 28 px images. Both these data sets contain only one class per image. The evaluation of GANomaly is done based on its efficiency in detecting anomalies on driving scene images from the dataset, Berkeley DeepDrive. We used the confusion matrix to evaluate its efficiency. We were able to follow the architecture of GANomaly [1] and the training method in the reference paper [1] and we successfully reproduced the work of GANomaly on the MNIST dataset with our modified settings. Moreover, we were also able to reconstruct, the driving scenes input images from Berkeley DeepDrive with high resolution and quality, using this method. However, when we applied GANomaly on the Berkeley DeepDrive for the task of anomaly detection, it suffered from low accuracy in the discriminating stage. The large number of classes in each image, the various angle of the view of each class in each image, different weather conditions, and light conditions in which images were taken, posed a great challenge for the method. In addition, the objects selected as unknown objects (trains, motorcycles, riders, bicycles) compared to known objects, occupied very fewer pixels space in the images. Another reason for the failure is, the method was able to reconstruct all the unknown objects that were not used during the training. This impacted the threshold metric used for classifying between anomaly and not anomaly images negatively. As a consequence, the threshold metric which is fundamentally based on mean squared error in the latent space, was not able to discriminate known images (not anomaly) from unknown images (anomaly). The method classified almost all the anomaly images in the test data, as normal (not anomaly) images, hence producing a high number of false negatives. So based on the results of our GANomaly approach replication on the BDD dataset, the current state-of-the-art methods for the task of anomaly detection in such images have not provided any indication that such methods could be directly used for anomaly detection in highly complex driving scenes. With our architecture modification and hyperparameters adjustment, we were able to reconstruct the complex images in such high resolution with high quality but at the discrimination level, the conducted approach with our modification was not able to discriminate the novel objects, contained in the driving scenes. In the future, other approaches will be explored in addition to more experiments will be applied to this approach towards finding the optimal adjustments for detecting the anomalies in such highly complex driving scenes.

References

1. Akcay, S., Atapour-Abarghouei, A., Breckon, T.P.: GANomaly: semi-supervised anomaly detection via adversarial training. In: Jawahar, C.V., Li, H., Mori, G., Schindler, K. (eds.) ACCV 2018. LNCS, vol. 11363, pp. 622–637. Springer, Cham (2019). https://doi.org/10.1007/978-3-030-20893-6_39
2. Alexander, A., et al.: Variational autoencoder for end-to-end control of autonomous driving with novelty detection and training de-biasing, pp. 568–575. IEEE (2018)
3. An, J., Cho, S.: Variational autoencoder based anomaly detection using reconstruction probability. Spec. Lect. IE 2(1), 1–18 (2015)
4. Aniculaesei, A., Grieser, J., Rausch, A., Rehfeldt, K., Warnecke, T.: Towards a holistic software systems engineering approach for dependable autonomous systems. In: Stolle, R., Scholz, S., Broy, M. (eds.) Proceedings of the 1st International Workshop on Software Engineering for AI in Autonomous Systems, pp. 23–30. ACM, New York (2018). https://doi.org/10.1145/3194085.3194091
5. Aniculaesei, A., Griesner, J., Rausch, A., Rehfeldt, K., Warnecke, T.: Graceful degradation of decision and control responsibility for autonomous systems based on dependability cages. In: 5th International Symposium on Future Active Safety Technology toward Zero, Blacksburg, Virginia, USA (2019)
6. Behere, S., Törngren, M.: A functional architecture for autonomous driving. In: Kruchten, P., Dajsuren, Y., Altinger, H., Staron, M. (eds.) Proceedings of the First International Workshop on Automotive Software Architecture, pp. 3–10. ACM, New York (2015). https://doi.org/10.1145/2752489.2752491
7. Behere, S., Törngren, M.: A functional reference architecture for autonomous driving. Inf. Softw. Technol. 73, 136–150 (2016). https://doi.org/10.1016/j.infsof.2015.12.008
8. Japkowicz, N., Myers, C., Gluck, M.: A novelty detection approach to classification. In: Proceedings of the 14th International Joint Conference on Artificial Intelligence, IJCAI 1995, vol. 1, pp. 518–523. Morgan Kaufmann Publishers Inc., San Francisco (1995)
9. Krizhevsky, A., Nair, V., Hinton, G.: CIFAR: learning multiple layers of features from tiny images. https://www.cs.toronto.edu/~kriz/cifar.html. Accessed 10 Feb 2023
10. LeCun, Y., Cortes, C., Burges, C., et al.: MNIST dataset of handwritten digits. https://www.tensorflow.org/datasets/catalog/mnist. Accessed 10 Feb 2023

11. Mahadevan, V., Li, W.X., Bhalodia, V., Vasconcelos, N.: Anomaly detection in crowded scenes. In: Proceedings of IEEE Conference on Computer Vision and Pattern Recognition, pp. 1975–1981 (2010)
12. Maurer, M., Gerdes, J.C., Lenz, B., Winner, H. (eds.): Autonomes Fahren. Springer, Heidelberg (2015). https://doi.org/10.1007/978-3-662-45854-9
13. Mauritz, M., Rausch, A., Schaefer, I.: Dependable ADAS by combining design time testing and runtime monitoring. In: FORMS/FORMAT 2014–10th Symposium on Formal Methods for Automation and Safety in Railway and Automotive Systems (2014)
14. Raulf, C., et al.: Dynamically configurable vehicle concepts for passenger transport. In: 13. Wissenschaftsforum Mobilität "Transforming Mobility - What Next", Duisburg, Germany (2021)
15. Rausch, A., Sedeh, A.M., Zhang, M.: Autoencoder-based semantic novelty detection: towards dependable AI-based systems. Appl. Sci. **11**(21), 9881 (2021). https://doi.org/10.3390/app11219881
16. Rushby, J.: Quality measures and assurance for AI software, vol. 18 (1988)
17. Sabokrou, M., Khalooei, M., Fathy, M., Adeli, E.: Adversarially learned one-class classifier for novelty detection. In: Proceedings of the IEEE Conference on Computer Vision and Pattern Recognition, pp. 3379–3388 (2018)
18. Salimans, T., Goodfellow, I., Zaremba, W., Cheung, V., Radford, A., Chen, X.: Improved techniques for training GANs. Adv. Neural Inf. Process. Syst. **29** (2016)
19. Schlegl, T., Seebőck, P., Waldstein, S.M., Schmidt-Erfurth, U., Langs, G.: Unsupervised anomaly detection with generative adversarial networks to guide marker discovery. CoRR abs/1703.05921 (2017). https://arxiv.org/abs/1703.05921
20. Sintini, L., Kunze, L.: Unsupervised and semi-supervised novelty detection using variational autoencoders in opportunistic science missions. In: BMVC (2020)
21. Wang, H., Li, X., Zhang, T.: Generative adversarial network based novelty detection using minimized reconstruction error. Front. Inf. Technol. Electron. Eng. **19**(1), 116–125 (2018). https://doi.org/10.1631/FITEE.1700786
22. Youtie, J., Porter, A.L., Shapira, P., Woo, S., Huang, Y.: Autonomous systems: a bibliometric and patent analysis. Technical report, Exptertenkommission Forschung und Innovation (2017)
23. Yu, F., Chen, H., Wang, X., Xian, W., et al.: BDD100K: a diverse driving dataset for heterogeneous multitask learning (2020). https://bdd-data.berkeley.edu/. Accessed 10 Feb 2023
24. Zenati, H., Foo, C.S., Lecouat, B., Manek, G., Chandrasekhar, V.R.: Efficient GAN-based anomaly detection. CoRR abs/1802.06222 (2018). https://arxiv.org/abs/1802.06222

Dynamic Prediction of Survival Status in Patients Undergoing Cardiac Catheterization Using a Joint Modeling Approach

Derun Xia[1], Yi-An Ko[1(✉)], Shivang Desai[2], and Arshed A. Quyyumi[2]

[1] Department of Biostatistics and Bioinformatics, Rollins School of Public Health, Emory University, Atlanta, GA, USA
{Derun.xia,yi-an.ko}@emory.edu
[2] Emory Clinical Cardiovascular Research Institute, Division of Cardiology, Department of Medicine, Emory University School of Medicine, Atlanta, GA, USA
{shivang.rajan.desai,aquyyum}@emory.edu

Abstract. Background: Traditional cardiovascular disease risk factors have a limited ability to precisely predict patient survival outcomes. To better stratify the risk of patients with established coronary artery disease (CAD), it is useful to develop dynamic prediction tools that can update the prediction by incorporating time-varying data to enhance disease management.

Objective: To dynamically predict myocardial infarction (MI) or cardiovascular death (CV-death) and all-cause death among patients undergoing cardiac catheterization using their electronic health records (EHR) data over time and evaluate the prediction accuracy of the model.

Methods: Data from 6119 participants were obtained from Emory Cardiovascular Biobank (EmCAB). We constructed the joint model with multiple longitudinal variables to dynamically predict MI/CV-death and all-cause death. The cumulative effect and slope of longitudinally measured variables were considered in the model. The time-dependent area under the receiver operating characteristic (ROC) curve (AUC) was used to assess the discriminating capability, and the time-dependent Brier score was used to assess prediction error.

Results: In addition to existing risk factors including disease history, changes in several clinical variables that are routinely collected in the EHR showed significant contributions to adverse events. For example, the decrease in glomerular filtration rate (GFR), body mass index (BMI), high-density lipoprotein (HDL), systolic blood pressure (SBP) and increase in troponin-I increased the hazard of MI/CV-death and all-cause death. More rapid decrease in GFR and BMI (corresponding to decrease in slope) increased the hazard of MI/CV-death and all-cause death. More rapid increase in diastolic blood pressure (DBP) and more rapid decrease in SBP increased the hazard of all-cause death. The time-dependent AUCs of the traditional Cox proportional model were higher than those of the joint model for MI/CV-death and all-cause death. The Brier scores of the joint model were also higher than those of the Cox proportional model.

Conclusion: Joint modeling that incorporates longitudinally measured variables to achieve dynamic risk prediction is better than conventional risk assessment models and can be clinically useful. The joint model did not appear to perform better than a Cox regression model in our study. Possible reasons include data

© The Author(s), under exclusive license to Springer Nature Switzerland AG 2023
D. Conte et al. (Eds.): DeLTA 2023, CCIS 1875, pp. 56–70, 2023.
https://doi.org/10.1007/978-3-031-39059-3_4

availability, selection bias, and quality uncertainty in the EHR. Future studies should address these issues when developing dynamic prediction models.

Keywords: Dynamic prediction · Longitudinal variable · Cardiovascular disease · Joint model · Risk prediction

1 Introduction

Globally, cardiovascular disease (CVD) is the major cause of death, accounting for 31% of deaths. CVD incidence and prevalence continue to climb in the United States, despite a drop in death rates [1]. By 2030, it is expected that 44% of American adults will suffer from at least one form of CVD [2]. CVD is associated with significant economic and health costs [3]. The development of precise risk assessment tools and cost-effective preventative and treatment strategies is an unmet need. Traditional cardiovascular disease risk factors only account for the likelihood of developing coronary artery disease (CAD), but they are less effective at predicting patient survival outcomes [4, 5]. To better stratify the risk of patients with established CAD, it is essential to develop dynamic diagnostic tools in order to enhance disease management.

Traditional models use baseline information but typically ignore longitudinal changes in risk markers, missing potential impact on risk assessment [6, 7]. Ideally, forecasting may be more precise if changes in marker values over time is also considered. Dynamic prediction utilizes time-dependent marker data obtained during a patient's follow-up to provide updated, more precise survival probability predictions. Electronic health record (EHR) data provides a rich source of clinical information on a large and diverse population, making it cost-effective and ideal for studying rare diseases and subpopulations. Meanwhile, EHR data also emphasizes the time-dependent characteristics of health events, as it records patient data longitudinally over time. This longitudinal data can provide valuable insights into disease progression, treatment response, and long-term outcomes, and can be used to identify patterns and trends in health outcomes over time. Therefore, EHR data is essential for healthcare professionals and researchers seeking to make accurate and informed dynamic predictions about future health outcomes.

Recently, many new methods have utilized longitudinal variables to dynamically predict the time-to-event, such as landmarking, joint model, functional principal component analysis (FPCA), and random forest [8–12]. However, these methods have their own disadvantages. By restricting the analysis to a subset of the data, landmarking can result in a loss of information and decreased statistical power; [13] Many machine learning algorithms are black-box models, making it difficult to understand the underlying relationships between the predictors and the outcome [14].

Joint models are suitable for dynamically predicting outcomes using EHR data [15, 16]. EHR data is often collected over time, and joint models can handle both time-varying and time-invariant variables, making them well suited for modeling these data. Moreover, longitudinal joint models can handle missing data frequently occurring in the EHR system in a principled way, making it possible to use all available information. The results of the joint model are also straightforward and can be represented graphically to illustrate the strength of the link between survival outcomes and longitudinal variables, such as the hazard ratio [17]. The predictive results of the combined model have the potential to help physicians make precise and timely medical decisions.

Our aim is to develop a joint model to dynamically predict the adverse events including MI/CV-death and all-cause death among patients undergoing cardiac catheterization using their EHR data over time and to evaluate the prediction accuracy of the model. We illustrate a method to develop individualized dynamic prediction models based on the progression of longitudinal variables. In addition, we will compare these results with a traditional Cox regression model that uses only baseline covariates.

2 Methods

2.1 Study Design and Participants

Data used in this analysis were obtained from Emory Cardiovascular Biobank (EmCAB), an ongoing prospective registry of patients undergoing cardiac catheterization, which was established to identify novel factors associated with the pathobiological process and treatment of cardiovascular disease. Detailed information on EmCAB study protocols, including participant inclusion and exclusion criteria have been described [18].

In our study, 6119 participants who underwent cardiac catheterization, enrolled 2004–2021, were included. At enrollment, patients are interviewed to collect information on demographic characteristics, medical history, detailed family history, medication usage, and health behaviors (alcohol/drug use) prior to cardiac catheterization. Each patients had a 1- and 5-year follow-up phone calls for any adverse events, including myocardial infarction (MI) and cardiovascular death (CV-death).

We selected longitudinal variables in the EHR data for which at least 90% of the patients had more than one observation, including blood pressure measurements, BMI, and labs (see below). Outliers and extreme values were reviewed and removed if necessary. In case of multiple measurements of the same variable within a day, we reduced the number of observations by using the median value for analysis.

The study was approved by the institutional review board (IRB) at Emory University (Atlanta, Georgia, USA) and is renewed annually. All participants provided written informed consent at the time of enrolment.

2.2 Statistical Analysis

Baseline characteristics of the study participants were summarized using mean ± standard deviation (SD) or median (interquartile range [IQR]) for continuous variables and frequencies and percentages for categorical variables.

We developed a joint longitudinal-survival modeling framework to focus on dynamic prediction of the future risk of MI or CV death and all-cause death. The joint model takes into account multiple longitudinal measures and their slopes, and the prediction of future risk can be updated based on multiple longitudinal measures as well as other baseline characteristics. The survival time was calculated from the enrollment to the time of MI/CV-death, all-cause death or censoring.

The joint model consists of two sub-models. The survival sub-model takes the form of a Cox proportional hazards model with baseline covariates including age, gender, race, education, and history of hypertension, smoking, diabetes, hypercholesterolemia, revascularization and heart failure. The longitudinal sub-model describes the evolution of the repeated measures over time with the main effects from observation time (in years), age, gender, and race. The longitudinal variables included estimated glomerular filtration rate (eGFR), body mass index (BMI), high-density lipoprotein (HDL), low-density lipoprotein (LDL), cardiac troponin-I, diastolic blood pressure (DBP), systolic blood pressure (SBP), and hemoglobin A1c (HbA1c). Random effects were used to capture the between-subject variation. For all longitudinal variables, the slope coefficients of observation time and intercepts vary randomly across individuals. We expanded the time effect in the longitudinal sub-model using a spline basis matrix to capture possibly the nonlinear subject-specific trajectories.

Let $y_{ij}(t)$ denote observation of the j-th measurement (j = 1, ..., n_i, where n_i is the number of observations for subject i) for the i-th subject (i = 1..., N) at time t. The following linear mixed model can be used to model a longitudinally measured variable:

$$y_{ij}(t) = m_{ij}(t) + \varepsilon_{ij}(t) = x_{ij}^{\mathbf{T}}(t)\beta + z_{ij}^{\mathbf{T}}(t)b_i + \varepsilon_{ij}(t) \tag{1}$$

$x_{ij}^{\mathbf{T}}(t_{ij})\beta$ is the fixed-effect and $z_{ij}^{\mathbf{T}}(t_{ij})b_i$ is the random-effects. $\varepsilon_{ij}(t_{ij})$ donates measurement error.

Given the eGFR as an example:

$$eGFR_i(t) = \mu + \theta_{0i} + (\beta_1 + \theta_{1i})B_n(t, \lambda_1) + (\beta_2 + \theta_{2i})B_n(t, \lambda_2) + (\beta_3 + \theta_{3i})B_n(t, \lambda_3) + \beta_4 * age_i + \beta_4 Gender_i + \beta_6 Black_i + \varepsilon_i(t) \tag{2}$$

$$\theta_i \sim N\left(0, \tau^2\right),$$

$$\varepsilon_i\left(t_{ij}\right) \sim N\left(0, \theta^2\right),$$

$$\theta_i \perp\!\!\!\perp \epsilon_{ij}$$

$$m_i(t) = \mu + \theta_{0i} + (\beta_1 + \theta_{1i})B_n(t, \lambda_1) + (\beta_2 + \theta_{2i})B_n(t, \lambda_2) + (\beta_3 + \theta_{3i})B_n(t, \lambda_3) + \beta_4 * age_i + \beta_4 Gender_i + \beta_6 Black_i \tag{3}$$

The $\mu + \theta_{0i}$ is the patient-specific intercept μ_0 is the overall intercept and θ_{0i} is the subject-specific difference from μ. The matrix represents a spline basis matrix for a natural cubic spline of time that has two internal knots, resulting in three degrees of freedom. These knots were placed at the 33.3% and 66.7% of the follow-up time points. The $\beta_u + \theta_{ui}$ is the subject specific slope for the u-th basic function of a spline with knots λ_u.

The hazard function is:

$$h_i(t) = h_0(t)\exp\left(\gamma^T\omega_i + \sum_{k=1}^{K} \alpha_k m_{ik}(t)\right) \tag{4}$$

$h_0(t)$ was the baseline hazard function. ω_i is the baseline covariate. We have K multiple longitudinal variables, the α_k linked the k-th (k = 1,..., K) linear mixed model and Cox regression model and assuming the hazard at time t was dependent on the longitudinal trajectory, $m_{ik}(t)$, through the estimated value at time t. When the α_k is significant, it indicated that there is an association between the k-th longitudinal variable and the longitudinal measures and time to event. And the $\exp(\alpha_k)$ was the hazard ratio for one unit increase in the $m_{ik}(t)$ at time t for k-th longitudinal variable. We also include the time-dependent slopes and the cumulative effects of longitudinal variables in the model.

The baseline hazard function is represented by $h_0(t)$. ω_i represents the baseline covariate. The α_k connects the linear mixed model and Cox regression model and assumes that the hazard at time t is dependent on the longitudinal trajectory, represented by $m_{ik}(t)$, through the estimated value at that time. If the α_k is significant, it indicates that there is a correlation between the longitudinal variable and time to event. The hazard ratio for a one unit increases in $m_i(t)$ at time t can be calculated as $\exp(\alpha)$. The model also considers the time-varying slopes and cumulative effects of the longitudinal variables. The cumulative effects $\frac{\int_0^t m_i(s)ds}{t}$ is the hazard of an event at t is associated with the area under the trajectory up to t.

Joint models for such joint distributions are of the following form. The θ_i is a vector of random effects that explains the interdependencies. $p(.)$ is the density function and $S(.)$ is the survival function.

$$p\left(y_{ij} \mid \theta_{ij}\right) = \prod_{k=1}^{n_{ij}} p\left(y_{ij,k} \mid \theta_{ij}\right) = \prod_j p\left(y_{ij} \mid \theta_{ij}\right) \tag{5}$$

$$p(y_i, T_i, \delta_i \mid \theta_i) = \prod_j p\left(y_{ij} \mid \theta_{ij}\right) p(T_i, \delta_i \mid \theta_i) \tag{6}$$

T_i is the observed event time for patient i and δ_i is the event indicator. The key assumption is that given the random effects, the repeated measurements in each outcome are independent, the longitudinal variables are independent of each other, and longitudinal outcomes are independent of the time-to-event outcome.

The Bayesian approach was adopted for model inference and for dynamic predictions. The key step in prediction for a new subject was to obtain samples of subject's random effects from the posterior distribution given the estimated parameters and previous longitudinal observations (at least one measure). The samples were then used to calculate the predictions for the longitudinal variables' future trajectories and risk of MI/CVdeath and all-cause death. Based on the general framework of joint models presented earlier, we are interested in deriving cumulative risk probabilities for a new subject j^* that has survived up to time point t and has provided longitudinal measurements $\mathcal{Y}_{kj^*}(t) = \{y_{kj^*}(t_{j^*l}); 0 \le t_{j^*l} \le t, l = 1, \ldots, n_{j^*}, k = 1, \ldots, K\}$, with K denoting the number of longitudinal outcomes. The probabilities of interest are:

$$\pi_{j^*}(u \mid t) = \Pr\left\{T_{j^*}^* \le u \mid T_{j^*}^* > t, \mathcal{Y}_{j^*}(t), \mathcal{D}_n\right\} = 1 - \iint \frac{S(u \mid b_{j^*}, \theta)}{S(t \mid b_{j^*}, \theta)} p\left\{b_{j^*} \mid T_{j^*}^* > t, \mathcal{Y}_{j^*}(t), \theta\right\} p(\theta \mid \mathcal{D}_n) db_{j^*} d\theta$$

$$\tag{7}$$

where $S(\cdot)$ denotes the survival function conditional on the random effects, and $\mathcal{Y}_{j^*}(t) = \{\mathcal{Y}_{1j^*}(t), \ldots, \mathcal{Y}_{Kj^*}(t)\}$. Combining the three terms in the integrand we can device a Monte Carlo scheme to obtain estimates of these probabilities, namely.

Firstly, we can sample a value $\tilde{\theta}$ from the posterior of the parameters $[\theta \mid \mathcal{D}_n]$ and sample a value $\tilde{b}_{j*}$ from the posterior of the random effects $\left[b_{j*} \mid T^*_{j*} > t, \mathcal{Y}_{j*}(t), \tilde{\theta}\right]$. We then compute the ratio of survival probabilities $S\left(u \mid \tilde{b}_{j*}, \tilde{\theta}\right) / S\left(t \mid \tilde{b}_{j*}, \tilde{\theta}\right)$. After replicating these steps L times, we can estimate the conditional cumulative risk probabilities by:

$$
1 - \frac{1}{L} \sum_{l=1}^{L} \frac{S\left(u \mid \tilde{b}_{j*}^{(l)}, \tilde{\theta}^{(l)}\right)}{S\left(t \mid \tilde{b}_{j*}^{(l)}, \tilde{\theta}^{(l)}\right)}
\tag{8}
$$

and their standard error by calculating the standard deviation across the Monte Carlo samples.

We calculated time-dependent areas under receiver-operating characteristics (ROC) curves (AUCs) and Brier score to assess the performance of the longitudinal marker at different time points over the follow-up period. We predicted the probabilities of MI and CV-death and all-cause of death occurring in the time frame (t, t + Δt], using all measures collected till time t. Then the AUCs were calculated to assess how well the longitudinal marker distinguished the status of patients at time t + Δt. The Brier score is a metric used to assess the precision of a predicted survival function at time t + Δt. It calculates the average squared difference between the observed survival status and the predicted survival probability, with a range of values from 0 to 1. Since the participants were reassessed approximately every year, we selected t at 2, 3, 4, 5 and 6 years, and Δt = 1, 2 (years). In general, higher AUCs indicate higher discrimination of the models and lower Brier score indicates worse precision of prediction. For comparison, we also fitted proportional hazards models (Cox model) with baseline measures. We then assessed the predictive performance of these models using time-dependent AUCs and Brier scores.

In addition, we applied the resulting joint models to predict the future longitudinal trajectories and risk of MI/CV-death for new participants. We selected 2 patients to demonstrated initialized dynamic prediction was updated over time as new clinical information became available. The joint model fitting and predictions were achieved using the R Jmbayes2 package [19].

3 Results

Table 1 summarizes the characteristics of the 6119 participants. The median follow-up time was 7.53 years (SD 4.16; range 0.09–13.73). The average age at baseline was 62.9 years (SD 12.7; range 18.7–99.6), 64.9% were women and 19.6% were black. 3984 (65.1%) patients had a smoking history. Among 6119 participants the average eGFR at baseline was 72.5 mL/min/1.73m^2 (SD 24.4; range 2.3–175.5). 76.9% of patients had a history of hypertension, 2143 (35.0%) patients had a history of diabetes mellitus, and 1827 (29.9%) patients had a history of hypercholesterolemia. Meanwhile, 1406 (23.0%) patients had a history of myocardial infarction, and 2246 (36.7%) have a history of heart failure. 2986 (48.8%) patients had a history of revascularization.

Table 1. Baseline characteristics of Emory Biobank participants stratified by gender.

Baseline variable	All (N = 6119)	Female (N = 2145)	Male (N = 3974)
Age (years)			
Mean (SD)	62.892 (12.662)	62.650 (13.522)	63.023 (12.172)
Range	18.645–99.627	18.645–98.795	20.879–99.627
Race			
Caucasian White	4661 (76.2%)	1510 (70.4%)	3151 (79.3%)
African American Black	1202 (19.7%)	563 (26.3%)	639 (16.1%)
Hispanic	51 (0.8%)	19 (0.9%)	32 (0.8%)
Asian	101 (1.7%)	23 (1.1%)	78 (2.0%)
Native American	7 (0.1%)	4 (0.2%)	3 (0.1%)
Pacific Islander	2 (0.0%)	0 (0.0%)	2 (0.1%)
Other	93 (1.5%)	25 (1.2%)	68 (1.7%)
Black			
Yes	1202 (19.6%)	563 (26.2%)	639 (16.1%)
History of Hypertension			
Yes	4704 (76.9%)	1650 (76.9%)	3054 (76.8%)
History of Diabetes Mellitus			
Yes	2143 (35.0%)	743 (34.6%)	1400 (35.2%)
History of hypercholesterolemia			
Yes	4292 (70.1%)	1424 (66.4%)	2868 (72.2%)
Ever smoker			
Yes	3984 (65.1%)	1231 (57.4%)	2753 (69.3%)
History of myocardial infarction			
Yes	1406 (23.0%)	396 (18.5%)	1010 (25.4%)
History of heart failure			
Yes	2246 (36.7%)	797 (37.2%)	1449 (36.5%)
Highest Level of Education			
Elementary or Middle School	191 (3.1%)	73 (3.4%)	118 (3.0%)
Some High School	565 (9.2%)	243 (11.3%)	322 (8.1%)
High School Graduate	1698 (27.7%)	664 (31.0%)	1034 (26.0%)
Some College	1417 (23.2%)	547 (25.5%)	870 (21.9%)
College Graduate	1242 (20.3%)	366 (17.1%)	876 (22.0%)
Graduate Education or Degree	1006 (16.4%)	252 (11.7%)	754 (19.0%)

(continued)

Table 1. (*continued*)

Baseline variable	All (N = 6119)	Female (N = 2145)	Male (N = 3974)
Significant CAD			
Yes	3237 (72.9%)	884 (63.1%)	2353 (77.4%)
N-Miss	1679	744	935
Normal catheterization			
Yes	691 (14.7%)	300 (19.6%)	391 (12.4%)
N-Miss	1424	615	809
eGFR			
Mean (SD)	72.514 (24.388)	72.298 (26.347)	72.630 (23.265)
Range	2.333–175.481	2.333–175.481	2.365–154.769
History of revascularization			
Yes	2986 (48.8%)	801 (37.3%)	2185 (55.0%)

eGFR, estimated glomerular filtration rate; Significant coronary artery disease (CAD) is defined as at least one artery with 50% or more stenosis based on angiogram findings.

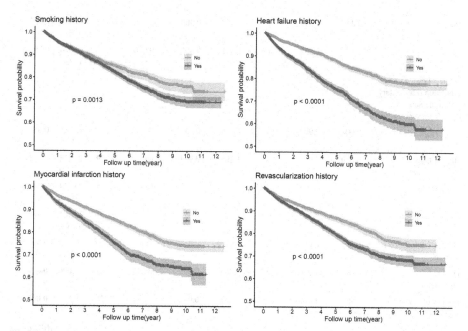

Fig. 1. Kaplan Meier plot of MI/CV-death stratified by smoking history, heart failure history, myocardial infarction history, and revascularization history.

Figure 1 shows Kaplan-Meier survival curves. The patients having the smoking history, history of heart failure, myocardial infarction, or a history of revascularization had a lower probability of MI/CV-death-free survival than the reference group during 12 years of follow-up.

Table 2 shows estimated hazard ratios from the joint models. Based on the results of the MI/CV-death joint models, age, gender, race, education, history of hypertension, history of myocardial infarction and history of heart failure (measured at baseline), eGFR, BMI, HDL, Troponin-I, and SBP (measured longitudinally) were all significant predictors of the hazard of MI/CV-death. For all-cause death joint model, age, gender, race, education, history of hypertension, history of diabetes mellitus, history of myocardial infarction, history of heart failure, history of revascularization (measured at baseline) and eGFR, BMI, HDL, Troponin-I, and SBP (measured longitudinally) were also significant predictors all-cause death. Compared with the MI/CV-death joint model, the slope of DBP, SBP and HbA1c were new significant predictors of all-cause death.

Table 2. Estimation of hazard ratio for joint model of MI/CV-death and all-cause death

	MI/CV-death				All-cause death			
	Coefficient	Hazard Ratio	95% CI	P-value	Coefficient	Hazard Ratio	95% CI	P-value
Baseline variables								
Age	0.012	1.012	(1.003, 1.021)	**0.009**	0.027	1.027	(1.019, 1.037)	**0.000**
Male	−0.440	0.644	(0.542, 0.763)	**0.000**	−0.366	0.693	(0.589, 0.816)	**0.000**
Black	0.289	1.335	(1.128, 1.601)	**0.002**	0.076	1.079	(0.914, 1.264)	0.353
Highest Level of Education (Elementary or Middle School as the reference group)	.	.	.	.				
Some High School	−0.008	0.992	(0.697, 1.434)	0.947	−0.027	0.974	(0.723, 1.335)	0.868
High School Graduate	−0.014	0.986	(0.717, 1.395)	0.904	−0.073	0.929	(0.713, 1.237)	0.585
Some College	0.011	1.011	(0.728, 1.43)	0.967	−0.169	0.845	(0.644, 1.131)	0.249
College Graduate	−0.352	0.703	(0.504, 1.005)	**0.052**	−0.287	0.751	(0.561, 1.012)	0.062
Graduate Education or Degree	−0.365	0.695	(0.487, 1.003)	**0.051**	−0.464	0.629	(0.469, 0.853)	**0.007**
History of Hypertension	0.223	1.250	(1.054, 1.497)	**0.009**	0.248	1.281	(1.096, 1.508)	**0.001**
History of Diabetes Mellitus	0.156	1.169	(0.999, 1.35)	**0.052**	0.205	1.228	(1.075, 1.408)	**0.002**

(*continued*)

Table 2. (*continued*)

	MI/CV-death				All-cause death			
	Coefficient	Hazard Ratio	95% CI	P-value	Coefficient	Hazard Ratio	95% CI	P-value
History of hypercholesterolemia	−0.027	0.973	(0.83, 1.138)	0.756	−0.226	0.798	(0.699, 0.914)	**0.001**
Ever smoker	0.128	1.136	(0.988, 1.314)	0.079	0.192	1.211	(1.063, 1.377)	**0.005**
History of myocardial infarction	0.220	1.246	(1.072, 1.454)	**0.004**	0.016	1.016	(0.879, 1.177)	0.818
History of heart failure	0.436	1.546	(1.349, 1.774)	**0.000**	0.515	1.673	(1.481, 1.887)	**0.000**
History of revascularization	0.202	1.224	(1.057, 1.431)	**0.007**	0.039	1.039	(0.914, 1.182)	0.550
Longitudinal variables								
eGFR (mL/min/1.73 m^2)	−0.017	0.984	(0.98, 0.987)	**0.000**	−0.018	0.982	(0.979, 0.986)	**0.000**
eGFR (slope) (mL/min/1.73 m^2/year)	−0.049	0.952	(0.921, 0.986)	**0.013**	−0.145	0.865	(0.799, 0.941)	**0.000**
log(BMI) (kg/m^2)	−1.422	0.241	(0.156, 0.371)	**0.000**	−1.827	0.161	(0.106, 0.243)	**0.000**
log(BMI) (slope) (kg/m^2/year)	−22.499	exp (−22.4991)	(exp(−29.4230), exp(−15.6880))	**0.000**	−95.336	0.000	(0, 0)	**0.000**
HDL (mg/dL)	−0.022	0.978	(0.97, 0.987)	**0.000**	−0.016	0.984	(0.976, 0.992)	**0.000**
HDL (slope) (mg/dL/year)	−0.815	0.443	(0.318, 0.618)	**0.000**	−1.694	0.184	(0.087, 0.417)	**0.000**
LDL (mg/dL)	0.000	1.000	(0.997, 1.004)	0.833	0.000	1.000	(0.996, 1.003)	0.932
LDL (slope) (mg/dL/year)	0.070	1.073	(0.998, 1.149)	0.060	0.187	1.205	(0.942, 1.52)	0.145
Troponin-I (ng/mL)	0.062	1.064	(1.05, 1.079)	**0.000**	0.043	1.044	(1.013, 1.076)	**0.006**
Troponin-I (slope) (ng/mL/year)	0.013	1.013	(0.975, 1.057)	0.519	0.061	1.063	(0.875, 1.302)	0.551
DBP (mmHg)	−0.005	0.995	(0.979, 1.012)	0.496	0.001	1.001	(0.984, 1.017)	0.934
DBP (slope) (mmHg/year)	0.003	1.003	(0.877, 1.153)	0.965	0.483	1.620	(1.077, 2.52)	**0.015**
SBP (mmHg)	−0.014	0.986	(0.978, 0.994)	**0.000**	−0.019	0.981	(0.974, 0.989)	**0.000**
SBP (slope) (mmHg/year)	0.023	1.023	(0.935, 1.115)	0.608	−0.299	0.742	(0.577, 0.949)	**0.019**
HbA1c (%)	−0.010	0.990	(0.894, 1.089)	0.860	−0.057	0.944	(0.839, 1.059)	0.337
HbA1c (slope) (%/year)	−0.808	0.446	(0.09, 1.953)	0.343	−9.308	0.000	(0, 0.004)	**0.000**

eGFR, estimated glomerular filtration rate; BMI, body mass index; HDL, high-density lipoprotein; LDL, low-density lipoprotein; DBP, diastolic Blood pressure; SBP, systolic blood pressure; HbA1c, hemoglobin A1c.

Table 3 presents the AUCs and Brier scores of Cox regression and the joint models. It shows that the AUCs of the joint models were lower than that of the Cox models. The Brier scores of joint models were higher than Cox regression models, which indicates that the prediction error of joint models was higher than the Cox regression model.

Table 3. Time-dependent AUC and Brier score of joint model and Cox model

		Δt (years)	Using information up to t follow-up (years)									
			2		3		4		5		6	
			Joint model	Cox model	Joint model	Cox model	Joint model	Cox model	Joint model	Cox model	Joint model	Cox model
AUC	MI/CV-death	2	0.616	0.698	0.628	0.694	0.638	0.670	0.631	0.686	0.635	0.657
		1	0.621	0.666	0.637	0.725	0.661	0.662	0.614	0.669	0.627	0.694
	All-cause death	2	0.641	0.693	0.611	0.701	0.651	0.701	0.648	0.706	0.639	0.703
		1	0.657	0.690	0.630	0.691	0.681	0.704	0.671	0.689	0.661	0.702
Brier Score	MI/CV-death	2	0.063	0.060	0.066	0.064	0.070	0.069	0.070	0.064	0.067	0.062
		1	0.034	0.033	0.033	0.032	0.037	0.037	0.039	0.037	0.035	0.032
	All-cause death	2	0.064	0.062	0.065	0.062	0.067	0.066	0.063	0.061	0.057	0.054
		1	0.034	0.034	0.034	0.033	0.034	0.034	0.037	0.037	0.030	0.028

t of the first row means the dynamic prediction uses the longitudinal measurements up to t years. Δt of the third column means the prediction interval of dynamic prediction. The third row represent the model used to dynamically predict the survival outcome. The AUCs were calculated to assess how well the longitudinal marker distinguished the status of patients at time $t + \Delta t$. The Brier score is a metric used to assess the precision of a predicted survival function at time $t + \Delta t$. Higher AUCs indicate higher discrimination of the models and lower Brier score indicates worse precision of prediction.

Figure 2 shows the dynamic prediction of all-cause death and MI/CV-death for two patients. As new longitudinal measurements were incorporated into the model, the linear mixed regression models were subsequently updated, and the risk function was simultaneously updated according to cumulative effect (the area under the model divided by the follow-up time). Lastly, the updated survival curve from the prediction time interval presented the predicted survival (event-free) probability.

The horizontal axis shows the number of years in follow-up, with a vertical dotted line indicating the time of longitudinal variables. The left-hand vertical axis displays eGFR, BMI, and HDL, with observed values denoted by stars and a solid line showing the longitudinal trajectory. The right-hand vertical axis presents the mean survival probability estimate and 95% confidence interval using dashed lines.

Table 3 presents the AUCs and Brier scores of Cox regression and the joint models. It shows that the AUCs of the joint models were lower than that of the Cox models. The Brier scores of joint models were higher than Cox regression models, which indicates that the prediction error of joint models was higher than the Cox regression model.

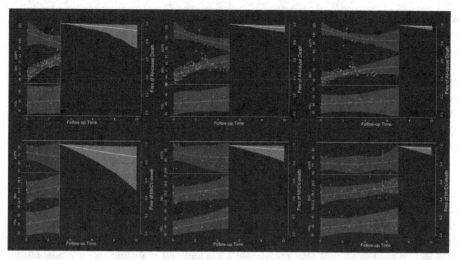

Fig. 2. The dynamic prediction of MI/CV-death and all-cause death probabilities for 2 different patients during follow-up

Figure 2 shows the dynamic prediction of all-cause death and MI/CV-death for two patients. As new longitudinal measurements were incorporated into the model, the linear mixed regression models were subsequently updated, and the risk function was simultaneously updated according to cumulative effect (the area under the model divided by the follow-up time). Lastly, the updated survival curve from the prediction time interval presented the predicted survival (event-free) probability.

4   Discussion

In this study, we built a joint model with multiple longitudinal variables to dynamically predict two survival outcomes, MI/CV-death and all-cause death, and found baseline and longitudinal variables from EHR that were significantly associated with survival outcomes. We also compared the discrimination power with the traditional Cox regression model based on time-dependent ROC of AUC. Results showed that the cumulative effect of variables such as eGFR, BMI, HDL and the slope over time was associated with survival outcomes. Based on AUC and Brier score, we did not find better discrimination power and prediction accuracy of joint model compared to Cox regression model.

Several approaches have been developed to accomplish dynamic prediction of conditional survival probabilities based on longitudinal and survival data, including joint models [20], landmark models [21], and random forests [10]. The landmark model takes into account only the most recent available measurement. For random forests, a limitation of RSF landmarking is that the predictions are not linked over time due to the use of independent RSF models at each landmark time result. If the longitudinal variables are extracted from EHR, monitoring of longitudinal variables may not always be organized as fixed follow-up intervals, so estimation of the model may introduce additional uncertainty and bias. The joint model has fewer restrictions on the longitudinal data, which is especially flexible for EHR. A rigidly specified follow-up plan is not required for joint modeling. These characteristics significantly increase the applicability of this joint model.

At each timepoint of longitudinal biomarkers measurement from EHR, our model can offer dynamic subject-specific predictions of MI/CV-death and all-cause of death. For patients who already have CAD, an accurate prediction model for their prognosis that is updated in real-time is crucial. This strategy may direct the frequency of tailored assessments and promote earlier diagnosis, hence improving prognosis and the timing of disease-modifying drug intervention, once accessible. The previous models [8, 22, 23] only considered the biomarkers measured at a few specific time points to the end of treatment. If these models are employed to forecast survival probability and stratify risk groups, individual variations in response to therapies will be disregarded. Although some joint models [24, 25] fulfill the dynamic prediction of cardiovascular risk, they only include one longitudinal biomarker in the model. Our model allows for a more comprehensive consideration of multiple factors associated with survival outcomes. Future prospective studies should also investigate a customized real-time therapy adaptation system to guide the treatment and health care based on the dynamic patterns of the longitudinal risk factors from EHR.

There are some limitations of our study. First, the joint model is computationally intensive, particularly for large datasets, leading to longer training and inference times. In our study, we have more than 6,000 subjects and 170,000 longitudinal records. More computation power is required for consideration of multiple longitudinal variables, their cumulative effects, and the trajectories over time. Second, the discrimination ability of the joint model did not show improvement compared with the Cox model. There could be several reasons. EHR data is not measured at regular intervals. Patients may have many measurements in a short period of time or no measurements for a very long time. As multiple variables are not measured simultaneously, large amount of missing data increases the computational effort, which may lead to inaccuracies in the estimation. Lastly, factors that are not accounted for by the proposed models may influence prediction and prediction performance.

References

1. Benjamin, E.J., et al.: Heart disease and stroke statistics-2019 update: a report from the American Heart Association. Circulation **139**(10), e56–e528 (2019)
2. Benjamin, E.J., et al.: Heart disease and stroke statistics-2017 update: a report from the American Heart Association. Circulation **135**(10), e146–e603 (2017)
3. Trogdon, J.G., Finkelstein, E.A., Nwaise, I.A., Tangka, F.K., Orenstein, D.: The economic burden of chronic cardiovascular disease for major insurers. Health Promot. Pract. **8**(3), 234–242 (2007)
4. D'Agostino, R.B., Sr., et al.: General cardiovascular risk profile for use in primary care: the Framingham Heart Study. Circulation **117**(6), 743–753 (2008)
5. Kannel, W.B., Dawber, T.R., Kagan, A., Revotskie, N., Stokes, J., 3rd.: Factors of risk in the development of coronary heart disease–six year follow-up experience: the Framingham Study. Ann. Intern. Med. **55**, 33–50 (1961)

6. Amor, A.J., et al.: Prediction of cardiovascular disease by the Framingham-REGICOR equation in the high-risk PREDIMED cohort: impact of the mediterranean diet across different risk strata. J. Am. Heart Assoc. **6**(3), e004803 (2017)
7. Chia, Y.C., Gray, S.Y., Ching, S.M., Lim, H.M., Chinna, K.: Validation of the Framingham general cardiovascular risk score in a multiethnic Asian population: a retrospective cohort study. BMJ Open **5**(5), e007324 (2015)
8. Sayadi, M., Zare, N., Attar, A., Ayatollahi, S.M.T.: Improved Landmark Dynamic Prediction Model to assess cardiovascular disease risk in on-treatment blood pressure patients: a simulation study and post hoc analysis on SPRINT data. BioMed Res. Int. **2020**, 2905167 (2020)
9. Keogh, R.H., Seaman, S.R., Barrett, J.K., Taylor-Robinson, D., Szczesniak, R.: Dynamic prediction of survival in cystic fibrosis: a landmarking analysis using UK patient registry data. Epidemiology **30**(1), 29–37 (2019)
10. Pickett, K.L., Suresh, K., Campbell, K.R., Davis, S., Juarez-Colunga, E.: Random survival forests for dynamic predictions of a time-to-event outcome using a longitudinal biomarker. BMC Med. Res. Methodol. **21**(1), 216 (2021). https://doi.org/10.1186/s12874-021-01375-x
11. Lin, X., Li, R., Yan, F., Lu, T., Huang, X.: Quantile residual lifetime regression with functional principal component analysis of longitudinal data for dynamic prediction. Stat. Methods Med. Res. **28**(4), 1216–1229 (2019)
12. Campbell, K.R., Martins, R., Davis, S., Juarez-Colunga, E.: Dynamic prediction based on variability of a longitudinal biomarker. BMC Med. Res. Methodol. **21**(1), 104 (2021). https://doi.org/10.1186/s12874-021-01294-x
13. Ferrer, L., Putter, H., Proust-Lima, C.: Individual dynamic predictions using landmarking and joint modelling: validation of estimators and robustness assessment. Stat. Methods Med. Res. **28**(12), 3649–3666 (2019)
14. Gong, X., Hu, M., Zhao, L.: Big data toolsets to pharmacometrics: application of machine learning for time-to-event analysis. Clin. Transl. Sci. **11**(3), 305–311 (2018)
15. Wulfsohn, M.S., Tsiatis, A.A.: A joint model for survival and longitudinal data measured with error. Biometrics **53**(1), 330–339 (1997)
16. Andrinopoulou, E.R., Rizopoulos, D., Jin, R., Bogers, A.J., Lesaffre, E., Takkenberg, J.J.: An introduction to mixed models and joint modeling: analysis of valve function over time. Ann. Thorac. Surg. **93**(6), 1765–1772 (2012)
17. Rizopoulos, D., Taylor, J.M., Van Rosmalen, J., Steyerberg, E.W., Takkenberg, J.J.: Personalized screening intervals for biomarkers using joint models for longitudinal and survival data. Biostatistics **17**(1), 149–164 (2016)
18. Ko, Y.A., Hayek, S., Sandesara, P., Samman Tahhan, A., Quyyumi, A.: Cohort profile: the emory cardiovascular biobank (EmCAB). BMJ Open **7**(12), e018753 (2017)
19. Rizopoulos, D., Papageorgiou, G., Miranda Afonso, P.: JMbayes2: extended joint models for longitudinal and time-to-event data. Version 0.3-0 (2022). https://CRAN.R-project.org/package=JMbayes2
20. Ibrahim, J.G., Chu, H., Chen, L.M.: Basic concepts and methods for joint models of longitudinal and survival data. J. Clin. Oncol. **28**(16), 2796–2801 (2010)
21. Liu, Q., Tang, G., Costantino, J.P., Chang, C.-C.H.: Landmark proportional subdistribution hazards models for dynamic prediction of cumulative incidence functions (2019). arXiv:1904.09002
22. Suchy-Dicey, A.M., et al.: Blood pressure variability and the risk of all-cause mortality, incident myocardial infarction, and incident stroke in the cardiovascular health study. Am. J. Hypertens. **26**(10), 1210–1217 (2013)
23. Haring, R., et al.: Association of sex steroids, gonadotrophins, and their trajectories with clinical cardiovascular disease and all-cause mortality in elderly men from the Framingham Heart Study. Clin. Endocrinol. **78**(4), 629–634 (2013)

24. Posch, F., Ay, C., Stoger, H., Kreutz, R., Beyer-Westendorf, J.: Longitudinal kidney function trajectories predict major bleeding, hospitalization and death in patients with atrial fibrillation and chronic kidney disease. Int. J. Cardiol. **282**, 47–52 (2019)

25. de Kat, A.C., Verschuren, W.M., Eijkemans, M.J., Broekmans, F.J., van der Schouw, Y.T.: Anti-Mullerian hormone trajectories are associated with cardiovascular disease in women: results from the Doetinchem cohort study. Circulation **135**(6), 556–565 (2017)

A Machine Learning Framework for Shuttlecock Tracking and Player Service Fault Detection

Akshay Menon[1]([envelope]) [iD], Abubakr Siddig[3] [iD], Cristina Hava Muntean[1] [iD],
Pramod Pathak[2] [iD], Musfira Jilani[1] [iD], and Paul Stynes[1] [iD]

[1] National College of Ireland, Mayor Street Lower, IFSC, Dublin 1, Ireland
akshaymenon81@yahoo.com
[2] Technological University Dublin, Grangegorman, Dublin, Ireland
[3] Griffith College, South Circular Road, Dublin 8, Ireland
https://www.ncirl.ie/, https://www.griffith.ie/,
https://www.tudublin.ie/

Abstract. Shuttlecock tracking is required for examining the trajectory of the shuttle-cock in badminton matches. Player Service Fault Detection identifies service faults during badminton matches. The match point scored by players is analyzed by the first referee based on the shuttlecock landing point and player service faults. If the first referee cannot decide, they use technology such as a third umpire system to assist. The current challenge with the third umpire system is based on the high number of marginal errors in predicting the match score. This research proposes a Machine Learning Framework to improve the accuracy of Shuttlecock Tracking and player service fault detection. The proposed framework combines a shuttlecock trajectory model and a player service fault model. The shuttlecock trajectory model is implemented using a pre-trained Convolutional Neural Network (CNN), namely Track-Net. The player service fault detection model uses Google MediaPipe Pose. A Random Forest classifier is used to classify the player's service faults. The framework is trained using the badminton world federation channel dataset. The dataset consists of 100000 images of badminton players and shuttlecock positions. The models are evaluated using a confusion matrix based on loss, accuracy, precision, recall, and F1 scores. Results demonstrate that the optimized TrackNet model has an accuracy of 90%, which is 5% more with 2.84% less positioning error compared to the current state of the art. The player service fault detection model can classify player faults with 90% accuracy using Google MediaPipe Pose, 10% more compared to the Openpose model. The machine learning framework for shuttlecock tracking and player service fault detection is of use to referees and the Badminton World Federation (BWF) for improving referee decision-making.

Keywords: CNN · TrackNet · MediaPipe · Shuttlecock tracking · Player service fault detection

1   Introduction

Badminton is one of the fastest sports in the world with the speed of the shuttlecock in the range of 480 to 493 km/h [1]. Hawkeye is a computer vision technology used in badminton sports to track the shuttlecock and identify badminton player's action and player's shot recognition [2]. The Hawkeye technology triangulates the position of the shuttlecock by cross matching images from several cameras placed around the court (see Fig. 1). One of the challenges in international badminton is the margin of error of 3.6 mm in computing the location of the shuttlecock from a graphic image when using Hawkeye technology [3]. The Badminton World Federation (BWF) are in the process of identifying technology which has errors of less than 3.6 mm.

The aim of this research is to investigate to what extent a machine learning framework for shuttlecock tracking and player service fault detection can accurately track the trajectory of the shuttlecock and detect player service faults. The major contribution of this research is a novel machine learning framework that combines a shuttlecock trajectory model and player service fault model to enhance match score analysis. A minor contribution is a Body Keypoint Landmark Coordinates dataset that consists of 94 images of player service faults with identified body landmarks which are present in Kaggle [4].

Fig. 1. Hawkeye Camera Position

2   Related Work

The use of computer vision technology has been used by referees in many sports, including badminton, to progress in the sports analysis sector [5]. The ball trajectory algorithm has been derived from various sports such as tennis and table tennis [6–8] which used temporal and spatial correlations to detect the fast motion of the ball using the Kalman filter, which was ineffective due to occlusion with 74% accuracy.

However, in the field of badminton specifically in [9], shuttlecock tracking methods such as YOLOv4 network [10], which has a precision of 88.39% when compared to SF-YOLOv4 [11] have been successfully used. Similarly, in the field of badminton robots, FTOC [12] used the AdaBoost algorithm for shuttlecock tracking using the OpenCV library. Whereas [13–15] used YOLOv3 where they achieved an average accuracy of 94.52% with 10.65 fps, the usage of the badminton robot is not cost-efficient as well as the probability of electrical error would be higher and challenging for match analysis. Hence, due to the fast image processing time, the solution would rely on a deep learning model using the convolutional neural network (CNN) algorithm independent of instrument usage. The SOTA (State of The Art Algorithm) [16] may not be an appropriate algorithm for shuttlecock tracking. Instead, our research will adopt the usage of CNN based framework, called TrackNet [17], a pre-trained deep learning model with a combination of VggNet-16 and a Deconvolutional Neural Network (Deconvnet) which provides a precision of 85%. The TrackNet model seems to fit into our research work, but we may need an enhanced model approach as it will be used in the real-time scenario to identify shuttlecock trajectory. The further research as referred to in [18] with computer virtuality on human posture estimation does not have a quantity robust approach when compared to research work in [19] which is focused on using HAR (human action recognition), which mostly used deep learning techniques based on CNN and LSTM. The usage of deep learning models such as CNN has been evolving in the field of badminton. The usage of such an algorithm in image processing does not need a processing phase as mentioned in the research paper by [20] as well as besides CNN, the second model for supervised deep learning is the recurrent neural network (RNN) model and one of the common RNN model is called Long Short-Term Memory (LSTM) with an accuracy of 92% and [21] suggest evaluation of different pre-trained deep CNN models such as GoogleNet model which has the highest classification and accuracy of 87.5% when compared to AlexNet, VggNet-16, and VggNet-19 in which justifies high accuracy for pose recognition with limitation of GPU usage which can affect processing time.

The images used from badminton matches in two classes hit and non-hit action make up the data inputs and trained using pre-trained AlexNet [22] CNN model obtained 98.7% accuracy of classification. The video analysis instead of a sensor-based approach mechanism is employed to dynamically generate a human skeleton topology map and a graph convolutional neural network-based skeleton action recognition method [23] to achieve action recognition. Experimental results demonstrate that the suggested method achieves 63%, 84%, and 92%, respectively, recognition accuracy on multiple benchmark datasets, enhancing the accuracy of human hitting action recognition. The player bounding box and skeleton are detected using YOLOv3 [12] and

OpenPose skeletal imaging [24] to predict player service fault detection which can be achieved when single or multiplayer pose estimation is recorded.

The Openpose package along with Yolov3 detects all human frames which is not required hence we have used Google MediaPipe pose as a model to detect service fault from a player in our research. Like our research, the MediaPipe model has been utilized in [25,26] martial art pose and Yoga poses classification with 95% accuracy using XgBoost.

In conclusion, the state of the art indicates that several models such as AlexNet, GoogleNet, VggNet-16, and VggNet-19 are used in the shuttlecock tracking out of which the TrackNet badminton algorithm has been improved and there is a need through experimentation to identify the optimal model for use in finding shuttle- cock trajectory. On the other hand, current research indicates that deep learning models can be optimized using CNN architecture for the detection of human activity recognition. Posing estimation is important for analyzing a player's service fault action on the badminton court. The 3D skeleton imaging is possible using a body landmark model which is a media Pipe and open pose package. But since we are aiming to detect only player action and remove another human pose from the frame due to which implementation of the Google MediaPipe model is suitable for our research work which generates 33 body keypoint landmarks for pose detection.

3   Research Methodology

The research methodology consists of five steps namely data gathering, data pre-processing, data transformation, data modelling, and evaluation (see Fig. 2).

The first step, *Data Gathering*, uses badminton match video data [27] from Yonex All England Open Championship BWF Super 1000 2018 of women singles between

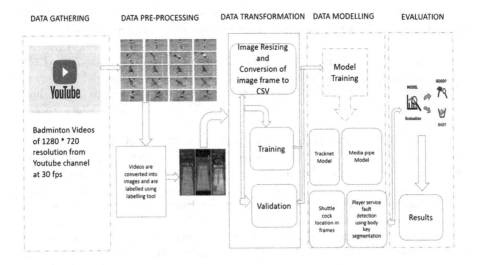

Fig. 2. Research Methodology

TAI Tzu Ying (TPE) vs Akane YAMAGUCHI (JPN). The video is scrapped for the first 7 min of the rally generating 12,180 frames with resolution of 1280 × 720 pixels at a frame rate of 30 fps. The seven-minute video was further divided into 12 videos each ranging from 5 sec to 20 sec. For training model, we have utilized 5 sec videos. Whereas the player service fault model uses video dataset [28] from BWF Development YouTube channel with relevant frame of 280.

The second step, *Data Pre-processing*, involves converting video to pictures for data generation. Images are resized from 1280 x 720 pixels to 512 x 288 pixels to enhance model performance and accelerate processing. In the case of shuttlecock trajectory, the resized images undergo labelling procedure and for the same this research have used Microsoft's visual object tagging tool which provides location of the shuttlecock in each frame and the details of the shuttlecock coordinates are recorded in CSV format which is used for data transformation and model training.

The data pre-processing of player service fault detection model entails skeletonizing the player pose in image dataset and then using the MediaPipe pose model to extract 33 3D landmark points in the x, y, and z axes. The 33 body keypoint landmark locations have x, y, and z coordinates served as the basis for estimating the player service fault pose. First, the OpenCV library was used to read the image. While MediaPipe pose requires RGB input, OpenCV can read data in BGR format. The images were initially pre-processed by changing it from BGR to RGB.

The third step, *Data Transformation*, involves labelled image data set of shuttlecock tracking and body keypoint landmark dataset of badminton player for service fault detection model. The dataset is in csv format which undergoes data transformation to form a train and test dataset. The shuttlecock tracking model generates a heat map (see Fig. 3) as ground truth which involves positioning of the shuttlecock in the image frames and results in 70% training data and 30% testing data. In total 156 images were generated, of which 109 training images and 47 test images were selected.

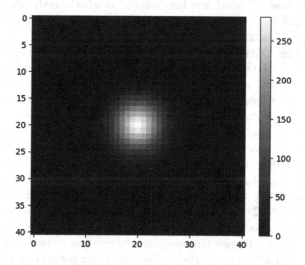

Fig. 3. Heat Map Detection

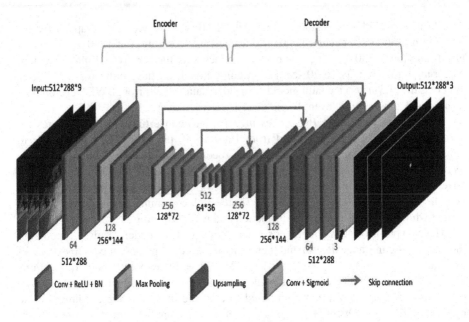

Fig. 4. TrackNet Architecture

Whereas, for player service fault model draws body landmark with key points which is in the form of X, Y, Z, and visibility where X, Y, Z are the coordinates of each pose with visibility as probability of landmark captured from each frame. The data collected has been converted to a network of keypoints for body posture, knee, foot, and ankle. The data transformation separates data into class variables and body landmarks. The class variable is estimation of player service fault which is categorized into "Not Foul" and "Foot Not Stationary" which was later label encoded. The body landmark consists of keypoints of each class. The data have been divided into 70:30 training and test data before applying classification.

The fourth step, *Data Modelling*, involves model training and implementation of the model. The TrackNet model architecture (see Fig. 4) for pre-trained transfer learning was employed in this study. It is a Fully Convolutional Network (FCN) model that uses VGG16 as an encoder to produce the feature map and DeconvNet as a decoder to decode utilizing the classification of the pixels. Multiple consecutive frames could be used as input for TrackNet, and the model will learn not only object tracking but also trajectory to improve its placement and recognition abilities. Gaussian heat maps centered on the shuttlecock were generated by TrackNet to show the ball's location. To calculate the difference between the heat map of the prediction and the ground truth, categorical cross-entropy was utilized as the loss function. TrackNet was trained with an image shape of (256,288,512). To predict the model weight to be used for retraining were trained with 30 epochs, categorical cross-entropy loss function, ReLU activation function, and SoftMax activation function, and tolerance was utilized. The models were optimized using Ada delta optimizer as well as some of the parameters such as learning rate, batch size, and epoch value are optimized during further retraining of the model.

The case of the player service fault detection model involves training the body key-point landmark data frame perceived through the MediaPipe pose model and evaluating through three classification model Random Forest, Support Vector Machine (SVM), and Decision Tree. The performance of the model was a good fit based on the evaluation matrix which was used for predicting player service faults.

The fifth step, *Evaluation*, involves evaluating the performance of the shuttlecock tracking model and player service fault detection model on the basis of accuracy, precision, and recall. To identify the optimal machine learning model, this research uses a pretrained TrackNet model [17]. The TrackNet model uses CNN based on heatmaps to precisely place the shuttlecock's location on the image frames. The player service fault detection model uses the Google MediaPipe Pose model [25] to identify the body landmark coordinates. The Google MediaPipe Pose model uses RGB video frames to infer 33 3D landmarks and a background segmentation mask on the entire body. Using RGB video frames, the machine learning technique infers 33 3D landmarks and a backdrop segmentation mask on the full body with random forest classifier model which seems to be suitable for player service fault prediction.

4   Design Specification

The machine learning framework combines shuttlecock tracking model and player service fault detection model (see Fig. 5). The components of the shuttlecock tracking model include a badminton YouTube video, image labeling, heat map image, and pretrained TrackNet model as discussed in Sect. 4.1 and components of the player service fault detection model as discussed in Sect. 4.2.

4.1   Shuttlecock Tracking Model

The shuttlecock tracking model was initiated with a dataset in the form of a badminton game video from BWF's YouTube channel. The initial resolution of the video was 1280 x 720 pixels, with a frame rate of 30 fps. The OpenCV library has been utilized to convert and read video frame images. The input images labelled generated coordinates of the shuttlecock in the frame. The labelled images are annotated by converting them into heat maps to find relative ground truth. After extracting the annotated images along with labelled images, the dataset undergoes train and test compilation to be applied to pre-trained CNN model known as TrackNet. The training of the TrackNet model with epoch, classes, load weight, and batch size creates saved weight which will be used to retrain the model and compare the two models. The model fit is evaluated based on accuracy, precision, recall, and PE in pixels.

4.2   Player Service Fault Detection Model

The player service fault detection model initiates through a video dataset from the BWF development YouTube channel. The video was downloaded with 1280 x 720 resolution running at 25.8 fps. The research work uses the OpenCV library to read each frame from the video. Each frame is converted from BGR to RGB and then it was ingested into the Google MediaPipe pose model. The pose model detects body landmarks with 33 keypoint. This player detection model focuses on service faults attempted by badminton players for which it calculates the angle between lower body parts such as the left leg using keypoints. The landmark coordinates detect two classes of pose estimation i.e.,"FOOT NOT STATIONARY" and "NOT FOUL".

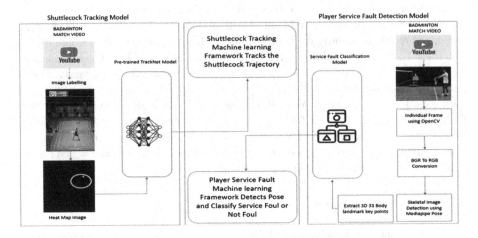

Fig. 5. Combined Machine Learning Framework

5   Implementation

The machine learning framework for shuttlecock tracking and player service fault detection was implemented on a Jupyter notebook along with a Google Colab notebook with Python version 3.9. The shuttlecock tracking model was implemented using following the Python libraries argparse, numpy, matplotlib, pillow, h5py, pydot, keras, tensorflow, opencv-python, CUDA 9.0 and cuDNN 7.0 for PyTorch (GPU) python libraries on Google colab notebook with GPU storage of 12 GB RAM and disk capacity of 78 GB as well as some parts of the script were implemented on Jupyter notebook with device configuration of 8GB RAM. The train and test images utilize keras library with tensorflow CPU backend to train TrackNet model. For convenience, TrackNet which takes one input frame is named model-I, and TrackNet which takes three consecutive input frames is named model-II. Both models were evaluated using accuracy, precision, and recall. A sample output of the TrackNet model-II of shuttlecock trajectory with three consecutive frames can be seen in (see Fig. 6). On the other hand, the player service fault detection machine learning model was implemented using MediaPipe pose model, OpenCV, Numpy, and Sklearn.

Fig. 6. Shuttlecock Trajectory on 1st Frame and 2nd Frame.

6   Evaluation and Results Analysis

This section evaluates the performance of the two-machine learning framework. Section 6.1 analyses the performance of the shuttlecock tracking model. Whereas Sect. 6.2 evaluates the performance of the player service fault detection.

6.1   Performance of the Shuttlecock Tracking Model

As per the first experiment, the TrackNet model was implemented on the first frame with a learning rate 1.0, batch size 2, steps per epoch 200, epochs 500, initial weights random uniform, and range of initial weights which detects shuttlecock trajectory with accuracy, precision, and recall as 85.0%, 57.7%, and 68.7% respectively. The researchers in [17] focused on training one image per frame which can cause occlusion. Hence, the Track-Net model-I using one frame to train which was used by previous research, and our model which was TrackNet model-II uses three consecutive images to train a TrackNet model with the accuracy of model-II having higher accuracy, precision, and recall than TrackNet model-I with 90%, 97%, and 92% respectively.

The prediction details of TrackNet model-I and model-II is shown in (see Fig. 7) with respect to the positioning error. TrackNet model-I and TrackNet model-II evaluation is based on TP, FP, TN, and FN which stand for true positive, false positive, true negative, and false negative, respectively. The "Visibility Class" is the key feature in predicting FP or FN in the image frame. The visibility class consists of VC1, VC2, and VC3 where FP justifies PE is greater than 7.5 pixels where FN justifies no shuttlecock detected or more than one shuttlecock detected when there is only one shuttlecock in the frame. The PE graph illustrates the percentage of error opted by two models while detecting the shuttlecock in the frame. The TrackNet model II seems to have less error positioning with 0.28% of the error margin which justifies the detection of occlusion in the frame and could use neighboring image frames for shuttlecock detection. The evaluation of the metrics is based on precision, recall, and F-1 score.

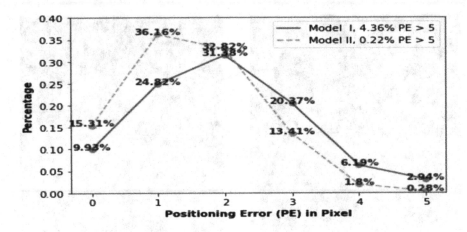

Fig. 7. TrackNet Model Positioning Error

6.2   Model Performance and Comparison of Player Service Fault Detection

The aim of the experiment on player service fault detection framework is to compare the performance of various machine learning algorithms on a 3D body landmark dataset. Three machine learning models namely decision tree, SVM, and random forest were analyzed. Comparison of the various classifier models based on the ROC curve. The AUC of the random forest classifier has the highest score when compared to the other two models (see Fig. 8). The accuracy score of random forest classifier stands out with 90% accuracy when compared to the other models followed by decision tree and SVC. The random forest classifier is used to develop and save models which will be used to predict the outcome of the player service fault detection model.

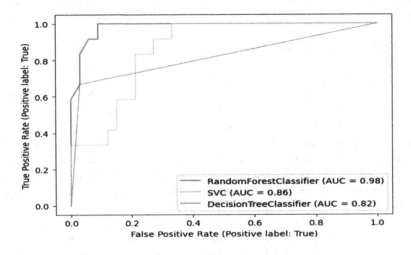

Fig. 8. ROC Curve Display of Model

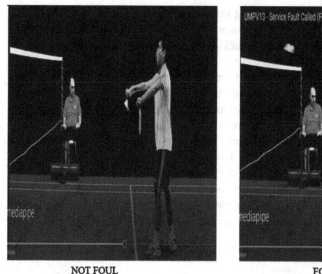

NOT FOUL   FOOT NOT STATIONARY

Fig. 9. Service Foul Classification

The prediction of the output post random forest classifier model produces predicted the output of the player attempting "FOOT NOT STATIONARY" service foul and has been detected with accuracy as Foul Classification (see Fig. 9). One of the major challenges faced during research work is to gather data which is relevant to the project. In the case of player service fault detection, the relevant dataset was unavailable hence we needed to create a dataset by scrapping video frames from YouTube channels and post-converting frames into CSV which has relevant training body landmarks used for predicting the model. Whereas, compared with the shuttlecock tracking model, occlusion in the video frame was a challenge which eventually prompted the use of multiple frames from training causing a high amount of CPU and GPU usage. In the real-time scenario, the shuttlecock tracking model needs to be tuned a bit further as well as the player service fault detection framework.

7   Conclusion and Future Work

This research aims to analyze to what extent a machine learning framework for shuttlecock tracking and player service fault detection can accurately track the trajectory of the shuttlecock and detect player service faults. The machine learning framework uses a shuttlecock tracking model that helps to predict the path and location of the shuttlecock from the projectile motion to the landing point on the badminton court. The badminton technical official currently uses Hawkeye camera technology to detect shuttlecock placement, but it has a marginal error that has challenged technical officials and players to re-analyze the outcome of the match. The shuttlecock Tracking model generated in our research known as TrackNet model II has been optimized by modifying the hyper-parameters and using three consecutive frames which resulted in an accuracy score of 90% which is better than other state of the art algorithm for shuttlecock tracking by 5%. On the Other hand, the Player Service Fault Detection model uses the Google MediaPipe Pose model to generate 3D body keypoint landmarks which were used for predicting service faults from video for which we trained data with a random forest classifier which in turn provided accuracy of 90% with a difference of 10% more compared to OpenPose model which detects multiple people including audience and referee in the frame which does not fit with our research requirement. The research to quantify badminton games such as shuttlecock tracking and player service fault detection has a scope of implementation in real-time scenarios as more research work in the future is needed. The further research work would enhance accuracy of the badminton match score as well as for player service fault detection multiple service fault with multiplayer action detection would be the scope of work which need to be worked upon and included in the next phase of the research.

References

1. Shuttlecock Speed. https://olympic.ca/2014/09/11/shuttlecock-and-balls-the-fastest-moving-objects-in-sport/
2. Hawk eye Analysis. https://bwfbadminton.com/news-single/2014/04/04/hawk-eye-to-determine-in-or-out
3. Hawk eye Error. https://bwfworldtour.bwfbadminton.com/newssingle/2021/11/21/bwf-infront-pan-asia-and-hawk-eye-statement
4. Body Keypoint Landmark. https://www.kaggle.com/datasets/axeireland/body-keypoint-landmark-coordinates-dataset
5. Jayalath, L.: Hawk eye technology used in cricket (2021)
6. Zhang, F., et al.: MediaPipe hands: On-device real-time hand tracking, arXiv preprint arXiv:2006.10214 (2020)
7. Thomas, G., Gade, R., Moeslund, T.B., Carr, P., Hilton, A.: Computer vision for sports: current applications and research topics. Comput. Vis. Image Underst. **159**, 3–18 (2017)
8. Zhao, Q., Lu, Y., Jaquess, K.J., Zhou, C.: Utilization of cues in action anticipation in table tennis players. J. Sports Sci. **36**(23), 2699–2705 (2018)
9. Shishido, H., Kitahara, I., Kameda, Y., Ohta, Y.: A trajectory estimation method for badminton shuttlecock utilizing motion blur. In: Klette, R., Rivera, M., Satoh, S. (eds.) PSIVT 2013. LNCS, vol. 8333, pp. 325–336. Springer, Heidelberg (2014). https://doi.org/10.1007/978-3-642-53842-1_28
10. Reno, V., Mosca, N., Marani, R., Nitti, M., D'Orazio, T., Stella, E.: Convolutional neural networks based ball detection in tennis games. In: Proceedings of the IEEE Conference on Computer Vision and Pattern Recognition Workshops, pp. 1758–1764 (2018)
11. Chen, Z., Li, R., Ma, C., Li, X., Wang, X., Zeng, K.: 3D vision based fast badminton localization with prediction and error elimination for badminton robot. In: 2016 12th World Congress on Intelligent Control and Automation (WCICA), pp. 3050–3055 (2016)
12. Cao, Z., Liao, T., Song, W., Chen, Z., Li, C.: Detecting the shuttlecock for a badminton robot: a yolo based approach. Expert Syst. Appl. **164**, 113833 (2021)
13. Chen, Y.-T., Yang, J.-F., Tu, K.-C.: Smart badminton detection system based on scaled-YOLOV4. In: International Symposium on Intelligent Signal Processing and Communication Systems (ISPACS), pp. 1–2. IEEE 2021 (2021)

14. Chen, W., et al.: Using FTOC to track shuttlecock for the badminton robot. Neurocomputing **334**, 182–196 (2019)
15. Vrajesh, S.R., Amudhan, A., Lijiya, A., Sudheer, A.: Shuttlecock detection and fall point prediction using neural networks. In: International Conference for Emerging Technology (INCET), pp. 1–6. IEEE 2020 (2020)
16. Lee, C.-L.: Badminton shuttlecock tracking and 3D trajectory estimation from video(2016)
17. Huang, Y.C., Liao, I.N., Chen, C.H., Ik, T.U., Peng, W.C.: TrackNet: a deep learning network for tracking high-speed and tiny objects in sports applications. In: 2019 16th IEEE International Conference on Advanced Video and Signal Based Surveillance (AVSS), pp. 1–8. IEEE (2019)
18. Zhu, C., Shao, R., Zhang, X., Gao, S., Li, B.: Application of virtual reality based on computer vision in sports posture correction. Wirel. Commun. Mob. Comput. **2022**, 1–15 (2022)
19. Host, K., Ivašić-Kos, M.: An overview of human action recognition in sports based on computer vision. Heliyon, e09633 (2022)
20. Rahmad, N.A., As'Ari, M.A., Ghazali, N.F., Shahar, N., Sufri, N.A.J.: A survey of video-based action recognition in sports. Indonesian J. Electr. Eng. Comput. Sci. **11**(3), 987–993 (2018)
21. binti Rahmad, N.A., binti Sufri, N.A.J., bin As' ari, M.A., binti Azaman, A.: Recognition of badminton action using convolutional neural network. Indonesian J. Electr. Eng. Inform. (IJEEI) **7**(4), 750–756 (2019)
22. Rahmad, N.A., As'ari, M.A., Ibrahim, M.F., Sufri, N.A.J., Rangasamy, K.: Vision based automated badminton action recognition using the new local convolutional neural network extractor. In: Hassan, M.H.A., et al. (eds.) MoHE 2019. LNB, pp. 290–298. Springer, Singapore (2020). https://doi.org/10.1007/978-981-15-3270-2_30
23. Liu, J., Liang, B.: An action recognition technology for badminton players using deep learning. Mob. Inf. Syst. **2022** (2022)
24. Cao, Z., Simon, T., Wei, S.E., Sheikh, Y.: Realtime multi-person 2D pose estimation using part affinity fields. In: Proceedings of the IEEE Conference on Computer Vision and Pattern Recognition, pp. 7291–7299 (2017)
25. Cabo, E.E., Salsalida, J.T., Alar, H.S.: Utilizing mediaPipe blazepose for a real-time pose classification of basic arnis striking and blocking techniques. SSRN 3992159
26. Sunney, J., Jilani, M., Pathak, P., Stynes, P.: A real-time machine learning framework for smart home-based yoga teaching system. In: 7th International Conference on Machine Vision and Information Technology (CMVIT 2023), Xiamen, China, 24–26 February 2023
27. YouTube Dataset Shuttlecock Tracking. https://youtu.be/PCyNbtMVkpI
28. YouTube Dataset Player Service Fault Detection. https://olympic.ca/2014/09/11/shuttlecock-and-balls-the-fastest-moving-objects-in-sport/

An Automated Dual-Module Pipeline for Stock Prediction: Integrating N-Perception Period Power Strategy and NLP-Driven Sentiment Analysis for Enhanced Forecasting Accuracy and Investor Insight

Siddhant Singh[1]([✉]) and Archit Thanikella[2]([✉])

[1] Memorial High School, Frisco, TX, USA
`singhsiddhantakshat@gmail.com`
[2] Panther Creek High School, Frisco, TX, USA
`archit.thanikella@gmail.com`

Abstract. The financial sector has witnessed considerable interest in the fields of stock prediction and reliable stock information analysis. Traditional deterministic algorithms and AI models have been extensively explored, leveraging large historical datasets. Volatility and market sentiment play crucial roles in the development of accurate stock prediction models. We hypothesize that traditional approaches, such as n-moving averages, may not capture the dynamics of stock swings, while online information influences investor sentiment, making them essential factors for prediction. To address these challenges, we propose an automated pipeline consisting of two modules: an N-Perception period power strategy for identifying potential stocks and a sentiment analysis module using NLP techniques to capture market sentiment. By incorporating these methodologies, we aim to enhance stock prediction accuracy and provide valuable insights for investors.

Keywords: Stock forecasting · Sentiment analysis · Automated pipeline · N-perception strategy · NLP · Market strategies · Predictive models · NBOS-OPT · Market sentiment polarity · N-observation period optimizer

1 Introduction

Investors actively participating in the stock market face the formidable challenge of selecting the most suitable stocks for their investments, given the presence of elusive parameters and complex market dynamics. Accurately forecasting a company's performance over an extended timeframe is further complicated by the existence of unknown variables that defy precise modeling, even with the aid of advanced AI algorithms. To make realistic predictions, a comprehensive understanding of various factors is essential, including an organization's service offerings, market competitiveness, consumer preferences, and other relevant information, all of which must be carefully integrated into the predictive models.

D. Conte et al. (Eds.): DeLTA 2023, CCIS 1875, pp. 84–100, 2023.
https://doi.org/10.1007/978-3-031-39059-3_6

The realm of stock market analysis and prediction has been a subject of intense research interest for many decades. Early methods, originating as far back as the 1800s, relied on candlestick patterns for analyzing stock behavior. Statistical analysis techniques were subsequently employed to identify and interpret these patterns, providing valuable signals to guide investors in their decision-making process. As computational capabilities advanced, a wide range of techniques emerged, encompassing classical variants of moving average techniques to more recent developments in deep learning, capable of predicting stock values at high frequencies, such as within seconds or minutes [1, 2]. While moving average techniques are commonly used to filter out noise from stock data, they may fall short in capturing the most recent and subtle price trends.

The Efficient Market Hypothesis (EMH) [3], a cornerstone theory in finance, postulates that all available information entering the market at any given time is rapidly incorporated into stock prices, rendering it virtually impossible to fully incorporate all relevant information using technical analysis. According to this hypothesis, The movement of prices in financial markets is characterized by a random walk pattern, indicating that they lack predictable patterns or trends. Furthermore, the absence of reliable indicators driving market volatility exacerbates the difficulty of forecasting stock prices with high precision [4].

In our pursuit of devising effective stock investment strategies, we adopt a flexible approach that considers two key two essential parameters in the market are volatility and market sentiment, which play significant roles in shaping the dynamics of financial markets. Even though these frameworks showcase a similarity, we treat them as distinct entities, each requiring a dedicated modeling step. In the initial step, we adopt a straightforward yet powerful approach by modeling volatility, enabling us to recognize the optimal number of stocks for investment. Rather than directly predicting stock prices, we utilize historical data and S&P indicators to analyze appropriate set of stocks. Subsequently, in the subsequent phase, we leverage investor sentiment to make well-informed decisions regarding the selected stock holdings. To track investor sentiment, we employ a cutting-edge attention-based models [5] that analyzes a vast collection of news headline data. This model allows us to classify news headlines into three categories (Buy, Sell, Neutral), and to facilitate the advancement of research in this domain, we have made our manually labeled dataset of 1,713 news headlines publicly available. Our approach has yielded promising results, outperforming baseline methods and successfully identifying optimal stocks for investment.

The remainder of the paper is structured as follows: Sect. 2 provides a comprehensive overview of the related work, highlighting key research contributions in the field of stock market analysis and prediction. In Sect. 3.1, we present the stock selection approach based on the N-Observation period, outlining the methodology employed to identify stocks with favorable investment potential [32]. Section 3.2 delves into our approach for discerning market sentiment, shedding light on the techniques utilized to extract valuable insights from news headlines. The datasets used in our research are described in Sect. 4, while Sect. 5 details the training approaches employed to optimize our models. Finally, in Sect. 6, we present the experimental results, offering a comprehensive evaluation of the effectiveness and performance of our proposed approach.

2 Related Work

The field of stock market analysis and prediction has garnered significant attention, leading to a wealth of research and diverse approaches. This section presents a comprehensive overview of related work, focusing on the identification of stock price trends, the utilization of machine learning algorithms, and the exploration of news sentiment correlation with stock prices.

To identify stock price trends, moving average techniques have been extensively employed. Studies such as [7, 8], and [9] have explored different variations of moving average methods, including exponential moving averages, weighted moving averages, and adaptive moving averages. These techniques aim to filter out noise and generate signals that aid in making informed investment decisions. Moreover, research [10] introduces advanced moving average strategies, such as the double crossover technique and the triple crossover technique, which consider multiple moving averages to provide enhanced trend analysis.

Machine learning algorithms have emerged as powerful tools in stock prediction. Various studies have employed linear models [11, 12], ensemble methods [13], support vector machines (SVM) [14], neural networks [15], and deep learning architectures [26, 27]. Linear regression models have been utilized to capture linear relationships between input features and stock prices [11], while ensemble methods like random forests and gradient boosting have leveraged the combination of multiple models for improved accuracy [13]. SVMs, known for their ability to handle high-dimensional data, have been applied to predict stock prices by considering various input factors [14]. Neural networks, including feedforward networks and recurrent neural networks (RNN), have been used to capture complex patterns and temporal dependencies in stock data [15]. Deep learning architectures, such as convolutional neural networks (CNN) and long short-term memory (LSTM) networks, have also exhibited promising results in stock prediction [16, 17].

The relationship between news sentiment and stock prices has been investigated extensively. Studies such as [18, 19], and [20] have explored the impact of news sentiment on stock market behavior. Sentiment analysis techniques, including lexicon-based approaches, machine learning-based approaches, and deep learning-based approaches, have been utilized to analyze news articles, social media posts, and financial reports. Moreover, natural language processing (NLP) methods, such as word embeddings and transformer models, have facilitated more accurate sentiment analysis and prediction of stock price movements [21, 22].

In addition to traditional approaches, recent advancements have explored the incorporation of alternative data sources for stock prediction. Research [23] investigates the utilization of social media data, such as tweets and sentiment analysis on Twitter, to predict stock market movements. Textual analysis of corporate financial reports and analyst reports has also been explored to gain insights into future stock performance [24, 25]. Furthermore, studies have explored the integration of alternative data sources, such as satellite imagery, web scraping, and geolocation data, to enhance stock prediction models [26–28].

It is worth noting that the development of robust stock prediction models necessitates careful feature engineering and model evaluation techniques. Studies such as [29, 30], and [31] have focused on feature selection and dimensionality reduction techniques, including principal component analysis (PCA), factor analysis, and autoencoders, to extract relevant features and reduce noise in stock data. Furthermore, research emphasizes the importance of model evaluation metrics, such as accuracy, precision, recall, and F1-score, to assess the performance of stock prediction models accurately.

In summary, a diverse range of approaches has been employed in the field of stock market analysis and prediction. Moving average techniques, machine learning algorithms, sentiment analysis, and the integration of alternative data sources have provided valuable insights for accurate stock price forecasting. Furthermore, feature engineering and model evaluation techniques play crucial roles in developing robust stock prediction models.

3 Architecture

In this section, we present the architecture of our proposed solution, NBOS-OPT, which aims to enhance the selection of hot stocks by optimizing the N-Observation period and incorporating a decision module that utilizes machine learning to generate signals guided by the opposition of investor sentiment. This NBOS-OPT conduit consists of two main modules: the Stock Picker and the Decision Module.

3.1 Stock Picker

The Stock Picker module plays a crucial role in the selection of stocks for investment. While moving average techniques have been extensively studied for their noise reduction capabilities in stock volatility, they may not promptly capture recent market changes. To address this limitation, we introduce the period optimizer, N-Observation period optimizer (NBOS-OPT), that optimizes investments within a specified observation period to make predictions for mid-range investment periods, typically ranging from 5 to 30 days.

The NBOS-OPT model leverages recent stock performance, assuming that if a stock has performed well in the recent past, it is likely to continue the trend in the short term. The performance of stocks over N observation periods and select the most optimal set based on their recent performance. To simplify the approach, the recent stock performance into N observation periods and use robust indicators to evaluate their effectiveness and make predictions about their future performance.

We compute the win ratio for each stock using the following function:

$$\psi_{\mathcal{P},s}^{OPT\_SCR} = \sum_{p_i \in \mathcal{P}} IS\_OPT\_STOCK(s, p_i)$$

Here, s represents a stock, From a given set of observation periods of interest P, IS_OPT_STOCK (s, p) determines if stock s can be declared a winner for the given observation period p [32]. We compare the performance of each stock to that of the S&P index and declare it a winner if it performs better than the S&P. The function returns the frequency of wins for stock s over the given observation periods [32].

Next, we calculate the win ratio ψ for all stocks and select those stocks that perform better than a predefined threshold. Through experimentation, we have found that a threshold of 0.8 provides an optimal balance between bucket size and forecast performance for the chosen period. The success of the algorithm is influenced by its hyperparameters, which consist of the length of the observation period (O), the number of observation periods (N), and the threshold win ratio (T) [32]. To understand the impact of these parameters on the algorithm's results, we conduct a sensitivity analysis [32].

Benchmark: To facilitate a comprehensive comparison, we propose a fundamental model referred to as OBS-OPT, which has been integrated into our algorithm. This model is distinctive in its usage of a singular observation period, designated by the length O * N. It utilizes the S&P indicator to pinpoint stocks that supersede the market during the defined observation timeframe.

Our algorithm incorporates the OBS-OPT protocol through an iterative process. For each stock in our dataset, we employ the IS_OPT_STOCK function, which takes into account sentiment analysis and the N-Perception Period Power Strategy. If a stock is classified as optimal, we delve further into sentiment analysis and power strategy data for that particular stock. These data are then combined and evaluated to reassess the stock's optimal status. If the stock remains optimal post-evaluation, it is added to the SelectedStocks list, along with its associated N-Perception Period Power Strategy and sentiment analysis data. Despite the simplicity of OBS-OPT leading to a typically extensive pool of shortlisted stocks, the model provides a robust benchmark for our study. The risk score for each stock in the SelectedStocks list is calculated, followed by sorting the list based on these risk scores. The most optimal stock is then selected and removed from the SelectedStocks list. Hence, OBS-OPT provides an effective groundwork that supports and strengthens the performance of our algorithm.

Both OBS-OPT and NBOS-OPT algorithms are evaluated in this study. OBS-OPT is an algorithm that determines which stocks are outperforming the S&P index over a given time span. NBOS-OPT builds upon the single-period approach by incorporating N observation periods to calculate a win ratio for the identified stocks.

In NBOS-OPT, users can configure the observation period (O), the number of observation periods (N), and the threshold score (T), which filters the stocks based on their computed win ratio. Achieving a near 1 win ratio requires consistent performance by the identified stocks over recent periods. Our results show that configuring the win ratio between 0.75 and 0.9 results in clearly defined stock buckets and a high degree of accuracy in prediction performance.

Algorithm 1: OBS-OPT

---

1. Input: Observation Period (O), Stocks (s1, s2, ..., sn), NLP techniques
2. Output: Selected Stocks (s1', s2', ..., sk') where k' is the number of selected stocks
3. Set SelectedStocks = []
4. for i = 1 to n do
5. isOptStock = IS_OPT_STOCK(s1, O, Sentiment Analysis, N-Perception Period Power Strategy)
6. if isOptStock then
7. SentimentData = ANALYSE_SENTIMENT(s1, NLP techniques)
8. PowerData=NPERCEPTION_PERIOD_POWER_STRATEGY(s1)
9. re-sult=COMBINE_SENTIMENT_AND_PERCEPTION_DATA(Sentiment Data, PowerData)
10. isOptStock = CHECK_OPT_RESULT(result)
11. if isOptStock then
12. SelectedStocks.add(s1, N-Perception Period Power Strategy, Sentiment Analysis)
13.   end if
14.   end if
15. end for
16. CalculateRiskScoreForSelectedStocks(SelectedStocks)
17. SortSelectedStocksByRiskScore(SelectedStocks)
18. SelectMostOptimalStock(SelectedStocks)
19. RemoveMostOptimalStock(SelectedStocks)
20. return SelectedStocks

Algorithm 2: NOBS-OPT

---

1. Input: Observation Period (O), Number of Periods (p), Stocks (s1, s2, ..., sn), Threshold Score (T), NLP techniques
2. Output: Selected Stocks (s1', s2', sk') where k' is the number of selected stocks
3. Set SelectedStocks = []
4. for i = 1 to n do
5. winRatio = 0
6. for j = 1 to p do
7. isOptStock = IS_OPT_STOCK(s1, Oj, Sentiment Analysis, N-Perception Period Power Strategy)
8. if isOptStock then
9. SentimentData = ANALYSE_SENTIMENT(s1, NLP techniques)
10. PowerData=N PERCEPTION_PERIOD_POWER_STRATEGY(s1)
11. result =COMBINE_SENTIMENT_AND_PERCEPTION_DATA(SentimentData, PowerData)
12. isOptStock = CHECK_OPT_RESULT(result)
13. winRatio += 1/p
14. end if
15. end for
16. if winRatio > T then
17. SelectedStocks.add(s1, N-Perception Period Power Strategy, Sentiment Analysis)
18. end if
19. end for
20. return SelectedStocks

3.2 AI Model

Our proposed solution incorporates sentiment analysis, a powerful natural language processing (NLP) technique, to determine the emotional tone of text and its relevance to stock investments. Sentiment analysis can be applied at different levels of granularity, such as documents, paragraphs, sentences, and phrases, and its effectiveness depends on the domain and the type of text being analyzed. In our solution, we employ an AI model that utilizes sentiment analysis to classify news headlines into three categories: Buy, Sell, or Neutral (Fig. 1).

The NBOS-OPT Stock Selection Algorithm leverages the latest advances in NLP to provide efficient and accurate stock selection strategies. The algorithm utilizes sentiment analysis, coupled with N-Perception period power strategies, to accurately identify optimal stocks within a given N-period Observation Window. This provides investors with the resources and insight to optimize their investments and maximize returns.

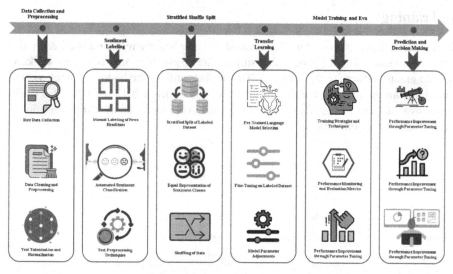

Fig. 1. Model architecture diagram

Transfer learning plays a crucial role in our AI model. Training transformers from scratch can be computationally intensive and impractical for many applications. However, transfer learning offers a practical approach by fine-tuning pre-trained transformer models with domain-specific data. This approach reduces the training time required for domain-specific data and enables the effective utilization of smaller datasets. It is important to note that transfer learning inherits any inherent biases present in the pre-trained models.

To determine the sentiment of news headlines and classify them as Buy, Sell, or Neutral, we employ a transformer model. This sentiment classification is instrumental in making informed investment decisions. Our approach involves fine-tuning a pre-trained transformer model using a carefully curated dataset comprising 1.7K labeled data points. This dataset includes news headlines related to various companies, collected from multiple sources.

The news headlines in our dataset are manually categorized as Buy, Sell, or Neutral to reflect the sentiment conveyed by the news. The resulting dataset is well-balanced, with 1046 samples labeled as Buy, 353 as Sell, and 314 as Neutral. It is important to maintain a balanced distribution of labeled samples across different sentiment categories to ensure representative coverage.

In the upcoming sections, we will delve into the details of our AI model, including the fine-tuning process, evaluation metrics, and the performance of the sentiment analysis module in classifying news headlines. By leveraging the power of sentiment analysis, we aim to provide valuable insights into market sentiment and enhance the decision-making process for stock investments.

4 Training

During the training phase, Our optimization module heavily relies on the S&P indicator to accurately identify stocks that surpass the threshold score and are most likely to yield greater returns in the long run. The module utilizes the historical stock values obtained from the dataset, which spans the time period from 2013 to 2020. This comprehensive dataset enables us to conduct rigorous experiments and accurately evaluate the performance of our model.

Table 1. Training and testing dataset

	Buy	Sell	Neutral
Total	2050	650	620
Training	1640	490	500
Testing	410	160	120

To enhance the effectiveness of our model, we collect news headlines from various sources that pertain to different companies. These news headlines are hand-labeled as buy, sell, or neutral, providing valuable sentiment information for our training process. The dataset consists of a total of 3,320 samples, with 2,050 labeled as buy, 650 as sell, and 620 as neutral. Table 1 displays the distribution of the dataset used for the train-test split (Fig. 2).

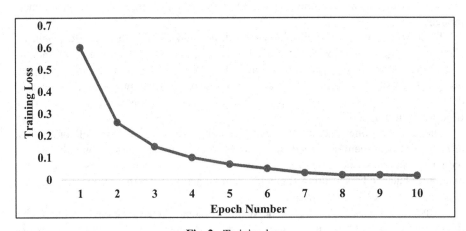

Fig. 2. Training loss

We incorporate this stock history and news sentiment data into the training process using the Transformers library in Python. The transfer learning technique of our model utilizes distilbert-base-uncased-finetuned-sst-2-english with hand-labeled data. We utilize the Transformers library, noted for its facile integration with MLflow, to facilitate rigorous experimental procedures as outlined in Reference [32].

Throughout the training iterations, the model undergoes a comprehensive training process for a span of ten epochs. It is discernible that the F1-Score approaches a plateau within this epoch range, suggesting that any furtherance of training does not contribute substantially towards the enhancement of the model's performance. It is important to note that the sentiment distribution in the dataset is imbalanced, with a higher proportion of buy and sell labels compared to neutral labels. To ensure fair comparison and evaluation of the results, we utilize the F1 score as a metric to represent the model's performance (Fig. 3).

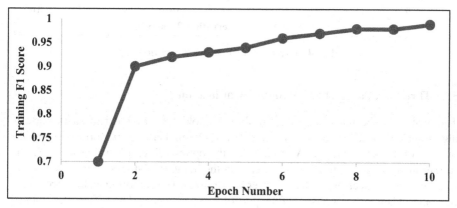

Fig. 3. F1 score

We evaluate the performance of the model on the test dataset, consisting of 410 positive, 160 negative, and 120 neutral data points. The model's performance on the test dataset is presented in Table 4, providing insights into its predictive capabilities and effectiveness in sentiment analysis for stock investment decision-making.

5   Experimental Results and Discussion

5.1   Observation Period Length and Performance Evaluation

In our study, We evaluated the performance of three models--the vanilla S&P indicator, the baseline model (OBS-OPT), and the NOBS-OPT model--over an 8-year period between 2003 and 2020, based on two factors: the bucket size and the performance of selected stocks over the forecasting period.

We found that the performance of the model is limited when observation is too long. In such cases, there are fewer periods available for evaluation, and information from more distant periods does not contribute significantly to identifying optimal stocks. Through a range of observation periods from 1 to 15, we determined that an observation period length of 10 consistently yielded the best results. Selecting an appropriate observation period length is crucial for achieving optimal performance (Fig. 4).

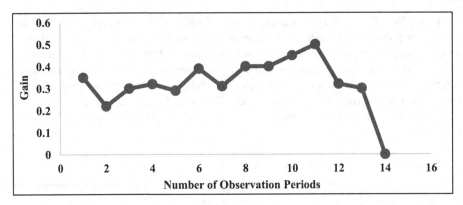

Fig. 4. Comparison of observation periods

5.2 Threshold Parameter and Stock Identification

The analysis indicated that increasing the threshold parameter led to an increase in algorithm performance. However, there is a trade-off between performance and the number of stocks identified. We found that the optimal threshold value was 0.8, and further increasing it resulted in a decrease in the number of stocks identified. Smaller threshold values, such as 0.7, led to larger bucket sizes. The algorithm yielded promising outcomes when the chosen size of the portfolio was between 17 and 22 stocks. Table 2 contains results comparison of baseline and average results over 8 years period and Table 3 contains results of different threshold values.

Table 2. Results comparison of baseline and average results over 8 years period

Year	NOBS-OPT Allocation Unit Size	NOBS-OPT Allocation Unit Size	Mean of S&P Index	Mean of OBS-OPT Index	Mean of NOBS-OPT Index
2013	238.09	21.93	0.35	0.31	0.43
2014	229.38	17.53	0.49	0.51	0.79
2015	264.69	19.72	0.23	0.21	0.18
2016	291.11	18.94	0.19	0.21	0.33
2017	238.23	21.54	0.29	0.28	0.51
2018	242.54	23.12	0.61	0.61	0.69
2019	239.51	19.26	0.51	0.52	0.49
2020	221.97	18.01	0.82	0.79	0.89

Table 3. Examining Profit Rate Result by Threshold & Allocation Unit Size Variation

Year	Threshold = 0.75		Threshold = 0.80		Threshold = 0.85	
	Allocation Unit Dimension	Profit Rate Result	Allocation Unit Dimension	Profit Rate Result	Allocation Unit Dimension	Profit Rate Result
2013	57.86	0.42	21.86	0.36	4.38	0.28
2014	51.59	0.63	22.62	0.37	3.08	0.46
2015	58.10	0.68	20.72	0.68	4.05	0.49
2016	63.08	0.75	25.86	0.48	5.19	0.42
2017	52.86	0.79	22.72	0.25	4.24	0.38
2018	65.59	0.75	18.24	0.78	4.78	0.71
2019	53.45	0.96	19.76	0.49	2.78	0.38
2020	59.21	0.46	21.29	0.85	2.54	1.21

5.3 News Sentiment Classification

For news sentiment classification, we approached it as a sentiment classification task with three labels: Buy, Sell, and Neutral. Assigned a Positive label to "Buy" and a Negative label to "Sell". Even with a compact dataset, our fine-tuning of the pre-trained model, bert-base-multilingual-uncased-sentiment, yielded encouraging results. This model demonstrated its adaptability to the market sentiment domain with limited training samples. Table 4 presents the sentimental results of transformer model on NEWZ headlines (Fig. 5).

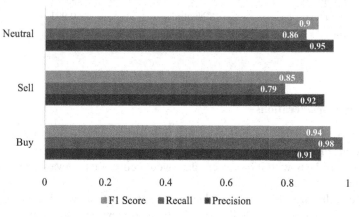

Fig. 5. Sentiment analysis results - precision, recall, and F1 score

Table 4. Sentimental results of transformer model

	Precision	Recall	F1 Score	Support
Buy	0.91	0.98	0.94	209
Sell	0.92	0.79	0.85	71
Neutral	0.95	0.86	0.9	63
Accuracy	–	–	0.92	343
Macro Avg	0.92	0.87	0.9	343
Weighted Avg	0.92	0.92	0.91	343

5.4 Performance Evaluation of the Model

The model we have proposed has demonstrated robust performance in addressing issues of label imbalance, consequently achieving elevated precision and recall metrics across all categories. The F1-scores pertaining to each class have been documented, serving as testament to the model's precision in discerning the contextual usage of words in varied situations. The trustworthy forecasts produced by this model offer valuable insights which can be deployed in the context of strategic investment decisions, leveraging the NOBS-OPT methodology (Table 5 and Figs. 6, 7).

Table 5. Our model's sample scores

Sentence	Label	Score
Apple incurred loses	Sell	0.973
Apple had huge profit	Buy	0.987
This is a random text	Neutral	0.991
Apple is a fruit	Neutral	0.919

5.5 Future Considerations

While our study provides valuable insights, there are certain limitations that need to be addressed. Firstly, the analysis is based on historical data from 2003 to 2020, and updating the analysis with more recent data would ensure its relevance in the current market conditions.

Secondly, the size of our curated dataset for sentiment analysis is relatively small. Expanding the dataset and including a wider range of news sources could enhance the model's performance and generalizability.

Additionally, exploring additional features and indicators, such as financial indicators, market trends, and macroeconomic factors, could further improve the accuracy of stock selection and sentiment analysis.

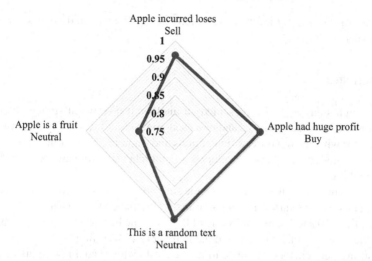

Fig. 6. Model's sentiment scores of sample sentences

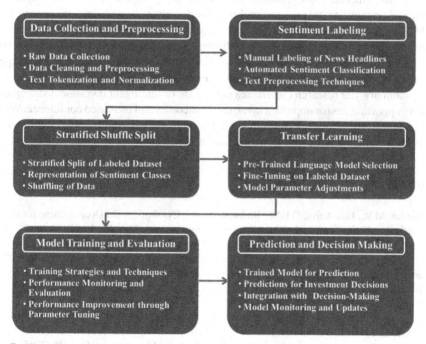

Fig. 7. Flow diagram illustrating the key steps in the research study, including data curation, data classification, stratified shuffle split, text task classification, model training, and prediction

In conclusion, our study demonstrates the effectiveness of combining the NOBS-OPT method with sentiment analysis for intelligent investment decision-making. As we continue to refine and enhance our approach, we believe it holds great potential for aiding investors in making informed decisions in the dynamic and complex world of stock markets.

6 Conclusion

In this research study, we have presented an intelligent investment strategy that combines historical stock data analysis and news sentiment classification. By integrating the S&P indicator with custom observation periods and training a transformer-based model for sentiment analysis, we have developed a comprehensive framework for informed decision-making.

Our experimental results demonstrate the superiority of our approach in identifying optimal stocks and capturing market sentiment. We have shown that an observation period length of 10 and a threshold value of 0.8 yield the best results, striking a balance between the quality of information and the number of observation periods.

The strong performance of our sentiment classification model highlights its ability to accurately classify news headlines into Buy, Sell, and Neutral categories, enabling strategic investments. These findings underscore the potential of AI technologies in revolutionizing investment practices and empowering investors with data-driven insights.

Moving forward, we plan to enhance our methodology by considering the time factor in observation periods, introducing an additional hyper-parameter for optimization. By refining and validating our framework, we aim to reshape the investment landscape, providing investors with valuable tools and improved decision-making processes.

In summary, our research contributes to the field of intelligent investment strategies, offering opportunities for improved investment outcomes and increased confidence. With further development, our approach has the potential to transform the way investments are made, leveraging advanced technologies for better financial decision-making.

References

1. Vargas, M.R., Dos Anjos, C.E.M., Bichara, G.L.G., Evsukoff, A.G.: Deep learning for stock market prediction using technical indicators and financial news articles. In: 2018 International Joint Conference on Neural Networks (IJCNN), pp. 1–8. IEEE (2018)
2. Thakkar, A., Chaudhari, K.: A comprehensive survey on deep neural networks for stock market: the need, challenges, and future directions. Expert Syst. Appl. **177**, 114800 (2021)
3. Sharpe, W.F.: Efficient capital markets: a review of theory and empirical work: discussion. J. Financ. **25**(2), 418–420 (1970)
4. Granger, C.W.J.: Long memory relationships and the aggregation of dynamic models. J. Econom. **14**(2), 227–238 (1980)
5. Vaswani, A., et al.: Attention is all you need. Adv. Neural Inf. Process. Syst. **30** (2017)
6. Xu, F., Yang, F., Fan, X., Huang, Z., Tsui, K.L.: Extracting degradation trends for roller bearings by using a moving-average stacked auto-encoder and a novel exponential function. Measurement **152**, 107371 (2020)
7. Liu, X., An, H., Wang, L., Jia, X.: An integrated approach to optimize moving average rules in the EUA futures market based on particle swarm optimization and genetic algorithms. Appl. Energy **185**, 1778–1787 (2017)

8. Rubi, M.A., Chowdhury, S., Rahman, A.A.A., Meero, A., Zayed, N.M., Islam, K.M.A.: Fitting multi-layer feed forward neural network and autoregressive integrated moving average for dhaka stock exchange price predicting. Emerg. Sci. J. **6**(5), 1046–1061 (2022)

9. Alam, T.: Forecasting exports and imports through artificial neural network and autoregressive integrated moving average. Decis. Sci. Lett. **8**(3), 249–260 (2019)

10. Engle, R.F., Granger, C.W.J.: Co-integration and error correction: representation, estimation, and testing. Econom.: J. Econom. Soc. 251–276 (1987)

11. Cakra, Y.E., Trisedya, B.D.: Stock price prediction using linear regression based on sentiment analysis. In: 2015 International Conference on Advanced Computer Science and Information Systems (ICACSIS), pp. 147–154. IEEE (2015)

12. Vachhani, H., et al.: Machine learning based stock market analysis: a short survey. In: Raj, J.S., Bashar, A., Ramson, S.R.J. (eds.) ICIDCA 2019. LNDECT, vol. 46, pp. 12–26. Springer, Cham (2020). https://doi.org/10.1007/978-3-030-38040-3_2

13. Xie, Y., Jiang, H.: Stock market forecasting based on text mining technology: a support vector machine method. arXiv preprint arXiv:1909.12789 (2019)

14. Moghar, A., Hamiche, M.: Stock market prediction using LSTM recurrent neural network. Procedia Comput. Sci. **170**, 1168–1173 (2020)

15. Oncharoen, P., Vateekul, P.: Deep learning for stock market prediction using event embedding and technical indicators. In: 2018 5th International Conference on Advanced Informatics: Concept Theory and Applications (ICAICTA), pp. 19–24. IEEE (2018)

16. Kumar, D., Sarangi, P.K., Verma, R.: A systematic review of stock market prediction using machine learning and statistical techniques. Mater. Today Proc. **49**, 3187–3191 (2022)

17. Tetlock, P.C.: Giving content to investor sentiment: the role of media in the stock market. J. Financ. **62**(3), 1139–1168 (2007)

18. Zhang, X., Fuehres, H., Gloor, P.A.: Predicting stock market indicators through twitter "I hope it is not as bad as I fear". Procedia Soc. Behav. Sci. **26**, 55–62 (2011)

19. Bollen, J., Mao, H., Zeng, X.: Twitter mood predicts the stock market. J. Comput. Sci. **2**(1), 1–8 (2011)

20. Pennington, J., Socher, R., Manning, C.D.: Glove: Global vectors for word representation. In: Proceedings of the 2014 Conference on Empirical Methods in Natural Language Processing (EMNLP), pp. 1532–1543 (2014)

21. Devlin, J., Chang, M.-W., Lee, K., Toutanova, K.: BERT: pre-training of deep bidirectional transformers for language understanding. arXiv preprint arXiv:1810.04805 (2018)

22. Aasi, B., Imtiaz, S.A., Qadeer, H.A., Singarajah, M., Kashef, R.: Stock price prediction using a multivariate multistep LSTM: a sentiment and public engagement analysis model. In: 2021 IEEE International IOT, Electronics and Mechatronics Conference (IEMTRONICS), pp. 1–8. IEEE (2021)

23. Mohan, S., Mullapudi, S., Sammeta, S., Vijayvergia, P., Anastasiu, D.C.: Stock price prediction using news sentiment analysis. In: 2019 IEEE Fifth International Conference on Big Data Computing Service and Applications (BigDataService), pp. 205–208. IEEE (2019)

24. Chiong, R., Fan, Z., Hu, Z., Adam, M.T.P., Lutz, B., Neumann, D.: A sentiment analysis-based machine learning approach for financial market prediction via news disclosures. In: Proceedings of the Genetic and Evolutionary Computation Conference Companion, pp. 278–279 (2018)

25. Deléglise, H., Interdonato, R., Bégué, A., d'Hôtel, E.M., Teisseire, M., Roche, M.: Food security prediction from heterogeneous data combining machine and deep learning methods. Expert Syst. Appl. **190**, 116189 (2022)

26. Bao, W., Yue, J., Rao, Y.: A deep learning framework for financial time series using stacked autoencoders and long-short term memory. PLoS ONE **12**(7), e0180944 (2017)

27. Ding, X., Zhang, Y., Liu, T., Duan, J.: Deep learning for event-driven stock prediction. In: Twenty-Fourth International Joint Conference on Artificial Intelligence (2015)

28. Hall, M.A.: Correlation-based feature selection of discrete and numeric class machine learning (2000)
29. Waqar, M., Dawood, H., Guo, P., Shahnawaz, M.B., Ghazanfar, M.A.: Prediction of stock market by principal component analysis. In: 2017 13th International Conference on Computational Intelligence and Security (CIS), pp. 599–602. IEEE (2017)
30. Lahmiri, S.: Minute-ahead stock price forecasting based on singular spectrum analysis and support vector regression. Appl. Math. Comput. **320**, 444–451 (2018)
31. Zhang, G.P.: Time series forecasting using a hybrid ARIMA and neural network model. Neurocomputing **50**, 159–175 (2003)
32. Agarwal, V., Madhusudan, L., Babu Namburi, H.: Method and apparatus for stock performance prediction using momentum strategy along with social feedback. In: 2nd International Conference on Intelligent Technologies (CONIT). IEEE (2022)

Machine Learning Applied to Speech Recordings for Parkinson's Disease Recognition

Lerina Aversano[1]([⊠])[iD], Mario L. Bernardi[1][iD], Marta Cimitile[2][iD], Martina Iammarino[1][iD], Antonella Madau[1][iD], and Chiara Verdone[1][iD]

[1] University of Sannio, Department of Engineering, Benevento, Italy
{aversano,bernardi,iammarino,chiverdone}@unisannio.it,
a.madau@studenti.unisannio.it
[2] Unitelma Sapienza University, Rome, Italy
marta.cimitile@unitelmasapienza.it

Abstract. Parkinson's disease is a common neurological condition that occurs when dopamine production in the brain decreases significantly due to the degeneration of neurons in an area called the substantia nigra. One of its characteristics is the slow and gradual onset of symptoms, which are varied and include tremors at rest, rigidity, and slow speech. Voice changes are very common among patients, so analysis of voice recordings could be a valuable tool for early diagnosis of the disease. This study proposes an approach that compares different Machine Learning models for the diagnosis of the disease through the use of vocal recordings of the vowel a made by both healthy and sick patients and the identification of the subset of the most significant features The experiments were conducted on a data set available on the UCI repository, which collects 756 different recordings. The results obtained are very encouraging, reaching an F-score of 95%, which demonstrates the effectiveness of the proposed approach.

Keywords: Parkinsons's disease · Speech recordings · Machine learning · Early diagnosis

1 Introduction

Parkinson's disease(PD) is a progressively evolving neurodegenerative disease that mainly involves some functions such as movement and balance control.

In PD, degeneration of nerve cells is observed in part of the basal ganglia, which are clusters of nerve cells located deep in the brain that help initiate and regulate voluntary muscle movements, and manage involuntary movements and changes in posture. When the nerve cells in the basal ganglia undergo degeneration, dopamine production decreases, and the number of connections between the nerve cells in the basal ganglia decreases. As a result, the basal ganglia cannot control muscle movements as they normally would, resulting in tremors, slow movements, a tendency to move less, problems with posture and walking, and a partial loss of coordination [23]. One of the most common problems in Parkinsonian patients is oral communication, symptoms of which include a weak monotone voice, wheezing, hoarse voice, and poor articulation. The patient may speak somewhat monotonously due to a loss of tone and modulation of the voice or because the voice is softer. There is a propensity to hasten sound output and mumble sentences when palilalia (repetition of syllables) arises. In recent years there has been growing interest in the development of telemonitoring systems for PD based on the measurement of motor system disturbances caused by the disease. In this regard, recent studies have proposed various approaches that exploit speech data, handwriting traces, and gait or cardiovascular measures for early diagnosis [3,5].

This study proposes an approach for the identification of PD, based on the use of Machine Learning models trained on features extracted from acoustic traces. The goal is twofold, to detect the disease by exploiting acoustic parameters of speech, such as voice intensity, frequency, or pressure level, to provide an objective and non-invasive measurement of symptomatic changes in Parkinson's disease and to identify the right subset of features. The experimentation was carried out on a dataset available in the UCI Repository[1], which collects information on 188 patients with PD and 64 healthy patients, making 756 vocal recordings available, each of which is composed of 754 features [22].

The document is structured as follows: Section 2 offers a brief discussion of related work, Sect. 3 details the proposed approach, and Sect. 4 describes the characteristics of the experiments conducted. Finally, the results are reported in Sect. 5, the discussion in Sect. 6, and the conclusions in Sect. 7.

2   Related Works

In the field of medicine, machine learning can be used to analyze large amounts of health data, such as clinical data, laboratory data, and imaging data, in order to identify patterns and trends that can be useful for decision support or biomedical research [24]. The possibility of having an early diagnosis of diseases already in the initial stages allows to improve clinical management of patients and to have a faster and more effective healing process. Through machine learning processes it is also possible to identify fundamental features present in patients with the same disease.

The classification and prognosis of various human diseases, including lung cancer [6], thyroid disease [2], and heart disease [4], have been the subject of numerous research. Similarly, much research has been conducted on the diagnosis and treatment of PD which differ in the methodologies applied and the data considered [7]. Little at. al [17], detecting hoarseness, presented an evaluation of the practical value of existing traditional and non-standard measures to discriminate people with PD from healthy ones. Specifically, they introduced intonation period entropy (PPE), a new measure of hoarseness. The work is based on a dataset containing 195 recordings belonging to 31 healthy people and 23 with PD. During the training of four ML models, the authors of the paper [16] conduct a study on audio data from the PPMI and UCI datasets. When Support Vector Machine (SVM), Random Forest, K-Nearest Neighbors (KNN), and Logistic

---

[1] https://urly.it/3t259.

Regression models are used to compare classification results, the Random Forest classifier emerges as the most accurate machine learning (ML) strategy for PD detection, achieving 91% accuracy.

The authors of the study [3] have developed a model based on the use of Neural Networks for the diagnosis of the disease on vocal recordings of different natures, made by both healthy and sick patients. The results are very interesting, highlighting the excellent performance of LSTM with an F-Score of 97%.

Sakar et al. [22] proposed, for the first time in the literature, the use of the Tunable Q-factor (TQWT) wavelet transform, which promises to have a higher frequency resolution than the classical discrete dyadic wavelet. The authors compare different classifiers, obtaining the maximum value of the F-score 84%, considering only the first 50 features among all those proposed.

Like the previous ones, our study aims to contribute to the early diagnosis of PD. In particular, we use the dataset presented in the paper [22], to improve the performance of the classification model and identify the ideal subset of features to consider to better distinguish healthy patients from those affected by PD.

3 Approach

This section details the approach, providing information on the dataset, the features, and the methodology adopted.

3.1 Dataset

The dataset used to conduct the experiments collects data from 252 patients, divided into healthy and PD patients, treated at the Department of Neurology, Cerrahpasÿa Faculty of Medicine, Istanbul University.

In detail, PD patients are 188, of which 107 men and 81 women aged between 33 and 87 years, while the control group of healthy individuals consists of 64 people, of which 23 men and 41 women aged between 41 and 82 years old. The data includes 756 voice recordings of the *a* vowel, collected with a microphone set to 44.1 KHz. There are 3 different records for each patient.

3.2 Features Model

The study considers 752 different features, whose detailed explanation was carried out by the authors in [22]. In short, the features extracted from voice recordings can be divided into six main macro groups:

– *Baseline*: this group comprises the most popular vocal characteristics used in PD studies [12,20], and includes Jitter (5), Shimmer (6), Fundamental Frequency Parameters (5), Harmony Parameters (2), Recurrence Period Density Entropy(RPDE) (1), Detrended Fluctuation Analysis (DFA)(1), and Pitch Period Entropy (PPE)(1). Acoustic analysis software Praat [8] was utilized to extract these parameters, except for RPDE, DFA, and PPE.

– *Time Frequency*: these characteristics are divided into three subgroups, Intensity parameters (3) relating to the strength of the vocal signal in dB, Formant frequencies (4) amplified by the vocal tract, and Bandwidth (4) relating to the frequency interval between the formant frequencies. As for the previous ones, these too have been extracted from the spectrograms of the speech signal using Praat.

– *Mel Frequency Cepstral Coefficients(MFCCs)*: for this group, 84 different features were considered concerning the mean and standard deviation of the 13 original MFCCS plus the logarithmic energy of the signal and their first-second derivatives. These characteristics closely resemble the efficient filtering capabilities of the human ear. The MFCC extraction approach produces narrow spectral sampling by combining cepstral analysis with spectral domain partitioning utilizing stacked triangular-shaped filter banks.

– *Wavelet Transform based*: this set includes 182 features related to the wavelet transform (WT), which obtained from the outline of the raw fundamental frequency F_0 of the vocal samples has also been employed as indicators of the UPDRS scale. The vocal samples are subjected to the 10-level discrete wavelet transform to obtain these features. After the vocal samples have been decomposed, the energy, Shannon and logarithmic energy entropy, and Teager-Kaiser energy of both the approximation and that of the detailed coefficients are then calculated.

– *Vocal fold*: this set includes four subsets, Glottis Quotient (GQ) (3), Glottal to Noise Excitation (GNE) (6), Vocal Fold Excitation Ratio(VFER) (7), and Empirical Mode Decomposition(EMD) (6). To measure their effectiveness and compare them to the suggested TQWT-based features, these characteristics are based on the vocal cord vibration pattern and the impacts of noise on the vocal cords.

– *Tunable Q-factor wavelet transform (TQWT)*: there are 432 different features involving the wavelet transform of the tunable Q factor(TQWT), which is a completely discrete and super complete WT. The TQWT consists of two banks of channel filters, which are applied iteratively, and in each iteration, the low-pass filter output is passed to the next iteration's low-pass/high-pass filters as inputs. There will be $j+1$ sub-bands emerging from the outputs of the high-pass filter j and a final low-pass filter after the decomposition phase, where j is the number of levels.

3.3 Proposed Methodology

Figure 1 shows the summary scheme of the phases involved in the proposed approach. In particular, as can be seen from the figure, five different experiments were conducted. The first classification was conducted on the entire original dataset (UCI Dataset), considering all the collected features. The second, on the other hand, envisaged only the use of the so-called baseline features (Dataset Baseline). The third was conducted after the first step of feature selection based on correlation (D1), and the fourth after the first and second steps of feature selection, therefore after the correlation and analysis of collinearity present in the data (D2). Finally, last classification experiment was conducted after the three steps of feature selection, correlation, collinearity analysis, and genetic algorithm, thus using only the minimum subset of features necessary to obtain good performance (Dataset Best Subset Features).

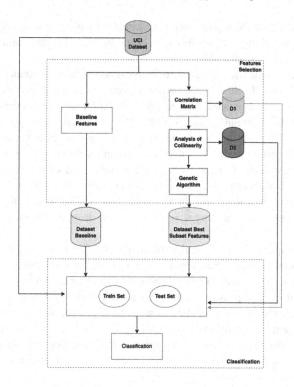

Fig. 1. Methodology Workflow

Feature Selection. The dataset used is characterized by a very large number of characteristics and this could be a problem, as it can introduce redundancy between features or noisy and irrelevant data that can lead to inaccurate classification, as well as slow down the training and classification process itself. Another issue related to many features is overfitting, which occurs when a model fits the training data exactly. The algorithm cannot work accurately with unknown data, as it starts learning noise or irrelevant information within the dataset.

In this regard, we applied various feature selection techniques to reduce the input variables to the classification model and identify the sufficient and necessary subset of features for the early diagnosis of the disease.

In particular, we proceed to a reduction of the dimensionality carried out in three different steps.

The first technique is feature selection based on the correlation matrix, a symmetric square matrix, which displays the correlations between variables in a data set. The correlation matrix, which shows how closely a change in one variable is related to a change in another variable, is specifically used to evaluate the degree of correlation between two or more variables. The correlation varies in a defined range between 1 and −1. If the correlation is negative, it means that the characteristics are related by an inversely proportional relationship. Conversely, if there is a positive correlation, they will have a directly proportional relationship. The dimensionality reduction is obtained by evaluating the correlation of the features with the target variable,that describe the patient's status. In particular, all the features having a correlation in absolute value lower than a pre-set threshold value are discarded.

The second technique used is based on the presence of the redundancy phenomenon in the considered data. In this regard, it is necessary to check for collinearity, which occurs when features in a dataset show a high correlation with each other. If the correlation matrix has off-diagonal elements with a high absolute value, we can talk about collinearity. If two features are collinear, they should not be considered together and only the most informative features should be considered. To identify collinear features, it is necessary to establish a threshold for the absolute value of the correlation coefficient. Then, pairs of all potential combinations of characteristics that show a correlation greater than the predetermined threshold are examined. The feature that has a lower correlation coefficient with the target variable is then dropped for each pair from the function.

The last technique is the genetic algorithm, a method of trait selection that uses the principles of natural evolution and genetic selection. Natural selection is simulated by genetic algorithms: organisms capable of adapting to environmental changes are able to survive, reproduce and pass on to the next generation. To address a problem, they essentially recreate the "survival of the fittest" among members of a future generation. The idea behind genetic algorithms is to create a population of random solutions, where each solution represents a set of characteristics. These solutions are evaluated on their ability to predict test data. In general, the best solutions are selected and used to create the next generation of solutions.

Classification. To classify PD patients and those belonging to the control group, different classifiers were compared to identify the best model. More specifically, both tree-based and augment-based models were considered, as well as a neural network-based model. The former applies the "divide and conquer" method and presents a hierarchical tree structure, consisting of a root node, branches, internal nodes, and leaf nodes. The Seconds are based on boosting and creating an ensemble model by sequentially combining many weak decision trees. The last model belongs to Deep learning which structures the algorithms to generate an artificial neural network. In detail we considered:

- *Decision Tree* (DT): builds a model based on a decision tree, where each node represents a decision to be made about the data and each branch represents a possible consequence of that decision [18].
- *Random Forest* (RF): based on the use of a combination of decision trees, in which each tree is trained on a random subset of data and characteristics, creating a model that is less susceptible to variations in the data than a single decision tree [11].
- *Extra Tree* (EXT): creates many decision trees, but the sampling for each tree is random, with no replacement. This creates a dataset for each tree with unique samples. A specific number of items, from the total set of items, are also randomly selected for each tree [15].

- *Gradient Boosting* (GB) : creates a sequence of weak models, each of which tries to correct the mistakes made by the previous models. Weak models have low prediction accuracy, but can be combined to get an overall model more accurate [14].
- *eXtreme Gradient Boosting* (XGB): sophisticated use of gradient boosting. It employs a framework for gradient augmentation [10].
- *CatBoost* (CB): builds a sequence of decision trees, all of the same type, in succession while training. Concerning earlier trees, each new tree is constructed with a smaller loss [1].
- *AdaBoost* (AB): it works by creating a sequence of weak models, each of which tries to correct the mistakes made by the previous models [13].
- *K-Nearest Neighbors* (KNN): classifies a data instance based on the classes of the instances closest to it in the data space [21].
- *Convolutional Neural Network* (CNN):a type of feed-forward neural network consisting of multiple stages in which each stage specializes in doing different things. It has an input block, one or more hidden blocks (hidden layers), which perform calculations via activation functions, and an output block which performs the actual classification [19].

4   Experimental Settings

This section specifically reports how the experiment on the previously presented dataset was conducted.

The first operation carried out on the dataset was to normalize the values of the features as these had a very wide range of possible values. We then proceeded to carry out a normalization of the min-max type in which the normalizer linearly rescales each feature to the interval $[0, 1]$.

Subsequently, three different steps of feature selection were applied:

- Feature selection based on correlation matrix by setting a threshold equal to 0.2, all features with lower correlation than the target variable are cut. After this feature selection, we obtain 168 features.
- Feature selection based on collinearity removal using a correlation threshold for choosing between considered feature pairs of 0.9. After this feature selection, we obtain 54 features.
- Feature selection based on genetic algorithm, we obtain 24 features using as scoring metric the accuracy and the following parameters: Random Forest Classifier as an estimator, cross-validation value equal to 5, max_features=30, initial population_size equal to 100, tournament_size (the best solution to the next generation) equal to 3 and 20 generations have been considered.

In the classification phase, the evaluation of the hyperparameters was done through GridSearch, but it was decided to use the default parameters because the performances did not improve significantly but the processing time greatly increased.

An appropriate dataset balancing procedure has been carried out because the initial dataset was unbalanced, with 564 occurrences for PD patients and only 192 for healthy controls. To solve this problem, we use the SMOTE (Synthetic Minority Oversampling

TEchnique) oversampling technique [9]; in particular, we oversample the cases in the minority class using SMOTE by randomly repeating them. It operates by generating synthetic data using the K-Nearest Neighbors method. Some samples of the minority class are initially randomly selected, after which the k nearest neighbors are determined for those observations. Linear interpolation is then used to construct the synthetic data between the random data and the chosen k-nearest neighbor at random.

We use the Cross-Validation technique as a validation method for this work. With this method, the input data is divided into k subsets. We choose the k-value equal to 10, so the procedure explained in the following is repeated 10 times. This procedure consisted of the training of the model on all subsets except one (k-1), and the subset that wasn't used for training is utilized for testing.

5   Experimental Results

This section reports the results of all the experiments conducted on the collected data.

Table 1. Results on UCI Dataset

Classifiers	Accuracy	Precision	Recall	F-score	Training Time
DT	0.8671	0.8663	0.857	0.859	7.20
RF	0.9317	0.9367	0.9351	0.9331	18.82
EXT	**0.9494**	**0.9553**	**0.9484**	**0.9493**	**6.04**
GB	0.9325	0.9332	0.9315	0.9287	285.81
XGB	0.9334	0.9384	0.9333	0.9331	447.32
CB	0.9467	0.9522	0.9466	0.9463	6524.28
AB	0.8954	0.8989	0.8952	0.8952	80.28
KNN	0.7925	0.8522	0.7924	0.7822	1.28
CNN	0.807	0.8299	0.807	0.8036	902.87

Table 1 shows the results obtained considering all the features, therefore using the UCI Dataset. In this case, the dataset has been subjected to normalization and balancing. Specifically, the Table shows the classifier in the first column, the validation metrics, such as Accuracy, Precision, Recall, and F-score, in the following columns, and the training time in the last column. As can be seen from the reported results, the classifier that best distinguishes PD patients from healthy ones is EXT, which reports an F-score of 95% and a training time of 6 s.

Table 2 reports the results obtained considering only the first group of features, the so-called baseline features, which amount to 22, and which are the most used in previous studies. In this case, the dataset has been subjected to normalization and balancing. As can be seen from the results shown, the classifier that best distinguishes PD patients from healthy ones is EXT, which reports an F-score of 87% and a training time among the lowest, equal to 4 s.

Table 2. Results on Dataset Baseline

Classifiers	Accuracy	Precision	Recall	F-score	Training Time
DT	0.7811	0.7882	0.7853	0.7881	0.65
RF	0.8413	0.8534	0.8447	0.8448	6.60
EXT	**0.8732**	**0.8913**	**0.8784**	**0.8661**	**4.07**
GB	0.7934	0.797	0.7941	0.7914	11.57
XGB	0.8546	0.863	0.8543	0.8536	33.04
CB	0.8395	0.8475	0.8394	0.8385	91.89
AB	0.7437	0.7475	0.7435	0.7424	4.10
KNN	0.7819	0.7913	0.7818	0.7796	0.43
CNN	0.636	0.6454	0.636	0.63	11.5

Table 3 reports instead the results obtained by training the classifiers, considering the characteristics obtained after the selection of the characteristics with the correlation matrix. In detail, using a threshold equal to 0.2 we consider 168 features, and we apply the balancing and the cross-validation. As shown in the table, in this case, the classifiers all work better, obtaining much better results than in the previous case. Also in this case EXT proved to be the best, obtaining an F-score of 94% and a training time of 5 s, which is the third lowest.

Table 3. Results on D1

Classifiers	Accuracy	Precision	Recall	F-score	Training Time
DT	0.8466	0.8638	0.8509	0.8568	3.68
RF	0.929	0.9368	0.9244	0.9323	13.12
EXT	**0.9379**	**0.9458**	**0.9422**	**0.9421**	**5.21**
GB	0.9131	0.9223	0.9121	0.9108	93.34
XGB	0.9388	0.9435	0.9387	0.9385	170.32
CB	0.9387	0.9445	0.9383	0.9383	883.76
AB	0.8785	0.8846	0.8783	0.8778	22.59
KNN	0.8599	0.8839	0.8597	0.8571	0.73
CNN	0.8991	0.8999	0.8991	0.8991	136.52

Table 4 shows the results obtained by applying the feature selection based on the correlation matrix and removal of collinearities. In this regard, 54 features were selected, applying balancing and cross-validation to the dataset. Compared to the previous case, only DT performs better, the others lose something in terms of evaluated performance. EXT, on the other hand, achieves the same results, always being the most suitable classifier for the intended objective, with an F-score of 94% and a training time of 4.5 s, compared to 5 s in the previous case.

Table 4. Results on D2

Classifiers	Accuracy	Precision	Recall	F-score	Training Time
DT	0.8679	0.8592	0.8615	0.8698	1.37
RF	0.929	0.9369	0.9315	0.9278	10.29
EXT	**0.9512**	**0.9484**	**0.9476**	**0.9422**	**4.54**
GB	0.8971	0.9016	0.8916	0.8894	31.64
XGB	0.9246	0.931	0.9244	0.9242	72.48
CB	0.929	0.9345	0.9289	0.9287	303.48
AB	0.8475	0.8519	0.8473	0.8568	7.79
KNN	0.851	0.8819	0.851	0.8476	0.60
CNN	0.8158	0.8221	0.8158	0.8149	19.76

Finally, Table 5 shows the results obtained by applying the feature selection using all three previously presented techniques: correlation matrix, collinearity removal, and genetic algorithm. These techniques lead to considering only 24 features chosen using the balanced dataset and with cross-validation. With this number of features, EXT reaches its best performance, with all validation metrics almost equal to 95% and a training time that is confirmed to be among the three fastest.

Table 5. Results on Dataset Best Subset Features

Classifiers	Accuracy	Precision	Recall	F-score	Training Time
DT	0.8369	0.8481	0.8383	0.8451	0.79
RF	0.9193	0.9269	0.9236	0.9138	7.48
EXT	**0.9521**	**0.95**	**0.9493**	**0.9475**	**4.92**
GB	0.8794	0.8807	0.8793	0.8797	15.03
XGB	0.9175	0.9224	0.9174	0.9172	37.07
CB	0.9246	0.9289	0.9245	0.9244	143.06
AB	0.8352	0.8404	0.8349	0.8345	4.65
KNN	0.8609	0.8793	0.8609	0.859	0.53
CNN	0.864	0.864	0.864	0.864	11.96

Finally, the results show that there are few cases in which the classifier makes mistakes in distinguishing patients with PD from healthy ones, but also that it is not necessary to collect a large amount of information on individuals but to identify the correct subset of the most indicative characteristics for disease detection.

6 Discussion

After conducting all the different experiments, the results show that EXT is the best-performing classifier on this data. More specifically, Fig. 2 shows the comparison of the F1 scores obtained from all the classifiers used on all the datasets considered.

The graph shows the score on the ordinate axis and the classifier on the abscissa axis, while the different ones represent the datasets. The results on the UCI Dataset are shown in blue, those of the Baseline Dataset in Orange, D1 in grey, D2 in yellow, and the dataset with the best features in light blue. It was chosen to start the y-axis scale from 0.7 to better highlight the results.

The figure confirms that EXT is the best classifier, which obtains the same performance, with an F1-score of 95% both on the dataset where we consider all the features, and on the one where we consider only 24 (Dataset Best Subset Features).

Another evident thing is that the Baseline features are not sufficient to train the classifiers to discern the disease, because the results are quite low.

Furthermore, KNN despite having the fastest-ever training times (<1 s), never matches the performance of the other classifiers.

It is also highlighted that the use of deep learning is not appropriate for the case study, as CNN takes a long time to train without obtaining comparable results to the others, this is because deep learning models require a much higher number of instances than the one present in the dataset in question.

Finally, Table 6 reports the features that have been selected after the entire feature selection process consisting of the three steps described, and which are contained in the Dataset Best Subset Features, on which the classifiers have achieved excellent performance, reducing times. The table shows the name and the group to which they belong. As can be seen, 70% of the selected features belong to the Tunable Q-factor wavelet

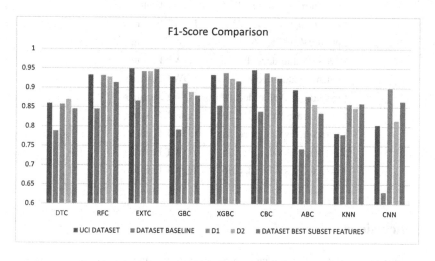

Fig. 2. F1-Score Comparison Across All Datasets

transform group, four to the Mel Frequency Cepstral Coefficient, two to the Vocal Fold, one to the Time-Frequency, and one to the Baseline.

This contribution is important because it demonstrates that it is not necessary to collect a large number of attributes to diagnose the disease, but it is sufficient to select the correct ones, and it is demonstrated that the features that have the greatest weight are precisely those introduced by the study by [22].

Table 6. Feature Selected

Name	Group
meanPeriodPulses	Baseline Features
f1	Formant Frequencies
VFER_NSR_SEO	Vocal Fold
IMF_SNR_entropy	Vocal Fold
std_4th_delta	MFCC
std_3rd_delta_delta	MFCC
std_6th_delta_delta	MFCC
std_10th_delta_delta	MFCC
tqwt_energy_dec_11	TQWT Features
tqwt_energy_dec_26	TQWT Features
tqwt_entropy_shannon_dec_8	TQWT Features
tqwt_entropy_shannon_dec_14	TQWT Features
tqwt_entropy_log_dec_16	TQWT Features
tqwt_entropy_log_dec_20	TQWT Features
tqwt_entropy_log_dec_28	TQWT Features
tqwt_entropy_log_dec_35	TQWT Features
tqwt_stdValue_dec_6	TQWT Features
tqwt_stdValue_dec_11	TQWT Features
tqwt_minValue_dec_8	TQWT Features
tqwt_kurtosisValue_dec_18	TQWT Features
tqwt_kurtosisValue_dec_20	TQWT Features
tqwt_kurtosisValue_dec_26	TQWT Features
tqwt_kurtosisValue_dec_35	TQWT Features
tqwt_kurtosisValue_dec_36	TQWT Features

7   Conclusions

PD is a neurodegenerative disease caused by progressive damage to certain parts of the brain and its diagnosis is essentially based on the study of the patient's symptoms and clinical history. Among the most common symptoms is that the voice of a person with

Parkinson's begins to change, often becoming much weaker and monotonous. In this regard, in this study, we use a public dataset of 756 voice recordings of voices recorded from both PD patients and a control group, for a total of 252 people. The proposed approach has a dual objective: classification to identify the best-performing classifier and feature selection to identify the most suitable subset of features for early diagnosis of the disease. Different algorithms based on trees, boosting and CNN were compared for the classification. The data set used is available on the UCI Machine Learning Repository and collects 752 different characteristics belonging to different groups, such as time-frequency, Wavelet-based transform, and Wavelet Tunable Q-factor transform. For the feature selection, three different techniques were used: correlation, multicollinearity, and genetic trees. Classification results are very interesting and show that the best performance is obtained with the EXT classifier which reaches an F-score of 95%, by making a feature selection based on all three proposed techniques. We further identify that the optimal subset of features consists of 24 features. These very encouraging results suggest that voice disturbances can be a valid tool to improve the diagnosis of PD, even in the early stage of the disease.

References

1. Al Daoud, E.: Comparison between XGBoost, lightGBM and CatBoost using a home credit dataset. Int. J. Comput. Inf. Eng. **13**(1), 6–10 (2019)
2. Aversano, L., et al.: Thyroid disease treatment prediction with machine learning approaches. Procedia Comput. Sci. **192**, 1031–1040 (2021). https://doi.org/10.1016/j.procs.2021.08.106, https://www.sciencedirect.com/science/article/pii/S1877050921015945. knowledge-Based and Intelligent Information and Engineering Systems: Proceedings of the 25th International Conference KES2021
3. Aversano, L., Bernardi, M.L., Cimitile, M., Iammarino, M., Montano, D., Verdone, C.: A machine learning approach for early detection of parkinson's disease using acoustic traces. In: 2022 IEEE International Conference on Evolving and Adaptive Intelligent Systems (EAIS), pp. 1–8. IEEE (2022)
4. Aversano, L., Bernardi, M.L., Cimitile, M., Iammarino, M., Montano, D., Verdone, C.: Using machine learning for early prediction of heart disease. In: 2022 IEEE International Conference on Evolving and Adaptive Intelligent Systems (EAIS), pp. 1–8 (2022). https://doi.org/10.1109/EAIS51927.2022.9787720
5. Aversano, L., Bernardi, M.L., Cimitile, M., Iammarino, M., Verdone, C.: Early detection of Parkinson's disease using spiral test and echo state networks. In: 2022 International Joint Conference on Neural Networks (IJCNN), pp. 1–8. IEEE (2022)
6. Aversano, L., Bernardi, M.L., Cimitile, M., Iammarino, M., Verdone, C.: An enhanced UNet variant for effective lung cancer detection. In: 2022 International Joint Conference on Neural Networks (IJCNN), pp. 1–8 (2022). https://doi.org/10.1109/IJCNN55064.2022.9892757
7. Aversano, L., Bernardi, M.L., Cimitile, M., Pecori, R.: Early detection of Parkinson disease using deep neural networks on gait dynamics. In: 2020 International Joint Conference on Neural Networks (IJCNN), pp. 1–8 (2020). https://doi.org/10.1109/IJCNN48605.2020.9207380
8. Boersma, P.: Praat: doing phonetics by computer (2007). http://www.praat.org/
9. Chawla, N.V., Bowyer, K.W., Hall, L.O., Kegelmeyer, W.P.: Smote: synthetic minority oversampling technique. J. Artif. Intell. Res. **16**, 321–357 (2002)

10. Chen, T., et al.: Xgboost: extreme gradient boosting. R package version 0.4-2 **1**(4), 1–4 (2015)
11. Cutler, A., Cutler, D.R., Stevens, J.R.: Random forests. ensemble machine learning: methods and applications, pp. 157–175 (2012)
12. Erdogdu Sakar, B., Serbes, G., Sakar, C.O.: Analyzing the effectiveness of vocal features in early telediagnosis of parkinson's disease. PLoS ONE **12**(8), e0182428 (2017)
13. Freund, Y., Schapire, R.E.: A decision-theoretic generalization of on-line learning and an application to boosting. J. Comput. Syst. Sci. **55**(1), 119–139 (1997)
14. Friedman, J.H.: Stochastic gradient boosting. Comput. Stat. Data Anal. **38**(4), 367–378 (2002)
15. Geurts, P., Ernst, D., Wehenkel, L.: Extremely randomized trees. Mach. Learn. **63**, 3–42 (2006)
16. Govindu, A., Palwe, S.: Early detection of Parkinson's disease using machine learning. Procedia Comput. Sci. **218**, 249–261 (2023). https://doi.org/10.1016/j.procs.2023.01.007, https://www.sciencedirect.com/science/article/pii/S1877050923000078. international Conference on Machine Learning and Data Engineering
17. Little *, M.A., McSharry, P.E., Hunter, E.J., Spielman, J., Ramig, L.O.: Suitability of dysphonia measurements for telemonitoring of Parkinson's disease. IEEE Trans. Biomed. Eng. **56**(4), 1015–1022 (2009). https://doi.org/10.1109/TBME.2008.2005954
18. Magee, J.F.: Decision Trees for Decision Making. Harvard Business Review, Brighton (1964)
19. O'Shea, K., Nash, R.: An introduction to convolutional neural networks. arXiv preprint arXiv:1511.08458 (2015)
20. Peker, M.: A decision support system to improve medical diagnosis using a combination of k-medoids clustering based attribute weighting and svm. J. Med. Syst. **40**(5), 116 (2016)
21. Peterson, L.E.: K-nearest neighbor. Scholarpedia **4**(2), 1883 (2009)
22. Sakar, C.O., et al.: A comparative analysis of speech signal processing algorithms for Parkinson's disease classification and the use of the tunable q-factor wavelet transform. Appl. Soft Comput. **74**, 255–263 (2019)
23. Shukla, L.C., Schulze, J., Farlow, J., Pankratz, N.D., Wojcieszek, J., Foroud, T.: Parkinson disease overview. GeneReviews®[Internet] (2019)
24. Sidey-Gibbons, J.A., Sidey-Gibbons, C.J.: Machine learning in medicine: a practical introduction. BMC Med. Res. Methodol. **19**, 1–18 (2019)

Vision Transformers for Galaxy Morphology Classification: Fine-Tuning Pre-trained Networks vs. Training from Scratch

Rahul Kumar[✉], Md Kamruzzaman Sarker, and Sheikh Rabiul Islam

Department of Computing Sciences, University of Hartford, West Hartford, CT, USA
{rkumar,sarker,shislam}@hartford.edu

Abstract. In recent years, the Transformer-based deep learning architecture has become extremely popular for downstream tasks, especially within the field of Computer Vision. However, transformer models are very data-hungry, making them challenging to adopt in many applications where data is scarce. Using transfer learning techniques, we explore the classic Vision Transformer (ViT) and its ability to transfer features from the natural image domain to classify images in the galactic image domain. Using the weights of models trained on ImageNet (a popular benchmark dataset for Computer Vision), we compare the results of two distinct ViTs: one base ViT (without pre-training) and another fine-tuned ViT pre-trained on ImageNet. Our experiments on the Galaxy10 dataset show that by using the pre-trained ViT model, we can get better accuracy compared to the ViT model built from scratch and do so with a faster training time. Experimental data further shows that the fine-tuned ViT model can achieve similar accuracy to the model built from scratch while using less training data.

Keywords: Deep learning · Vision transformer (ViT) · Transfer learning · Pre-trained · Image classification

1 Introduction

The formation of galaxies has always been of particular interest to astronomers and astrophysicists throughout history. Known as galaxy morphology, examining galactic structures can help us understand the processes that go into star creation and indicate potential areas for the formation of life outside the Milky Way [2]. With the launch of the James Webb Space Telescope in December 2021, the observatory will allow us to understand the history of the universe through the use of infrared light. The James Webb Space Telescope can peer billions of years into the past to uncover the first stars and galaxies ever created by magnifying the infrared light with its 18 gold-plated mirrors while simultaneously blocking off the sun's heat rays using an expansive heat shield. Using this advanced technology, images taken of the oldest galaxies in the universe can offer valuable insight into the evolution of galaxies from the beginning of time to the

---

This research is conducted with the support of the University of Hartford through the Vincent Coffin Grant (ID: 398131).

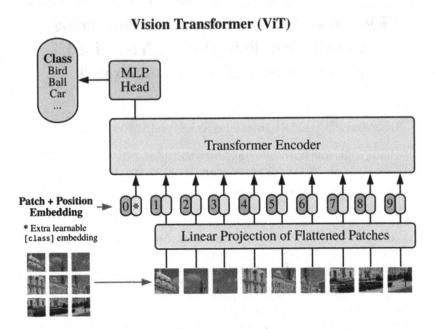

Fig. 1. Vision Transformer Architecture from Dosovitskiy et al.'s An Image is Worth 16 × 16 Words: Transformers For Recognition At Scale

present day. However, with the life expectancy of this mission being around five to ten years due to fuel capacity, collecting image data on just galaxies is limited due to the endless reservations of astronomers needing the observatory for research purposes [10].

One solution to address this issue of limited data involves leveraging transfer learning. Transfer Learning is a subset of machine learning which uses pre-trained models to transfer knowledge from one domain to a new problem. We aim to determine whether we can improve the classification process of galaxies using transfer learning and the Vision Transformer architecture to save time and money on data collection. For the classifier model, we choose the Vision Transformer (ViT) [5] due to it's improved performance compared to the CNN model in recent years. In order to train our model to classify galaxies, we use the Galaxy10 [7]. To evaluate the transfer capability of the ViT in galaxy classification, we devise two hypothesis: 1: Can we achieve a better accuracy by using a pre-trained model? 2: How much data does the pre-trained model need to obtain a similar performance to the model built from scratch?

Below, we outline the dataset along with how it is pre-processed, the steps that went into pre-training and fine-tuning the models, as well as the results of their performance when fine-tuned on galactic data. In Sect. 3, we discuss the data being used in this paper. Section 4 discusses the experimental approach and the methods for pre-training, fine-tuning, and calculating accuracy. In Sect. 5, we discuss the results from each of these experiments. Finally, the paper ends with a discussion and a brief conclusion in Sect. 6.

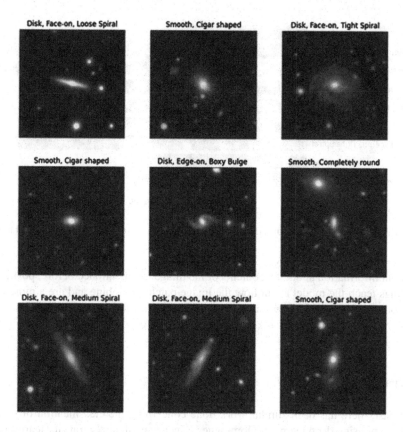

Fig. 2. One image from each of the nine class of the Galaxy10 dataset. Note that we used images from nine classes instead ten classes, as we found one class have significantly lower number of images.

2 Related Work

2.1 Vision Transformer

We start the discussion with the introduction of Vision Transformer. Introduced by Dosovitskiy et al., the Vision Transformer (ViT) [5] serves as the backbone of this paper with its popularity in recent years. First, the ViT breaks up an input image into equal-sized patches. These patches are then "flattened" into number vectors and attached with a positional embedding, since the model can only accept arrays of numbers. The purpose of the positional encoding is to help the model understand the relative position of each patch within the image. Finally, the model calculates the likelihood of each class in the final layer and outputs the most likely label for the image. Architecture of the ViT is shown in Fig. 1. Despite Convolutional Neural Networks (CNN) being the de-facto model for imaging tasks, vision transformers have slightly outperformed CNNs due to two main features; the attention mechanism and positional encoding [13]. Attention in Deep Learning directs a machine to general locations within an image to extract

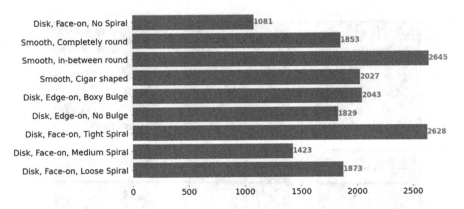

Fig. 3. Distribution of the modified Galaxy10 DECals dataset for each classification label.

important information about the contents of the image. As mentioned before, positional encodings teach the model about how the image patches combine and are a more efficient solution compared to passing the whole image into the model. This architecture will serve as the backbone for the experiments that are conducted and presented in this paper.

3   Transfer Learning

In the past, there has been similar work done concerning the classification of galaxy morphology using deep learning. Domínguez Sánchez et al. apply transfer learning for galaxy classification using three datasets: The Sloan Deep Sky Survey (SDSS), the Dark Energy Survey (DES), and the Dark Energy Camera Legacy Survey (DECaLS) [4]. The data contained images with one of three labels: Smooth/Disk, Edge-on, and Bar Sign. Consisting of three different types of hidden layers, Sanchez and her team created a standard deep neural network architecture to investigate the transferability of one image survey to another. To demonstrate the effects of transfer learning, models were trained and evaluated under unique and specific conditions. One model was trained on SDSS data and immediately transferred to DES data, another model was trained on SDSS data with some model fine-tuning, a model trained on SDSS data with added layers for DES data, and a model trained on DES data from scratch. Ultimately, the team achieved results of >90% accuracy using this strategy but required more than 300k prepared galaxy classifications in the process.

Tonkes et al. also investigates the application of transfer learning by comparing state-of-the-art CNN architectures with Vision Transformers in the artistic domain [11]. By pre-training these models on ImageNet, a very popular benchmark dataset consisting of images of common objects, Tonkes' evaluation examines how well these architectures perform on data outside of these common objects; in this case, art. Also, his work includes investigations into transformer architectures, unlike past research into this domain which has extensively explored transfer learning with CNNs. The dataset consisted of digitized artwork that falls into three categories: Type, Material, and Artist. To classify these images, eight different architectures (four CNNs, and four ViTs) were implemented to test how transferable each model is when extracting prominent features from images. Each of these models was tested with an on-the-shelf approach (OTS) and a fine-tuned approach (FT). Similar to Domínguez Sánchez et al., the OTS approach consisted of directly applying the model to a new dataset without any fine-tuning of the weights in the model while the FT approach altered the model to fit the new dataset. Ultimately, the Swin Transformer (a variant of the ViT) after fine-tuning provided the best results overall with an average of 92% accuracy for each art classification, followed closely by the ConvNext CNN Architecture. Our paper is similar to the work done in this paper but will be different in two ways. First, we use a different dataset (Galaxy10 dataset) from what Tonkes [11] used. Secondly, we evaluate the required dataset size where the pre-trained ViT model demonstrates similar results to that of the scratch ViT model using 100% of the dataset.

Table 1. Comparing the results of training from scratch (ViT reported in [5]) to classification after fine-tuning a pre-trained ViT with varying dataset sizes. The number in the trial name represents the percentage of the Galaxy10 dataset that the respective ViT was fine-tuned on. The best results are highlighted in bold. The colored cells represent the models trained on 100% of the dataset.

Scheme	Trial	Validation Accuracy	Validation Loss
Train from scratch	base_vit	69.49%	0.9018
Fine-tuning	ft_vit_40	67.13%	0.92
Fine-tuning	ft_vit_50	69.60%	0.8878
Fine-tuning	ft_vit_60	71.28%	0.8541
Fine-tuning	ft_vit_70	71.75%	0.8129
Fine-tuning	ft_vit_80	73.33%	0.7735
Fine-tuning	ft_vit_90	73.33%	0.7797
Fine-tuning	ft_vit_95	75.20%	0.7548
Fine-tuning	ft_vit_100	75.53%	0.7225

4 Dataset

The data used to fine-tune the ViT architectures was the Galaxy10 dataset, which consists of 17,736 images across 10 galaxy classifications[1]. This dataset is a subset of the Galaxy10 DECals and Galaxy Zoo datasets. The Galaxy Zoo Data Release 2 (GZ DR2) had volunteers classify $\sim 270k$ images from the Sloan Digital Sky Survey (SDSS) across 10 broad classes [12]. From there, the GZ DR2 combined with the DESI Legacy Imaging Surveys (DECals) to create the Galaxy10 DECals dataset. resulting in 441k unique galactic images in one dataset. For the purposes of this paper, we utilize a smaller

---

[1] https://github.com/henrysky/Galaxy10.

version of the Galaxy10 DECals dataset, which contains $\sim 18k$ images across the same 10 galaxy classes as GZ DR2. Some example images from the dataset are displayed in Fig. 2. Due to the disproportionate size of the Disk, Edge-on, Rounded Bulge class with only 300 images, we delete this class from our current dataset. The modified version of this dataset can be seen in Fig. 3 below. Each image has a shape of 256×256 pixels and contains 3 color channels. The Galaxy10 dataset is publicly available online through the astroNN Python library.

4.1   Fine-Tuning Preparation

Using a pre-trained network and applying it to other domains has yielded better performing models than training from scratch. However, when fine-tuning a deep network, the optimal setup varies between applications [13]. For this paper, we create a data pipeline that performs various transforms on the images for the ViT models. These transforms include resizing images to 224×224, converting the image to a tensor, as well as normalizing them with a mean of $[0.485, 0.456, 0.406]$ and a standard deviation of $[0.229, 0.224, 0.225]$ [8]. Images are then loaded into training and validation Torch dataloaders with an 80/20 split and batch sizes of 32.

5   Experiment Details

Each trial was run using the Vision Transformer architecture as provided in the Torchvision package in PyTorch [9]. Utilizing a pre-trained network, we then analyze the effectiveness of the network when using it just as a feature extractor versus training the same model from scratch. We also change the fine-tuning dataset to determine the dataset size where the pre-trained model yields similar results to the model with no initialized weights.

5.1   Model Selection and Pre-training

Within the Torchvision PyTorch package, there are many models and pre-trained weights that are pubicly available at[2]. Out of the various Vision Transformer variants, we chose the vit_b_16 model since this is the exact architecture that was proposed in Dosovitskiy et al.'s An Image is Worth 16×16 Words: Transformers For Recognition At Scale. The b within the name refers to the"Base" version of the model and 16 refers to the 16×16 input patch size. With authenticity in mind, we chose the ViT_B_16_Weights.IMAGENET1K_V1 weights to use for our pre-trained model because these weights were retrieved from a model trained exclusively on ImageNet. Specifically, the vit_b_16 was run on ImageNet images for 300 epochs using the AdamW optimizer with a learning rate of 0.003. For this paper, we also use AdamW as our optimizer while using the CrossEntropyLoss loss function and a learning rate of 0.0001. All the experimental source code is available at Github[3].

---

[2] https://torchvision.mlverse.org.

[3] https://github.com/kumarrah2002/ViT_Pretrained_vs_Scratch.

Table 2. Comparing the time duration for training from scratch versus fine-tuning a pre-trained ViT with varying dataset sizes. The colored cells represent the models trained on 100% of the dataset.

Trial	Total Epochs	Average Epoch Duration	Total Training Time
base_vit	26	302 s	131 min.
ft_vit_40	16	109 s	29 min.
ft_vit_50	23	123 s	47 min.
ft_vit_60	19	147 s	47 min.
ft_vit_70	22	161 s	59 min.
ft_vit_80	27	178 s	80 min.
ft_vit_90	42	190 s	133 min.
ft_vit_95	36	198 s	119 min.
ft_vit_100	50	206 s	172 min

5.2 Pre-trained ViT as a Feature Extractor

We use the pre-trained network as a feature extractor and only modify the classification layer. To accomplish this, the last layer in the Sequential block of the pre-trained ViTs is changed to have a different output shape. Specifically, the out_features within the Linear layer was edited from 1000 classes for ImageNet to 9 classes for the Galaxy10 dataset [3]. We also prevent the weights within every layer of the ViT from being updated except for the last Linear layer. Finally, we conduct nine different trials: one trial trained a ViT with no weights using 100% of the dataset while the other eight used pre-trained ViTs. Of these eight models, the dataset sizes that were used for fine-tuning were 40%, 50%, 60%, 70%, 80%, 90%, 95%, and 100%. We present our findings in Sect. 5.

5.3 Accuracy and Loss Calculation

We use the standard equations of calculating accuracy and loss throughout the trials. The accuracy calculations utilized the following formula:

$$Accuracy = \frac{TP + TN}{TP + TN + FP + FN}$$

The loss, L_δ, is defined as

$$L_\delta = -\sum_{c=1}^{M} y_{o,c} \log(p_{o,c})$$

where the number of classes, M, must be $M > 2$. Here, we have $M = 9$ since we have nine classes within the Galaxy10 dataset.

5.4   Hardware and Software

Both experiments were performed on a NVIDIA GeForce RTX 3060 TI GPU. To activate this GPU, PyTorch, an open-source machine learning framework [9], is used for all trials. The Vision Transformer architecture and its pre-trained weights were taken from the Torchvision library within PyTorch. We use Matplotlib to visualize our results [6].

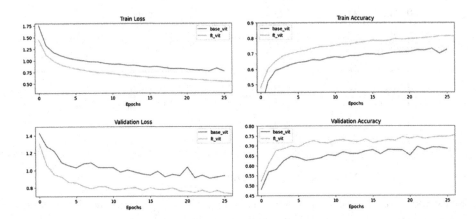

Fig. 4. Loss and accuracy curves of the fine-tuned model (ft_vit in orange) and the base model (base_vit in blue) (Color figure online)

6   Results

We now present the main findings of our study and report results for the eight trials as seen in Table 1. It is important to note that for both experiments, an early stopping approach was used to avoid the models from overfitting. To do this, the lowest loss value, min_loss_value, is set to the first value that the model outputs (oftentimes the largest value throughout the training). We also initialize a counter variable to 0, a threshold variable(δ), and a patience variable. These variables are explained further in the following paragraph.

After each epoch, min_loss_value is compared to the current epoch; if the current value is smaller, then min_loss_value would become the current value. If the current epoch had a loss greater than min_loss_value + δ, then we increment the counter. Once the counter is equivalent to the patience, then the model stops training. To acquire the best results without overfitting, we set $\delta = 0.03$ and $patience = 3$.

6.1   Training from Scratch vs. Fine-Tuning

In this section, we analyze the performance of both the ViT baseline as well as the ViT that was fine-tuned on 100% of the Galaxy10 dataset. When it comes to galaxy morphology classification, we can see that the baseline ViT and fine-tuned ViT achieved

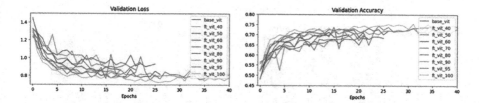

Fig. 5. The validation loss and accuracy curves obtained from both experiments. The number next to the model name represents the dataset size percentage that the model was fine-tuned on.

final accuracies of $\sim 69\%$ and $\sim 76\%$, respectively. After the first three epochs of this trial, the fine-tuned ViT (ft_vit) had an average accuracy increase of 5% compared to the baseline ViT (base_vit) as seen in Fig. 4. This result was not unexpected, as the pre-trained weights from ImageNet allowed ft_vit to recognize patterns within the galaxy images more accurately than training base_vit with no initialized weights. Additionally, ft_vit had an average training time of 201 seconds/epoch whereas base_vit trained at 302 seconds/epoch as seen in Table 2, resulting in a 31% speed-up from the fine-tuned ViT to the base ViT. It is important to note that base_vit began to show signs of overfitting after 25 epochs while ft_vit began to overfit at 50 epochs. This experiment clearly confirms the benefits of transfer learning and its ability to increase accuracy, reduce training time, and delay overfitting.

6.2 Fine-Tuning with Varying Dataset Sizes

Now that we have compared the fine-tuned ViT with the base ViT, we report our findings on the second part of this paper: evaluating the dataset size required for the pretrained model. While it is clear in Fig. 4 that pre-trained weights demonstrate a considerable boost in performance over training from scratch, this begs the question: At what dataset size does a fine-tuned model yield similar results to a baseline model using 100% of the dataset? We approach this problem by creating seven additional trials to those proposed in section Training from Scratch vs. Fine-Tuning.

To modify the dataset for fine-tuning, we created a function that implemented the train_test_split function from the Sci-Kit Learn library [1]. After splitting the original dataset by our desired dataset size, each ViT was trained using the early stopping technique described earlier. The results of these trials are shown in Fig. 5. Upon examination, it is clear that ViT_80 − 100 perform very well compared to their counterparts as they reach an accuracy of $\sim 72\%$–75%. The highest accuracies and lowest loss values for each trial in Fig. 5 can be found in Figs. 6 and 7 below.

Additionally, we can see that as the dataset size decreases, the overall training time decreases. This is clearly shown in Table 2 where the trials with smaller dataset sizes have lower epochs and training times than those of higher dataset sizes. In response to the question proposed at the beginning of this section, we can conclude that a pre-trained model only requires around 50% of the images and 40% of the time used to train the baseline model in order to get similar results when classifying galaxy morphology on the Galaxy10 dataset using the Vision Transformer.

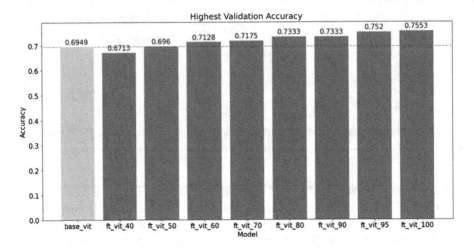

Fig. 6. Visualization of the highest validation accuracy for each trial. The red dotted-line represents the best accuracy from the base ViT with no pre-training. (Color figure online)

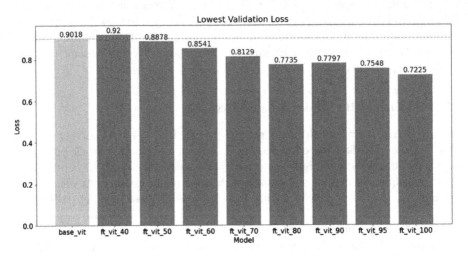

Fig. 7. Visualization of the lowest validation loss for each trial. The red dotted-line represents the best loss from the base ViT with no pre-training. (Color figure online)

7   Conclusion

In this paper, we assess the transferability of the Vision Transformer (ViT) for galaxy morphology classification. We evaluate scratch and pre-trained versions of the ViT with two main experiments: comparing a fine-tuned ViT with a baseline ViT using the same dataset size and comparing multiple pre-trained ViTs with varying dataset sizes against the baseline ViT. Based on the experimental results, we can see that the pre-trained and fine-tuned ViTs were able to outperform the scratch ViT until the dataset percentage was below 50%. Additionally, we can see the differences in training duration from scratch to pre-trained models. Specifically, we see a 32% decrease in training time when comparing base_vit and ft_vit_100, hence demonstrating the implications on time saved when using pre-trained models. When applying computer vision to problems within industry, a reduction of 50% in dataset size and 32% in training time allows for businesses to spend less money and time on data acquisition and preprocessing.

While we focused on the classic ViT architecture for this study, we aim to experiment with other models, such as CNN variants and the latest Transformer architectures like SWin Transformers (SWin) and Data-Efficient Image Transformers (DEiT), for the future. As the world shifts its focus towards space exploration, galaxy classification will allow us to better understand the development of distant galaxies. Specifically, we will gain more insight into the origins of the Universe based on a galaxy's shape and composition, hence solving questions that have gone unanswered for thousands of years. This research is a step in the direction for classifying galaxy morphology where data is scarce.

References

1. Buitinck, L., et al.: Api design for machine learning software: experiences from the scikit-learn project. arXiv preprint arXiv:1309.0238 (2013)
2. Buta, R.J.: Galaxy Morphology (2011)
3. Deng, J., Dong, W., Socher, R., Li, L.J., Li, K., Fei-Fei, L.: Imagenet: a large-scale hierarchical image database. In: 2009 IEEE Conference on Computer Vision and Pattern Recognition, pp. 248–255. IEEE (2009)
4. Domínguez Sánchez, H., et al.: Transfer learning for galaxy morphology from one survey to another. Mon. Not. R. Astron. Soc. **484**(1), 93–100 (2019)
5. Dosovitskiy, A., et al.: An image is worth 16x16 words: transformers for image recognition at scale (2021)
6. Hunter, J.D.: Matplotlib: a 2D graphics environment. Comput. Sci. Eng. **9**(03), 90–95 (2007)
7. Lintott, C.J., et al.: Galaxy Zoo: morphologies derived from visual inspection of galaxies from the sloan digital sky survey. Monthly Not. R. Astron. Soc. **389**(3), 1179–1189 (2008). https://doi.org/10.1111/j.1365-2966.2008.13689.x
8. Liu, Y., Sangineto, E., Bi, W., Sebe, N., Lepri, B., Nadai, M.: Efficient training of visual transformers with small datasets. In: Advance in Neural. Information Processing System, vol. 34, pp. 23818–23830 (2021)
9. Paszke, A., et al.: Pytorch: An imperative style, high-performance deep learning library. In: Advances in Neural Information Processing Systems, vol. 32 (2019)
10. Robertson, B.E.: Galaxy formation and reionization: key unknowns and expected breakthroughs by the James webb space telescope. Ann. Rev. Astron. Astrophys. **60**, 121–158 (2022)
11. Tonkes, V., Sabatelli, M.: How well do vision transformers (VTs) transfer to the non-natural image domain? an empirical study involving art classification. In: Karlinsky, L., Michaeli, T., Nishino, K. (eds.) ECCV 2022. LNCS, vol. 13801, pp. 234–250. Springer, Cham (2022). https://doi.org/10.1007/978-3-031-25056-9_16
12. Walmsley, M., et al.: Galaxy zoo decals: detailed visual morphology measurements from volunteers and deep learning for 314 000 galaxies. Mon. Not. R. Astron. Soc. **509**(3), 3966–3988 (2022)
13. Zhao, Y., Wang, G., Tang, C., Luo, C., Zeng, W., Zha, Z.J.: A battle of network structures: An empirical study of CNN, transformer, and MLP (2021)

A Study of Neural Collapse for Text Classification

Jia Hui Feng[✉][iD], Edmund M.-K. Lai[iD], and Weihua Li[iD]

School of Engineering, Computer and Mathematical Sciences,
Auckland University of Technology, Auckland, New Zealand
jiahui.feng@autuni.ac.nz, {edmund.lai,weihua.li}@aut.ac.nz

Abstract. In this study, we explore the phenomenon of neural collapse (NC) in text classification using convolutional neural networks (CNNs) applied to the AG News dataset [23]. Initially, our findings indicate the occurrence of NC, which initially underperforms compared to a non-collapsed CNN. However, upon closer examination, we uncover an intriguing insight: certain data points converge towards an unknown cluster during NC. Further analysis reveals that this additional cluster represents an additional topic within the dataset, challenging the initial assumption of four distinct classes in AG News. This significant discovery suggests a promising research direction, where NC can serve as a tool for cluster discovery in semi-supervised learning scenarios.

Keywords: Neural networks · Natural language processing · Deep learning · Semi-supervised learning · Neural collapse · Text classification

1 Introduction

In the last decade or so, using deep learning [3], where the ANN models have many hidden layers, for classification has become standard practice. A lot of experience and insights into training such networks have been gained. It is generally not advisable to over-train an over-parametrized model such as a neural network. This is to prevent *overfitting* the model to the training data and therefore affects its generalization ability [6,8].

Recently, an unusual phenomenon has been discovered empirically when a network is over-trained. This occurs when training persists beyond the point where the error vanishes. At this point, if training continues to drive the loss towards zero, then the network will enter the so-called Neural Collapse (NC) state [17]. The network is said to be in the terminal phase of training (TPT) when the error is very small or zero. Surprisingly, such an over-trained network is able to achieve high levels of classification performance. Experimental results in [17] showed that NC networks achieves a performance that is better than a non-collapsed neural network. This is because in the neural collapsed state, the last layer activations concentrates towards a symmetric structure, making the job of the classifier relatively simple. It has also been shown to be a model that has generalizable and transferable learning properties [2].

Many subsequent works have extended the research on NC to further understand this phenomenon and deep learning in general. Some works contributed theoretic understanding using unconstrained features model [9,15]. Further geometric insights are provided by [26,28]. Others studied the effects of the loss function on NC [5,12,27].

The effects of imbalanced training have also been considered [19]. It has also been shown that an NC model has generalizable and transferable learning properties [2,11]. There are also potential improvements proposed to make NC more efficient and effective [22]. There has also been research looking into the limitations of NC. One limitation researchers have discovered is that NC can only occur with weight decay [18]. Also, it has been found that NC does not occur on imbalanced datasets [19]. However, such limitations can be overcome [25]. There are more questions related to NC that need to be investigated. In [7], the authors noted that the NC phenomenon occurs only with the training data. They found that when separate test data are used with the NC network, does not apply and hence they concluded that NC is primarily an optimization phenomenon.

1.1 Our Contributions

In this paper, we seek to answer some further questions regarding NC. First, since all previous research on NC have been conducted using image data, we would like to know if NC occurs for text classification. Since the deep learning network is, in a sense, agnostic to the type of data, we hypothesize that NC also applies. Secondly, although the NC network may not perform well for unseen test data, what can we learn from the misclassified data? For the misclassified test data, the activation of the penultimate layer will not be in the region of those that are correctly classified. Is there any topological structure for these activations?

Our research shows that NC does occur for text classification. Furthermore, the classification accuracy for unseen test data is much lower compared to one that has not been trained to collapse. We make use of topological data analysis to determine the topological structure of activations for the misclassified data. It shows that apart from clustering around the classes, there is an additional cluster formed for the misclassified training data. Examination of these misclassified texts gave us insight into the reasons why they are misclassified. These insights into the training data are very useful in helping us understand the training data better. We postulate that NC can be utilized for semi-supervised learning that help discover new classes or labels.

The rest of this paper is organized as follows. Section 2 presents our experiments and findings when we train a convolutional neural network (CNN) toward NC. A topological data analysis (TDA) of the activation in the penultimate layer of the CNN is presented in Sect. 3. This is followed by a discussion, in Sect. 3.2, regarding the misclassified texts using the insight gained through TDA. Finally, Sect. 4 concludes the paper with suggestions for further research.

2 Neural Collapse in Text Classification

2.1 Characteristics of Neural Collapse

In the seminal paper on NC [17], the authors listed four inter-related characteristics exhibited by NC:

NC1: Collapse of Variability. In the penultimate layer of the network, the data samples that belong to the same class moves towards their class mean and the variability of the intra-class features is lost as they collapse to their class mean. That is, the feature within-class variance. To quantify this collapse, we can use the covariance matrix, denoted by Σ_W, which captures the within-class variance. The covariance matrix measures the relationships between pairs of features and can provide insights into the spread and correlations within the data.

The equation for the covariance matrix is given by:

$$\Sigma W = \underset{i,c}{\text{Ave}}\,(hi, c - \mu c)\,(hi, c - \mu_c)^\top,\tag{1}$$

Here, $h_{i,c}$ represents the feature vector of the i-th sample belonging to class c. The symbol μ_c denotes the class mean, which is the average of all feature vectors belonging to class c. The operator $\underset{i,c}{\text{Ave}}$ indicates the averaging operation, which computes the average covariance matrix over all samples and classes.

By examining the covariance matrix Σ_W, we can observe how the within-class variance changes throughout the network. In the case of the collapse of variability, as the network progresses, the values in the covariance matrix tend to approach zero, indicating a significant reduction in the variability of the intra-class features.

NC2: Preference Towards a Simplex Equiangular Tight Frame (EFT). The class means of the final layer moves towards forming a simplex ETF, which can be mathematically represented by the following:

$$\frac{\langle \mu_c - \mu_G, \mu_{c'} - \mu_G \rangle}{\|\mu_c - \mu_G\|_2 \|\mu_{c'} - \mu_G\|_2} \to \begin{cases} 1, & c = c' \\ \frac{-1}{C-1}, & c \neq c' \end{cases}\tag{2}$$

$$\|\mu_c - \mu_G\|_2 - \|\mu_{c'} - \mu_G\|_2 \to 0 \quad \forall c \neq c'$$

where

$$\mu_G = \underset{i,c}{\text{Ave}}\, h_{i,c},\tag{3}$$

is the feature global mean.

In these equations, μc represents the mean feature vector of class c, while $\mu c'$ represents the mean feature vector of class c'. The symbol μ_G denotes the global mean of the features, which is computed as the average of all feature vectors across all classes. The angle brackets $\langle \cdot, \cdot \rangle$ represent the dot product, and the notation $|\cdot|_2$ denotes the Euclidean norm.

The first equation expresses the cosine similarity between the differences of the class means and the global mean, normalized by the product of their Euclidean norms. As the network progresses, this similarity tends towards specific values: 1 when comparing the same class means ($c = c'$), and $-\frac{1}{C-1}$ when comparing different class means ($c \neq c'$), where C is the total number of classes.

The second equation states that the difference in Euclidean norms between any pair of class means ($c \neq c'$) tends towards zero. In other words, the magnitudes of the class mean vectors become approximately equal.

Together, these equations indicate the network's preference towards a simplex equiangular tight frame (EFT) formation in the final layer, where the class means exhibit specific angular relationships and similar magnitudes. This preference suggests a structured representation of the classes, potentially aiding in better discriminability and classification performance.

NC3: Convergence to Self-duality

There is also a convergence to self-duality observed in the vectors of the last layer, represented mathematically as follows:

$$\frac{w_c}{|w_c|_2} - \frac{\mu_c - \mu_G}{|\mu_c - \mu_G|_2} \to 0 \tag{4}$$

In this equation, w_c represents the weight vector associated with class c in the last layer of the network. The term $\mu_c - \mu_G$ represents the difference between the class mean μ_c and the global mean μ_G of the features.

The equation states that as the network progresses, the difference between the normalized weight vector $\frac{w_c}{|w_c|_2}$ and the normalized difference between the class mean and the global mean $\frac{\mu_c - \mu_G}{|\mu_c - \mu_G|_2}$ tends towards zero. In other words, the weight vectors become increasingly aligned with the difference between the class mean and the global mean, in a normalized sense.

NC4: Simplification to the Nearest Class Mean

In the last layer of the network, the classifier examines the features from the penultimate layer of a test point and assigns a label to the test point based on the closest train-class mean. This process is similar to a 1-Nearest Neighbour classifier [10]. Mathematically, this can be expressed as

$$\arg\max c' \langle wc', h \rangle + b_{c'} \to \arg\min c' |h - \mu c'|_2 \tag{5}$$

where wc' is the weight vector associated with the class c' in the last layer of the network. $b_{c'}$ denotes the bias term associated with class c'. h denotes the feature vector of the test point, which is extracted from the penultimate layer. The term $\mu c'$ represents the class mean of the training samples for class c'.

The equation indicates that the classifier calculates the dot product between the weight vectors and the feature vector h, adds the corresponding bias terms, and selects the class c' that maximizes this expression. This process effectively assigns the test point to the class with the highest similarity between its weight vector and the test point's feature representation. Simplifying this equation, we can rewrite it as selecting the class c' that minimizes the Euclidean distance between the feature vector h and the class mean $\mu_{c'}$. This approach is analogous to the nearest neighbor classification, where the test point is assigned the label of the closest training sample based on the Euclidean distance. This simplification suggests that the network is learning to map the features of a test point to the class means in a manner that resembles a nearest neighbor approach. It implies that the network's decision-making process is driven by the similarity between the test point and the class prototypes represented by the class means.

A visualization of NC with three labels is shown in Fig. 1.

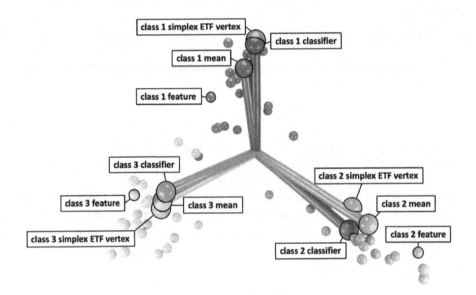

Fig. 1. Visualization of NC with labels [5]

2.2   Experimental Setup

The experiments for text classification is set up as follows.

Dataset. We use AG's corpus of news articles [23] containing textual data of news articles and their corresponding class label. Each class label represents a news category. There are four categories – world, sports, business and science/technology. In our experiments, 50 samples for each label are randomly chosen for each dataset. 45 of these samples are used for training and 5 for testing, i.e. a 90/10 split.

Network Architecture. The same CNN architecture as that used in [24] is employed here. Its architecture is shown in Fig. 2. It works well in detecting local and positional invariant patterns [14]. In addition, weight decay is introduced into the training as it has been shown that it can lead to neural collapse [18]. But the baseline model does not use weight decay. The specific details of the network are:

- Embedding dimension: 300
- Filters: (100, 100, 100)
- Activation: ReLU
- Pooling: 1-max pooling
- Dropout rate: 0.5
- Word Embedding: fastText
- Weight Decay: $5e - 4$

fastText is used as the word embedding as it has shown promising results for CNN in text classification [20].

Training. The baseline model is trained for 20 epoch. The NC model is trained until neural collapse occurs. In this case, it takes 500 epochs.

2.3 Experimental Results

Figure 3a shows that the training error decreases although it is not able to reach zero. The four characteristics of NC described in Sect. 2.1 are shown in Figs. 3b (NC1), 3c (NC2), 3d (NC3), and 3e (NC4) respectively. NC1, NC3, and NC4 show convergence. Although there are still fluctuations in NC2, the variation is small. Therefore, we can say that the network reaches the NC state.

The classification accuracies of the baseline CNN and the NC network are shown in Table 1. While the baseline model achieves a classification accuracy of over 92% after being trained for 20 epochs, the NC model's accuracy is only around 55%. This indicates that the NC network does not generalize well for the test data.

Table 1. Classification accuracies of baseline CNN and NC network

Model	Acc.
Baseline	$92.3\% \pm 0.2\%$
NC CNN	$55.2\% \pm 0.5\%$

Next, we examine the output confidence of the two models using specific texts from the test data. Table 2 shows the output confidence for a sample text taken from the topic of science/technology:

A company founded by a chemistry researcher at the University of Louisville won a grant to develop a method of producing better peptides, which are short chains of amino acids, the building blocks of proteins

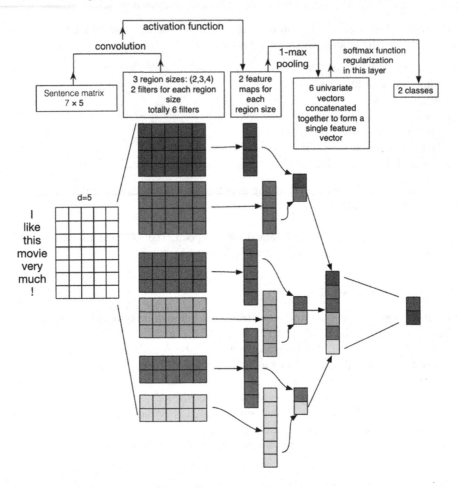

Fig. 2. CNN Architecture [24]

For the baseline CNN, the highest confidence for this input is for the *science/technology* label (79.3%), which is the correct label. However, with the NC network, the confidences of all four labels are similar, with the highest being 26.29% for the *world* label. It seems that after NC, the prediction for each label for each misclassified data sample is very similar. Hence, the NC model's prediction cannot choose any of the classes with confidence.

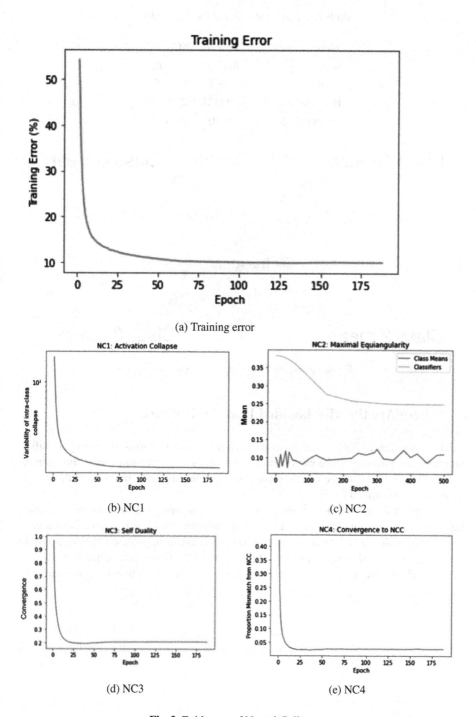

(a) Training error

(b) NC1

(c) NC2

(d) NC3

(e) NC4

Fig. 3. Evidences of Neural Collapse

Table 2. Output confidence of the two models

Topic	Baseline	NC
World	5.64%	**26.29%**
Sports	2.08%	25.69%
Business	12.98%	23.02%
Science/Technology	**79.30%**	25.00%

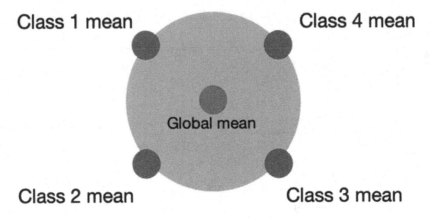

Fig. 4. Geometric Distribution of Class Means

3   Where Are the Misclassified Data Activations?

The results given in Table 2 seems to indicate that the activations of the misclassified data may be clustered. Since each of the four classes form an equiangular tight frame, it is reasonable to postulate that misclassified samples are clustered around the global mean, as demonstrated in Fig. 4.

A graph of the global mean is shown in Fig. 5. Comparing this with the misclassified data mean as shown in Fig. 9, it is clear that the misclassified sample activations do not reside around the global mean. Even though this is the case, it could still be possible that the misclassified samples are still clustered together since their output confidences are similarly distributed. In order to confirm this, we resort to using topological data analysis.

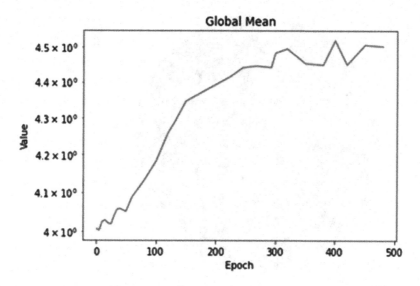

Fig. 5. Global Class Mean during NC training

3.1 Topological Data Analysis

Topological data analysis (TDA) is a statistical method, utilizing algebraic topological concepts to find structure in data. It has been used in a variety of applications. More recently, it is used for cluster analysis in machine learning [21].

Persistent Homology. Persistent homology is a mathematical tool in algebraic topology [1] that enables the study of the topological properties of data from a point cloud in a metric space such as $\mathbb{R}^d$. Where $\mathbb{R}^d$ represents a Euclidean space of dimension d.

In persistent homology, topological spaces are represented by simplicial complexes. A simplicial complex is a collection of simplices, where a simplex σ is a subset of vertices and let V be a set of vertices, typically denoted as $V = 1, \ldots, |V|$, where $|V|$ represents the total number of vertices in the complex. For example, if we have a simplicial complex with three vertices, we can represent it as $V = 1, 2, 3$. Each subset of vertices $\sigma \subseteq V$ corresponds to a simplex.

A simplicial complex K on the vertex set V is a collection of simplices σ, where each $\sigma \subseteq V$. Additionally, for any simplex τ that is a subset of a simplex σ in K, we have $\tau \in K$. This means that if, for example, we have a simplex $1, 2$ in K, then its subsets 1 and 2 must also be included in K. The dimension n of a simplex σ is given by $n = |\sigma| - 1$, where $|\sigma|$ represents the number of vertices in σ. For instance, if $\sigma = 1, 2, 3$, then $n = 3 - 1 = 2$ (Fig. 6).

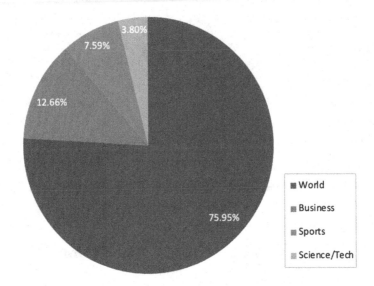

Fig. 6. Distribution of (incorrect) labels by model

A filtration of a simplicial complex is a function $f : K \rightarrow \mathbb{R}$ that assigns a real value to each simplex in the complex. This function satisfies the property that for any simplex τ that is a subset of another simplex σ in the filtration, the assigned value of τ is less than or equal to the assigned value of σ. In mathematical terms, this can be expressed as $f(\tau) \leq f(\sigma)$ whenever $\tau \subseteq \sigma$. The filtration function f allows us to order the simplices based on their assigned values, providing an ordering of the simplices in the complex.

Persistent homology enables the visualization of the formation and destruction of loops and voids in the data space. It can be visualized using persistent diagrams and barcodes. A persistent diagram shows the filtration at which a cluster is born and when it dies. It consists of points in a plane, where each point represents a cluster and its coordinates indicate the filtration values at which the cluster is born and dies. A barcode representation provides the same information but in the form of bar graphs, where the length of each bar represents the lifespan of a cluster.

Example of a series of filtrations are shown in Fig. 7.

Very often, the data points exist in high dimension and therefore are difficult to visualize. Persistent homology can be visualized through the use of persistent diagrams and what is commonly known as barcodes. Persistent diagrams show us the filtration at which a cluster is born and when it dies. Barcodes show the same information but in the form of bar graphs.

TDA on Textual Dataset. Using the Python library *gudhi* [13], we analyzed the activations of the penultimate layer of the CNN for the dataset that is used for training and testing. The persistent diagram and barcode are shown in Fig. 8. The top end of the barcode (Fig. 8b indicates that there are more than 4 major clusters. The correspond-

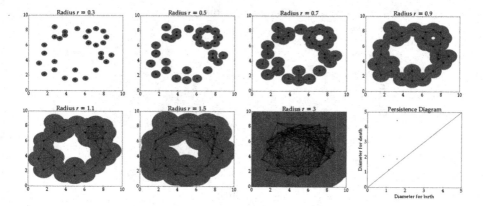

Fig. 7. Example of a series of filtrations on a set of point cloud data [16]

ing points on Fig. 8a are the two data points between 7.5 and 10 on the death axis are actually two sub-classes that were combined into one in our labelling of the data classes.

3.2 Discussions

The distribution of the misclassified labels is shown in Fig. 9. This means that over 70% of the misclassified samples are classified as *world* by the NC network. A small sample of texts from the topics of *Sports, Business*, and *Science/Technology* that have been misclassified as *World* are shown in Table 3. It can be observed that they all contain geographical locations in their texts. For example, under the topic of *sports*, the texts contain words such as "Atlanta", "Minnesota". Also, under the topic of *business*, there are "New York" and "London". Under *science/technology* there are "Southern California", and also the word "Canadian". It seems that the model relates these geographic names to the topic of *world*. However, in the context of the dataset, this classification is incorrect.

Now consider some examples from the topic *world* that have been misclassified are shown in Table 4. These texts also contain geographic locations. So it cannot be the deciding factor here.

We notice that in both tables, there are text examples about oil (petrol) that was classified wrongly by the NC model. In Table 3, under the *business* category, the second text example talks about oil prices reported in London and discusses its impact on other countries. But the model has incorrectly labeled it as *business*. Furthermore, in Table 4, when it talks about oil in New York, the model has classified it as *business*. In terms of news context, one discusses more about the global impact of oil prices which would be closer to *world*, and the other is more of local impact of oil prices.

Based on this analysis, it seems that the label *world* can be split into two: one is **global** and the other is **local**. Both of them combined and the major topic would be **contextual impact**, instead of *world*.

What this seems to indicate is that we can make use of NC to help analyze the characteristics of the dataset when it comes to text classification.

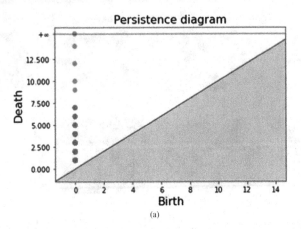

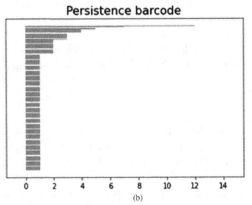

Fig. 8. Persistent diagram and barcode of activations

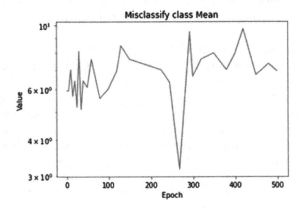

Fig. 9. Mislabeled Class Mean during NC training

Table 3. Samples Misclassified as Topic 1 (World)

Correct Topic	Misclassified Examples
Sports	Outfielder J.D. Drew missed the Atlanta Braves' game against the St. Louis Cardinals on Sunday night with a sore right quadriceps
	The Cleveland Indians pulled within one game of the AL Central lead by beating the Minnesota Twins, 7–1, Saturday night with home runs by Travis Hafner and Victor Martinez
Business	NEW YORK (Reuters) - U.S. Treasury debt prices slipped on Monday, though traders characterized the move as profit-taking rather than any fundamental change in sentiment
	LONDON (Reuters) - Oil prices struck a fresh record above $48 a barrel on Thursday, spurred higher by renewed violence in Iraq and fresh evidence that strong demand growth in China and India has not been slowed yet by higher energy costs
Sci/Tech	Southern California's smog-fighting agency went after emissions of the bovine variety Friday, adopting the nation's first rules to reduce air pollution from dairy cow manure
	Vermont's Republican governor challenged the Bush administration's prescription drug policy in federal court yesterday, marking the first time a state has chosen a legal avenue in the expanding battle over Canadian imports

Table 4. Incorrect Classification for Class 1 (World) Samples

NC Label	Text Examples (World)	Title
Business	The price of oil has continued its sharp rise overnight, closing at a record high. The main contract in New York, light sweet crude for delivery next month, has closed at a record $US46.75 a barrel - up 70 cents on yesterday	Oil prices bubble to record high
Sports	Students at the Mount Sinai School of Medicine learn that diet and culture shape health in East Harlem	Future Doctors, Crossing Borders

3.3 Possible Further Applications

Text classification models are traditionally trained by supervised learning (SL) techniques. However, the process of labelling data by human annotators can be time-consuming and expensive. When limited labelled data are available, semi-supervised learning (SSL) is used. SSL is a combination of both supervised and unsupervised learning. It allows the model to integrate the data that has not been categorized. The goal of SSL is to maximize the learning performance through pre-labeled and newly discovered labels and these newly-labeled examples [4].

The experimental process in this paper could be utilized to uncover hidden clusters. A neural collapsed model trained on labelled textual data. Unlabelled data could then be used as test data to this model. Some of them would be clustered to the existing class means. So these samples are then automatically labelled. The remaining samples could be analyzed using TDA on the activation pattern of the penultimate layer. Those forming new clusters could be assigned new labels.

4   Conclusions and Future Works

In this paper, it has been verified that the neural collapse phenomenon also occurs in text classification. However, the NC model performed very poorly in the classification of test data. We performed analysis using TDA on the activation pattern of the penultimate layer of CNN for misclassified data. This led to the discovery of the possible deficiencies in labeling of the original dataset that contributes to poor classification accuracies.

We were able to find a hidden cluster that some of the data points were converging towards. This hidden cluster turned out to be an additional topic. Specifically for the AG News dataset, the NC model has been able to identify additional topics which can label the news as an impact on *local* or a *global* context. To provide additional verification of this finding, the textual content of the dataset labeled world can be relabeled as global and local. Then followed by an experiment using a text classification technique such as CNN and/or NC. If the performance results demonstrate a significant improvement, it will confirm the presence of the hidden topic.

Overall, this indicates that NC can be used for cluster discovery in semi-supervised learning situations. This is an area of future research.

References

1. Edelsbrunner, H., Harer, J.: Persistent homology - a survey. In: Surveys on Discrete and Computational Geometry, vol. 453, p. 257. American Mathematical Society (2008)
2. Galanti, T., György, A., Hutter, M.: On the role of neural collapse in transfer learning, January 2022. https://doi.org/10.48550/arXiv.2112.15121. http://arxiv.org/abs/2112.15121. arXiv:2112.15121 [cs]
3. Goodfellow, I., Bengio, Y., Courville, A.: Deep Learning. MIT Press, Cambridge (2016). Google-Books-ID: omivDQAAQBAJ
4. Hady, M.F.A., Schwenker, F.: Semi-supervised learning. In: Bianchini, M., Maggini, M., Jain, L.C. (eds.) Handbook on Neural Information Processing. ISRL, vol. 49, pp. 215–239. Springer, Heidelberg (2013). https://doi.org/10.1007/978-3-642-36657-4_7
5. Han, X.Y., Papyan, V., Donoho, D.L.: Neural collapse under MSE loss: proximity to and dynamics on the central path, May 2022. https://doi.org/10.48550/arXiv.2106.02073. http://arxiv.org/abs/2106.02073. arXiv:2106.02073 [cs, math, stat]
6. He, F., Tao, D.: Recent advances in deep learning theory, March 2021. https://doi.org/10.48550/arXiv.2012.10931. http://arxiv.org/abs/2012.10931. arXiv:2012.10931 [cs, stat]
7. Hui, L., Belkin, M., Nakkiran, P.: Limitations of neural collapse for understanding generalization in deep learning, February 2022. http://arxiv.org/abs/2202.08384. arXiv:2202.08384 [cs, stat]

8. Jabbar, H.K., Khan, R.Z.: Methods to avoid over-fitting and under-fitting in supervised machine learning (comparative study). In: Computer Science, Communication and Instrumentation Devices, pp. 163–172. Research Publishing Services (2014). https://doi.org/10.3850/978-981-09-5247-1_017. http://rpsonline.com.sg/proceedings/9789810952471/html/017.xml

9. Ji, W., Lu, Y., Zhang, Y., Deng, Z., Su, W.J.: An unconstrained layer-peeled perspective on neural collapse, April 2022. http://arxiv.org/abs/2110.02796. arXiv:2110.02796 [cs, stat]

10. Kothapalli, V., Rasromani, E., Awatramani, V.: Neural collapse: a review on modelling principles and generalization, June 2022. http://arxiv.org/abs/2206.04041. arXiv:2206.04041 [cs]

11. Li, X., et al.: Principled and efficient transfer learning of deep models via neural collapse, January 2023. http://arxiv.org/abs/2212.12206. arXiv:2212.12206 [cs, eess, stat]

12. Lu, J., Steinerberger, S.: Neural collapse under cross-entropy loss. Appl. Comput. Harmonic Anal. **59**, 224–241 (2022). https://doi.org/10.1016/j.acha.2021.12.011. https://www.sciencedirect.com/science/article/pii/S1063520321001123

13. Maria, C.: Persistent cohomology user manual - gudhi documentation (2016). https://gudhi.inria.fr

14. Minaee, S., Kalchbrenner, N., Cambria, E., Nikzad, N., Chenaghlu, M., Gao, J.: Deep learning-based text classification: a comprehensive review. ACM Comput. Surv. **54**(3), 1–40 (2022). https://doi.org/10.1145/3439726

15. Mixon, D.G., Parshall, H., Pi, J.: Neural collapse with unconstrained features. Sampling Theory Sig. Process. Data Anal. **20**(2), 11 (2022). https://doi.org/10.1007/s43670-022-00027-5

16. Munch, E.: A user's guide to topological data analysis. J. Learn. Anal. **4**(2), 47–61 (2017). https://doi.org/10.18608/jla.2017.42.6. https://learning-analytics.info/index.php/JLA/article/view/5196

17. Papyan, V., Han, X.Y., Donoho, D.L.: Prevalence of neural collapse during the terminal phase of deep learning training. Proc. Natl. Acad. Sci. **117**(40), 24652–24663 (2020). https://doi.org/10.1073/pnas.2015509117

18. Rangamani, A., Banburski-Fahey, A.: Neural collapse in deep homogeneous classifiers and the role of weight decay. In: ICASSP 2022–2022 IEEE International Conference on Acoustics, Speech and Signal Processing (ICASSP), pp. 4243–4247, May 2022. https://doi.org/10.1109/ICASSP43922.2022.9746778. ISSN: 2379-190X

19. Thrampoulidis, C., Kini, G.R., Vakilian, V., Behnia, T.: Imbalance trouble: revisiting neural-collapse geometry, August 2022. https://doi.org/10.48550/arXiv.2208.05512. http://arxiv.org/abs/2208.05512. arXiv:2208.05512 [cs, stat]

20. Umer, M., et al.: Impact of convolutional neural network and FastText embedding on text classification. Multimed. Tools Appl. **82**(4), 5569–5585 (2023). https://doi.org/10.1007/s11042-022-13459-x

21. Wasserman, L.: Topological data analysis, September 2016. https://doi.org/10.48550/arXiv.1609.08227. http://arxiv.org/abs/1609.08227. arXiv:1609.08227 [stat]

22. Yaras, C., Wang, P., Zhu, Z., Balzano, L., Qu, Q.: Neural collapse with normalized features: a geometric analysis over the riemannian manifold, September 2022. http://arxiv.org/abs/2209.09211. arXiv:2209.09211 [cs, eess, math, stat]

23. Zhang, X., Zhao, J., LeCun, Y.: Character-level convolutional networks for text classification. In: Advances in Neural Information Processing Systems, vol. 28. Curran Associates, Inc. (2015). https://papers.nips.cc/paper/2015/hash/250cf8b51c773f3f8dc8b4be867a9a02-Abstract.html

24. Zhang, Y., Wallace, B.: A sensitivity analysis of (and practitioners' guide to) convolutional neural networks for sentence classification, April 2016. http://arxiv.org/abs/1510.03820. arXiv:1510.03820 [cs]

25. Zhong, Z., et al.: Understanding imbalanced semantic segmentation through neural collapse (2023)

26. Zhou, J., Li, X., Ding, T., You, C., Qu, Q., Zhu, Z.: On the optimization landscape of neural collapse under MSE loss: global optimality with unconstrained features. In: Proceedings of the 39th International Conference on Machine Learning, pp. 27179–27202. PMLR, June 2022. https://proceedings.mlr.press/v162/zhou22c.html. ISSN: 2640-3498
27. Zhou, J., et al.: Are all losses created equal: a neural collapse perspective, October 2022. https://doi.org/10.48550/arXiv.2210.02192. https://arxiv.org/abs/2210.02192v2
28. Zhu, Z., et al.: A geometric analysis of neural collapse with unconstrained features (2021)

Research Data Reusability with Content-Based Recommender System

M. Amin Yazdi(✉) [ID], Marius Politze [ID], and Benedikt Heinrichs [ID]

IT Center, RWTH Aachen University, Seffenter Weg 23, Aachen, Germany
{yazdi,politze,heinrichs}@itc.rwth-aachen.de

Abstract. The use of content-based recommender systems to enable the reusability of research data artifacts has gained significant attention in recent years. This study aims to evaluate the effectiveness of such systems in improving the accessibility and reusability of research data artifacts. The study employs an empirical study to identify content-based recommender systems' strengths and limitations for recommending research data-collections (repositories). The empirical study involves developing and evaluating a prototype content-based recommender system for research data artifacts. The literature review findings reveal that content-based recommender systems have several strengths, including providing personalized recommendations, reducing information overload, and enhancing retrieved artifacts' quality, especially when dealing with cold start problems. The results of the empirical study indicate that the developed prototype content-based recommender system effectively provides relevant recommendations for research data repositories. The evaluation of the system using standard evaluation metrics shows that the system achieves an accuracy of 79% in recommending relevant items. Additionally, the user evaluation of the system confirms the relevancy of recommendations and enhances the accessibility and reusability of research data artifacts. In conclusion, the study provides evidence that content-based recommender systems can effectively enable the reusability of research data artifacts.

Keywords: Recommender systems · Content-based · RDF graph data models · Research data management

1 Introduction

Research Data Management (RDM) has become an essential practice for researchers to store, manage, and share their data effectively. With the exponential growth of digital data, managing research data has become increasingly complex, and more researchers have realized the importance of RDM. As a result, RDM has been gaining attention from researchers and institutions worldwide.

Recent studies show that the number of researchers applying RDM for their data artifacts has been growing steadily. For example, in a survey conducted in 2019 found that 73% of researchers agreed that data management is vital for their research [18]. For

D. Conte et al. (Eds.): DeLTA 2023, CCIS 1875, pp. 143–156, 2023.
https://doi.org/10.1007/978-3-031-39059-3_10

instance, Coscine[1] is an RDM platform that uses SHACL[2] and RDF[3] to build knowledge graphs that can be validated according to defined metadata profiles (also known as application profiles) to maintain meta information and controlled vocabularies regarding every data item. These metadata profiles facilitate applying good scientific practices to boost research data Findability, Accessibility, Interoperability, and Reusability (FAIR) [7]. Accordingly, domain experts and scholars in a scientific discipline define the Application Profiles to ensure that the metadata accurately describes every data artifact. However, enabling the reusability of relevant research data can be challenging, especially with the exponential growth of digital data [21]. For example, a recent study found that over 90% of researchers reported difficulties finding and accessing relevant data [16]. This difficulty can be attributed to factors such as inadequate metadata or lack of standardization in metadata descriptions. As a result, many researchers spend a significant amount of time reproducing findings or results from their fellow researchers, resulting in unnecessary duplication of effort and resources [20].

To address this challenge, content-based recommendation systems have been proposed to improve the discoverability and reuse of research data. These systems use machine learning algorithms to analyze the research data content and make recommendations based on similarities in the content. Several studies have explored the use of content-based recommendation systems in the context of research data management. For example, Färber et al. [6] proposed a system for recommending research datasets based on their content and metadata. The system uses machine learning algorithms to identify similar datasets and recommends them to researchers. Similarly, in a study by Nair et al., [11] authors proposed a content-based recommendation system for research papers using the semantic relationships between papers. However, applying content-based recommendation systems to research data management requires a structured metadata profile that accurately describes the research data. This is where RDM platforms such as Coscine could play a crucial role in enabling content-based recommendations. By providing a standardized metadata profile using SHACL and RDF graph data models, Coscine encourages acquiring an accurate description of research data, making it easier for content-based recommendation systems to operate effectively.

In summary, while the demand for RDM continues to grow, the discoverability and reuse of research data remain a significant challenge for researchers. Content-based recommendation systems offer a promising solution to this challenge by leveraging the relationships between research data items and their metadata. However, applying these systems to research data management requires a structured metadata profile such as that provided by RDM platforms like Coscine.

Overall, this paper aims to answer the following research questions: *How can a content-based recommender system enable the reusability of research data repositories using RDF-based graph data models?* By answering this question, we hope to contribute to the growing body of literature on RDM and provide insights into how researchers can improve the discoverability and reuse of research data.

---

[1] www.coscine.de.

[2] https://www.w3.org/TR/shacl/.

[3] https://www.w3.org/TR/rdf11-concepts/.

2   Preliminaries

Table 1. An example dataset curated according to metadata profile defined by a research group.

Repository Name	File Name	*MillingParameters*	*LaserPulseEnergy*	*LiftOutRegion*	*PulseFrequency*	...
Repository-1	F1.csv	pA2.0/0.1 um 40 pA	60	Pt protection layer $14 \times 4 \times 4$ m Trench milling (2x)	125	...
Repository-1	F2.jpg	0.43 nA/1.5 m 0.43	100	fracture cross-section $10 \times 1.5 \times 0.5$ m	125	...
Repository-2	F3.csv	0.43 nA/0.75 m 0.23	100	Pt protection layer $14 \times 4 \times 4$ m Trench	250	...
Repository-2	F4.csv	pA2.0/0.1 um 40 pA	100	cross-section $3 \times 1.5 \times 0.5$ m	250	...
...	...	...	...	...	...	

RDM practices are a crucial aspect of scientific research that facilitates long-term data storage, preservation, and sharing. To ensure efficient and effective RDM, specialized metadata profiles have been developed to describe data in a scientific context. The data management platform Coscine employs these specialized discipline specific metadata profiles, allowing domain experts to define their own set of metadata and vocabulary to explain data in their context. However, despite the availability of search engines to locate data, discovering relevant data artifacts can be challenging without prior knowledge of their existence [19].

Coscine's metadata profiles are based on the SHACL model, where information is stored in the Subject, Predicate, and Object formats [13]. This allows for more detailed and flexible descriptions of data, where the subject points at the data, the Predicate defines the relationship to its Object, and the Object contains meta information explaining the Subject. For instance, Table 1 lists a portion of the acquired sample dataset, where *Repository* Name and *File Name* jointly stand for the Subject, while *Milling-Parameters, LaserPulseEnergy, LiftOutRegion,* and *PulseFrequency* are the Predicates; every value below Predicates are Objects. Objects are commonly unstructured texts manually noted by users in a field [19].

In this study, we acquired metadata provided for files uploaded and provided by experts over one year in the domain of Material Science and Chemistry. Manual metadata entry for every file is a tedious task for researchers and is often neglected or postponed, making these relatively small datasets very expensive and valuable [17]. Figure 1 presents the distribution of files to their data repositories (collections). In total, we have acquired 543 entries from 25 repositories. In this figure, we can observe the natural distribution of files in repositories, where some repositories contain about 55 files and some with as few as 1 or 2 files.

Figure 2 illustrates the data sparsity for every Predicate or column. Although the RDM system under study does not allow the uploading of files without providing mandatory meta information, with employing process mining techniques, it becomes apparent that some users bypass this requirement by uploading their research data to

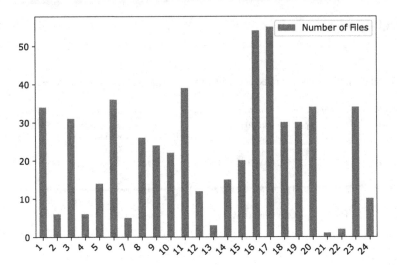

Fig. 1. Distribution of files per repository. X-axis represents repository id, and the y-axis is the number of files.

their data repositories using clients [22]. Thus, addressing data imputation for the missing values is a crucial step in our recommender system preprocessing pipeline. Our proposed recommender system suggests relevant repositories by building profiles from files' metadata. Furthermore, the recommender system recommends repositories' Persistent Identifiers (PIDs) as links, which users can utilize to contact the PID owner and request access to data owners while adhering to data protection laws. We use the scikit-learn[4] machine learning library for Python programming language to implement the proposed methodology.

3   Approach

consistency of the data features, datatype information, semantic definition in the future may allow to consider overlapping profiles. In this section, we elaborate on the recommender system pipeline with six crucial steps as represented in Fig. 3.

3.1   Sparsity Check and Imputation

The imputation step is crucial for a recommender system pipeline as it ensures we have a complete dataset with minimal missing values. The constructed dataset is then used for further analysis and evaluation. Previous studies have shown that imputing missing values can improve the performance of recommender systems [12, 14]. Therefore, our approach to handling missing values is based on best practices from previous studies.

---

[4] www.scikit-learn.org.

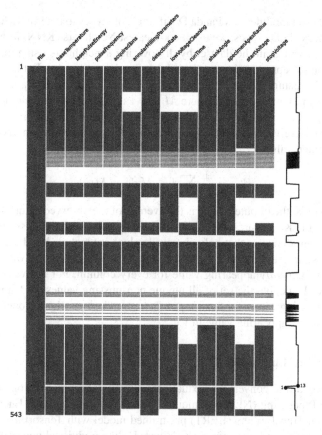

Fig. 2. Data sparsity per feature. Files and repositories columns jointly act as Subjects, feature headers are Predicates, and white spaces represent missing Objects/values.

In this step, let D denote the input dataset. We evaluate the input dataset D against data sparsity before imputing missing values for every feature and row. Generally, according to the legacy study by Bennett et al., imputing missing data is considered statistically a reasonable approach when the amount of missing data is less than 10% of the total data [3]. However, due to the cold start problem and limited sample data availability, we increase the threshold for data imputation to 30%, beyond which features and rows will be dropped for further evaluation or imputation.

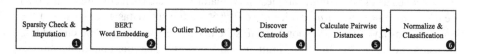

Fig. 3. The general overview of the approach.

The imputation module uses Panda Dataframe data types to determine features data type. For numerical data types, we use the k-Nearest Neighbors (KNN) imputer to predict the missing values in the dataset by discovering k neighbors using the Euclidean distance metric to represent the most similar to a selected feature. Let $d_{i,j}$ denote the value of the jth feature of the ith instance in D, and let NaN denote a missing value. Let M be the binary mask matrix, where $M_{i,j} = 1$ if $d_{i,j}$ is observed (not missing) and $M_{i,j} = 0$ if $d_{i,j}$ is missing. Let $\mathcal{N}_i$ be the set of $K = 3$ nearest neighbors of the ith data point (excluding itself) based on the Euclidean distance between the observed features. Then, the imputed value $\hat{X}_{i,j}$ for the missing value $X_{i,j}$ is given by:

$$\hat{d}_{i,j} = \frac{1}{K} \sum k \in \mathcal{N}_i d_{k,j} \cdot M_{k,j} \tag{1}$$

In other words, the imputed value is the average of the observed values for the same feature among the K=3 nearest neighbors of the data point with the missing value. The binary mask matrix ensures that only observed values are included in the average.

We use a simple imputer for the object data type where missing values are filled with the *most frequently* appearing value for every column. For a given feature j of object type, let V_j denote the set of all unique non-missing values for feature j. Then, we fill in any missing values for feature j as $d_{i,j} = \text{mode}(V_j)$, where $\text{mode}(V_j)$ denotes the *most frequently* appearing value in V_j.

3.2 BERT Word Embedding

In this step, we aim to convert the textual data into vector representations using Natural Language Processing (NLP) techniques. We employ Bidirectional Encoder Representations from Transformers (BERT) pre-trained model with TensorFlow to generate contextualized word and sentence embeddings. Unlike traditional approaches such as TF-IDF, BERT leverages a deep learning approach and is trained on vast amounts of text data to generate rich representations of words and sentences that capture their context and meaning. To facilitate cross-lingual sentence understanding, we use a multi-lingual version of BERT known as the "bert_multi_cased" [5]. This model is designed to pre-train deep bidirectional representations from an unlabeled text corpus by joint conditioning on both left and right contexts, preserving the distinction between lower and upper case.

Every column with textual data is embedded into a 768-dimensional vector using the "bert_multi_cased" model. The size of the resulting data is proportional to the number of columns with textual data and the vector length, plus every column with numerical data. We process the data in chunks to avoid overflowing the processing power [15]. Hence, if D be a dataset with m samples and n features, where each column j has a data type $dt_j \in numeric, string$, and let $D_s \in \mathbb{R}^{m \times k}$ be the sub-dataset consisting of the string columns and their corresponding BERT embeddings, where k is the embedding dimension (i.e., $k = 768$ for "bert_multi_cased" model). Then, the final dimension of the processed data is given by:

$$D_f \in \mathbb{R}^{m \times (k \cdot \forall dt_j = string + \forall dt_j = numeric)} \tag{2}$$

Overall, this step generates rich vector representations of the textual data, which can be further utilized for downstream tasks.

3.3 Outlier Detection

This step aims to identify outlier data points or buckets that exhibit significant dissimilarity from other data points or do not follow the expected behavior of the majority of points. Distance-based methods are the most commonly used techniques to detect outliers [2, 4]. These algorithms are based on computing the distance between observations, where a point is considered an outlier if it is significantly far from its neighbors. In our approach, we first calculate the average pairwise distance of repositories, and then estimate the corresponding z-score for each bucket. A threshold factor of 3 is used to identify and filter out repositories that need to be excluded from the recommendation task.

3.4 Discover Centroids

In this step, we aim to discover the centroids of the clusters. K-means clustering is a widely used unsupervised learning technique for clustering data points into K clusters based on similarity. In our study, we use the K-means clustering algorithm to create a cluster per repository, where each point in the cluster is a file represented in a fixed number of feature dimensions derived from the BERT word embeddings discussed in Sect. 3.2.

The K-means clustering algorithm starts with an initial random assignment of data points to clusters and calculates the initial cluster centroids. Then, it iteratively assigns each data point to the nearest centroid and updates the centroids until convergence. The algorithm stops when the cluster assignments no longer change, or a maximum number of iterations is reached.

We use the scikit-learn implementation of K-means clustering with the default settings. After clustering, we obtain k clusters with their respective centroids corresponding to the number of repositories after outlier detection, which are used as the basis for the final step of the recommendation task.

3.5 Calculate Pairwise Distances

Pairwise distances between centroids are calculated using either Euclidean or Cosine metrics. This allows for constructing a confusion matrix that captures the distances between every pair of repository's centroids. Using either of these metrics, we can create a confusion matrix showing the distances between each repository. This allows us to assess the similarity between items and ultimately generate end-user recommendations.

3.6 Normalization and Classification

This section outlines the methodology employed to scale and classify similarity results in our analysis. The final step of the analysis involves normalizing the confusion matrix to fall within the range of 0 to 1, where 0 represents the closest distance, and 1 represents the farthest distance between any two clusters. The normalization process is achieved by dividing the confusion matrix's raw values by the matrix's maximum value.

After normalization, the resulting values are further recasted to facilitate the evaluation process. Specifically, the normalized values are mapped to an equal-length multiclass scale with five classes. Thus, the range of 0 to 1 is divided into five equal intervals, with the lowest interval representing the least similarity and the highest interval representing the highest similarity between cluster pairs. This casting method is chosen due to its natural mapping of start-like ratings, where five stars represent most liked, and one star expresses dislikes [8]. The resulting classification scheme is represented as follows:

$$C_{i,j}^{class} = \begin{cases} 1 & 0.8 \leq C_{i,j}^{norm} \leq 1.0 \\ 2 & 0.6 \leq C_{i,j}^{norm} < 0.8 \\ 3 & 0.4 \leq C_{i,j}^{norm} < 0.6 \\ 4 & 0.2 \leq C_{i,j}^{norm} < 0.4 \\ 5 & 0.0 \leq C_{i,j}^{norm} < 0.2 \end{cases} \tag{3}$$

where $C_{i,j}^{class}$ represents the resulting class of the normalized confusion matrix. It is important to note that the classification scheme functions as an instrument for evaluating the similarity between cluster pairs. Therefore, a cluster pair assigned to class 1 is considered the least similar, while a pair assigned to class 5 is considered the most similar.

4   Results and Analysis

In this section, we present the empirical studies conducted using the proposed methodology to provide recommendations for repositories and assess its performance. The dataset used for the evaluation is described in Sect. 2 and was obtained from the Coscine platform. To select a set of appropriate features, a domain expert was consulted to select a list of twelve features (Predicates) that best describe artifacts within repositories. The columns (except File) in Fig. 2 demonstrate the corresponding independent variables. According to the user, the metadata for the selected set of features can assist domain scientists in distinguishing between repositories without requiring them to analyze the data artifacts individually.

The evaluation pipeline was executed twice, once for each distance metric, to ensure the robustness and consistency of the results. Figures 4a and 4b illustrate the pairwise similarity of repositories as heatmaps. To evaluate the performance of the pipeline, a ground truth matrix was created by asking the domain expert to heuristically evaluate and rank the similarity of pairwise repositories and provide a similarity score between 1 and 5 based on the discussed classification earlier. This approach has been widely used in previous studies to evaluate the performance of distance metrics [1].

Table 2 lists the Precision, Recall, and F1 score for each class per distance metrics. We observe that the F1 score provides the best results for class 1 regardless of its metric. This may be attributed to the imbalanced distribution of the dataset across classes, as indicated by the support values.

Table 2. Comparing the recommender system results for Cosine and Euclidean distance metrics on each class.

	Precision		Recall		F1		Support
	Cos.	Euc.	Cos.	Euc.	Cos.	Euc.	
Class 1	0.95	0.87	0.70	0.93	0.81	0.90	177
Class 2	0.47	0.70	0.29	0.88	0.36	0.78	153
Class 3	0.42	0.76	0.49	0.56	0.45	0.64	111
Class 4	0.26	0.76	0.32	0.68	0.29	0.72	82
Class 5	0.44	0.97	1.00	0.70	0.61	0.81	53
Macro Avg	0.51	0.81	0.56	0.75	0.50	0.77	576
Weighted Avg.	0.58	0.80	0.52	0.79	0.53	0.78	576

Table 3. Overall performance of Cosine and Euclidean distance metrics to suggest correct items.

	Silhouette	RMSE	MAE	AUC	Accuracy
Cosine	0.68	0.83	0.54	0.68	0.52
Euclidean	0.74	0.54	0.24	0.68	0.79

Table 3 shows the record of the overall performance of our recommender system. We use the Silhouette score to assess the soundness of our predicted ratings and ensure separation between the discovered groups of repositories. This assists us in avoiding overfitting or underfitting our model and guarantees a reasonable distance between the clusters in the predicted dataset is achieved. Accordingly, the Silhouette scores suggest reasonable distances between clusters. We also use the Root Mean Squared Error (RMSE) and Mean Absolute Error (MAE) to measure the difference between predicted and actual ratings. Our results indicate high error rates for both RMSE and MAE while using the Cosine distance metric.

Furthermore, the Area Under the Curve (AUC) scores across all classifications are reported in Table 3. The AUC score of 0.68 suggests that the recommender system can reasonably distinguish between true positive and false negative items, as demonstrated in Fig. 5 by each class's Receiver Operating Characteristic (ROC) curve. Finally, we note that the Euclidean distance metric yields a promising recommender accuracy of 0.79, whereas Cosine returns an unsatisfactory accuracy of 0.52.

In addition to quantitative evaluation, we conducted a qualitative evaluation of the recommender system. We have converted our code-base into a publicly available python package[5] where a repository id (pivot) and a set of pre-selected features are given as package input, and it returns the top five similar items. A domain expert evaluated the system by picking the independent variables as parameters. The study was conducted twice using the Euclidean distance factor. During the 30 min long assessment, the domain expert reviewed files and their metadata for the recommended set and heuristically judged the similarity and ranking of findings. The evaluation involved

---

[5] https://pypi.org/project/DA4RDM-RecSys-ContentBased/.

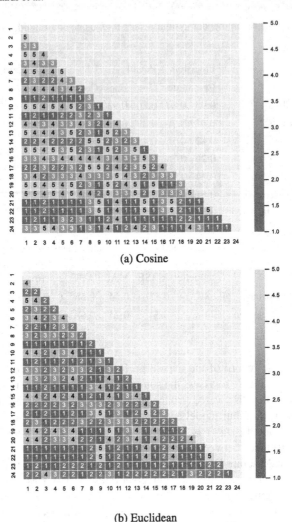

(a) Cosine

(b) Euclidean

Fig. 4. Heatmap of predicted pairwise ratings for 24 repositories. 5 is the most similar, and 1 represents dissimilarity.

a detailed review of each recommended item. The user agreed with the recommendations by stating, "*these two repositories seems to be really alike, maybe they should be merged.*" suggesting that project members have unknowingly created separate repositories for similar studies. Additionally, the user was able to mentally construct a mind map of the relations between the recommended items and understand the reasoning behind the recommendations without the need for explicit justifications by mentioning "*I think I can see why this one is not the first [recommended collection in the list of top five].*".

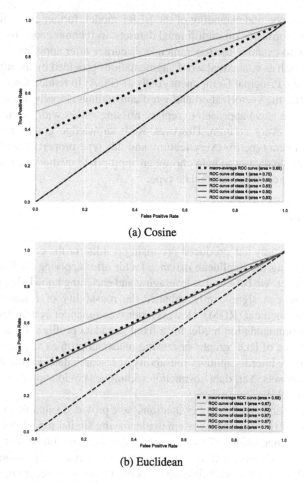

(a) Cosine

(b) Euclidean

Fig. 5. AUC Curve for the overall and class-specific performance of the system to distinguish between TPR and FPR.

5   Discussion and Future Work

The literature review findings reveal that content-based recommender systems have several strengths, including providing personalized recommendations, reducing information overload, and enhancing retrieved artifacts' quality, especially when dealing with cold start problems. However, while content-based recommendation relies on its metadata, limitations of this category of recommender systems manifest in their reliance on sample data quality.

We plan to extend our study for future work to increase accuracy and applicability toward a broader range of disciplines. One important aspect is to evaluate the sensitivity of the BERT pre-trained model toward the multilingual dataset, given that our sample dataset contains manually entered values in German and English. Despite the cross-lingual sentence understanding of the BERT model, further evaluation is needed to consider its performance on multilingual datasets, as recommended by [10].

We also plan to examine the recommender accuracy after applying data augmentation techniques such as contextual word augmentations provided by the *nlpaug*[6] python library or applying Principal Component Analysis (PCA) to reduce the dimensionality of the dataset after the vectorization and word embeddings. Finally, for non-numerical columns, we used a naive approach to replace missing values with the most frequently seen value within every column. However, RDF knowledge graphs provide precise information on a data type for every feature, and this type-property could be imported as part of the sample dataset to help choose an appropriate method for data imputation or vectorization concerning feature data type.

6   Conclusion

In conclusion, our study has produced promising results for the content-based recommender system using the Euclidean distance factor after applying NLP to an unstructured sample dataset. We believe that leveraging and enriching knowledge graphs across different domains can significantly enhance the reusability of research artifacts in research data management (RDM). However, our recommender system has a limitation in building a recommendation model for a fixed metadata profiles. Therefore, we propose the inheritance of RDF graphs or reusing metadata profiles as potential solutions to gain access to the interdisciplinary reusability of research data despite its potential to generate large datasets. Yet, data abstraction techniques could support us in achieving promising results [23].

Our qualitative and quantitative evaluations have provided valuable insights into the effectiveness of our recommender system in identifying similar repositories and providing relevant recommendations. These findings align with previous research, emphasizing the importance of incorporating domain expertise in evaluating recommender systems [9]. In addition, our work builds upon recent research on recommender systems, such as the study by Zhang et al. [24] on deep learning-based recommendation models.

We position the contribution of our work at the intersection of Recommender Systems, Ontology-based data graphs, and research data management fields, where our work opens new opportunities for the reusability of research data.

References

1. Aggarwal, C.C., Hinneburg, A., Keim, D.A.: On the surprising behavior of distance metrics in high dimensional space. In: Van den Bussche, J., Vianu, V. (eds.) ICDT 2001. LNCS, vol. 1973, pp. 420–434. Springer, Heidelberg (2001). https://doi.org/10.1007/3-540-44503-X_27
2. Aggarwal, C.C., Yu, P.S.: Outlier detection for high dimensional data. In: Proceedings of the 2001 ACM SIGMOD International Conference on Management of Data, pp. 37–46 (2001)
3. Bennett, D.A.: How can i deal with missing data in my study? Aust. NZ. J. Public Health **25**(5), 464–469 (2001)

---

[6] https://nlpaug.readthedocs.io.

4. Boukerche, A., Zheng, L., Alfandi, O.: Outlier detection: methods, models, and classification. ACM Comput. Surv. (CSUR) **53**(3), 1–37 (2020)
5. Devlin, J., Chang, M.W., Lee, K., Toutanova, K.: Bert: pre-training of deep bidirectional transformers for language understanding. arXiv preprint arXiv:1810.04805 (2018)
6. Färber, M., Leisinger, A.K.: Recommending datasets for scientific problem descriptions. In: CIKM, pp. 3014–3018 (2021)
7. Heinrichs, B.P.A., Politze, M., Yazdi, M.A.: Evaluation of architectures for FAIR data management in a research data management use case. In: Proceedings of the 11th International Conference on Data Science, Technology and Applications (DATA 2022), SCITEPRESS - Science and Technology Publications, Setúbal (2022). https://doi.org/10.5220/0011302700003269
8. Ho-Dac, N.N., Carson, S.J., Moore, W.L.: The effects of positive and negative online customer reviews: do brand strength and category maturity matter? J. Market. **77**(6), 37–53 (2013)
9. Jones, A.M., Arya, A., Agarwal, P., Gaurav, P., Arya, T.: An ontological sub-matrix factorization based approach for cold-start issue in recommender systems. In: 2017 International Conference on Current Trends in Computer, Electrical, Electronics and Communication (CTCEEC), pp. 161–166. IEEE (2017)
10. Kenton, J.D.M.W.C., Toutanova, L.K.: Bert: Pre-training of deep bidirectional transformers for language understanding. In: Proceedings of naacL-HLT, pp. 4171–4186 (2019)
11. Nair, A.M., Benny, O., George, J.: Content based scientific article recommendation system using deep learning technique. In: Suma, V., Chen, J.I.-Z., Baig, Z., Wang, H. (eds.) Inventive Systems and Control. LNNS, vol. 204, pp. 965–977. Springer, Singapore (2021). https://doi.org/10.1007/978-981-16-1395-1_70
12. Phung, S., Kumar, A., Kim, J.: A deep learning technique for imputing missing healthcare data. In: 2019 41st Annual International Conference of the IEEE Engineering in Medicine and Biology Society (EMBC), pp. 6513–6516. IEEE (2019)
13. Politze, M., Bensberg, S., Müller, M.S.: Managing discipline-specific metadata within an integrated research data management system. In: ICEIS (2), pp. 253–260 (2019)
14. Revathy, V.R., Anitha, S.P.: Cold start problem in social recommender systems: state-of-the-art review. In: Bhatia, S.K., Tiwari, S., Mishra, K.K., Trivedi, M.C. (eds.) Advances in Computer Communication and Computational Sciences. AISC, vol. 759, pp. 105–115. Springer, Singapore (2019). https://doi.org/10.1007/978-981-13-0341-8_10
15. Rogers, A., Kovaleva, O., Rumshisky, A.: A primer in BERTology: What we know about how BERT works. Trans. Assoc. Comput. Linguist. **8**, 842–866 (2021)
16. Tenopir, C., et al.: Academic librarians and research data services: attitudes and practices. IT Lib: Inf. Technol. Libr. J. Issue 1 (2019)
17. Tenopir, C., et al.: Data sharing, management, use, and reuse: practices and perceptions of scientists worldwide. PLoS ONE **15**(3), e0229003 (2020)
18. Ünal, Y., Chowdhury, G., Kurbanoğlu, S., Boustany, J., Walton, G.: Research data management and data sharing behaviour of university researchers. In: Proceedings of ISIC: The Information Behaviour Conference, vol. 3, p. 15 (2019)
19. Vardigan, M., Donakowski, D., Heus, P., Ionescu, S., Rotondo, J.: Creating rich, structured metadata: lessons learned in the metadata portal project. IASSIST Q. **38**(3), 15–15 (2015)
20. Yazdi, M.A.: Enabling operational support in the research data life cycle. In: Proceedings of the First International Conference on Process Mining (ICPM), Doctoral Consortium, pp. 1–10 (2019)
21. Yazdi, M.A., Politze, M.: Reverse engineering: the university distributed services. In: Arai, K., Kapoor, S., Bhatia, R. (eds.) FTC 2020. AISC, vol. 1289, pp. 223–238. Springer, Cham (2021). https://doi.org/10.1007/978-3-030-63089-8_14

22. Yazdi, M.A., Schimmel, D., Nellesen, M., Politze, M., Müller, M.S.: Da4rdm: data analysis for research data management systems. In: 13th International Conference on Knowledge Discovery, Knowledge Engineering and Knowledge Management, (KMIS), pp. 177–183 (2021). https://doi.org/10.5220/0010678700003064

23. Yazdi, M.A., Ghalati, P.F., Heinrichs, B.: Event log abstraction in client-server applications. In: 13th International Conference on Knowledge Discovery and Information Retrieval (KDIR), pp. 27–36 (2021). https://doi.org/10.5220/0010652000003064

24. Zhang, S., Yao, L., Sun, A., Tay, Y.: Deep learning based recommender system: a survey and new perspectives. ACM Comput. Surv. (CSUR) 52(1), 1–38 (2019)

MSDeepNet: A Novel Multi-stream Deep Neural Network for Real-World Anomaly Detection in Surveillance Videos

Prabhu Prasad Dev, Pranesh Das[(⊠)], and Raju Hazari

Department of Computer Science and Engineering, NIT Calicut,
Kozhikode, Kerala, India
{prabhu_p210052cs,praneshdas,rajuhazari}@nitc.ac.in

Abstract. Anomaly detection in real time surveillance videos is a challenging task due to the scenario dependency, duration and multiple occurrences of anomalous events. Typically, weakly supervised video anomaly detection that involves video-level labels is expressed as a multiple instance learning (MIL) problem. The objective is to detect the video clips containing abnormal events, while representing each video as a collection of such clips. Existing MIL classifiers assume that the training videos only have anomalous events of short duration. However, this may not hold true for all real-life anomalies and it cannot be dismissed that there may be multiple occurrences of anomalies in the training videos. In order to detect such anomalies, a novel multi-stream deep neural network (MSDeepNet) is proposed by employing spatio-temporal deep feature extractors along with weakly supervised temporal attention module (WS-TAM). The features extracted from the individual streams are fed to train the modified MIL classifier by employing a novel temporal loss function. Finally, a fuzzy fusion method is used to aggregate the anomaly detection scores. To validate the performance of the proposed method, comprehensive results have been performed on the large-scale benchmark UCF Crime dataset. The suggested multi-stream architecture outperforms state-of-the-art video anomaly detection methods with the frame-level AUC score of 84.72% for detecting anomalous events and lowest false alarm rate of 0.9% for detecting normal events.

Keywords: Multiple instance learning · Anomaly detection · Fuzzy fusion · Weakly supervised learning · Surveillance videos

1 Introduction

In recent years, there has been a notable increase in the deployment of surveillance cameras in public areas such as crowded streets, educational institutions, banks, transportation hubs, shopping malls, religious congregations, and other similar locations. However, the ability of law enforcement agencies to monitor the situation has not kept up with the times. As a consequence, more surveillance

© The Author(s), under exclusive license to Springer Nature Switzerland AG 2023
D. Conte et al. (Eds.): DeLTA 2023, CCIS 1875, pp. 157–172, 2023.
https://doi.org/10.1007/978-3-031-39059-3_11

cameras remain unutilized and it is very difficult for humans to manually monitor the cameras 24/7. In order to avoid the effort and time of human monitoring, it becomes imperative to analyze and detect anomalous activities automatically by using intelligent computer vision algorithms. Video anomalies can be interpreted as the phenomenon of unusual video patterns in usual circumstances or of usual video patterns in unusual circumstances. For instance, it is unusual for two people to fight in a crowded street considered to be an anomaly event, but two boxers fighting inside a ring is considered to be a normal event. The primary objective of an anomaly detection system in real-world surveillance videos is to recognize abnormal events and identify the exact time frames of their occurrence. Nevertheless, video anomaly detection remains challenging due to the complexities involved and the scenario dependency of unusual events. Furthermore, the probability of the occurrences of abnormal events is relatively lower than the probability of the occurrences of normal events, resulting in imbalanced datasets [18].

A widely used paradigm of anomaly detection in surveillance videos is one-class classification, alternatively known as unary classification, which encodes usual video patterns extracted from normal (non-anomalous) videos and detects the videos as anomalies if there are distinctively encoded video patterns. However, it will be challenging to collect all normal training samples in the datasets, making it impossible to encode the usual video patterns universally. Consequently, few normal samples may exhibit deviations from the encoded usual video patterns resulting in a high false alarm rate to detect anomalous events. In order to solve the challenges with unary classification, several researchers used a binary classification paradigm that encodes usual (normal) features extracted from normal videos and unusual (anomalous) features extracted from abnormal videos [16]. Sultani et al. [20] divided the training dataset into bags of normal and anomalous segments and labeled them as normal or abnormal by leveraging the MIL classifier. In order to improve the performance of anomaly detection techniques, Zhu et al. [29] incorporated a temporal context into the MIL classifier through an attention block and also introduced an augmented network to learn the motion-aware feature. Several strategies for handling the anomaly detection problem have been proposed in recent years [1,3,4,6,10,11,13,21,23,25].

The task of detecting anomalies in videos without supervision, also known as unsupervised video anomaly detection (uVAD), involves training models with normal videos and using them to identify abnormal videos that exhibit patterns different from those of normal videos. Xu et al. [24] introduced a novel unsupervised deep learning framework by utilizing stacked denoising autoencoders to learn both spatial (appearance) and temporal (motion) features. In [24], multiple one-class Support Vector Machines (SVMs) are employed by integrating early and late fusion strategies to predict the anomaly scores. Hasan et al. [5] proposed a fully connected autoencoder framework by leveraging conventional handcrafted spatiotemporal local features. In order to address the challenges of anomaly detection for real-time video surveillance, Nawaratne et al. developed an incremental spatiotemporal deep learning approach by incorporating fuzzy aggregation that distinguishes between dynamically evolving normal and abnormal behavior. Nguyen et al. [17] suggested a fusion model that is comprised of two streams: the first stream is a reconstruction network that assists in the reconstruction of the appearance features and the second stream is a U-Net network that anticipates the instant motion based on an input frame. Zaheer et al. [25] developed a generative corporate learning (GCL) framework by incorporating cross-supervision for a training generator that reconstructs the available normal feature representations as well as for a training discriminator that evaluates the probability of a video instance being anomalous.

Weakly supervised-based VAD (wVAD) methods aim to detect anomalous events in the video by leveraging annotations at the video level instead of the frame level. Existing state-of-the-art VAD approaches can be categorized into two types. The first approach i.e. Encoder-agnostic (classifier-centered) approaches focus only on training the classifier. Sultani et al. [20] proposed a deep learning framework by incorporating the MIL classifier to detect anomalies in video. In addition, they have introduced a large-scale weakly-supervised VAD dataset, UCF-Crime, which has attracted the research community for the purpose of being used as a benchmark dataset. Zhang et al. [27] extended the MIL framework by utilizing the temporal convolution network with inner bag loss, which takes into account the lowest score in an anomalous instance as well as the highest score in the bag. Following their approach, Wan et al. [22] designed two novel losses: the first is a dynamic MIL loss, which learns features that are more easily separable, and the second is a center loss, which learns the feature center of each class and penalizes the distance between the feature representations and their corresponding class centers. An attention-based framework was proposed by Ma et al. [15], and it is composed of three modules: the anomaly attention module (AAM), which measures the abnormal degree of each frame; the discriminative anomaly attention module (DAAM), which improves AAM by optimizing the video-level video classification; and the generative anomaly attention module (GAAM), which uses a conditional variational autoencoder to model the likelihood of each frame given the attention for refining AAM. As a consequence, AAM is able to provide greater anomaly scores for abnormal frames while producing lower anomaly scores for typical frames. The second approach i.e. Encoder-based (feature-centered) methods focus on training both the classifier and feature extractor. Zhu et al. [29] proposed the attention block by incorporating temporal context into MIL and obtained the motion-aware features by utilizing the temporal augment network. Zhong et al. [28] reformulated the weakly supervised problem as a binary classification problem under noise labels. Furthermore, they have applied the Graph Convolutional Network (GCN) to correct the noisy predictions. Recent research works [12,14,26] have focused on using weak and noisy labels together to enhance the quality of noisy labels.

Despite the fact that the above approaches used a binary classification paradigm, the ability to discriminate between normal and abnormal videos is inadequate. First, there is an insignificant difference between normal and abnormal events in different situations. Second, they are incapable of capturing discriminating spatiotemporal information to detect subtle anomalies like shoplifting, robbery, stealing, shooting. In order to correctly recognize anomalous behavior, it is essential to extract robust temporal features at the video level and encode robust features at the training level. Existing works such as MIL [20], graph convolutional network [28] utilize either features based on images or features learned via deep learning to detect anomalies. Regardless of the depth of the networks or feature dimension, features based on volume with integrated motion information generally exhibit superior performance compared to those based on images. Combining irregular activity patterns with motion-aware features can improve the effectiveness of detecting anomalous events. Moreover, it's desirable for the detection algorithm to be independent of any prior information about the event. With these considerations in mind, the authors are motivated to introduce a novel MSDeepNet approach to address the above challenges in anomaly detection.

In a nutshell, the major contributions of this paper are summarized as follows:

- A novel multi-stream deep neural network (MSDeepNet) is proposed to detect anomalous events in surveillance videos.
- This paper focuses on combining spatial information by utilizing Inflated 3-Dimensional (I3D) network [2] and motion information by employing Flow-Former optical flow method [7].
- In order to obtain more robust temporal features in the proposed MSDeepNet framework, a WS-TAM is proposed. An enhanced Multiple Instance Learning (MIL) classifier is developed by employing a novel temporal loss function to handle feature variations.
- Experiments conducted on a challenging real-world video anomaly dataset, namely UCF Crime, demonstrate that the proposed approach outperforms the state-of-the-art anomaly detection methods.
- To improve the performance of the proposed MSDeepNet anomaly detection model, a fuzzy fusion method is employed by aggregating the anomaly scores of multiple independent feature streams.

The rest of this paper is structured as follows. Section 2 presents the details of the proposed methodology. Comprehensive experiments are conducted in Sect. 3 to validate the effectiveness of the proposed method and finally this paper is concluded in Sect. 4.

2   Proposed Work

This section begins by defining the problem statement, followed by a description of the proposed feature extraction network. Subsequently, the proposed framework (MSDeepNet) is presented, along with the depiction of the temporal attention MIL loss function in Fig. 1.

2.1   Problem Formulation

Let $X_{normal} = \{x_i\}_{i=1}^{N_{normal}}$ denote the normal video set which contains N_{normal} videos, where x_i is the ith video in X_{normal}, and let $X_{anomaly} = \{x_j, y_j\}_{j=1}^{N_{anomaly}}$

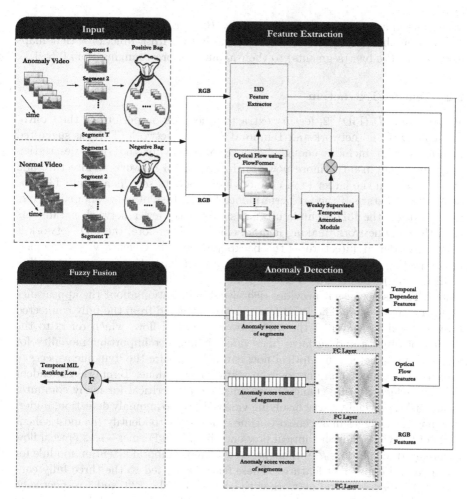

Fig. 1. Architecture of the proposed anomaly detection approach (MSDeepNet)

denote the anomaly video set which contains $N_{anomaly}$ videos, where x_j is the jth video in $X_{anomaly}$ and $y_j = [y_{j;1}, y_{j;2}, ..., y_{j;C}]$; C is the total number of abnormal event categories. If the x_j contains the kth abnormal event category, where $k \in 1, 2, ...C$, then $y_{j,k} = 1$, otherwise $y_{j,k} = 0$. Then, the proposed model is trained on X_{normal} and $X_{anomaly}$ with the guidance of the overall objective function. Note that videos in both the X_{normal} and $X_{anomaly}$ are similarly processed in the feature extraction and prediction steps. Thus, each video v is taken as an example to describe the process in the feature extraction and prediction. In addition, during the training, each video in X_{normal} and $X_{anomaly}$ is split into several non-overlapping video snippets (segments). MIL-based methods assume each video as a bag and each snippet as an instance. A positive video is considered as a positive bag denoted by $B_a = x_j = \{x_{j1}, x_{j2}, ..., x_{jT}\}$, and a negative video is considered as a negative bag denoted by $B_n = x_i = \{x_{i1}, x_{i2}, ..., x_{iT}\}$, where T is the number of segments. The objective is to learn a function l that maps instances of the bag (segments) to their anomaly scores, ranging from 0 to 1.

2.2   Feature Extraction

The Inflated 3D (I3D) [2] feature extraction network outperformed the Convolutional 3D (C3D) network and Temporal Segment Network (TSN) as suggested in literature. It includes enough image information for the video description task, since they use the more powerful Inception-V3 network architecture and were trained on the larger (and cleaner) Kinetics data set. As a result, I3D has demonstrated state-of-the-art performance for the task of video anomaly detection. Further, the I3D architecture has significantly fewer network parameters than the C3D network, making it more efficient. Therefore, the I3D Network is adopted in our architecture as the backbone network. Given a video segment, the extraction of RGB, Optical flow and temporal dependent features is needful. RGB (spatial) feature is used as the primary modality for anomaly detection in videos. This is because it provides rich visual information about the appearance of objects and scenes. The RGB features are extracted from the fully connected layer ('mixed_5c') layer of the I3D network. Optical flow, which refers to the motion of objects and surfaces in a video, is another important modality for video anomaly detection. Optical flow can help capture the dynamic aspects of a scene that cannot be inferred from static images alone. Temporal-dependent features, which capture changes over time, are also critical for many computer vision tasks, especially those involving videos. In video anomaly detection, understanding the changes in a scene over time is necessary to identify the most salient parts of the video. For the optical flow encoding, FlowFormer++ as optical flow is employed to speed up the whole process [7] and temporal attention module for the temporal dependent features. These features are fed to the three fully connected (FC) layers neural network with layer nodes 512, 128, and 1 respectively. The combination of extracted visual data from I3D and encoded flow data from FlowFormer++ aided the network in learning spatial (appearance) and temporal (motion) features. Each individual stream is independently trained with distinct features.

2.3   Weakly Supervised-Temporal Attention Module (WS-TAM)

When presented with a large number of frames, people often choose to focus on only a few of them in order to correctly identify anomalies. It indicates that the majority of anomalous segments can be captured by a small number of significant frames. Motivated by this characteristic, a weakly-supervised temporal attention module (WS-TAM) is developed which depicted in Fig. 2.

To determine temporal weights, the WS-TAM analyzes the temporal information present in each frame and applies spatial average pooling to compress the spatial information of input features. This approach enables it to concentrate exclusively on detecting anomalies in the temporal information for anomaly

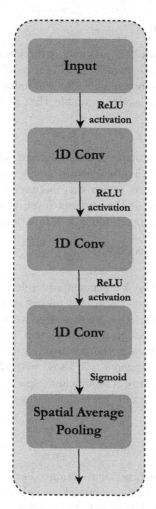

Fig. 2. Weakly Supervised Temporal Attention Module (WS-TAM)

detection purposes. In order to obtain attention scores, the initial two layers implement a Rectified Linear Unit (ReLU) activation function, while the final layer utilizes a sigmoid activation function. The 1D convolutional layers capture the temporal relationships between adjacent segments using filters of various sizes. Finally in order to detect anomalous events, the temporally refined vectors undergo analysis by three fully-connected layers. The utilization of WS-TAM enables the network to allocate more attention towards segments that are deemed to be more important.

2.4 Anomaly Detection

Inspired by the previous work [20], anomaly detection is formulated as a regression problem. In contrast to segments found in a regular video, it is anticipated that segments present in an anomalous video will exhibit higher anomaly scores. If annotations at the segment level is available, then a ranking loss can be applied as indicated in Eq. (1).

$$\ell(V_a) > \ell(V_n) \tag{1}$$

where (V_a) and (V_n) are segments of anomaly and normal videos. Here, $\ell(V_a)$ and $\ell(V_n)$ are the functions that is responsible for converting a video segment into a predicted anomaly score, which vary between 0 and 1. However, we are unable to access annotations at the segment level, thus the Eq. (1) is modified to Eq. (2) as a multiple instance ranking loss function:

$$\max_{k \in B_a} \ell(V_a^k) > \max_{k \in B_n} \ell(V_n^k) \tag{2}$$

where, (V_a^k) and (V_n^k) are the kth segments of relatively anomaly and normal video respectively. Also, max is taken over all instances (segments) in each bag (video). The ranking loss function proves effective because the negative bag contains no anomalies. As a result, anomalous true positive segments (i.e. those with the highest anomaly score in the positive bag) are typically ranked higher than normal false positive segments (i.e. those with the highest anomaly score in the negative bag). To ensure a significant separation between positive and negative instances, the authors in [20] have employed the hinge-loss function as outlined in Eq. (3).

$$Hloss(B_a, B_n) = max(0, 1 - \max_{k \in B_a} \ell(V_a^k) + \max_{k \in B_n} \ell(V_n^k)) \tag{3}$$

However, the ranking loss function mentioned earlier fails to take into account the temporal structure inherent in anomalous videos. Because, some segments in the normal video look most similar to the anomalous segments. Thus to mitigate the above limitations, authors in [20] have incorporated the smoothness and sparsity constraints on anomaly scores. Since, multiple anomalous instances may be present in anomaly video, enforcing smoothness constraints between adjacent instances on anomaly scores can be unreasonable. To alleviate this problem, the temporal context (attention module) have been incorporated that is capable of capturing the aggregated anomaly score of a video. The intuition is the aggregated anomaly score of instances contained within an anomalous video should exceed that of instances present in a normal video. Therefore, the ranking loss of Eq. (3) is modified to Eq. (4) as;

$$loss(B_a, B_n) = max(0, 1 - \sum_{k \in B_a} w_k \ell(V_a^k) + \sum_{k \in B_n} w_k \ell(V_n^k)) \tag{4}$$

where w_k represents the end-to-end learned attention weights within the network. Moreover, the sparsity constraints in loss function is utilized because anomalies often occur rarely for a short duration. Consequently, the final ranking loss function is formulated and presented in Eq. 5:

$$loss(B_a, B_n) = max(0, 1 - \sum_{k \in B_a} w_k \ell(V_a^k) + \sum_{k \in B_n} w_k \ell(V_n^k)) + \lambda \underbrace{\sum_{k \in B_a} w_k \ell(V_a^k)}_{(1)}$$

$$(5)$$

The sparsity constraints constitute the first term, while the smoothness factor for these constraints is denoted by λ.

2.5   Fuzzy Fusion of Anomaly Scores

Anomalies are common in real-world situations but tend to occur briefly, resulting in sparse segment scores in the positive bag. Furthermore, it is expected that the anomaly score will display a smooth transition between adjacent segments. Therefore, it is recommended to combine various streams by utilizing a normalized anomaly score. The present section discusses a fusion process that utilizes fuzzy logic to fuse the anomaly scores of the localized segments. We have used fuzzy logic for video anomaly detection because it can capture the uncertainty and imprecision that often exists in video data. Anomalies in video data can take many forms and may not always be easily defined or detectable with traditional binary logic. Fuzzy logic can provide a more flexible and nuanced approach to modeling these anomalies. While neural networks and ensemble approaches can also be used in video anomaly detection, they may be more computationally expensive and require more training data than fuzzy logic. In addition, ensemble approaches may be more complex to implement and tune. As illustrated in Fig. 1, the proposed architecture comprises three integrated streams of features, which are processed through separate MIL classifiers to obtain the detection scores for every segment. At last, triangular membership function of fuzzy logic is utilized to fuse these scores.

3   Experiments and Results

3.1   Dataset Details and Evaluation Metrics

The effectiveness of the proposed framework is evaluated through experiments on a large-scale public dataset i.e. UCF-Crime dataset with a total duration of 128 h [20]. It includes 1900 surveillance videos of public areas, displaying anomalies, normal events, and crime scenes. It covers 13 different anomalous event categories including Robbery related anomalies (Burglary, Stealing, Robbery, Shoplifting, and Shooting), Human related anomalies (Abuse, Fighting, Arrest, Shooting, Assault), Fire related anomalies (Arson and Explosion), and Road Accident anomaly. In the experiments, the official training and testing splits of the dataset are utilized, which respectively contain 810 anomalous and 800 normal videos in the training set and 140 anomalous and 150 normal videos in the testing set. From previous literature [20, 29], and [22], frame-level Area Under the Curve (AUC) of the Receiver Operating Characteristics (ROC) for performance evaluation is leveraged. A higher AUC suggests that the model is better at detecting abnormal events. In addition, a false alarm rate (FAR) is used for performance evaluation on normal videos only. The lower the FAR, the better the model detects normal events.

3.2   Implementation Details

Each video is partitioned into 32 non-overlapping temporal segments, with T being set to 32. Before computation of feature extraction, each video frame is resized into 240×320 pixels with a frame rate of 30 frame per second (fps). The visual RGB features are extracted from the mixed5c layer of the I3D network [2], which had been pre-trained on the Kinetics dataset [8]. FlowFormer++ [7] is utilized as the optical flow for the flow features, while the WS-TAM is employed for the temporal dependent features. These features are then fed into a three-layer fully connected neural network with layer nodes of 512, 128, and 1, respectively. Following each of these FC layers, a ReLU activation function with a dropout rate of 0.8 are applied. Adam optimizer [9] is employed with a learning rate of 0.001, a weight decay of 0.0005 and a batch size of 64. By taking these hyper-parameter settings, the proposed model achieves best performance. The proposed model is trained by randomly selecting a mini-batch size of 32 positive and 32 negative bags. All experiments are performed on an NVIDIA GeForce RTX 3080 graphics card using the PyTorch framework [19].

3.3   Result Analysis

The qualitative results of the proposed method are presented on a sample of four videos and compared with the best two existing methods as depicted in Fig. 3. The proposed approach enables the timely identification of anomalous events by assigning a high anomaly detection score to the frames that exhibit anomalies. Throughout the anomaly score, the proposed approach generates a low anomaly score in normal videos, resulting the best FAR.

Table 1 presents the comparison of AUC values for quantitative evaluation. The results show that the proposed MSDeepNet significantly outperforms than the SOTA approaches i.e. Deep feature encoding [5], dictionary based encoding [14], MIL method [20], MIL with inner bag loss [27], attention-aware feature encoding [29], and graph convolutional neural network with TSN [28]. It can be observed that the proposed MSDeepNet achieves the highest AUC (83.29%) as compared to all approaches outlined in Table 1. By incorporating the weakly supervised temporal attention module (WS-TAM), the performance is improved by 1.43% (83.29 → 84.72). Since the AUC of the suggested method is 2% higher than that of the best method currently available, it has a greater chance of identifying true positive anomalies which is extremely valuable in various surveillance scenarios as shown in Fig. 4. In terms of FAR, the proposed approach achieves FAR (1.2%) without using WS-TAM and FAR (0.9%) using WS-TAM which

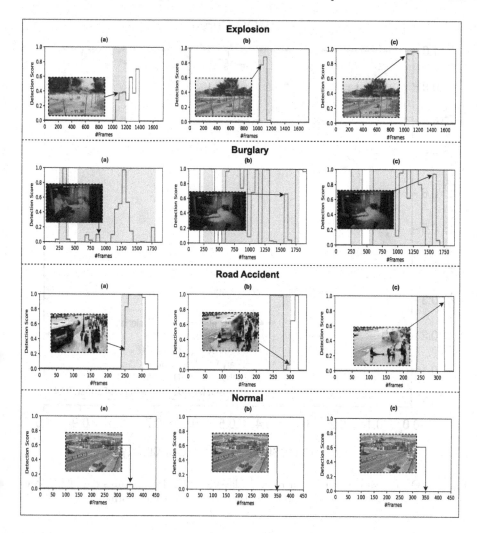

Fig. 3. Qualitative comparison results of the existing best two methods vs proposed method. X-axis and Y-axis represents the frame number and anomaly detection scores respectively. The areas shaded in pink represent the ground truth. (a) presents the anomaly detection score achieved by [20] (in red color). (b) presents the results achieved by [26] (in green color). (c) presents the results achieved by fusion of I3D + Optical flow with FlowFormer and TAM (in blue color) (Color figure online)

is significantly 0.5% lesser than the best available method (1.4%) as shown in Table 2. The lower the FAR, the better the model detects normal events.

As previously mentioned, the MIL does not take into account the limitation that utilizes the temporal smoothness constraints. The necessity of temporal smoothness can be avoided if motion information is incorporated into the training process. However, to enforce temporal smoothness, the proposed method

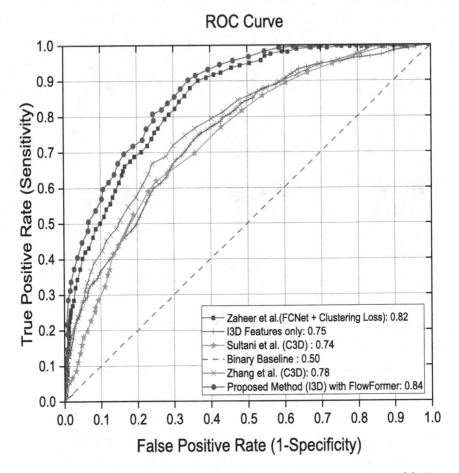

Fig. 4. ROC curves comparison of Binary Classifier, [20], I3D features only [2], Zhang et al. [27], Zaheer et al. [26], Proposed Method (I3D) with FlowFormer

incorporates the constraint (λ) determined in Eq. (5). Furthermore, the default values of λ is set to 8×10^{-5}. The efficiency of λ at different scales is shown in Table 3. Yet, there is no noticeable improvement in performance. We compare the computational time (in milliseconds) of the proposed MSDeepNet with other state-of-the-art methods [5, 14, 20, 28]. From the Table 4, it is clear that the computational time of the proposed method is lower than for the suggested baseline methods.

Table 1. AUC comparison of various approaches on UCF Crime dataset

Supervision	Method	AUC (%)
OCC	Conv-AE	50.60
	ST-Graph	72.71
Unsupervised	Binary Classifier	50.00
	Deep feature encoder [5]	65.51
	Dictionary-based encoding [14]	65.60
Weakly Supervised	C3D RGB + MIL loss [20]	75.41
	I3D RGB + MIL loss [20]	77.92
	C3D RGB + IBL [27]	78.66
	Motion aware features [29]	79.00
	GCN + TSN RGB [28]	81.08
	FC Net + Clustering loss [26]	82.67
	Proposed method (I3D + Optical Flow with Flowformer)	**83.29**
	Proposed method (I3D + Optical Flow with Flowformer + WS-TAM)	**84.72**

Table 2. False alarm rate comparison of various approaches on normal videos of UCF Crime dataset

Supervision	Method	FAR (%)
OCC	Conv-AE	–
	ST-Graph	–
Unsupervised	Binary Classifier	–
	Deep feature encoder [5]	27.2
	Dictionary-based encoding [14]	3.1
Weakly Supervised	C3D RGB + MIL loss [20]	1.9
	I3D RGB + MIL loss [20]	1.7
	C3D RGB + IBL [27]	–
	Motion aware features [29]	–
	GCN + C3D RGB [28]	2.8
	FC Net + Clustering loss [26]	1.4
	Proposed method (I3D + Optical Flow with Flowformer)	**1.2**
	Proposed method (I3D + Optical Flow with Flowformer + WS-TAM)	**0.9**

Table 3. AUC comparison of different smoothness factor value (λ) on UCF Crime dataset

Sl. No	Smoothness factor (λ)	Proposed Method
1	8×10^{-2}	84.19
2	8×10^{-3}	84.31
3	8×10^{-4}	84.64
4	8×10^{-5}	84.72
5	8×10^{-6}	84.47

Table 4. Computational complexity comparison of various approaches on normal videos of UCF Crime dataset

Supervision	Method	Computational Time (ms)
OCC	Conv-AE	–
	ST-Graph	–
Unsupervised	Binary Classifier	–
	Deep feature encoder [5]	1560
	Dictionary-based encoding [14]	2319
Weakly Supervised	C3D RGB + MIL loss [20]	548
	I3D RGB + MIL loss [20]	382
	C3D RGB + IBL [27]	–
	Motion aware features [29]	–
	GCN + C3D RGB [28]	278
	FC Net + Clustering loss [26]	–
	Proposed method (I3D + Optical Flow with Flowformer)	**184**
	Proposed method (I3D + Optical Flow with Flowformer + WS-TAM)	**156**

4   Conclusion

In this paper, a multi-stream deep neural network (MSDeepNet) is proposed to detect anomalous events in surveillance videos by combining spatial visual features using I3D, temporal features using FlowFormer and WS-TAM. Besides, a novel loss function is developed to learn the temporal dependency features. Comprehensive experiments on large scale benchmark UCF Crime dataset demonstrates the effectiveness of the proposed approach. Despite the fact that the suggested method provides better performance than existing mentioned state-of-the-arts methods, the normal-abnormal feature encoding method is nevertheless susceptible to the subjective characterization of an anomalous occurrence. The performance of anomaly detection can be improved by incorporating noise filtering techniques and additional bag-loss techniques that have been introduced in [27] to handle complex anomalies which can be considered as a future research directions.

References

1. Barbalau, A., et al.: SSMTL++: revisiting self-supervised multi-task learning for video anomaly detection. Comput. Vis. Image Underst. **229**, 103656 (2023)
2. Carreira, J., Zisserman, A.: Quo vadis, action recognition? A new model and the kinetics dataset. In: Proceedings of the IEEE Conference on Computer Vision and Pattern Recognition, pp. 6299–6308 (2017)
3. Feng, J.C., Hong, F.T., Zheng, W.S.: MIST: multiple instance self-training framework for video anomaly detection. In: Proceedings of the IEEE/CVF Conference on Computer Vision and Pattern Recognition, pp. 14009–14018 (2021)
4. Hao, Y., Li, J., Wang, N., Wang, X., Gao, X.: Spatiotemporal consistency-enhanced network for video anomaly detection. Pattern Recogn. **121**, 108232 (2022)
5. Hasan, M., Choi, J., Neumann, J., Roy-Chowdhury, A.K., Davis, L.S.: Learning temporal regularity in video sequences. In: Proceedings of the IEEE Conference on Computer Vision and Pattern Recognition, pp. 733–742 (2016)

6. Huang, C., et al.: Self-supervised attentive generative adversarial networks for video anomaly detection. IEEE Trans. Neural Netw. Learn. Syst. (2022)
7. Huang, Z., et al.: FlowFormer: a transformer architecture for optical flow. In: Avidan, S., Brostow, G., Cissé, M., Farinella, G.M., Hassner, T. (eds.) ECCV 2022. LNCS, vol. 13677, pp. 668–685. Springer, Cham (2022). https://doi.org/10.1007/978-3-031-19790-1_40
8. Kay, W., et al.: The kinetics human action video dataset. arXiv preprint arXiv:1705.06950 (2017)
9. Kingma, D.P., Ba, J.: Adam: a method for stochastic optimization. arXiv preprint arXiv:1412.6980 (2014)
10. Le, V.T., Kim, Y.G.: Attention-based residual autoencoder for video anomaly detection. Appl. Intell. **53**(3), 3240–3254 (2023). https://doi.org/10.1007/s10489-022-03613-1
11. Li, D., Nie, X., Li, X., Zhang, Y., Yin, Y.: Context-related video anomaly detection via generative adversarial network. Pattern Recogn. Lett. **156**, 183–189 (2022)
12. Li, N., Zhong, J.X., Shu, X., Guo, H.: Weakly-supervised anomaly detection in video surveillance via graph convolutional label noise cleaning. Neurocomputing **481**, 154–167 (2022)
13. Li, S., Liu, F., Jiao, L.: Self-training multi-sequence learning with transformer for weakly supervised video anomaly detection. In: Proceedings of the AAAI Conference on Artificial Intelligence, vol. 36, pp. 1395–1403 (2022)
14. Lu, Z., Fu, Z., Xiang, T., Han, P., Wang, L., Gao, X.: Learning from weak and noisy labels for semantic segmentation. IEEE Trans. Pattern Anal. Mach. Intell. **39**(3), 486–500 (2016)
15. Ma, H., Zhang, L.: Attention-based framework for weakly supervised video anomaly detection. J. Supercomput. **78**(6), 8409–8429 (2022). https://doi.org/10.1007/s11227-021-04190-9
16. Nayak, R., Pati, U.C., Das, S.K.: A comprehensive review on deep learning-based methods for video anomaly detection. Image Vis. Comput. **106**, 104078 (2021)
17. Nguyen, T.N., Meunier, J.: Anomaly detection in video sequence with appearance-motion correspondence. In: Proceedings of the IEEE/CVF International Conference on Computer Vision, pp. 1273–1283 (2019)
18. Pang, G., Shen, C., Cao, L., Hengel, A.V.D.: Deep learning for anomaly detection: a review. ACM Comput. Surv. (CSUR) **54**(2), 1–38 (2021)
19. Paszke, A., et al.: PyTorch: an imperative style, high-performance deep learning library. In: Advances in Neural Information Processing Systems, vol. 32 (2019)
20. Sultani, W., Chen, C., Shah, M.: Real-world anomaly detection in surveillance videos. In: Proceedings of the IEEE Conference on Computer Vision and Pattern Recognition, pp. 6479–6488 (2018)
21. Thakare, K.V., Sharma, N., Dogra, D.P., Choi, H., Kim, I.J.: A multi-stream deep neural network with late fuzzy fusion for real-world anomaly detection. Expert Syst. Appl. **201**, 117030 (2022)
22. Wan, B., Fang, Y., Xia, X., Mei, J.: Weakly supervised video anomaly detection via center-guided discriminative learning. In: 2020 IEEE International Conference on Multimedia and Expo (ICME), pp. 1–6. IEEE (2020)
23. Wang, L., Tian, J., Zhou, S., Shi, H., Hua, G.: Memory-augmented appearance-motion network for video anomaly detection. Pattern Recogn. **138**, 109335 (2023)
24. Xu, D., Ricci, E., Yan, Y., Song, J., Sebe, N.: Learning deep representations of appearance and motion for anomalous event detection. arXiv preprint arXiv:1510.01553 (2015)

25. Zaheer, M.Z., Mahmood, A., Khan, M.H., Segu, M., Yu, F., Lee, S.I.: Generative cooperative learning for unsupervised video anomaly detection. In: Proceedings of the IEEE/CVF Conference on Computer Vision and Pattern Recognition, pp. 14744–14754 (2022)
26. Zaheer, M.Z., Lee, J.H., Astrid, M., Mahmood, A., Lee, S.I.: Cleaning label noise with clusters for minimally supervised anomaly detection. arXiv preprint arXiv:2104.14770 (2021)
27. Zhang, J., Qing, L., Miao, J.: Temporal convolutional network with complementary inner bag loss for weakly supervised anomaly detection. In: 2019 IEEE International Conference on Image Processing (ICIP), pp. 4030–4034. IEEE (2019)
28. Zhong, J.X., Li, N., Kong, W., Liu, S., Li, T.H., Li, G.: Graph convolutional label noise cleaner: train a plug-and-play action classifier for anomaly detection. In: Proceedings of the IEEE/CVF Conference on Computer Vision and Pattern Recognition, pp. 1237–1246 (2019)
29. Zhu, Y., Newsam, S.: Motion-aware feature for improved video anomaly detection. arXiv preprint arXiv:1907.10211 (2019)

A Novel Probabilistic Approach for Detecting Concept Drift in Streaming Data

Sirvan Parasteh[✉] and Samira Sadaoui

Computer Science Department, University of Regina, 3737 Regina, Wascana Parkway, Canada
{sirvan.parasteh,Samira.Sadaoui}@uregina.ca

Abstract. Concept drift, which indicates data-distribution changes in streaming scenarios, can significantly reduce predictive performance. Existing concept drift detection methods often struggle with the trade-off between fast detection and low false alarm rates. This paper presents a novel concept drift detection algorithm, called SPNCD*, based on probabilistic methods, particularly Sum-Product Networks, that addresses this challenge by offering high detection accuracy and low mean lag time. Based on three benchmark datasets, the proposed method is evaluated against state-of-the-art algorithms, such as DDM, ADWIN, KSWIN, and HDDM_A. Our experiments demonstrate that SPNCD* outperforms the existing algorithms in terms of true positive rate, recall, precision, and mean lag time while improving the performance of the base classifier. The SPNCD* algorithm provides a reliable solution for detecting concept drift in real-time streaming data, enabling practitioners to maintain their machine learning models' performance in dynamic environments.

Keywords: Concept drift detection · Probabilistic methods · Sum-product networks · Synthetic datasets · Real concept drift · Virtual concept drift

1 Introduction

Concept drift (CD) is a prevalent challenge in machine learning (ML) that arises when the statistical properties of data evolve over time [11]. This challenge becomes particularly significant in streaming environments, where data is continuously generated at a high volume and speed. The non-stationary nature of such data hinders the ability of ML models to maintain accurate performance in real-world applications. For instance, fluctuations in user preferences can considerably influence the purchasing behavior of users [19]. The COVID-19 pandemic exemplified this phenomenon, drastically altering consumers' purchasing patterns and rendering models trained on pre-pandemic data ineffective.

Failing to identify and adapt to CD leads to degraded performance and obsolete patterns [12]. Traditional ML algorithms consist of data training and prediction [22], and the underlying assumption in this process is that incoming data follows the same patterns or distribution as the historical data [9]. However, this assumption becomes invalid in environments where CD is present. Consequently, it is vital to equip ML

© The Author(s), under exclusive license to Springer Nature Switzerland AG 2023
D. Conte et al. (Eds.): DeLTA 2023, CCIS 1875, pp. 173–188, 2023.
https://doi.org/10.1007/978-3-031-39059-3_12

algorithms with proper mechanisms for detecting CD [4], such as continuously monitoring the model performance and/or statistical characteristics of the input stream [6]. Figure 1 illustrates the general idea of incorporating a concept drift handling mechanism in machine learning models.

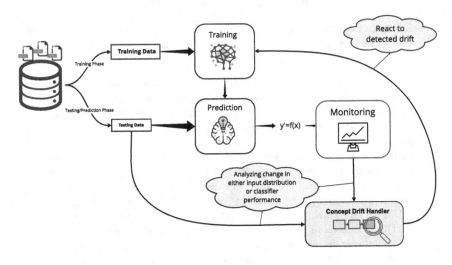

Fig. 1. Equipping ML model with CD handler to provide a reaction to concept drift

CD detection methods are mainly two folds: data distribution-based and learner performance-based (error rate-based) [4,22]. The appropriate method to use depends on the specific data characteristics and the changes in the problem at hand. Performance-based methods primarily focus on observing the classifier's performance; nonetheless, they may result in sub-optimal outcomes owing to the delayed detection of concept drift [11]. To overcome this issue, it is necessary to consider changes in the data distribution. Monitoring changes at the distribution level can provide a more comprehensive understanding of CD [15,22]. Nevertheless, distribution-based methods may not always detect drifts. Additionally, unsupervised methods, commonly used in distribution-based approaches, can lead to a high rate of false alarms [13].

This research proposes a hybrid approach for CD detection, which combines the advantages of data distribution-based and performance-based methods. Our approach uses a probabilistic method, specifically the Sum-Product Networks (SPN), to learn the joint probability distribution of the incoming data in a tractable way. By directly learning the joint distribution, our approach can capture both changes in the data distribution and decision boundary, providing a more comprehensive understanding of CD. Our method can perform probabilistic inference on the joint distribution, such as the conditional, marginal, and likelihood distributions, making detecting drifts in real-time feasible. Through our experiments on synthetic datasets, we demonstrate the effectiveness of our approach in detecting both real and virtual CD. We also compare our new approach to numerous well-known CD detection methods. We observe a significant improvement in the False Alarm rate, Precision and Recall compared to state-of-the-art methods.

The remainder of this paper is organized as follows. Section 2 provides a theoretical background on the CD notion and its types. Section 3 highlights numerous CD detection methods and their implementation. Section 4 introduces the key concepts behind our new approach for CD detection using SPN-based probabilistic inferences. Section 5 presents the experimental results and the comparison with existing methods. Section 6 concludes our work and discusses future research directions.

2   Probabilistic Definition of Concept Drift

We follow the probabilistic CD definitions introduced in [11, 16, 32]. Data is often modeled as a fixed set of observations, represented as a set of feature vectors, x, and their corresponding target variables, y. In a streaming scenario, data arrives continuously over time and is modeled as a sequence of instances, $D = \{(x_1, y_1), (x_2, y_2), \cdots \}$. The goal of a classifier in a streaming environment is to learn the underlying relationship between the features and target variable and make predictions for each incoming instance, $\hat{y}_t = f(x_t)$ [11]. In the Bayesian Decision Theory, $y \in \{0, 1, .., c\}$, where c is equal to the size of the class labels, the goal is predicting $\hat{y}$, computed by the following equation [7, 16]:

$$\hat{y} = \arg \max_{y \in 0,1,..,c} P(y|\mathbf{x}) \tag{1}$$

where $P(y|\mathbf{x})$ is the posterior probability of class y given the input features. The posterior probabilities can be computed using the Bayes' theorem:

$$P(y|\mathbf{x}) = \frac{P(\mathbf{x}|y)P(y)}{P(\mathbf{x})} \tag{2}$$

where $P(\mathbf{x}|y)$ is the likelihood of the features given the class, $P(y)$ is the prior probability of the class, and $P(\mathbf{x})$ is the marginal probability of the features. Thus, the classification problem is modeled based on the joint probability distribution of the class and the feature vector, $P(y, \mathbf{x})$ using the Bayesian approach.

In the circumstances of CD, the prior distribution $P(y)$ and likelihood distribution $P(\mathbf{x}|y)$ can change over time, leading to a change in the posterior distribution [9]. This makes the Bayesian approach a popular choice for CD detection, as it captures the changes in the joint distribution of the features and class labels over time and makes predictions accordingly [16]. Following the notations in [11, 32], the arriving samples up to time t can be defined as [32]:

$$Concept_{[0,t]} = P_{[0,t]}(\mathbf{x}, y) \tag{3}$$

Considering that the concept can change over time, thus we can identify the drift at time u, u is a time after t, when the distributions are changed:

$$Concept_t \neq Concept_u, P_t(\mathbf{x}, y) \neq P_u(\mathbf{x}, y) \tag{4}$$

In other words, we assume a concept will remain stable for a period of time and then it may turn into another concept at time u. Following these definitions, two different CD types are proposed in the literature:

- **Virtual CD** means the change happens in the distribution of the observations $P_t(X)$, which has no impact on the conditional distribution $P_t(y \mid \mathbf{x})$ [22]:

$$P_t(\mathbf{x}) \neq P_{t+1}(\mathbf{x}) \quad such \ that \ P_t(y \mid \mathbf{x}) = P_{t+1}(y \mid \mathbf{x}) \tag{5}$$

- **Real CD** means the conditional distribution of the target variable $P(y \mid \mathbf{x})$ changes, while there no change in the distribution of the input features [22]:

$$P_t(y \mid \mathbf{x}) \neq P_{t+1}(y \mid \mathbf{x}) \quad such \ that \ P_t(\mathbf{x}) = P_{t+1}(\mathbf{x}) \tag{6}$$

- **Mixed CD:** a mixture of both Virtual and Real drifts, where both $P(y \mid X)$ and $P(X)$ change. In real-world situations, this case is the most probable one as a change in any of the distributions with a high probability affects the other distribution [22].

$$P_t(y \mid \mathbf{x}) \neq P_{t+1}(y \mid \mathbf{x}) \quad such \ that \ P_t(\mathbf{x}) \neq P_{t+1}(\mathbf{x}) \tag{7}$$

3   Related Works

Detecting CD is not a newly revealed challenge. Researcher, since mid-1990, has shown their interest in this topic [18, 33], and a good amount of methods for detecting changes in data stream applications have been published. Among these methods, error-based methods have gained significant attention. These methods are based on the observation that the prediction error of a classifier increases when the CD occurs. Table 1 lists various CD detection algorithms, including their implementation, if any, year, and popularity regarding the citation number. We investigate and report the existing implementation of the models using three main resources: (1) **GitHub** for the public repositories, (2) **Code** for the provided link to the source code, and (3) **Multiflow or River** for the implementation on Scikit-multiflow or River, which are Python-based packages for ML for streaming environments.

Examining various error-based CD detection techniques reveals a chronological progression in their development. The foundational Page-Hinkley (PH) method in 1954 [24] marked the beginning of this progression, which has since evolved to include increasingly sophisticated approaches such as DDM [10], EDDM [1], and ADWIN [5]. Recent methods, like KSWIN [29], demonstrate continued innovation in the field. While some techniques have garnered significant attention regarding the citation count, others have remained relatively less influential. Nonetheless, the diverse landscape of error-based concept drift detection methods offers researchers a wide array of techniques to choose from based on their specific application needs.

A comprehensive study on the performance of error-based detection algorithms was conducted in [14] and [3]. These studies analyzed various CD algorithms and found that no single method outperformed the others in all cases. The results showed that the choice of method depends on the application's specific requirements. The findings in these studies highlight the limitations of the existing methods, such as sensitivity to parameter tuning, high computational costs, and difficulty in handling complex and non-stationary data streams. These limitations emphasize the need for more robust and efficient methods to handle different data types and adapt to changing concepts in real time. The results of these studies suggest that there is a need for further research to overcome these limitations and develop advanced and efficient solutions.

Table 1. Error-based CD Detection Methods.

CD Detection Methods			
Method	Implementation	Year	#citation
DDM [10]	MultiFlow, River	2004	1596
EDDM [1]	MultiFlow, River	2006	868
STEPD [23]	Git	2007	286
ADWIN [5]	MultiFlow, River	2007	1549
ECDD [30]	Git	2012	341
HDDM [8]	MultiFlow, River, Git	2014	271
PH [24]	MultiFlow, River	1954	6563
SEED [17]	code	2014	68
FHDDM [26]	Git	2016	134
FTDD [20]	Git (Fisher test)	2018	90
RDDM [2]	Git	2017	130
SeqDrift [25]	Git	2013	46
MDDM [27]	Git	2018	55
EWMA [30]	Git	2012	382
KSWIN [29]	MultiFlow	2020	58

4   Methodology

The majority of existing CD detection methods primarily focus on tracking the performance of models to detect real drifts but fall short of providing insight into changes in the data distribution. To address this gap, we present a new probabilistic approach for identifying both Real and Virtual CD types by understanding the distribution of the data. Our approach leverages SPN to learn the joint probability distribution of the features, thereby enabling tractable probabilistic inference. The trained SPN model then will be used to categorize the learner's prediction into three regions: real drift, virtual drift, and no drift. A decision-making strategy is then applied to determine whether to fire a drift alarm or not. In the following sections, we introduce SPN and explain our proposed algorithm.

4.1 Sum-Product Networks

We utilize SPNs to learn the joint probability distribution of the input data and target variable, $P(x, y)$. SPNs are a class of deep probabilistic models widely used for various ML tasks due to their ability to handle missing data and perform tractable inference. The SPN model represents the joint probability distribution of the random variables in a compact manner that enables efficient evaluation of probabilistic queries, such as marginal probability, conditional probability, and the most probable explanation (MPE), in linear time [31]. The formulation of an SPN can be defined as follows [28]:

$$P(x) = \sum_{i=1}^{n} w_i \prod_{j=1}^{d_i} P_{i,j}(x_{S_{i,j}})$$ (8)

where x is the random variable with dimensions D, n is the number of sum nodes, d_i is the number of children of the ith sum node, w_i is the weight of the ith sum node, and each product node, $P_{i,j}(x_{S_{i,j}})$, is the probability function corresponding to the jth child of the ith sum node, and $S_{i,j}$ is the subset of variables within the scope of that product node [28]. This formulation represents the hierarchical (recursive) structure of the network, which is composed of sum nodes and product nodes because $P_{i,j}(x_{S_{i,j}})$ could itself be another sum-product network. The sum nodes represent the marginal probabilities of the variables, and the product nodes capture the dependencies between variables. The recursive nature of this formula allows it to model complex relationships between variables and capture multiple layers of sum and product nodes implicitly.

4.2 Detecting CD Using SPN

Our method utilizes SPN to detect CD by categorizing samples into no-drift, real-drift, and virtual drift regions based on the Bayesian formulation for CD (Table 2). To detect and categorize drifts, we compute two terms: the posterior probability $P(y|\mathbf{x})$, corresponding to the classification function, and the marginal distribution of input features, $P(\mathbf{x})$, which is the normalization term in Bayes' formula:

$$P(\mathbf{x}) = \sum_{y \in Y} P(\mathbf{x}|y)$$ (9)

The marginal distribution, $P(\mathbf{x})$, represents the input feature distribution irrespective of target variables and plays a crucial role in drift detection.

Table 2. Probabilistic drift trace in data distribution

$P(y|x)$	$P(x)$	Decision	Explanation
✓	⇓	Virtual drift	Classifier's prediction is correct, the new sample does not follow the learned distribution
✗	⇑	Real drift	Classifier's prediction is incorrect, the new sample belongs to the learned distribution

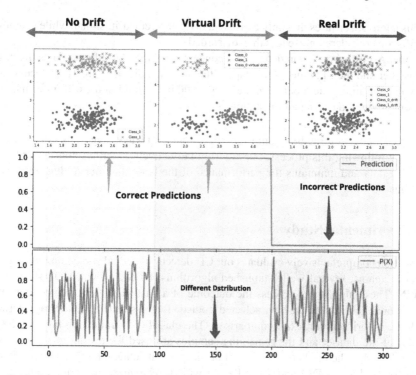

Fig. 2. Model outputs for all three types of data samples: No-drift, Virtual, and Real drifts, with respect to Table 2.

Using SPN, we categorize samples into drift regions based on $P(y|X)$ and $P(X)$. Figure 2 illustrates the results of applying our method to a synthetic dataset, validating the hypothesis.

Our detection algorithm is built upon the hypothesis described earlier, utilizing the SPN model for learning and computing probability queries. The algorithm operates as follows:

1. **Initial Training**: The algorithm first trains a base classifier and an SPN model on a portion of the data, which is considered the historical dataset. This step establishes the initial models used for classifying incoming samples.
2. **Classification and Probability Evaluation**: As new samples arrive, the algorithm uses the trained base classifier and the SPN model to classify them. It then calculates the posterior probability $P(y|\mathbf{x})$ and the marginal distribution $P(\mathbf{x})$ for each sample to identify the drift type according to the conditions outlined in Table 2.
3. **Drift Assessment**: The algorithm evaluates the consistency of the detected drifts using a scoring and aging mechanism. When a sample is detected as drifting in the second step, its corresponding drift object's score is increased. The updated score is then compared with a predefined threshold to determine if it should lead to a drift alarm. If a drifting object is not observed for a while, the aging mechanism gradually reduces its score. Once the score reaches zero, the drift object is removed.

This approach ensures that only consistent drifts trigger a drift signal, while sporadic drifts are considered noise and then discarded.

4. **Model Adaptation**: Upon detecting a concept drift, the algorithm updates the base classifier and the SPN model to adapt to the new data distribution. This process can involve retraining the models on the latest data, incorporating the drift information, or employing other adaptation strategies.

Our CD detection method comprises these four sequential stages, as illustrated in Fig. 3. By following this process, our algorithm effectively identifies concept drifts in streaming data and maintains the performance of the base classifier in dynamic environments.

5   Experimental Study

This section comprehensively evaluates our CD detection method and compares its performance against several well-established algorithms: DDM, Adwin, HDDM_A, and KSWIN. These algorithms process the outcome of a base classifier to detect concept drifts. In our experiments, we have selected Logistic Regression (LR) as the base learner for all the algorithms for a fair comparison. The classifier is trained on the first 2000 samples of the dataset and the remaining samples are used for testing in a streaming manner. Whether the prediction is correct or incorrect, its outcome is fed into the algorithms for analysis. DDM and ADWIN are widely recognized as some of the most seminal methods for detecting CD, whereas HDDM_A is considered one of the top algorithms in comparison studies conducted in [3]. KSWIN, on the other hand, is a more recently published method that demonstrated superior performance compared to the conventional methods [29].

We implemented all the algorithms in Python using the SPflow library for training the Sum-Product Networks (SPN) and the River package for online classification tasks, specifically utilizing LR as the base classifier. The experimental evaluation was conducted on a Linux-based system with a Core i9 processor and 64 GB of RAM. The key steps involved in implementing the proposed method included pre-processing the data, which consisted of normalization and simulating the data stream, training the SPN, training the LR model, and performing drift detection while processing the streaming data.

In the following subsections, we discuss the selected datasets and evaluation metrics, the experimental results and provide a detailed analysis and discussion of the findings.

5.1   Selected Datasets and Evaluation Metrics

In order to evaluate the performance of our proposed drift detection method, we utilized three synthetic datasets where the locations of drifts were known. This approach is commonly used in the literature since real-world datasets do not usually provide this information. The benchmark datasets used in our experiments were Mixed, Sine, and

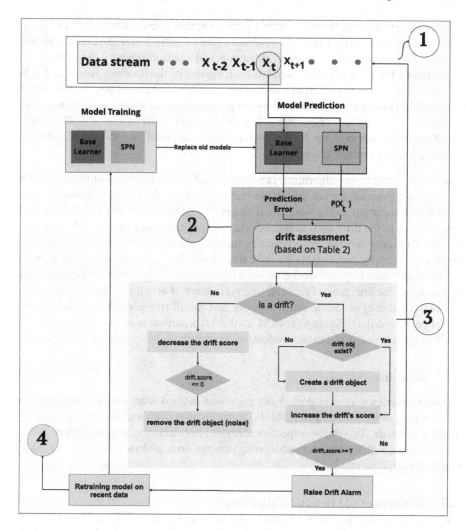

Fig. 3. Proposed Probabilistic Concept Drift Detection Algorithm.

Random Tree (RT), which were previously published on the Harvard Dataverse [21]. These datasets contain 40,000 samples generated from four different concepts and used for binary classification tasks. The drifts occur abruptly every 10,000 samples. By utilizing these benchmark datasets, our results can be compared and validated with other studies in the literature, leading to more reliable conclusions. The datasets are characterized as follows [21]:

- **Mixed**: This synthetic dataset, generated by the Mixed stream generator, follows a sequence of classification functions in the reverse order of 1-0-1-0. It contains four numerical features and a balanced binary class label.

- **Sine**: Created using the Sine stream generator, this dataset adheres to a series of classification functions with the reversed order of SINE2-SINE2-reversed SINE1-SINE1. It consists of two numerical features and a balanced binary class label.
- **Random Tree (RT)**: This artificial dataset, created by the Random Tree stream generator, follows a series of tree random state functions. The parameters for the deepest tree level, smallest leaf depth, and the proportion of leaves at each level were configured to 6, 3, and 0.15, respectively. The arrangement of the functions is 8873-9856-7896-2563, known as the F1 configuration. It consists of two numerical attributes and a balanced binary class label.

The effectiveness of the algorithms can be assessed by the precision of their decision-making. Our experiments categorize incoming samples as drift or no drift based on the known drift locations in the synthetic datasets. We consider 2% of the length of each new concept as the drift area, containing the true positive cases suggested by [14]. We use the following metrics: True Positives (detecting drift within the drift area), False Positives (detecting drift outside the drift area), False Negatives (missing drift signal in a drift area), Mean-Lag (difference between the location of a true positive signal and the beginning of the drift area), Precision (the percentage of actual positive signals among the total number of predicted positive ones), and Recall (the percentage of recognized actual positive drifts). Figure 4 provides a visual representation of the drift regions and labeling system as defined in our evaluation procedure.

5.2 Discussion

In this section, we have evaluated our proposed method from two perspectives: the performance of the drift detector and the base classifier, with and without applying the detector methods. This dual-perspective analysis offers a comprehensive understanding of our method's effectiveness in addressing concept drift, underscoring its potential for real-world applications.

A. Performance of CD Detection Algorithms
The results of the experiments demonstrate the effectiveness of the proposed method for detecting concept drift in data streams. In this section, we discuss the findings from the experiments and evaluate the performance of the proposed method against other state-of-the-art concept drift detection algorithms. We provide a detailed analysis of the results obtained from the experiments and highlight the strengths and weaknesses of the proposed method. We also discuss the implications of the findings and provide recommendations for future research in this area.

Table 3 shows the comparison results for the proposed method with other concept drift detection algorithms on the Mixed dataset. The results indicate that our method, along with KSWIN, achieved perfect precision and recall by detecting all true positive instances of concept drift without reporting any false positives. Furthermore, the mean lag of our method is significantly lower than that of other algorithms, indicating that our method can detect concept drift more quickly. These findings suggest that our method, in combination with KSWIN, has the dominant performance among the tested algorithms for detecting concept drift in the Mixed dataset.

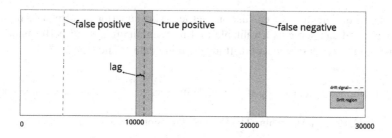

Fig. 4. Drift region and labeling logic

Table 3. Performance Comparison of Concept Drift Detection Algorithms on the Mixed Dataset

Detection Algorithm	TP	FP	FN	Precision	Recall	Mean Lag
DDM	3	4	0	42.86	100.0	85.00
ADWIN	3	1	0	75.00	100.0	135.00
KSWIN	3	0	0	100.00	100.0	16.00
HDDM_A	3	1	0	75.00	100.0	50.67
SPNCD*	3	0	0	100.00	100.0	0.00

The results presented in Table 4 highlight the superior performance of our proposed method in detecting concept drift in the RT dataset. Our method achieved a perfect true positive rate and recall of 100% by accurately detecting all instances of concept drift with only one false positive instance reported, which did not significantly affect the overall performance. In contrast, the other benchmark algorithms, such as ADWIN, had poor detection rates and failed to detect any instances of concept drift. While KSWIN exhibited a lower Mean Lag than our method, our method still outperformed KSWIN in all other metrics, showcasing its effectiveness in identifying concept drift instances. The significantly better performance of our method in precision and recall, as well as the relatively competitive Mean Lag, suggests that it provides a more balanced and robust approach to concept drift detection.

Overall, these results demonstrate that the proposed method is highly effective in detecting concept drift for the RT dataset, consistently outperforming other detection

Table 4. Performance Comparison of Concept Drift Detection Algorithms on the RT Dataset

Detection Algorithm	TP	FP	FN	Precision	Recall	Mean Lag
DDM	1	4	2	20.00	33.33	114.0
ADWIN	0	3	3	0.00	0.00	N/A
KSWIN	1	12	2	7.69	33.33	57.0
HDDM_A	1	3	2	25.00	33.33	115.0
SPNCD*	3	1	0	75.00	100.00	79.0

algorithms in the majority of evaluation metrics. This superior performance indicates that our method can serve as a valuable tool in maintaining system performance and minimizing the impact of concept drift in streaming data applications.

Table 5. Performance Comparison of Concept Drift Detection Algorithms on the Sine Dataset

Detection Algorithm	TP	FP	FN	Precision	Recall	Mean Lag
DDM	4	1	0	80.00	100.00	95.50
ADWIN	1	2	2	33.33	33.33	47.00
KSWIN	3	0	0	100.00	100.00	20.67
HDDM_A	3	0	0	100.00	100.00	18.67
SPNCD*	3	0	0	100.00	100.00	1.60

The comparison results in Table 5 for the Sine dataset demonstrate that three out of the five methods, including our proposed method (SPNCD*), achieved perfect precision and recall of 100%. Notably, our proposed method had the lowest mean lag compared to all the other algorithms, indicating a shorter delay between detecting concept drift and taking appropriate action. In contrast, ADWIN had a significantly lower true positive rate and recall of only 33.33%. These results suggest that our proposed method consistently outperformed other benchmark detection algorithms across the different datasets, underlining its effectiveness in detecting concept drift with high accuracy and low mean lag.

The average results presented in Table 6 further highlight the superiority of our proposed method (SPNCD*), as it achieved the highest true positive rate, recall, and precision while having the lowest false negative and false positive rates among the tested algorithms. Although KSWIN had the overall lowest mean lag, our method detected drifts notably faster in two out of the three datasets. Additionally, our method did not miss any drifts in the RT dataset, while KSWIN missed two. Consequently, our method demonstrated the best mean lag performance without missing any drifting points. These results emphasize the effectiveness of our proposed method in detecting concept drift, ensuring accurate and timely detection, which is essential for maintaining system performance and avoiding costly errors.

B. Performance of the Base Classifier
The comparison results of the learner performance, as shown in Tables 7, 8, and 9, demonstrate the effectiveness of the proposed SPNCD algorithm in detecting concept drift while maintaining learner performance. For conciseness and clarity, we have only included the results for the best-performing methods in the tables and excluded ADWIN and DDM, as their performance was consistently worse. The tables provided compare the performance of different concept drift detection algorithms in combination with the learner for three datasets. The results show that incorporating a reliable concept

Table 6. Comparison of Average Performance Metrics for Concept Drift Detection Algorithms Across Datasets.

Detection Algorithm	TP	FN	FP	Precision	Recall	Mean Lag
DDM	2.67	0.67	3.00	47.62	77.78	98.17
ADWIN	1.33	1.67	2.00	36.11	44.44	N/A
KSWIN	2.33	0.67	4.00	69.23	77.78	23.58
HDDM_A	2.33	0.67	1.33	66.67	77.78	61.45
SPNCD*	3.00	0.00	0.33	91.67	100.00	26.87

drift detection algorithm can significantly improve the performance of the learner. Our proposed method outperformed other algorithms in terms of model accuracy and F1 score across two out of three datasets, indicating its effectiveness in detecting concept drifts while the results in this aspect could be affected by the learner's capabilities.

The SPNCD* algorithm achieved the highest score in terms of model accuracy and F1 score for two datasets, namely the Sine and RT datasets. However, in the Mixed dataset, our proposed method achieved the second-best performance, with KSWIN having slightly better accuracy and F1 score. It is worth noting that all algorithms had perfect concept drift detection performance on the Mixed dataset, implying that the results may be limited by the base learner's performance.

In summary, the results suggest that incorporating our proposed method or other reliable concept drift detection algorithms can enhance the performance of the learner and maintain system performance by avoiding costly errors.

Table 7. Comparison of Learner Performance with Different Detection Algorithms for the Mixed Dataset.

Detection Algorithm	C_Precision	C_Recall	F1	Accuracy
No Detection	48.8	48.83	48.8	48.8
KSWIN	91.12	90.85	90.98	91.00
HDDM_A	90.92	90.27	90.59	90.63
SPNCD*	90.97	90.80	90.89	90.89

These tables showcase the learner's performance when equipped with different concept drift detection algorithms, including a No-Detection method to show the results of the learner without a detection mechanism and not getting updated. As seen in the tables, our proposed method, SPNCD*, demonstrated better performance in terms of accuracy and F1 score in the Sine and RT datasets, while achieving the second-best performance in the Mixed dataset. This demonstrates the advantages of incorporating a reliable concept drift detection algorithm in maintaining learner performance and avoiding costly errors.

Table 8. Comparison of Learner Performance with Different Detection Algorithms for the RT Dataset.

Detection Algorithm	C_Precision	C_Recall	F1	Accuracy
No Detection	59.20	59.73	58.67	59.19
KSWIN	77.47	65.03	70.71	73.87
HDDM_A	75.80	68.28	71.84	74.05
SPNCD*	75.86	70.99	73.34	74.98

Table 9. Comparison of Learner Performance with Different Detection Algorithms for the Sine Dataset.

Detection Algorithm	C_Precision	C_Recall	F1	Accuracy
No Detection	49.07	47.59	48.32	49.1
KSWIN	89.92	90.07	89.99	89.98
HDDM_A	89.82	90.28	90.05	90.02
SPNCD*	90.00	90.58	90.29	90.25

Overall, the results indicate that our proposed SPNCD* algorithm is effective in detecting concept drift while preserving the learner's performance. It highlights the importance of using a reliable concept drift detection algorithm in real-world scenarios where data distributions change over time, which is crucial for maintaining system performance and preventing costly mistakes.

6   Conclusion and Future Work

In this study, we presented a novel concept drift detection algorithm, SPNCD*, and demonstrated its effectiveness in identifying concept drift in three datasets. Our method consistently outperformed benchmark detection algorithms regarding the true positive rate, recall, precision, and mean lag. Additionally, we showed that incorporating our proposed method or other reliable concept drift detection algorithms can enhance the learner's performance and maintain system performance by avoiding costly errors. Our findings significantly impact the development of an ML model that can effectively respond to changing data distributions. By incorporating our proposed method, such systems can ensure accurate and timely detection of concept drift, which is essential for maintaining high performance and avoiding costly mistakes.

Despite the success of our proposed method, there is still room for improvement and further exploration. As part of future work, the following directions could be considered:

1. **Hybrid Approaches:** Investigate the potential of combining SPNCD* with other drift detection algorithms to create a hybrid approach that capitalizes on the strengths of multiple techniques, potentially leading to even better performance in detecting concept drift.

2. **Adaptive Window Sizes:** Examine the impact of adaptive window sizes on the performance of our proposed method, enabling the algorithm to respond better to varying degrees of concept drift.
3. **Different Base Learners:** Evaluate the performance of SPNCD* in combination with other types of base learners, including deep learning models, to assess the generalization capability of our approach across a wider range of learning algorithms.
4. **Online Learning:** Extend our approach to online learning settings, where data is processed one sample at a time, and investigate its performance in real-time detection of concept drift.
5. **Application to Real-world Problems:** Apply our proposed method to real-world problems in various domains, such as finance, healthcare, and manufacturing, to assess its practical utility and performance in diverse real-world scenarios.

References

1. Baena-García, M., del Campo-Ávila, J., Fidalgo, R., Bifet, A., Gavalda, R., Morales-Bueno, R.: Early drift detection method. In: Fourth International Workshop on Knowledge Discovery from Data Streams, vol. 6, pp. 77–86 (2006)
2. Barros, R.S., Cabral, D.R., Gonçalves, P.M., Jr., Santos, S.G.: RDDM: reactive drift detection method. Expert Syst. Appl. **90**, 344–355 (2017)
3. Barros, R.S.M., Santos, S.G.T.C.: A large-scale comparison of concept drift detectors. Inf. Sci. **451**, 348–370 (2018)
4. Bayram, F., Ahmed, B.S., Kassler, A.: From concept drift to model degradation: an overview on performance-aware drift detectors. Knowl.-Based Syst. **245**, 108632 (2022). https://doi.org/10.1016/j.knosys.2022.108632
5. Bifet, A., Gavalda, R.: Learning from time-changing data with adaptive windowing. In: Proceedings of the 2007 SIAM International Conference on Data Mining, pp. 443–448. SIAM (2007)
6. Demšar, J., Bosnić, Z.: Detecting concept drift in data streams using model explanation. Expert Syst. Appl. **92**, 546–559 (2018)
7. Fieguth, P.: An Introduction to Pattern Recognition and Machine Learning. Springer, Cham (2022). https://doi.org/10.1007/978-3-030-95995-1
8. Frias-Blanco, I., del Campo-Ávila, J., Ramos-Jimenez, G., Morales-Bueno, R., Ortiz-Diaz, A., Caballero-Mota, Y.: Online and non-parametric drift detection methods based on Hoeffding's bounds. IEEE Trans. Knowl. Data Eng. **27**(3), 810–823 (2014)
9. Gama, J., Castillo, G.: Learning with local drift detection. In: Li, X., Zaïane, O.R., Li, Z. (eds.) ADMA 2006. LNCS (LNAI), vol. 4093, pp. 42–55. Springer, Heidelberg (2006). https://doi.org/10.1007/11811305_4
10. Gama, J., Medas, P., Castillo, G., Rodrigues, P.: Learning with drift detection. In: Bazzan, A.L.C., Labidi, S. (eds.) SBIA 2004. LNCS (LNAI), vol. 3171, pp. 286–295. Springer, Heidelberg (2004). https://doi.org/10.1007/978-3-540-28645-5_29
11. Gama, J., Žliobaitė, I., Bifet, A., Pechenizkiy, M., Bouchachia, A.: A survey on concept drift adaptation. ACM Comput. Surv. (CSUR) **46**(4), 1–37 (2014)
12. Garnett, R., Roberts, S.J.: Learning from data streams with concept drift. Technical report, Dept. of Engineering Science (2008)
13. Gemaque, R.N., Costa, A.F.J., Giusti, R., Dos Santos, E.M.: An overview of unsupervised drift detection methods. Wiley Interdiscip. Rev. Data Min. Knowl. Discov. **10**(6), e1381 (2020)

14. Gonçalves, P.M., Jr., de Carvalho Santos, S.G., Barros, R.S., Vieira, D.C.: A comparative study on concept drift detectors. Expert Syst. Appl. **41**(18), 8144–8156 (2014)

15. Gözüaçık, Ö., Büyükçakır, A., Bonab, H., Can, F.: Unsupervised concept drift detection with a discriminative classifier. In: Proceedings of the 28th ACM International Conference on Information and Knowledge Management, pp. 2365–2368 (2019)

16. Hoens, T.R., Polikar, R., Chawla, N.V.: Learning from streaming data with concept drift and imbalance: an overview. Prog. Artif. Intell. **1**, 89–101 (2012)

17. Huang, D.T.J., Koh, Y.S., Dobbie, G., Pears, R.: Detecting volatility shift in data streams. In: 2014 IEEE International Conference on Data Mining, pp. 863–868. IEEE (2014)

18. Klinkenberg, R., Renz, I.: Adaptive information filtering: learning in the presence of concept drifts. Learn. Text Categ. 33–40 (1998)

19. Kuncheva, L.I.: Combining Pattern Classifiers: Methods and Algorithms. John Wiley & Sons, Hoboken (2014)

20. de Lima Cabral, D.R., de Barros, R.S.M.: Concept drift detection based on fisher's exact test. Inf. Sci. **442**, 220–234 (2018)

21. López Lobo, J.: Synthetic datasets for concept drift detection purposes. Harv. Dataverse (2020)

22. Lu, J., Liu, A., Dong, F., Gu, F., Gama, J., Zhang, G.: Learning under concept drift: a review. IEEE Trans. Knowl. Data Eng. **31**(12), 2346–2363 (2018)

23. Nishida, K., Yamauchi, K.: Detecting concept drift using statistical testing. In: Corruble, V., Takeda, M., Suzuki, E. (eds.) DS 2007. LNCS (LNAI), vol. 4755, pp. 264–269. Springer, Heidelberg (2007). https://doi.org/10.1007/978-3-540-75488-6_27

24. Page, E.S.: Continuous inspection schemes. Biometrika **41**(1/2), 100–115 (1954)

25. Pears, R., Sakthithasan, S., Koh, Y.S.: Detecting concept change in dynamic data streams. Mach. Learn. **97**(3), 259–293 (2014). https://doi.org/10.1007/s10994-013-5433-9

26. Pesaranghader, A., Viktor, H.L.: Fast hoeffding drift detection method for evolving data streams. In: Frasconi, P., Landwehr, N., Manco, G., Vreeken, J. (eds.) ECML PKDD 2016. LNCS (LNAI), vol. 9852, pp. 96–111. Springer, Cham (2016). https://doi.org/10.1007/978-3-319-46227-1_7

27. Pesaranghader, A., Viktor, H.L., Paquet, E.: Mcdiarmid drift detection methods for evolving data streams. In: 2018 International Joint Conference on Neural Networks (IJCNN), pp. 1–9. IEEE (2018)

28. Poon, H., Domingos, P.: Sum-product networks: a new deep architecture. In: 2011 IEEE International Conference on Computer Vision Workshops (ICCV Workshops), pp. 689–690. IEEE (2011)

29. Raab, C., Heusinger, M., Schleif, F.M.: Reactive soft prototype computing for concept drift streams. Neurocomputing **416**, 340–351 (2020)

30. Ross, G.J., Adams, N.M., Tasoulis, D.K., Hand, D.J.: Exponentially weighted moving average charts for detecting concept drift. Pattern Recognit. Lett. **33**(2), 191–198 (2012)

31. Sánchez-Cauce, R., París, I., Díez, F.J.: Sum-product networks: a survey. IEEE Trans. Pattern Anal. Mach. Intell. **44**(7), 3821–3839 (2021). IEEE

32. Webb, G.I., Lee, L.K., Petitjean, F., Goethals, B.: Understanding concept drift. arXiv preprint arXiv:1704.00362 (2017)

33. Widmer, G., Kubat, M.: Learning in the presence of concept drift and hidden contexts. Mach. Learn. **23**(1), 69–101 (1996)

Explaining Relation Classification Models with Semantic Extents

Lars Klöser[1]([✉])(ID), Andre Büsgen[1](ID), Philipp Kohl[1](ID), Bodo Kraft[1],
and Albert Zündorf[2]

[1] Aachen University of Applied Sciences, 52066 Aachen, Germany
{kloeser,buesgen,p.kohl,kraft}@fh-aachen.de
[2] University of Kassel, 34109 Kassel, Germany
zuendorf@uni-kassel.de

Abstract. In recent years, the development of large pretrained language models, such as BERT and GPT, significantly improved information extraction systems on various tasks, including relation classification. State-of-the-art systems are highly accurate on scientific benchmarks. A lack of explainability is currently a complicating factor in many real-world applications. Comprehensible systems are necessary to prevent biased, counterintuitive, or harmful decisions.

We introduce semantic extents, a concept to analyze decision patterns for the relation classification task. Semantic extents are the most influential parts of texts concerning classification decisions. Our definition allows similar procedures to determine semantic extents for humans and models. We provide an annotation tool and a software framework to determine semantic extents for humans and models conveniently and reproducibly. Comparing both reveals that models tend to learn shortcut patterns from data. These patterns are hard to detect with current interpretability methods, such as input reductions. Our approach can help detect and eliminate spurious decision patterns during model development. Semantic extents can increase the reliability and security of natural language processing systems. Semantic extents are an essential step in enabling applications in critical areas like healthcare or finance. Moreover, our work opens new research directions for developing methods to explain deep learning models.

Keywords: Relation classification · Natural language processing · Natural language understanding · Explainable artificial intelligence · Trustworthy artificial intelligence · Information extraction

1 Introduction

Digitalizing most areas of our lives lead to constantly growing amounts of available information. With digital transformation comes the need for economic and social structures to adapt. This adaption includes the structuring and networking of information for automated processing. Natural language is a significant source of unstructured information. While comfortable and efficient for humans, processing natural language remains challenging for computers. Structuring and networking the amount of text created daily in sources like online media or social networks requires incredible manual effort.

D. Conte et al. (Eds.): DeLTA 2023, CCIS 1875, pp. 189–208, 2023.
https://doi.org/10.1007/978-3-031-39059-3_13

The field of *information extraction* (IE) is a subfield of *natural language processing* (NLP) and investigates automated methods to transform natural language into structured data. It has numerous promising application areas; examples are the extraction of information from online and social media [12], the automatic creation of knowledge graphs [8], and the detailed analysis of behaviors in social networks [4].

Modern deep-learning-based NLP systems show superior performance on various benchmarks compared to humans. For example, on average deep learning systems like [33] score higher than humans on the *SuperGLUE*[1] [27] benchmark. [19] discusses that such results do not directly imply superior language skills. A primary reason is that the decision-making processes are not intuitively comprehensible. NLP systems are optimized on datasets and therefore run the risk of learning counterintuitive patterns from the data [10, 11]. This research focuses on *relation classification*, a central task in IE, and aims to compare the central decision patterns of human and deep-learning-based task-solving.

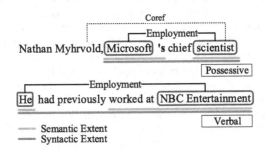

Fig. 1. Two relation mentions from the ACE 05 corpus. The first example shows a possessive formulation. The second example shows a verbal formulation. In the first example, *scientist* is preferable to *Nathan Myhrvold*. *scientist* is part of a predefined class of formulations (possessive), while *Nathan Myhrvold* is not.

In more detail, relation classification describes classifying the relation for a given pair of entity mentions, the relation candidate, in a text. Figure 1 shows two examples for the task. We assume that one of the possible relations applies to the relation candidate. A real-world application requires additional steps like detecting entity mentions or numerating relation candidates. We analyze a deep-learning NLP model on the ACE 05 dataset [7]. Our primary focus is the comparison between human and deep-learning decision patterns. We introduce the concept of *semantic extents* to identify human decision patterns.

[7] introduced the related concept of *syntactic classes* and *syntactic extents*. Natural language offers various syntactical ways to formulate relationships between mentions of certain entities in text. Each involves different syntactical aspects of a sentence. For example, as shown in Fig. 1, a relationship can be expressed through a possessive formulation or a verbal construction. In these examples, *possessive* and *verbal* are the

---

[1] https://super.gluebenchmark.com/.

relation's syntactic classes. The syntactic extent contains both arguments and the text between them. If necessary for the syntactic construction implied by the syntactic class, it contains additional text outside the arguments. The ACE corpus contains only relations that fit into one of the given syntactic classes.

Syntactic extents neglect the fact that the semantics of a decision can strongly influence decision patterns. Semantic extents are minimum necessary and sufficient text passages for classification decisions concerning a specific *decider*. Humans, as well as NLP models, can be deciders. We introduce a tool for practical human annotation of semantic extents. The semantic extents for humans and models show significant differences. Our results indicate that human decisions are more context-dependent than model decisions. Models tend to decide based on the arguments and ignore their context. We provide an analysis based on adversarial samples to support this assumption. Finally, we compare semantic extents to other existing explainable *artificial intelligence* (AI) techniques. The results of this study have the potential to reveal incomprehensible and biased decision patterns early on in the development process, thus improving the quality and reliability of IE systems.

We provide reproducible research. All source code necessary to reproduce this paper's results are available via GitHub[2]. The paper aims to investigate and compare the decision patterns of human and deep-learning-based models concerning relation classification. Specifically, we analyze the concept of semantic extents and compare them concerning human and model decisions. The paper also explores existing explainable AI techniques for relation classification. The primary research question can be summarized as follows:

What are the decision patterns of human and deep-learning-based models when solving the task of relation classification, and how do they compare?

Our main contributions are:

1. the introduction and analysis of semantic extents for relation classification,
2. the first analysis of explainable AI techniques like input reductions for relation classification, and
3. a novel preprocessing pipeline based on a state-of-the-art python tool stack for the ACE 05 dataset.

2 Related Work

This section surveys different approaches and techniques proposed to interpret and explain the decisions made by NLP models and discusses their strengths and limitations.

The best information extraction approaches on scientific benchmarks finetune large pretrained language models like BERT [6] and LUKE [29]. Various studies try to understand the internal processes that cause superior performance on various NLP benchmarks.

---

[2] https://github.com/MSLars/semantic_extents https://github.com/MSLars/ace_preprocessing.

One direction of research focuses on analyzing how models represent specific meanings. Besides the superior performance on benchmarks, pretrained text representations may be the source of harmful model behaviors. [2] shows that static word embeddings, such as Word2Vec, tend to be biased toward traditional gender roles. Approaches, such as [17,30,32], tried to remove these biases. [3] investigate the origins of biases in these vector representations. Similarly, [13] showed that pretrained transformer language models suffer from similar biases. These results show that public data sources are biased. This property could carry over to finetuned models.

[16] focused on the internal knowledge representation of transformer models and showed that they could change specific factual associations learned during pretraining. For example, they internalized the fact that *the Eifel tower is in Rome* in a model.

Another direction of research focuses on the language capabilities of pretrained language models. [5] showed that BERT's attention weights correlate with various human-interpretable patterns. They concluded that BERT models learn such patterns during pretraining. *Probing* is a more unified framework for accessing the language capabilities of pretrained language models. [28] provided methods for showing that language models learn specific language capabilities. [24] showed that latent patterns related to more complex language capabilities tend to occur in deeper network layers. Their findings suggest that pretrained transformer networks learn a latent version of a traditional NLP pipeline.

The previously mentioned approaches investigate the internal processes of large pretrained language models. Another research direction is analyzing how finetuned versions solve language tasks in a supervised setting. [15] created diagnostic test sets and showed that models learn heuristics from current datasets for specific tasks. Models might learn task-specific heuristics instead of developing language capabilities. [18] provided a software framework for similar tests and revealed that many, even commercial systems, fail when confronted with so-called diagnostic test sets or adversarial samples.

A different way to look at the problem of model interpretability is the analysis of single predictions. [21] introduced gradient-based *saliency maps*. [22,23] adopted the concept for NLP. Each input token is assigned a saliency score representing the influence on a model decision. *Input reductions* remove supposedly uninfluential tokens. They interpret the result as a semantic core regarding a decision. This concept is similar to semantic extents, except that we extend and they reduce input tokens. Similarly, [9] rearrange the input. If perturbation in some area does not influence a model's decision, the area is less influential. Besides gradient information from backpropagation, researchers analyze the model components' weights directly. [5] visualizes the transformer's attention weights. Their results indicate that some attention heads' weights follow sequential, syntactical, or other patterns.

[20] introduce an explainability approach for relation extraction with distant supervision. They use a knowledge base to collect text samples, some expressing the relationship between entities and others not. The model then predicts the relationship between these samples with methods that assess the model's relevance for each sample.

Meanwhile, [1] brings us a new development in the field with their interpretable relation extraction system. By utilizing deep learning methods to extract text features and combining them with interpretable machine learning algorithms such as decision trees, they aim to reveal the latent patterns learned by state-of-the-art transformer models when tasked with determining sentence-level relationships between entities.

The field of explainable NLP covers various aspects of state-of-the-art models. We are a long way from fully understanding decisions or even influencing them in a controlled manner. Various practical applications, especially in sensitive areas, require further developments in this field.

3 Dataset

The ACE[3] 05 dataset is a widely used benchmark dataset for NLP tasks such as entity, event, and relation extraction. It was a research initiative by the United States government to develop tools for automatically extracting relevant information from unstructured text. The resources are not freely available.

The dataset consists of news articles from English-language newswires from 2003 to 2005. This paper analyzes explainability approaches for relation classification models using the ACE 05 dataset. This dataset includes annotations for various entities, relations, and events in 599 documents. Entities, relations, and events may have various sentence-level mentions. We focus on entities and relations and leave out the event-related annotations.

Our analysis focuses on the sentence-level relation mentions annotated in the ACE 05 dataset. The annotations have additional attributes, such as tense and syntactic classes, often ignored by common preprocessing approaches for event [31] and relation [14] extraction. We publish a preprocessing pipeline on top of a modern Python technology stack. Our preprocessing approach makes this meta-information easily accessible to other researchers. We refer to the original annotation guidelines for further information about the relation types and syntactic classes[4]. Our preprocessing pipeline is available via a GitHub repository.

3.1 Preprocessing

Our preprocessing pipeline includes all data sources provided in the ACE 05 corpus. In contrast, [14] excludes all Usenet or forum discussions and phone conversation transcriptions. However, real-world NLP systems might encounter informal data from social media and comparable sources. We see no reason to exclude these relations. Our train-dev-test split expands [14]. We extended each split with random samples from the previously excluded data sources and kept the ratio between all three splits similar.

The currently most widespread pipelines use a combination of Java and Python implementations. The sentence segmentation in [14] uses a Standford NLP version from 2015. Our implementation uses the Python programming language and the current version of spaCy[5]. All source codes are open for further community development since the technologies used are common in the NLP community.

---

[3] Short for Automated Content Extraction.

[4] https://www.ldc.upenn.edu/sites/www.ldc.upenn.edu/files/english-relations-guidelines-v5.8.3.pdf.

[5] https://spacy.io/, version 3.5.

We offer a representation of the data structures from the original corpus in a modern Python environment. In contrast to existing approaches, we can access cross-sentence coreference information. Users can access the coreference information between *Entities*, *Relations*, and *Events*. Our preprocessing enables researchers to use the ACE 05 corpus for further tasks such as document-level relation extraction or coreference resolution.

We use spaCy's out-of-the-box components for tokenization and sentence segmentation. Each *relation mention* forms a relation classification sample if both arguments are in one detected sentence. Additionally, one sentence may occur in multiple relation classification samples if it contains multiple *relation mentions*.

3.2 Task Definition

We define the relation classification task as classifying the relationship between two argument spans in a given sentence. Given an input sentence $s = x_1, ..., x_n$ that consists of n input tokens. We refer to sentences in the sense of ordinary English grammar rules. In relation classification, two argument spans, $a_1 = x_i, ..., x_j$ and $a_2 = x_k, ..., x_l$ with $1 \le i < j < k < l \le n$, are defined as relation arguments. The model's task is to predict a label from a set of possible labels $l \in R$, where R is the set of all possible relation labels. It is important to note that the model only considers samples where one of the relations in R applies to the pair of arguments.

Our definition of relation classification differs from relation extraction. Relation extraction typically involves identifying entity mentions in the text and then analyzing their syntactic and semantic relationships. Relation classification, conversely, involves assigning predefined relationship types to given pairs of entities in text. In this task, we provide the relation's arguments, and the goal is to classify each pair of entities into the correct predefined relationship category.

3.3 Dataset Statistics

By adding Usenet or forum discussions and phone conversation transcriptions as data sources, we increase the semantical and syntactical scope of the dataset. Figure 2 shows the distribution of relation labels and syntactic classes. While the more informal sources contain more *Family* relations, the other sources contain more business-related *Employer* relations. Similarly, other sources contain less *Possessive* and *Verbal* formulations. The results indicate that adding the new sources extends the semantic and syntactic scope of the dataset since added samples have different characteristics.

4 Semantic Extent

To answer the main research question, we formalize the concept of *decision patterns*. Decision patterns are text areas contributing significantly to decisions in the context of an NLP task. We refer to these significantly influential text areas as the *semantic extent*. The semantic extent contains the relation's arguments for the relation classification task.

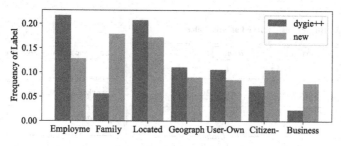

(a) Most frequent relation labels.

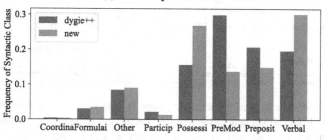

(b) Most frequent syntactic classes.

Fig. 2. The histograms show the semantic and syntactic properties of different parts of the ACE 05 dataset. *dygie++* refers to all samples in [14,25]. The name originates from the implementation we used. *new* refers to the excluded samples.

We need four additional concepts to specify the scope of semantic extents. Given an input sentence $s = x_1, ..., x_n$, each ordered subset $s_{cand} \subset s$ that contains the arguments can be interpreted as a *semantic extent candidate*. A *candidate extension*, $s_{ext} = s_{cand} \cup x_i$ for some $x_i \notin s_{cand}$, is the addition of one token.

An extension is *consistent* if the added token does not substantially reduce the influence of other tokens of the candidate. If no extension of a candidate that causes the classification decision to change exists, then the candidate is *significant* and referred to as a semantic extent.

Figure 3 shows how to apply the previous definitions to determine a semantic extent. We aim to explain why the sentence *He had previously worked at NBC Entertainment* expresses an *Employer* relation between *He* and *NBC Entertainment*.

Are the arguments themselves a valid semantic extent? Our first candidate is *He NBC Entert.* One consistent extension could be *He visits NBC Entertainment*. It suggests a *Located* relation. Therefore, *He NBC Entertainment* is not a significant candidate nor a semantic extent. To create the next semantic extent candidate, we add *at*. As the second column in Fig. 3 suggests, we can still create consistent extensions that cause a different classification decision. Finally, *He works at NBC Entertainment* does not allow consistent extensions that change the classification label. *He works at NBC Entertainment* is the semantic extent.

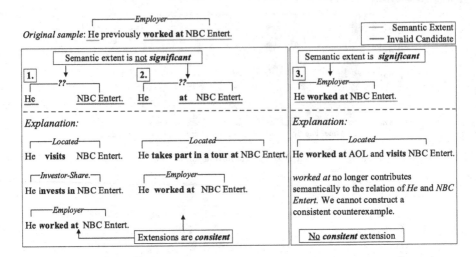

Fig. 3. Example determination of the semantic extent The top row contains suggestions for the semantic extent. Each column contains extensions that allow a different interpretation. The last column contains the valid semantic extent for the sample *He had previously worked at NBC Entertainment*. The top row shows different semantic extent candidates. Each column list possible extensions that change the relation label.

However, extensions that change the classification decision are still possible. For example, *He works at AOL and visits NBC Entertainment. works at*, as part of the semantic extent, does not influence the classification decision. *visits* is this extension's most significant context word. We do not consider this a consistent extension and assume that no consistent extension exists.

Generally, it is impossible to prove the absence of any consistent extension changing the relation label. Nevertheless, we argue that the concepts of *interpretability* and *explainability* suffer similar conceptual challenges. We apply a convenient extension rule: if an average language user cannot specify a convenient extension, we assume such does not exist. Our evaluation uses the additional concept of *semantic classes*. Semantic classes group different kinds of semantic candidate extensions. The following section motivates and describes the concept in detail.

4.1 Semantic Extent for Humans

This section proposes a framework and the corresponding tooling for human semantic extent annotations. We explain and justify the most relevant design decisions. In general, many other setups and configurations are possible. A comparative study is out of the scope of this research.

A possible approach would be to show annotators the complete text and task them to mark the semantic extent directly. With this approach, annotators manually reduce the input presented to them. We assume that annotators would annotate to small semantic extents because the additional information biases them. We, therefore, define an extensive alternative to such reductive approaches.

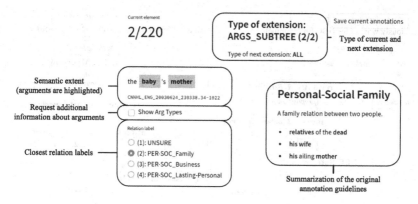

Fig. 4. Screenshot of our annotation tool's central area. Annotators use the keyboard for label selection, navigation between samples, and extension of the semantic extent.

We start with confronting annotators with only the argument text spans. If they are sure to determine the relation label, they can annotate the relation and jump to the following sample. If not, they can request further tokens. This procedure continues until they classify a sample or reject a classification decision. Figure 4 shows a screenshot of our annotation tool.

The extension steps in Fig. 3 are examples of such an iterative extension. At first, an annotator sees the arguments *He* and *NBC Entertainment*. The first extension reveals an *at*, and the final extension reveals the verb *works*. We implemented a heuristic for the priority of words in this process.

Our priority can be summarized as follows. The initial semantic extent candidate contains only arguments (OA). The second stage contains the syntactic subtrees of the arguments (AS). The third stage shows the verbs on the syntactic path between the arguments (VOP). Afterward, we show all tokens between the arguments (BA). Finally, we show the full semantic extent annotated in the ACE 05 corpus (E) and all remaining tokens (A) if necessary. Some of these stages may contain multiple or even none tokens. Figure 5 illustrates our prioritization for two examples. We refer to the category of the final extension as *semantic class*. We use this class extensively during the evaluation in Sect. 5.

We implemented some simplifications to increase practicability and reduce the necessary human annotation effort.

1. Annotators had to check the annotation guidelines to select possible relation candidates. The dataset contains 18 different relation labels. Adequate familiarization with the annotation scheme is too time-consuming. We implemented a label preselection. Annotators choose between the three most likely labels (concerning our deep learning model). The preselection reduces the need to know every detail of the annotation guidelines but makes the semantic challenge comparable. Additionally, we provide a summary of the annotation guidelines.
2. Many expressions needed to be looked up by the annotators. For example, the ACE 05 corpus contains many sources about the war in Iraq. Names of organizations or

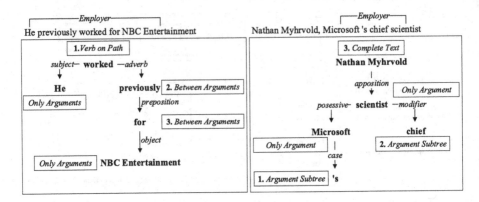

Fig. 5. Example of semantic extension types in combination with the dependency parse tree for two samples. The indices show the priority index during the expanding annotation process.

locations were unknown. We offered annotators the possibility to check the fine-grained argument entity types. This possibility usually avoids additional look-ups and speeds up the annotation process.

3. We implement the previously mentioned priority heuristic about which tokens are candidates for extensions of the semantic extent. In general, this is necessary for our extensive setting. However, since annotators cannot annotate custom semantic extents, unnecessary tokens might be part of the semantic extent. We use the extension categories for our analysis. For each annotation, we can interpret these semantic classes as statements like *As annotator, I need the verb to classify the sample.*

We had to distinguish between practicability and the danger of receiving biased annotations. However, we assume that our adoptions did not influence or bias the annotations in a significant way. The source code for the annotation process is available in our GitHub repository, so the results are reproducible with other design decisions.

4.2   Semantic Extent for Models

This section describes how we determine semantic extents for deep learning models. Finetuned transformer networks are fundamental for most state-of-the-art results in NLP approaches. We define a custom input format for relation classification and finetune a pretrained transformer model, *RoBERTa*, for relation classification. Figure 6 illustrates our deep learning setup.

We tokenize the inputs and use the model-specific *SEP* tokens to indicate the argument positions. To receive *logits*, we project the final representation of the *CLS* token to the dimension of the set of possible relation labels. Logits are a vector representation $h \in \mathbb{R}^{|L|}$ assigning each label a real-valued score. We apply the softmax function to interpret these scores as probabilities

$$P(l) = \frac{\exp(h_l)}{\sum_{k \in L} \exp(h_k)}, \ \forall l \in L$$

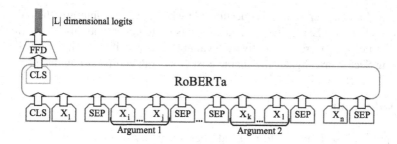

Fig. 6. Illustration of our deep learning model architecture. We special *SEP* (separator) tokens to indicate the argument positions.

and calculate the cross-entropy loss to optimize the model.

Our approach requires a single forward pass to a transformer network. The computational complexity is comparable to many current state-of-the-art approaches.

As described in the previous sections, the semantic extent is a mixture of reductive and expanding input variations. We use this as a guide and define two methods for determining text areas interpretable as semantic extent for deep learning models.

4.3   Expanding Semantic Extent

Compared to human annotation, the expanding semantic extent is determined similarly (Subsect. 5.1). At first, we calculate the model prediction l_{all} on a complete sentence. Then we create semantic extent candidates. The initial candidate contains only the arguments. We expand the extent by adding words concerning the priority described in Subsect. 5.1 and compute a model prediction $l_{reduced}$ on the candidate. If $l_{all} = l_{reduced}$ and $P(l_{reduced}) > \theta$ for some threshold θ, we interpret the current candidate as semantic extent. We interpret the category of the final extension as the semantic class.

4.4   Reductive Semantic Extent

We apply gradient-based *input reductions* [10] to determine a reductive semantic extent. Like Subsect. 4.3, input reductions are a minimal subset of the input with an unchanged prediction compared to the complete input. The main difference is that input reductions remove a maximal number of tokens. Our implementation uses the input reduction code in [26].

We apply a beam search to reduce the input as far as possible while keeping the model prediction constant. We compute an approximation for the token influence at each reduction step based on gradient information. Let p be the probability distribution predicted by a model and l_{pred} be the predicted label. In the next step we compute a loss score $L_{inter}(l_{pred}, p)$. We use the gradient $\frac{\partial L_{inter}}{\partial x_i}$, for each x_i in the current input, to approximate the influence of each token on the model prediction. During the beam search, we try to exclude the least influential tokens.

Reductive semantic extents may be sentence fragments without grammatical structure and unintuitive for humans. If the essential structure deviates too much from the data set under consideration, it is questionable whether we can interpret the reductions as decision patterns. Expanding semantic extents try to get around this by creating sentence fragments that are as intuitively meaningful as possible through prioritization. Following Sect. 5 investigates the characteristics of both approaches in detail.

5   Evaluation

In the previous section, we formalized decision patterns for the relation classification task by introducing *semantic extents*. This section compares semantic extents for humans and deep learning models. We reveal similar and dissimilar patterns related to human and deep learning-based problem-solving.

5.1   Human Annotation

Preparing for the previously mentioned comparison, we analyze human decision patterns by determining and analyzing semantic extents. Due to human resource constraints, we could not annotate semantic extents on all test samples. The available resources are sufficient to prove an educated guess about human decision patterns. Humans generally require additional context information to the relation arguments to make a classification decision. We assume that *possessive, preposition, pre-modifier, coordination, formulaic,* and *participial*[6] formulations require a syntactically local context around arguments. In contrast, *verbal* and *other* formulations require more *sentence-level* information. This section refers to both categories as *local* and *sentence-level* samples. We sampled 220 samples with an equal fraction of local and sentence-level samples to prove our claim.

We asked two annotators to annotate semantic extents on these 220 samples using the previously introduced application (Subsect. 5.1). Table 1 shows central characteristics of the annotations. The first annotator made a more accurate prediction concerning F1 scores, but both annotators showed high agreement on the relation labels. While the agreement on fine-grained semantic classes is moderate, annotators agree more clearly on the coarse classes. Regarding the number of tokens in the semantic extents, both annotators show similar numbers with a comparably high standard deviation. Annotators reported that, in some cases, arguments have large subtrees. Large subtrees require many extensions to see the verb of other sentence-level tokens.

Figure 7 shows the distribution of semantic classes for local and sentence-level samples. The semantic classes OA and AS indicate that annotators made decisions based on tokens that are children of the relation arguments in a dependency parse tree. The semantic classes VOP, BA, E, and A suggest that annotators needed more sentence-level information for their decision. The syntactic classes *verbal* and *other* also suggest a higher influence of sentence-level contexts. For sentence-level samples, the annotators often need tokens outside the syntactical subtree of either of the arguments. For local samples, the local context suffices for deciding relation labels. Only some cases require

---

[6] These refer to the syntactic classes for relation-mention formulations introduced in the ACE 05 annotation guidelines.

Table 1. A summary showing the human annotations' results on 220 test samples. All F1 scores refer to the relation labels, and *label agreement* (LA) indicates the agreement on these annotations. SC refers to *semantic class*. *fine* refers to all semantic classes, while *coarse* distinguishes only arguments and argument subtree from the others.

	Annotator 1	Annotator 2
F1 micro	.85	.78
F1 macro	.84	.78
LA	.76	
SC coarse	.67	
SC fine	.46	
SC size	5.2 ± 5.5	5.2 ± 5.9

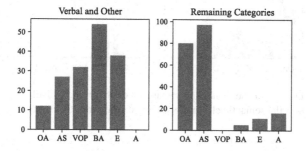

Fig. 7. Semantic class of human-annotated semantic extents for different syntactic classes. *Verbal* and *Other* correlate with context-dependent semantic classes. The remaining categories do not.

additional context information. These additional contexts differ from the previously described sentence-level context. We suppose that annotators had problems classifying these samples and tried to get information from the extensions to avoid excluding examples.

In conclusion, the amount of context information needed for decisions depends on the type of formulation. For the task of relation classification, humans recognize distinguishable patterns. We revealed that humans decide between patterns based on contexts syntactically local to the arguments and contexts that span entire sentences. Most decisions are not possible completely without any context.

5.2 Model Performance

The previous section revealed two human decision patterns. Humans either focus on local contexts around the arguments or on sentence-level contexts. This section similarly investigates latent model decision patterns. Besides traditional metrics such as micro and macro F1 scores, we assign decisions to different behavioral classes and investigate the implications on model language capabilities and real-world applications. We use the complete test dataset for the following analyses unless otherwise stated.

Table 2. Model and human performance on the ACE 05 relation classification corpus. (*) on a subset of 220 samples. (**) on the complete test set.

	F1-micro	F1-macro	#Parameters
Annotator 1*	.846	.843	
Annotator 2*	.779	.779	
RoBERTa base**	.832	.678	123 million
RoBERTa large**	.855	.752	354 million

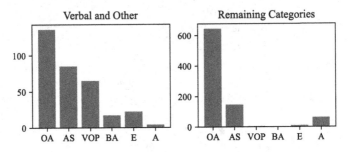

Fig. 8. Semantic classes for the expanding semantic extents for the RoBERTa-base model. The two histograms show the semantic classes depending on the syntactic classes.

Table 2 presents the transformer models' overall performance compared to humans. We trained two model variants, one with the base version of RoBERTa and the other with the large version. F1-micro scores for annotator 1 and both models are in a similar range[7]. The F1-macro scores reveal that model performances differ across relation labels. An investigation of F1 scores per label shows that RoBERTa-base recognizes neither *Sports Affiliation* nor *Ownership* relation. RoBERTa-large suffers from similar problems in a slightly weaker form. We selected a subset of 220 samples for human annotation. Some effects might be unlikely for underrepresented classes at this sample size. However, concerning the F1-micro score, the results show that transformer-based NLP models solve the sentence-level relation classification task (Subsect. 3.2) comparably well as humans.

5.3 Extensive Semantic Extents

The models' semantic extents hint at differences in model and human decision patterns. Figure 8 shows the classes of expanding semantic extents (Subsect. 4.3). Confronted with only the arguments, models tend to make confident decisions similar to their decision on complete samples. In comparison, Fig. 7 indicates that humans cannot make similar decisions based on the arguments. This observation may have different reasons. Models are optimized to reproduce annotations on concrete datasets. These

---

[7] Since Annotator 1 performs better than annotator 2, we take his scores as a careful estimation of human performance.

annotations may contain shortcuts and lead models towards heuristics diverging from language capabilities we would expect from humans. Another reason may be that we force models to make confident decisions in common deep-learning setups. Argument texts allow good heuristic solutions, and our setup offers no way to express uncertainty for models. In combination, that may cause overconfident model decisions if certain argument text correlate with particular labels in the dataset.

Table 3 relates the semantic class with model confidences and the correctness of predictions. We distinguish two categories of decisions. The first summarizes samples for which the model makes the final decision based on arguments alone. Secondly, we focus on samples where the model needs all tokens to make its final decision. Even slight changes in the input change the model prediction for these samples.

Predictions that fall in the first category are significantly more likely to be correct and show higher confidence with less deviation. The results indicate two things. First, in this data set, arguments are often a strong indication of the relation label. Second, models recognize such correlations and use them for decisions. Our experiments indicate that people solve the relation classification task fundamentally differently. This model behavior is beneficial to increase metrics on benchmark datasets. This may result in wrong predictions in real-world applications.

Figure 9 shows examples giving intuition that models may encounter cases where sentences express different relations between identical argument texts. Our previous results indicate that obvious correlations between argument texts and certain relations may cause models to internalize specific shortcuts. If model decisions do not react to context changes in these cases, this causes vulnerabilities and counterintuitive decision processes.

Table 3. Performance and confidence of the RoBERTa base model on samples with different semantic classes.

	Confidence	F1 Micro	F1-Macro
Complete dataset	.88 ± .16	.82	.66
Only Arguments	.94 ± .13	.86	.69
Non-Only Arguments	.82 ± .16	.77	.57
All tokens in extent	.74 ± .16	.72	.60
Not all tokens in extent	.90 ± .14	.82	.65

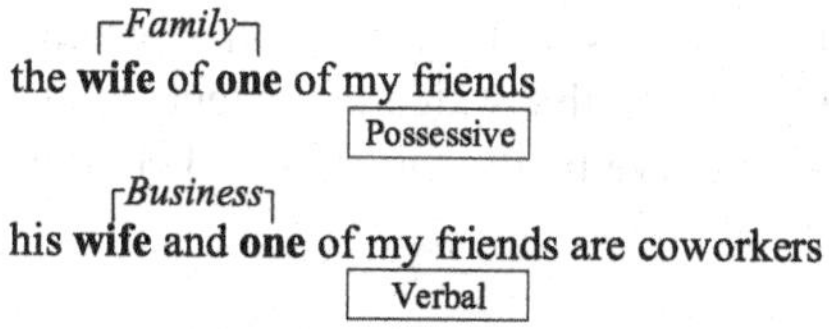

Fig. 9. Samples with identical argument text and different relation labels. Correct decisions on both samples require including context information.

Table 4. Evaluation on adversarial samples. We create multiple adversarial samples from each original sample. The model prediction is considered *correct* for an adversarial sample if the prediction changes from the original to the adversarial sample.

	Adv. *Only Arguments*	Adv. *Other*
# Arg Pairs	12	12
# Samples	120	120
Accuracy	.41 ± .31	.84 ± .20
Confidence	.88 ± .19	.85 ± .18

[11] shows that deep learning models have problems reacting appropriately to specific context changes. We transfer the concept to the relation classification task introduced in Subsect. 3.2. For a given sample, we keep the argument texts and automatically change the context to change the relation between arguments with Chat GPT[8]. We refer to these as *adversarial samples*. Table 4 shows how the RoBERTa base model performs on these samples. The first column refers to samples where the model predicts the final label from only the arguments with high confidence. For these samples, we suggest that models might ignore relevant context information. The second column refers to adversarial samples created from other instances. For these other instances, the model refers to the context for its decision.

The accuracy indicated the fraction of times when the adversarial adoptions lead to a changed model prediction. While the F1 score is higher on samples with semantic class *Only Arguments*, the model reacts to adversarial samples less often than samples with different semantic classes. Our results indicate that deep learning models learn correlations from the dataset. These correlations may be unintuitive for humans. This causes model decisions that differ from human decision patterns for relation classification.

5.4   Reductive Semantic Extents

The previous results show how comparing extensive semantic extents for models, and humans can reveal spurious decision patterns. Determining extensive semantic extents requires a task-specific prioritization of possible candidate expansions (Subsect. 5.1). Subsection 4.4 shows we can interpret input reductions as reductive semantic extents.

Our previous results require identifying samples where models focus mainly on the arguments and ignore the context. In theory, input reduction methods could reduce all input tokens from such samples and create similar results compared to extensive semantic extents. Figure 10 shows that reductive extents tend to contain more tokens than extensive extents. We observe hardly any cases where only the arguments remain in input reductions.

---

[8] https://openai.com/blog/chatgpt.

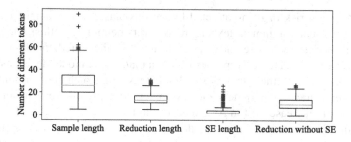

Fig. 10. Average number of tokens in samples extensive and reductive (input reductions) semantic extents.

6 Limitations

This research formalized the concept of an *explanation* for relation classification. With that, we are looking at a fundamental NLP task. The concept of *semantic extents* reveals latent decision patterns for this essential task. Real-world applications confront models with scenarios that are structural and conceptual more challenging. Our results demonstrate the fundamental applicability of semantic extents. Our methodology is a basis for further developments in explainable and trustworthy NLP.

We compared semantic extents for both humans and models. The comparison hints toward spurious behavioral patterns. To prove such suppositions, we applied a further analysis that includes the creation of adversarial samples. We applied Chat GPT to reduce the manual effort. The methodology presented is a tool to identify potential weaknesses. Eliminating these vulnerabilities requires a combination of semantic extents with different approaches.

Our results showed that a deep-learning model's focus on relation arguments might be too high. While our research uncovered this spurious decision pattern, there may be other spurious patterns that were not the focus of our study. One example might be various biases. Since many data sources describe the war in Iraq from an American perspective, the classification decisions might be biased if arguments indicate that people, locations, or organizations have certain geopolitical connections.

7 Conclusion

The novel concept of *semantic extents* formalizes explanations for the relation classification task. We defined a procedure to determine semantic extents for humans and deep learning models. A novel annotation tool supports human annotations, and a methodical framework enables the determination of semantic extents for NLP models.

We train and evaluate our models on the ACE 05 corpus. In contrast to most current applications of this dataset, we extensively use provided metadata to compare human and model decision patterns. Many previous approaches use preprocessing pipelines with no convenient access to this metadata. We implement a preprocessing pipeline on top of a uniform state-of-the-art python technology stack.

Datasets are at risk of being affected by various biases. For example, our results suggest that particular argument texts correlate overwhelmingly with specific relation labels. NLP models recognize such patterns and base their predictions on them. If these patterns do not relate to properties of human language, model decisions become counterintuitive. Our methodology is suitable for revealing such counterintuitive ties. We want to encourage practitioners and researchers to apply our concepts or develop methodologies to increase trust in AI and NLP solutions.

Novel developments like Chat GPT draw the public's attention to NLP and AI in general. This results in new opportunities to transfer scientific innovations into practical applications. We know many weaknesses of current deep learning models and must show ways to recognize and fix them in developed approaches. Semantic extents are an essential step towards explainable information extraction systems and an important cornerstone to increasing the acceptance of AI solutions.

References

1. Ayats, H., Cellier, P., Ferré, S.: A two-step approach for explainable relation extraction. In: Bouadi, T., Fromont, E., Hüllermeier, E. (eds.) IDA 2022. LNCS, vol. 13205, pp. 14–25. Springer, Cham (2022). https://doi.org/10.1007/978-3-031-01333-1_2
2. Bolukbasi, T., Chang, K.W., Zou, J., Saligrama, V., Kalai, A.: Man is to computer programmer as woman is to homemaker? Debiasing word embeddings (2016)
3. Brunet, M.E., Alkalay-Houlihan, C., Anderson, A., Zemel, R.: Understanding the origins of bias in word embeddings. ArXiv (2018)
4. Büsgen, A., Klöser, L., Kohl, P., Schmidts, O., Kraft, B., Zündorf, A.: Exploratory analysis of chat-based black market profiles with natural language processing. In: Proceedings of the 11th International Conference on Data Science, Technology and Applications, pp. 83–94. SCITEPRESS - Science and Technology Publications, Lisbon (2022). https://doi.org/10.5220/0011271400003269
5. Clark, K., Khandelwal, U., Levy, O., Manning, C.D.: What does BERT look at? An analysis of BERT's attention (2019). https://doi.org/10.48550/arXiv.1906.04341
6. Devlin, J., Chang, M.W., Lee, K., Toutanova, K.: BERT: pre-training of deep bidirectional transformers for language understanding. arXiv:1810.04805 [cs] (2019)
7. Doddington, G., Mitchell, A., Przybocki, M., Ramshaw, L., Strassel, S., Weischedel, R.: The automatic content extraction (ACE) program tasks, data, and evaluation. In: Proceedings of the Fourth International Conference on Language Resources and Evaluation (LREC 2004). European Language Resources Association (ELRA), Lissabon (2004)
8. D'Souza, J., Auer, S., Pedersen, T.: SemEval-2021 task 11: NLPContributionGraph - structuring scholarly NLP contributions for a research knowledge graph. In: Proceedings of the 15th International Workshop on Semantic Evaluation (SemEval-2021), pp. 364–376. Association for Computational Linguistics, Online (2021). https://doi.org/10.18653/v1/2021.semeval-1.44
9. Ebrahimi, J., Rao, A., Lowd, D., Dou, D.: HotFlip: white-box adversarial examples for text classification. arXiv:1712.06751 [cs] (2018)
10. Feng, S., Wallace, E., Ii, A.G., Iyyer, M., Rodriguez, P., Boyd-Graber, J.L.: Pathologies of neural models make interpretations difficult. Undefined (2018)
11. Gardner, M., et al.: Evaluating models' local decision boundaries via contrast sets. In: Findings of the Association for Computational Linguistics: EMNLP 2020, pp. 1307–1323. Association for Computational Linguistics, Online (2020). https://doi.org/10.18653/v1/2020.findings-emnlp.117

12. Klöser, L., Kohl, P., Kraft, B., Zündorf, A.: Multi-attribute relation extraction (MARE) - simplifying the application of relation extraction. In: Proceedings of the 2nd International Conference on Deep Learning Theory and Applications, pp. 148–156 (2021). https://doi.org/10.5220/0010559201480156

13. Li, B., et al.: Detecting gender bias in transformer-based models: a case study on BERT. ArXiv (2021)

14. Li, Q., Ji, H.: Incremental joint extraction of entity mentions and relations. In: Proceedings of the 52nd Annual Meeting of the Association for Computational Linguistics (Volume 1: Long Papers), pp. 402–412. Association for Computational Linguistics, Baltimore (2014). https://doi.org/10.3115/v1/P14-1038

15. McCoy, R.T., Pavlick, E., Linzen, T.: Right for the wrong reasons: diagnosing syntactic heuristics in natural language inference (2019)

16. Meng, K., Bau, D., Andonian, A., Belinkov, Y.: Locating and editing factual associations in GPT (2023)

17. Nissim, M., van Noord, R., van der Goot, R.: Fair is better than sensational: man is to doctor as woman is to doctor. Comput. Linguist. **46**(2), 487–497 (2020)

18. Ribeiro, M.T., Wu, T., Guestrin, C., Singh, S.: Beyond accuracy: behavioral testing of NLP models with CheckList (2020)

19. Schlangen, D.: Targeting the benchmark: on methodology in current natural language processing research. arXiv:2007.04792 [cs] (2020)

20. Shahbazi, H., Fern, X., Ghaeini, R., Tadepalli, P.: Relation extraction with explanation. In: Proceedings of the 58th Annual Meeting of the Association for Computational Linguistics, pp. 6488–6494. Association for Computational Linguistics, Online (2020). https://doi.org/10.18653/v1/2020.acl-main.579

21. Simonyan, K., Vedaldi, A., Zisserman, A.: Deep inside convolutional networks: visualising image classification models and saliency maps. CoRR (2013)

22. Smilkov, D., Thorat, N., Kim, B., Viégas, F., Wattenberg, M.: SmoothGrad: removing noise by adding noise. arXiv:1706.03825 [cs, stat] (2017)

23. Sundararajan, M., Taly, A., Yan, Q.: Axiomatic attribution for deep networks. arXiv:1703.01365 [cs] (2017)

24. Tenney, I., Das, D., Pavlick, E.: BERT rediscovers the classical NLP pipeline (2019)

25. Wadden, D., Wennberg, U., Luan, Y., Hajishirzi, H.: Entity, relation, and event extraction with contextualized span representations. In: Proceedings of the 2019 Conference on Empirical Methods in Natural Language Processing and the 9th International Joint Conference on Natural Language Processing (EMNLP-IJCNLP), pp. 5784–5789. Association for Computational Linguistics, Hong Kong (2019). https://doi.org/10.18653/v1/D19-1585, https://aclanthology.org/D19-1585

26. Wallace, E., Tuyls, J., Wang, J., Subramanian, S., Gardner, M., Singh, S.: AllenNLP interpret: a framework for explaining predictions of NLP models. arXiv:1909.09251 [cs] (2019)

27. Wang, A., et al.: SuperGLUE: a stickier benchmark for general-purpose language understanding systems (2020)

28. Wu, Z., Chen, Y., Kao, B., Liu, Q.: Perturbed masking: parameter-free probing for analyzing and interpreting BERT (2020). https://doi.org/10.18653/v1/P18-1198

29. Yamada, I., Asai, A., Shindo, H., Takeda, H., Matsumoto, Y.: LUKE: deep contextualized entity representations with entity-aware self-attention. In: Proceedings of the 2020 Conference on Empirical Methods in Natural Language Processing (EMNLP), pp. 6442–6454. Association for Computational Linguistics, Online (2020). https://doi.org/10.18653/v1/2020.emnlp-main.523

30. Zhang, B.H., Lemoine, B., Mitchell, M.: Mitigating unwanted biases with adversarial learning. In: Proceedings of the 2018 AAAI/ACM Conference on AI, Ethics, and Society, pp. 335–340 (2018). https://doi.org/10.1145/3278721.3278779

31. Zhang, T., Ji, H., Sil, A.: Joint entity and event extraction with generative adversarial imitation learning. Data Intell. 1(2), 99–120 (2019)
32. Zhao, J., Wang, T., Yatskar, M., Ordonez, V., Chang, K.W.: Men also like shopping: reducing gender bias amplification using corpus-level constraints. In: Proceedings of the 2017 Conference on Empirical Methods in Natural Language Processing, pp. 2979–2989. Association for Computational Linguistics, Copenhagen (2017). https://doi.org/10.18653/v1/D17-1323
33. Zhong, Q., et al.: Toward efficient language model pretraining and downstream adaptation via self-evolution: a case study on SuperGLUE (2022)

Phoneme-Based Multi-task Assessment of Affective Vocal Bursts

Tobias Hallmen[✉], Silvan Mertes, Dominik Schiller, Florian Lingenfelser, and Elisabeth André

Chair for Human -Centered Artificial Intelligence, University of Augsburg, Augsburg, Germany
{tobias.hallmen,silvan.mertes,dominik.schiller, florian.lingenfelser,elisabeth.andre}@uni-a.de
https://hcai.eu

Abstract. Affective speech analysis is an ongoing topic of research. A relatively new problem in this field is the analysis of affective vocal bursts, which are non-verbal vocalisations such as laughs or sighs. The current state of the art in the analysis of affective vocal bursts is predominantly based on wav2vec2 or HuBERT features. In this paper, we investigate the application of the wav2vec2 successor data2vec and the extension wav2vec2phoneme in combination with a multi-task learning pipeline to tackle different analysis problems at once, e.g., type of burst, country of origin, and conveyed emotion. Finally, we present an ablation study to validate our approach. We discovered that data2vec appears to be the best option if time and lightweightness are critical factors. On the other hand, wav2vec2phoneme is the most appropriate choice if overall performance is the primary criterion.

Keywords: data2vec · wav2vec2 · wav2vec2phoneme · Vocal bursts · Affective vocal bursts

1 Introduction

The human voice is a fundamental means of communication. While it can be used to produce spoken language, it can also carry an enormous amount of information on its own. Especially in the field of affect, non-verbal patterns are often even more important than linguistic content [21]. This becomes particularly apparent when listening to *vocal bursts*, which are short and intense vocalizations, often expressing strong emotions. The fact that vocal bursts can effectively communicate affective information without using verbal language makes them an interesting object of research. However, computational analysis of affective vocal bursts still remains a challenging topic [8,10,23,24]. As such, it is surprising that the current state-of-the-art approaches for affective vocal burst analysis rely on

© The Author(s), under exclusive license to Springer Nature Switzerland AG 2023
D. Conte et al. (Eds.): DeLTA 2023, CCIS 1875, pp. 209–222, 2023.
https://doi.org/10.1007/978-3-031-39059-3_14

wav2vec2 [5] or HuBERT [14] models that were trained on speech data, which has a substantially different structure than non-verbal vocal bursts. Therefore, in this paper, we examine the use of a successor and an extension of wav2vec2:

- First, we study if *data2vec* [4], a more generic version of wav2vec2, can be used to effectively infer various characteristics from vocal bursts.
- Second, we conduct various experiments using *wav2vec2phoneme* [28], which, instead of being trained on raw audio data, makes use of a phoneme vocabulary. As vocal bursts can, similar to speech, also be seen as a series of phonemes, we examine if using that intermediate representation additionally can improve automatic affect analysis pipelines.

We evaluate both architectures in a multi-task setting using HUME-VB dataset [9], a dataset of vocal bursts annotated regarding 5 different tasks. To further our understanding of how to build a successful system for analysing affective voice breaks, we subject our approach to an ablation study. Therefore, we investigate how different aspects of our training pipeline contribute to the performance of the analysis pipeline.

2 Related Work

Multi-task learning for vocal bursts recently became a popular research topic, partly because it was addressed in multiple conference challenges. E.g., in the *ExVo2022* challenge, participants were asked to predict the expression of 10 emotions along with the age and native country of the speaker at the same time [7]. [25] approached the task by experimenting with various encoder frontends as well as handcrafted features. They found that using the HuBERT model [14], which is closely related to the wav2vec architecture and training approach, as a backbone yielded the best performance. Purohit et al. [22] compared various embeddings that have been either trained using self-supervision or directly in a task-dependent manner. They found that overall, the self-supervised embeddings are outperforming the task-dependent ones, which, supports the choice of data2vec for our experiments. Anuchitanukul and Specia [1] also rely on wav2vec and HuBERT backbones to extract embeddings for their multi-task training system. They further utilise an adversarial training approach to disentangle the input representations into shared, task-specific ones. Their experiments showed that the wav2vec-based model performs best, but using ensemble techniques to combine multiple variations of their wav2vec and HuBERT models can achieve even higher performance.

Another challenge that addressed similar tasks was the *ACII-VB* challenge [6]. Here, participants of the challenge had to assess the type, valence/arousal, intensity of the emotion type, and the emotional type specific to certain countries. Again, the majority of contributions made use of either HuBERT or wav2vec2 models [2,3,11,15,19,26,27].

All those works indicate that self-supervision in general and wav2vec specifically are building a good foundation for the task at hand and confirming our choice of data2vec and wav2vec2phoneme as the successor of wav2vec.

3   Dataset

For our experiments, we utilised the HUME-VB dataset [9], which consists of emotional non-linguistic vocal bursts. Overall, there are roughly 37 h of recorded audio clips at a 16 kHz sampling rate spread over 59,201 files. The data has been recorded in 4 countries (China, South Africa, U.S. and Venezuela) representing different cultures, totalling in 1702 speakers with ages from 20 to 39 years. Each vocal burst was rated on average by 85 raters from the same country as the vocal burst's origin. For our experiments, we use the *Train* and *Validation* splits provided by the authors of dataset. The corpus provides multiple annotations for each sample that we use to train and evaluate our multi-task learning architecture:

- *High* refers to the intensity of 10 different emotions: *Awe, Excitement, Amusement, Awkwardness, Fear, Horror, Distress, Triumph, Sadness, Surprise.*
- *Country* labels inform about the origin of the person a vocal burst was recorded from.
- *Culture* labels provide the country-specific annotations of the 10 different emotions. As such, for each country, a 10 different emotion gold standard is given that was derived from annotators of the same country, resulting in $4 \cdot 10 = 40$ dimensions.

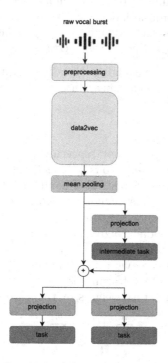

Fig. 1. Overview of the data2vec architecture.

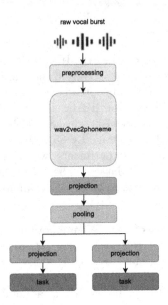

Fig. 2. Overview of the wav2vec2phoneme architecture.

- *Two* refers to the two-dimensional continuous emotion representation of the samples, i.e., valence/arousal labels.
- *Type* annotations are given to divide the samples into 8 different expression classes, i.e., *Gasp, Laugh, Cry, Scream, Grunt, Groan, Pant* and *Other*.

High, Culture, and *Two* are multi-label regressions with each label ranging from 0 to 1. *Type* and *Country* are classifications, having 8 respectively 4 classes.

4   Methodology

data2vec. Since vocal bursts are not "speech made out of words", but rather "speech made out of vocalised emotions", the not fine-tuned version of data2vec is used for our first experiments. Instead of handcrafted features which are often specifically engineered for spoken language, data2vec uses dataset-specific representations learnt in an unsupervised way.

To follow data2vec's modality-agnostic approach (i.e., it can be applied to either audio, video, or textual data) no or as few assumptions as possible are posed for the downstream supervised fine-tuning. To use all the provided labels of task $i = 1, \ldots, n$, multi-task learning is applied in a self-learning way [16], which approximates optimal task weights by the learning task uncertainty σ_i:

$$L = \sum_i^n \frac{1}{\sigma_i^2} L_i + \sum_i^n \log \sigma_i \tag{1}$$

In Fig. 1, the network architecture is depicted. The raw audio is fed into the pretrained data2vec model, including its preprocessor. Variable sequence lengths are zero-padded to the longest seen sequence. Because of the attention mask, the extracted features vary in length. Therefore, these are mean-pooled and fed into downstream projection layers. To investigate the question of whether knowing certain tasks before predicting other tasks is helpful, the intermediate tasks are separated and their prediction is fed along with the extracted features to the remaining tasks.

The projection layers reduce the output dimension of data2vec (768 base, 1024 large) to 256, apply GELU [12], and further reduce the dimension down to the five task's required dimensions (high: 10, culture: 40, two: 2, type: 8, country: 4). The tasks' layers then apply softmax for classification and compute the loss via cross-entropy using inversely proportional class weights. For regression, we apply sigmoid and compute the loss using the concordance correlation coefficient. The losses are then linearly combined through the learnt optimal uncertainty-based weights. This architecture serves as a starting point for the ablation study, where the design is iteratively improved.

wav2vec2phoneme. Using the insights drawn from our experiments with data2vec (see Sect. 5), the architecture was slightly altered for wav2vec2phoneme, as seen in Fig. 2 - the intermediate tasks were removed. With the projection of wav2vec2's 1024-dimensional feature vector down to a 392-dimensional phoneme vocabulary, the resulting sequence had to be aggregated to a fixed length for downstream tasks.

Training. The whole architecture has a size of 360 MB for the base and 1.2 GB for the large version of data2vec, and 1.4 GB for wav2vec2phoneme. They are trained in a two-stage matter. First, we freeze the net and train only the tasks with their projection layers. Second, we unfreeze the net and fine-tune the whole architecture. Both stages are trained for a maximum of 30 epochs with early stopping using a patience of 2 on the validation split to avoid overfitting on the training data. We use a batch size of 32 for the base version of data2vec, 24 for the large version, and 16 for wav2vec2phoneme. As optimiser, we used AdamW [18]. For the first stage, the optimiser was initialised with default parameters and a learning rate of $1 \cdot 10^{-3}$. For the second stage, the learning rate is set to $4 \cdot 10^{-5}$ and follows a cosine schedule with a warmup of 1 epoch.

The training runs for our experiments were done on a single Nvidia A40 GPU and took 5 to 6.5 h for the base version of data2vec and 5.5 to 8.5 h for the large version. For wav2vec2phoneme, training took between 10 to 13 h. Replacing the optimiser with ASAM [17], which we did for some experiments as described in Sect. 5, increased the training times by a factor of 2.5 to 3.5.

Table 1. Results for the experiment sets on *Intermediate Tasks* and *Task Loss*.

Method	High CCC	Culture CCC	Two CCC	Type UAR	Country UAR	All HMean
Baseline	.564	.436	**.499**	.412	–	.472
2/3	.639	.625	.252	.562	.631	.476
1/4	.628	.614	.245	.542	.603	.463
0/5	.650	.636	.254	.564	.633	.481
MSE	.624	.598	.244	.553	.591	.460
MAE	.640	.573	.251	.575	.655	.474
-High	–	.608	.244	.543	.600	.432
-Culture	.625	–	.255	.539	.599	.442
-Two	.651	.637	–	.574	.645	**.625**
-Type	.649	.634	.265	–	**.665**	.476
-Country	**.659**	**.642**	.266	**.577**	–	.467

Table 2. Results for the experiment sets on *Weighting, Fine-Tuning Splits* and *Network Size*.

Method	High CCC	Culture CCC	Type UAR	Country UAR	All HMean
-CW	.645	.631	.276	.632	.480
+SW	.643	.629	.319	.644	.511
-SM	.654	.639	**.584**	**.657**	**.632**
B10m	.656	**.642**	.577	.655	.631
B100h	.639	.624	.552	.634	.610
B960h	**.658**	**.642**	.567	.647	.626
L	.635	.622	.554	.603	.602
L10m	.641	.624	.555	.607	.605
L100h	.497	.601	.540	.578	.551
L960h	.613	.617	.540	.539	.575

5   Experiments and Results

data2vec. In our experiments, we varied certain details of the data2vec network architecture as listed below. The results for the validation set are shown in Tables 1 and 2, additionally to the official baseline results that were published by the authors of the dataset [6]. Note that the single experiments were conducted iteratively, carrying the best configuration of the previous set of experiments over to the next set of experiments. Since we use a multi-task approach, comparing different approaches can be difficult if not all tasks perform better or worse than in the experiment being compared to. Therefore, we calculate the harmonic mean of the tasks' metrics to provide an overall comparison. The results of the first set of experiments are listed in Table 1.

Intermediate Tasks. Following the dataset's motivation that vocal bursts depend on the country of origin to assess the conveyed emotions like a rater of the same origin would, we tried three different variants: (i) *Country* and *Type* as intermediate tasks before predicting the remaining three (*2/3*). (ii) Only *Country* as an intermediate task (*1/4*), assuming the type depends less on the country. (iii) No intermediate tasks, i.e., predicting all tasks simultaneously (*0/5*). (iii) performed best.

Task Loss. To investigate why task *Two* performed so poorly, we experimented with replacing the CCC loss by (i) mean squared error (*MSE*) and (ii) mean absolute error (*MAE*). Further, to investigate if there is an issue with a single task, we removed each task from training once (iii) - (vii) (e.g. *-Task*, *-High*). (v), i.e., removing task *Two*, turned out to be the best option. As removing that task improves the harmonic mean for the other tasks by large margins, we decided to drop it for the following experiments. Results of the second set of experiments are reported in Table 2.

Weighting. Since the cross-entropy losses in tasks *Country* and *Type* have inverse proportional class weights while the other tasks do not, we experimented with (i) removing these class weights from the training (*-CW*) so that every sample's tasks are unweighted, (ii) adding sample weights inversely proportional to *Country* (*+SW*), thereby weighting all tasks, and (iii) keeping the class weights but removing the sigmoid activation from the last layer (*-SM*) and clamping the linear output to $[0, 1]$ in the regression tasks *High* and *Culture* instead. We could marginally boost the network's performance by applying (iii).

Fine-Tuning Splits. To investigate if fine-tuning the self-supervised learnt audio representations using labeled speech improves the pipeline's performance, we experimented with different versions of the base network. Each version was fine-tuned on word labels after pretraining using connectionist temporal classification (CTC) loss on a different amount of LibriSpeech [20] data: (i) 10 min (*B10m*), (ii) 100 h (*B100h*), (iii) 960 h (*B960h*). Version (i) performed best overall.

Network Size. At last, we replaced the base network with (i) the large version (*L*) and (ii) the respective fine-tuning splits (*L10m*, *L100h*, *L960h*). None of the large versions could outperform the former base version experiments. For the tasks *High* and *Culture*, experiment *B960h* worked best on the validation set. For the tasks *Type* and *Country*, experiment *-SM* performed best on the validation set.

Table 3. Results for the experiment sets on *Aggregation, Loss adjustments, Features, Combination, Optimiser*, and *Loss Revision*.

Method	High CCC	Culture CCC	Type UAR	Country UAR	All HMean
LSTM-1	.549	.558	.468	.253	.412
Mean-1	.665	.652	.574	.660	.635
Count	.053	.055	.125	.250	.082
Regularisation	.041	.039	.125	.250	.064
Weighting	.267	.160	.166	.250	.200
LSTM-2	.673	.656	.580	.562	.614
Mean-2	.327	.642	.581	**.703**	.516
Separate	.670	.638	.595	.688	.646
Concat-1	.675	.650	.596	.689	**.650**
Mean-3	.664	.649	.583	.696	.645
ASAM-1	**.683**	**.667**	**.602**	.624	.642
ASAM-2	.673	.645	.588	.699	.649
DRUW-1	.570	.643	.584	.692	.619
DRUW-2	.680	**.667**	.593	.561	.621
DRUW-3	.666	.650	.562	.675	.635

wav2vec2phoneme. The insights on data2vec so far are:

- No intermediate tasks are needed.
- Removing task *Two* greatly benefits the other tasks.
- Removing the sigmoid activation in the last layer improves the performance.
- Using larger versions of the model does not improve performance.
- No fine-tuning on word labels is needed.

With those findings as a starting point, we investigated if using another vocabulary than words made out of letters to describe vocal bursts improves the performance. For this, we use wav2vec2phoneme, which transcribes audio using a phoneme vocabulary. Intuitively, it is conceivable that this allows for a better vocal burst description. The results of the following experiments are reported in Table 3.

Aggregation. The next set of experiments targets the aggregation of the varying-length sequence of the transcribed 392-dimensional phoneme vocabulary. We did this by (i) using a bi-directional two-layer LSTM [13] with a feature size of 768 (*LSTM-1*), the same feature size as data2vec's base version. Further, we (ii) applied mean pooling (*Mean-1*), and (iii) simply counted the occurrences of each phoneme (*Count*). (ii) performed best. The experiments also showed that experiment (iii), by aiming to learn constant predictions, artificially induced the loss to decrease – the net converges towards chance level.

Loss Adjustments. (i) To avoid the regularisational term from becoming negative, a lower bound was introduced to the uncertainties in Eq. 1 (*Regularisation*):

$$L = \sum_i^n \frac{1}{\sigma_i^2} L_i + \sum_i^n \log(1 + \sigma_i) \qquad (2)$$

(ii) additionally to the above adjustment, the loss is weighted by the sequence length (*Weighting*). Both experiments did not improve the results. As such, the former best configuration remains the starting point for the next set of experiments.

Features. To investigate if the transcription to human-readable 392 different phonemes or the preceding 1024-dimensional layer should be passed on, the latter is (i) fed to a bi-directional two-layer LSTM with a hidden size of 1024 (*LSTM-2*), and (ii) mean pooled (*Mean-2*). While for both the harmonic mean decreased compared to the former best, the LSTM improved the regression, while mean pooling improves the classification tasks.

Combination. In these experiments, we tried to find a combination of LSTM and mean pooling to improve the harmonic mean, thereby overall performance, partly at the cost of a higher dimension. In order to do so, we (i) separately used the LSTM features for regression and the mean-pooled features for classification (*Separate*). In (ii), both were concatenated (*Concat-1*), and last, in (iii), they were averaged to keep the dimension the same (*Mean-3*). (ii) performed best.

Optimiser. To investigate if the loss issues are caused by the optimiser, it is extended by applying adaptive sharpness-aware minimisation (ASAM) [17]. For parameter ρ, we experimented with (i) $\rho = .5$ (*ASAM-1*) and (ii) $\rho = .05$ (*ASAM-2*). Both slightly reduced the harmonic mean – they decrease some tasks' performance and increase the others'. As such, no overall improvement could be observed.

Loss Revision. By applying dynamic restrained uncertainty weighting (DRUW) [15], we tried to tackle the loss issues through further adjustments to Eq. 2:

$$L = \sum_i^n (\frac{1}{\sigma_i^2} + \lambda_i) L_i + \sum_i^n \log(1 + \log \sigma_i^2)$$
$$+ |\phi - \sum_i^n |\log \sigma_i|| \qquad (3)$$

with the dynamic weights λ_i being:

$$\lambda_i = n \frac{\exp(\tau \frac{L_{i,t-1}}{L_{i,t-2}})}{\sum_i^n \exp(\tau \frac{L_{i,t-1}}{L_{i,t-2}})} \qquad (4)$$

Table 4. Transfer of best setup *Concat-1* to other wav2vec2-architectures. Upper half ex- and lower half includes task *Two* while applying the same configuration.

Method	High CCC	Culture CCC	Two CCC	Type UAR	Country UAR	All HMean
w2v2-B	.642	.630	-	.563	.661	.622
w2v2-B960h	.293	.285	-	.125	.250	.211
w2v2-L	.651	.642	-	.579	**.711**	.642
w2v2-L960h	.366	.636	-	.584	.636	.527
Concat-1	**.675**	**.650**	-	.596	.689	**.650**
w2v2-B	.652	.636	.137	.581	.669	.367
w2v2-B960h	.198	.273	.232	.125	.250	.200
w2v2-L	.665	**.650**	**.263**	.588	.706	.502
w2v2-L960h	.138	.163	.137	.493	.592	.205
Concat-2	.669	.563	.255	.593	.695	.485

Using the configuration parameters proposed by [15], i.e., $\tau = 1$ for temperature and $\phi = 1$ as regularisation constant, DRUW is applied to *Concat-1* and both *ASAM-1/2* experiments, resulting in experiments (i) - (iii) (*DRUW-1/2/3*). No overall performance improvement was observed. However, (ii) managed to maintain the performance for task *Culture*.

Transfer to wav2vec2. To investigate if phonemes are really more suited than words, the experiments that worked best were transferred to the base and large versions of wav2vec2 (*w2v2-B, w2v2-L*). The same modifications were applied to the respective fine-tuned models (*w2v2-B960h, w2v2-L960h*). Additionally, all of those four experiments were run with the inclusion of task *Two* in order to validate if the aforementioned negative interferences of that task still occur for the wav2vec2 model. Results are shown in Table 4. *Concat-1* remains the overall best choice.

6   Discussion

Recapitulating the conducted experiments and considering the results, the following insights can be drawn:

Intermediate Tasks. Determining the country of origin before assessing the conveyed emotions like a rater of that country would may be beneficial to a human rater in order to detect and adjust emotional biases. However, it is disadvantageous for our pipeline – the extracted features already encompass these biases and need not be handcrafted into.

Task Loss. Revising the task losses, adding further regularisational terms and applying a sharpness-aware optimiser did not improve the poor performance on task *Two*. Since the baseline shows double the performance here, neither of the self-supervised learnt audio or word-/phoneme-based representations are suited for estimating valence and arousal in these architectures. Therefore, the model learns to predict a rather constant output for *Two*. As such, the uncertainty in this task is artificially reduced, minimising the penalising uncertainty term, but also maximising the task weight in the computation of the MTL loss. The weight can become so large that it substantially degrades the other tasks' performance. Our experiments have shown that excluding task *Two* greatly benefits the assessment of the remaining tasks.

Weighting. These experiments investigated different weighting techniques to counter imbalance in the training data. Removing the inversely proportional class weights in the calculation of the cross-entropy losses greatly reduces the performance in *Type*. Extending the cross-entropy weights over the whole sample to inversely proportional intra-batch weights depending on *Country* alleviates this only slightly, despite having the annotations made from raters of the same country as the vocal burst's utterer. Inversely proportional class weights both in *Type* and *Country* to counter class imbalance combined with the removal of sigmoid activation, is the best approach, as it increases the net's sensitivity to samples close to the boundaries of the value ranges.

Fine-Tuning Splits. Here, we examined the assumption that vocal bursts are not a word-based "language". As such, we observed a decline in performance when fine-tuning the pretrained model using CTC on (English) word-based labels. Consequently, this decline is only slight when using 10 min of fine-tuning data, but more so for 100 h. More than 50% performance degradations are visible when using all 960 h of data for fine-tuning. Therefore, sticking with the pretrained network, or using only sensible fine-tuning, i.e. phonemes, seems to be the best option.

Feature Summarisation. When processing a sequence of phonemes, i.e. listening, one intuitively expects a recurrent neural network (RNN) to be best suited, since those type of architectures consider chronology. However, *LSTM-1* and *Mean-1* showed that simply computing the distribution of phonemes, regardless of the time of occurence, outperforms a RNN. Since *Count* performed poorly even with loss adjustments, it does not matter how often (or how long) different phonemes were uttered - only the proportion is relevant. Interestingly, when dropping the human-readable phoneme transcription and directly using wav2vec2phoneme's features, *LSTM-2* and *Mean-2* show an equal overall performance. However, each of them were better in either both regression or both classification tasks. Therefore, providing both summarisations by concatenating them (*Concat-1*) leads to the best overall performance.

Feature Representations. In these last experiments, we evaluated if our initial assumption, e.g., a phoneme-based feature representation, indeed has the ability to outperform the more traditional wav2vec2 approach. Therefore, we used the configuration that worked best in the preceding experiments and applied them to different versions of wav2vec2. The observation that the performance of neither data2vec nor wav2vec2phoneme could be matched supports our claim that using a phoneme-based feature representation can be a valid choice for the task at hand.

7   Conclusion

In this work, we have shown that a single network for multi-task affective vocal burst assessment is a valid choice. Per-task ensembling and large structures are not necessarily needed. Task weighting can be done automatically via task uncertainty approximation. Although being pretrained on English-only speech in a self-supervised manner, data2vec is able to assess vocal bursts originating out of different (non-English) countries after fine-tuning it for fiveish hours. Furthermore, by substituting the data2vec architecture with wav2vec2phoneme, a larger and phoneme-based net, we could further boost the pipeline's performance, while only doubling the required time for training. If time and lightweightness are of essence, data2vec seems to be the better choice. If overall performance is the most important criterion, wav2vec2phoneme fits best. Applying a sharpness-aware optimiser can yield even better results for specific subtasks, but comes with the cost of a decreased overall performance. By comparing our best configuration to a word-based wav2vec2, we can conclude that phonemes are better suited for the assessment of affective vocal bursts than words.

Acknowledgements. This work was partially funded by the KodiLL project (FBM2020, Stiftung Innovation in der Hochschullehre), project TherapAI (DFG, German Research Foundation, grant number 493169211) and project Panorama (DFG, German Research Foundation, grant number 442607480).

References

1. Anuchitanukul, A., Specia, L.: Burst2vec: an adversarial multi-task approach for predicting emotion, age, and origin from vocal bursts. arXiv preprint arXiv:2206.12469 (2022)
2. Atmaja, B.T., Sasou, A.: Predicting affective vocal bursts with finetuned wav2vec 2.0. arXiv preprint arXiv:2209.13146 (2022)
3. Atmaja, B.T., Sasou, A., et al.: Jointly predicting emotion, age, and country using pre-trained acoustic embedding. arXiv preprint arXiv:2207.10333 (2022)
4. Baevski, A., Hsu, W.N., Xu, Q., Babu, A., Gu, J., Auli, M.: Data2vec: a general framework for self-supervised learning in speech, vision and language. arXiv preprint arXiv:2202.03555 (2022)
5. Baevski, A., Zhou, Y., Mohamed, A., Auli, M.: wav2vec 2.0: a framework for self-supervised learning of speech representations. In: Advances in Neural Information Processing Systems vol. 33, pp. 12449–12460 (2020)

6. Baird, A., Tzirakis, P., Batliner, A., Schuller, B., Keltner, D., Cowen, A.: The ACII 2022 affective vocal bursts workshop and competition: Understanding a critically understudied modality of emotional expression. arXiv preprint arXiv:2207.03572v1 (2022). https://doi.org/10.48550/arXiv.2207.03572

7. Baird, A., et al.: The ICML 2022 expressive vocalizations workshop and competition: Recognizing, generating, and personalizing vocal bursts. arXiv preprint arXiv:2205.01780v3 (2022). https://doi.org/10.48550/ARXIV.2205.01780

8. Cordaro, D.T., Keltner, D., Tshering, S., Wangchuk, D., Flynn, L.M.: The voice conveys emotion in ten globalized cultures and one remote village in Bhutan. Emotion **16**(1), 117 (2016)

9. Cowen, A., et al.: The Hume vocal burst competition dataset (H-VB) — raw data [exvo: updated 02.28.22] [data set]. Zenodo (2022). https://doi.org/10.5281/zenodo.6308780

10. Cowen, A.S., Elfenbein, H.A., Laukka, P., Keltner, D.: Mapping 24 emotions conveyed by brief human vocalization. Am. Psychol. **74**(6), 698 (2019)

11. Hallmen, T., Mertes, S., Schiller, D., André, E.: An efficient multitask learning architecture for affective vocal burst analysis (2022)

12. Hendrycks, D., Gimpel, K.: Gaussian error linear units (GELUs). arXiv preprint arXiv:1606.08415 (2016)

13. Hochreiter, S., Schmidhuber, J.: Long short-term memory. Neural Comput. **9**(8), 1735–1780 (1997)

14. Hsu, W.N., Bolte, B., Tsai, Y.H.H., Lakhotia, K., Salakhutdinov, R., Mohamed, A.: HuBERT: self-supervised speech representation learning by masked prediction of hidden units. IEEE/ACM Trans. Audio Speech Lang. Proc. **29**, 3451–3460 (2021)

15. Karas, V., Triantafyllopoulos, A., Song, M., Schuller, B.W.: Self-supervised attention networks and uncertainty loss weighting for multi-task emotion recognition on vocal bursts. arXiv preprint arXiv:2209.07384 (2022)

16. Kendall, A., Gal, Y., Cipolla, R.: Multi-task learning using uncertainty to weigh losses for scene geometry and semantics. In: Proceedings of the IEEE Conference on Computer Vision and Pattern Recognition, pp. 7482–7491 (2018)

17. Kwon, J., Kim, J., Park, H., Choi, I.K.: ASAM: adaptive sharpness-aware minimization for scale-invariant learning of deep neural networks. In: International Conference on Machine Learning, pp. 5905–5914. PMLR (2021)

18. Loshchilov, I., Hutter, F.: Decoupled weight decay regularization. arXiv preprint arXiv:1711.05101 (2017)

19. Nguyen, D.K., Pant, S., Ho, N.H., Lee, G.S., Kim, S.H., Yang, H.J.: Fine-tuning wav2vec for vocal-burst emotion recognition. arXiv preprint arXiv:2210.00263 (2022)

20. Panayotov, V., Chen, G., Povey, D., Khudanpur, S.: LibriSpeech: an ASR corpus based on public domain audio books. In: 2015 IEEE International Conference on Acoustics, Speech and Signal Processing (ICASSP), pp. 5206–5210. IEEE (2015)

21. Phutela, D.: The importance of non-verbal communication. IUP J. Soft Skills **9**(4), 43 (2015)

22. Purohit, T., Mahmoud, I.B., Vlasenko, B., Doss, M.M.: Comparing supervised and self-supervised embedding for exvo multi-task learning track. arXiv preprint arXiv:2206.11968 (2022)

23. Scherer, K.R.: Expression of emotion in voice and music. J. Voice **9**(3), 235–248 (1995)

24. Schröder, M.: Experimental study of affect bursts. Speech Commun. **40**(1–2), 99–116 (2003)

25. Sharma, R., Vuong, T., Lindsey, M., Dhamyal, H., Singh, R., Raj, B.: Self-supervision and learnable STRFs for age, emotion, and country prediction. arXiv preprint arXiv:2206.12568 (2022)
26. Syed, M.S.S., Syed, Z.S., Syed, A.: Classification of vocal bursts for ACII 2022 A-VB-Type competition using convolutional network networks and deep acoustic embeddings. arXiv preprint arXiv:2209.14842 (2022)
27. Trinh, D.L., Vo, M.C., Kim, S.H., Yang, H.J., Lee, G.S.: Self-relation attention and temporal awareness for emotion recognition via vocal burst. Sensors **23**(1), 200 (2022)
28. Xu, Q., Baevski, A., Auli, M.: Simple and effective zero-shot cross-lingual phoneme recognition. arXiv preprint arXiv:2109.11680 (2021)

Using Artificial Intelligence to Reduce the Risk of Transfusion Hemolytic Reactions

Maya Trutschl[1], Urska Cvek[2(✉)], and Marjan Trutschl[2]

[1] Caddo Parish Magnet High, 1601 Viking Dr, Shreveport, LA 71101, USA
[2] Louisiana State University Shreveport, 1 University Pl, Shreveport, LA 71115, USA
ucvek@lsus.edu

Abstract. The monocyte monolayer assay is a cellular assay, an in-vitro procedure that mimics extravascular hemolysis. It was developed to predict the clinical significance of red blood cell antibodies in transfusion candidates with intent to determine whether the patient needs to receive the expensive, rare, antigen-negative blood to avoid an acute hemolytic transfusion reaction that could lead to death. The assay requires a highly trained technician to spend several hours evaluating a minimum of 3,200 monocytes on a glass slide under a microscope in a cumbersome process of repetitive counting. Using the YOLO neural network model, we automate the process of identifying and categorizing monocytes from slide images, a significant improvement over the manual counting method. With this technology, blood bank technicians can save time and effort while increasing accuracy in the evaluation of blood transfusion candidates, leading to faster and better medical diagnosis. The trained model was integrated into an application that can locate, identify, and categorize monocytes, separating them from the background and noise on the images acquired by an optical microscope camera. Experiments involving a real-world data set demonstrate that F1-score, mAP scores, precision and recall are above 90%, indicating that this workflow can ease and accelerate the medical laboratory technician's repetitive, cumbersome, and error-prone counting process, and therefore contributes to the accuracy of medical diagnosis systems.

Keywords: Machine learning · Object detection · Classification · Health informatics · Blood · Decision support · Medical diagnosis system

1 Introduction

Red blood cell (RBC) transfusion remains an important medical procedure for rapid increase of hemoglobins in patients to ensure sufficient oxygenation levels in the blood. This procedure is commonly documented for patients with trauma, surgical complications, pregnancy, cardiovascular diseases, and autoimmune hemolytic anemias. Hemolytic transfusion reactions are caused by antibodies produced by the recipient's host immune response against transfused RBCs and in some cases can result in life-threatening reactions.

Monocytes are agranulocytes that are produced in the bone marrow and travel to the organs in the body. When they enter the connective tissue, they differentiate into

D. Conte et al. (Eds.): DeLTA 2023, CCIS 1875, pp. 223–234, 2023.
https://doi.org/10.1007/978-3-031-39059-3_15

phagocytic and antigen-presenting cells, which have major roles in clearing apoptotic cells, removing debris, ingesting pathogens, and mediating immune responses. When a patient needs a blood transfusion, pre-transfusion testing requires blood typing, antibody screening, and crossmatching in the laboratory before the blood product can be dispensed from the blood bank. Antibody reactions can cause decreased (hemolysis) or shortened survival of transfused RBCs carrying the corresponding antigen (i.e., acute or delayed hemolytic transfusion reaction). Thus, the recipient's plasma is tested for the presence of unexpected antibodies before a RBC transfusion. This is of particular importance if the intended recipient has autoantibodies [7]. If a patient has a clinically significant antibody, the transfusion service selects and reserves the appropriate red cell components that do not carry the corresponding antigen.

1.1 Background

Identifying the patient's red cell antibodies and crossmatching them with the appropriate red cell components can take hours or even days, depending on the antibody or antibodies found. The Monocyte Monolayer Assay (MMA) is a laboratory test [8, 15] used to predict the risk of a clinically significant reaction at the time of blood transfusion, and results in the calculation of the Monocyte Index (MI).

The MMA is a lengthy, tedious, and subjective evaluation process that takes a trained technician several hours to complete as they use the patient's serum or plasma, fresh monocytes (from a volunteer), and setup tests with antigen positive and antigen negative RBCs. The test is loaded onto an eight-chamber glass slide, 25 mm × 75 mm in size, with $0.98\,cm^2$ working volume per chamber. The glass slide is placed under a microscope in order to identify individual monocytes and classify if they are free, ingested, or adhered [3]. At least 400 monocytes have to be counted in each of the eight chambers, thus counting and classifying at least 3,200 monocytes using manual hand counters while looking through the microsope. This process is not only time consuming and costly, but it is also complex due to the variations in light, location, and overlap of the monocytes on the slide.

After the monocytes are located and predicted either as free or ingested and adhered, we can calculate the MI based on the formula in Eq. 1. The formula for the calculation of MI makes a distinction of the two cutoffs that were confirmed based on retrospective data from twenty years of studies [3]. A negative MMA with MI at or below 5% indicates that incompatible blood can be given without risk of a hemolytic transfusion reaction.

$$MI = \frac{\text{Count of ingested/adhered monocytes}}{\text{Count of all monocytes}} \times 100 \qquad (1)$$

Although artificial intelligence and machine learning algorithms represent an intense area of research in radiology and pathology for automated or computer-aided diagnosis, MMA and MI automation has only been proposed once recently [14], but not as a viable computer-based solution due to its performance.

In this paper we offer a novel medical diagnosis system approach with which we detect, locates, and classify monocytes based on the YOLO algorithm. Section 2 provides a short overview of previous work in machine learning as related to monocyte detection and classification. Section 3 explains the proposed solution in detail. Section 4 presents the results of experiments conducted to test the efficacy of the model, and Sect. 5 contails our conclusions.

The result is an integrated solution that results in MI calculation, which can help the blood bank quickly and efficiently evaluate Monocyte Monolayer Assays, resulting in an improved medical diagnosis system. Section 2 will show previous related work. Section 3 gives a detailed explanation of our proposed solution. Section 4 provides results of our experiments and Sect. 5 contains our conclusions.

2 Related Work

Machine learning and deep neural networks have significantly improved in accuracy and speed in the past few years. Target detection algorithms in deep learning include one and two-stage algorithms that have been designed and applied to target counting, tracking, and other upper-layer applications. Although improved versions of R-CNN [11], one of the first and most commonly utilized algorithms were proposed to speed up the inference, two-stage architectures limit the processing speed. Unlike the R-CNN, YOLO reduces processing time by performing classification and bounding box regression at the same time, in subsequent further improvements through different versions of the YOLO algorithm.

YOLOv5 is a family of compound-scaled object detection models trained on the Common Objects in Context (COCO) data set, which is a large-scale object detection, segmentation, and captioning data set [13]. A network trained with the YOLO algorithm identifies objects in a given image. Its output consists of bounding boxes around the detected objects, the class of each detected object together with the individual item's confidence score. YOLOv5 is one of the fastest and most accurate object detectors based on the YOLO paradigm, based on Microsoft COCO [1], a commonly used, general-purpose object detection dataset.

3 Methodology

In this paper we develop a deep learning algorithm for monocyte detection, localization, and classification based on the YOLO [17] object detection. "You only look once" version 5 or YOLOv5 algorithm [17] from Ultralytics' GitHub was selected after researching and comparing the performance of several algorithms. Models were trained repeatedly using Jupyter Notebook with code written in Python in the Google Colab [6] environment. Roboflow [10] was used for organization, curation, pre-processing, and data augmentation of the images. Tensorflow [2] and Weights and Biases [5] were used to track hyperparameters, system metrics and performance evaluation.

3.1 Dataset Description and Methodology

The original 109 data files used to train and test the model were collected at LifeShare Blood Center in Shreveport. They were captured using a ZEISS Axiocam 208 color/202 mono microscope camera and stored in JPEG format at resolution of 3840 × 2160 pixels. Annotation files include 109 files in JavaScript Object Notation (JSON) format for a one-to-one correspondence with the images. The annotation was done manually to include the coordinates for each of the monocytes in an individual image, outlining a polygonal region of interest (ROI). Each monocyte is also labeled as one of two classes; Free (labeled as FRMonocyte) or Ingested/Adhered (labeled as IAMonocyte), which is the information required for the calculation of MI. Additional 12 data files were acquired using the same microscope and were used only in the experimental demonstration phase, as they were not annotated.

Figure 1 shows two examples of original images with an overlay of the annotations, which localize the monocyte positions, as well as determine their class. Only binary classes are utilized, and every monocyte is classified as *Free* or *Ingested or Adhered.*

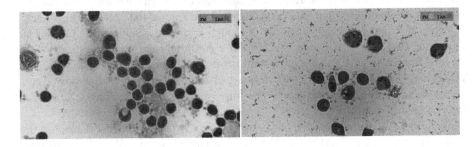

Fig. 1. Polygonal annotations of monocytes in representative original images. The polygonal objects were drawn manually, and green outlines mark the *Free* monocytes (FM) and red outlines mark the *Ingested/Adhered* monocytes (IAM). (Color figure online)

3.2 Image Preprocessing

Original 109 images were imported into Roboflow [10], which was used for organization, curation, and processing of images. Hold-out method was used for processing of the data into random sets for training, validation, and testing at 70-20-10 distribution, respectively, resulting in 76 images for training, 22 images for validation, and 11 images for testing. This process was repeated in order to train the model and fine tune the hyper-parameters.

Each of the images used for model optimization and prediction was resized in Roboflow framework to 640 × 640 pixels. Since the width and height of the original images are not the same, rescaling was performed with aspect ratio of 6:1 along the width, and 3.375:1 along the height of the image. Additionally, each of the ROI regions was reshaped from a polygon to a rectangle by identifying the min_x, max_x, min_y, and max_y of the bounding rectangle. To match the YOLOv5 data format, the coordinates of the ROI rectangle were reformatted to a "*class x_{center} y_{center} width height*" string. Furthermore, the values of x_{center}, y_{center}, width, and height were normalized between 0.0

and 1.0 concerning the width and height of the image, allowing the image to be resized independently of the ROI annotation.

Figure 2 shows the samples of augmented images with an overlay, representing new bounding box annotations in YAML format, which is required for YOLOv5. Yellow boxes outline the free monocytes, while the cyan boxes outline the ingested or adhered monocytes. The algorithm identifies the monocytes and is able to ignore the background noise in these images.

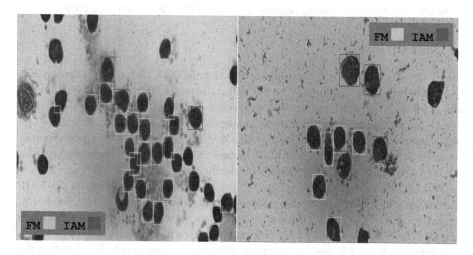

Fig. 2. Rectangular annotations of monocytes in the augmented sample images. Rectangular bounding boxes were drawn to replace the polygons, and yellow outlines the *Free* monocytes (FM) and magenta outlines mark the *Ingested/Adhered* monocytes (IAM). (Color figure online)

Data augmentation can artificially expand the data set, increase the diversity of the data, and improve the robustness of the model [9]. Augmentation was applied to the training images in order to increase the diversity of the learning examples for the model. Each of the training images received up to 3 random augmentations, such as flipping horizontally or vertically, rotating 90° clockwise, counter-clockwise, or upside down. Each time the random selection of the 70-20-10 distribution was applied, 76 training images were augmented, resulting in 216 images that were used for training. This is in addition to the 22 images that were used for validation, and 11 for testing.

3.3   Development Environment

Eighty-class COCO data set was used as a baseline and transfer learning was applied to our two-class data set. Models were trained repeatedly using Jupyter Notebook interactive computing platform with code written in Python programming language in the Google Colab environment [6]. The YOLOv5s (small model) and YOLOv5m (large model) were ran and hyperparameters were adjusted repeatedly until the best performance was achieved on each. YAML [4] was used as the format for specifying the configurations and file locations for the model.

Other tools imported into Jupyter included Pytorch [16], an open-source machine learning framework, and NumPy [12], a framework package that was used for its support for large arrays, matrices, and mathematical functions. Tensorflow [2], an end-to-end machine learning platform, and Tensorboard, a visualization, and tooling for model performance evaluation, utilized together with Weights and Biases [5] as a central dashboard for keeping track of all hyperparameters, system metrics, and predictions.

3.4 Model Training

Image sets were used to train, validate, and test the YOLOv5 model (Version 6.1, released February 22, 2022), varying the training between YOLOv5s and YOLOv5m, which are based on different network structures due to different widths and depths. YOLOv5s, which has the smallest width and depth, but the fastest speed, was tested first and compared to YOLOv5m, with resulting models compared for their performance. Since the speed of YOLOv5s was much greater at the negligibly lower performance than YOLOv5m, YOLOv5s was adopted as the model of choice.

YOLOv5s was trained repeatedly on random subsets of the images at 70-20-10 distribution. The hyperparameters were adjusted, including the learning rate, batch size, image size, and epochs (iterations of the training process with a randomized set of batch images). The final model used a learning rate of 0.01, a batch size of 16, a 640×640-pixel image size, and it was executed for 200 epochs. One epoch is defined as the number of training iterations in which the model has completed a full pass of the whole training set. The choice of 200 epochs was made based on the empirical observation that in all of these experiments the learning converged well. Google Colab [6] utilized a Tesla T4 Graphic Processing Unit (GPU) with 40 processors and trained for 17 min and 32 s in a shared environment. The model's best prediction was achieved at epoch 189 out of 200 epochs (0 through 199).

The best performing model from epoch 189 was adopted as the model to identify free and ingested/adhered monocytes and corresponding confidence values and locations were marked in the images. This final model was loaded onto a Raspberry Pi and a small script written to allow connections through a web interface. New images acquired through the microscope can be uploaded to the model and a list of predictions of monocyte locations, followed by confidence levels and class predictions (free or ingested/adhered monocyte) is returned, together with the calculated MI.

4   Results and Discussion

We developed a deep learning algorithm for monocyte detection, localization, and classification. The best prediction on the training data was achieved during epoch 189. The model detected 273 monocytes on one set of 22 sample validation images. Furthermore, out of those 273 monocytes, 214 were free and 59 were ingested/adhered. Figure 3 shows the summary of the best model's performance metrics on the training and validation data sets (epoch 189). Precision across both classes (free and ingested/adhered) was 0.942, which means that 94.2% of the monocyte class predictions were correctly detected. The recall or sensitivity was measured at 0.977, which means that 97.7% of the true positives in the two classes of all positives were correct.

Class	Images	Labels	P	R	mAP@.5	mAP@.5:.95
all	22	273	0.942	0.977	0.989	0.841
FRMonocyte	22	214	0.951	0.987	0.989	0.840
IAMonocyte	22	59	0.934	0.966	0.989	0.842

Fig. 3. Detailed architecture of the YOLOv5s model.

As shown in Fig. 4, objectness loss on the training and validation sets was 0.04, with classification loss at 0.004 and 0.002, respectively. Box loss was 0.02 in both of the sets. The model improved swiftly in terms of precision, recall, and mean average precision, before plateauing after about 120 epochs. The box, objectness, and classification losses of the validation and training data showed a rapid decline until around epoch 50. The mAP@.5 precision for both classes was at 0.989 and mAP@.5:.95 was at 0.841.

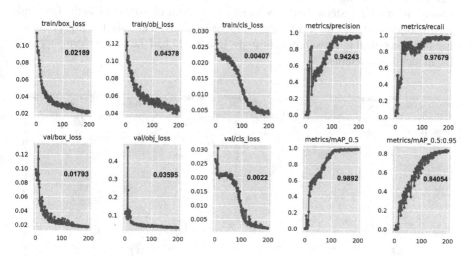

Fig. 4. Performance metrics for the best model built at epoch 189. The top row shows the training set metrics, and the bottom row shows the validation set metrics in the first three images (L to R). Remaining two columns display precision, recall, and mAP scores during the training.

The F1 score (Fig. 5) was 0.96 at 0.653 confidence, which provides the information on the model's balanced ability to both capture positive cases (recall) and be accurate with respect to the cases it does capture (precision).

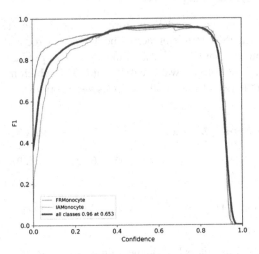

Fig. 5. F1 score for the trained model with overall 0.96 F1 score across both classes.

Examples in Fig. 6 show a sample image with predicted bounding boxes for the monocytes and their subsequent classifications in epochs 39 (left image) and 189 (right image). Some bounding boxes have a very low confidence score at epoch 39 and are discarded in the later steps. Some bounding boxes are classified as both, free and ingested/adhered class, but at different confidence levels, and only the higher confidence box is retained in the subsequent steps. Epoch 189 had the best performance and is the model that is retained.

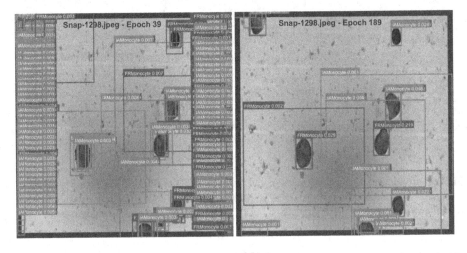

Fig. 6. Predicted bounding boxes for monocytes on one sample image at epochs 39 (left) and 189 (right) showing the improvement in the learning process.

```
{
  "predictions": [
    {
      "x": 415.5,
      "y": 186.5,
      "width": 35,
      "height": 25,
      "confidence": 0.891,
      "class": "FRM"
    },
    {
      "x": 45.5,
      "y": 70,
      "width": 25,
      "height": 32,
      "confidence": 0.88,
      "class": "FRM"
    },
    {
      "x": 171,
      "y": 133.5,
      "width": 32,
      "height": 31,
      "confidence": 0.874,
      "class": "FRM"
    },
    {
      "x": 364.5,
      "y": 214.5,
      "width": 33,
      "height": 31,
      "confidence": 0.868,
      "class": "FRM"
    },
    {
      "x": 494,
      "y": 189,
      "width": 34,
      "height": 26,
      "confidence": 0.866,
      "class": "FRM"
    },
    {
      "x": 348.5,
      "y": 161,
      "width": 31,
      "height": 30,
      "confidence": 0.859,
      "class": "FRM"
    },
    {
      "x": 70.5,
      "y": 65.5,
      "width": 29,
      "height": 31,
      "confidence": 0.857,
      "class": "FRM"
    },
    {
      "x": 249.5,
      "y": 67.5,
      "width": 33,
      "height": 33,
      "confidence": 0.852,
      "class": "FRM"
    },
    {
      "x": 468.5,
      "y": 201.5,
      "width": 29,
      "height": 27,
      "confidence": 0.832,
      "class": "FRM"
    },
    {
      "x": 114.5,
      "y": 75,
      "width": 27,
      "height": 22,
      "confidence": 0.823,
      "class": "FRM"
    },
    {
      "x": 269,
      "y": 279.5,
      "width": 30,
      "height": 17,
      "confidence": 0.819,
      "class": "FRM"
    },
    {
      "x": 138.5,
      "y": 67,
      "width": 27,
      "height": 26,
      "confidence": 0.816,
      "class": "FRM"
    },
    {
      "x": 460.5,
      "y": 17.5,
      "width": 33,
      "height": 29,
      "confidence": 0.809,
      "class": "FRM"
    },
    {
      "x": 351.5,
      "y": 110.5,
      "width": 27,
      "height": 23,
      "confidence": 0.805,
      "class": "FRM"
    },
    {
      "x": 505.5,
      "y": 128,
      "width": 13,
      "height": 30,
      "confidence": 0.638,
      "class": "FRM"
    },
    {
      "x": 217.5,
      "y": 164,
      "width": 31,
      "height": 34,
      "confidence": 0.875,
      "class": "IAM"
    }
  ]
}
```

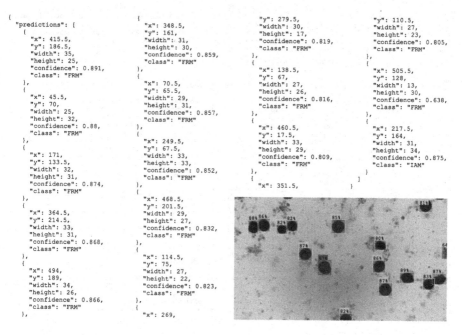

Fig. 7. Monocyte location and class predictions for one test image. Locations include the x and y position, width, and height of the bounding box together with prediction confidence.

The trained model was used to make predictions for new and unseen images and process them with the algorithm. The example in Fig. 7 shows that the algorithm detected both classes of monocytes (free and ingested/adhered) with high accuracy. In this image there were 16 monocytes, one of which was detected at 64% confidence, as it was clipped on the edge when the image was taken. Remaining fifteen monocytes were detected with the confidence level of 80 to 89%. Only one of the monocytes in this image was identified as ingested/adhered (it is marked with a cyan box), and was detected with 87% confidence. The remaining detected monocytes were free monocytes. The algorithm worked very efficiently to ignore the background noise.

Figure 8 shows that the model took less than 1 s to process the 27 sample images. Average processing time was 271 ms per image and the number of monocytes present in the images does not seem to be related to the time required to process an image. Overall, there were 344 free monocytes and 56 ingested/adhered monocytes detected across the 27 images, with the confidence threshold of at least 0.5 or 50%. Post-processing calculates the MI value based on the 27 images as 14% using Eq. 1. This MI value can be shared with clinicians requesting the MMA, allowing them to provide personalized medicine to the patient.

```
detect: weights=['/content/yolov5/best.pt'], source=/content/yolov5/testImages/, data=data/coco128.yaml,
imgsz=[640, 640], conf_thres=0.5, iou_thres=0.45, max_det=1000, device=, view_img=False, save_txt=True,
save_conf=True, save_crop=False, nosave=False, classes=None, agnostic_nms=False, augment=False,
visualize=False, update=False, project=runs/detect, name=exp, exist_ok=False, line_thickness=3,
hide_labels=False, hide_conf=False, half=False, dnn=False, vid_stride=1
YOLOv5 🚀 v7.0-59-gfdc35b1 Python-3.8.16 torch-1.13.0+cu116 CPU

Fusing layers...
custom_YOLOv5s summary: 232 layers, 7249215 parameters, 0 gradients

image  1/27 /content/yolov5/testImages/Snap-01.jpeg      8 FRMonocytes  9 IAMonocytes 280.1ms
image  2/27 /content/yolov5/testImages/Snap-02.jpeg      3 FRMonocytes 11 IAMonocytes 269.8ms
image  3/27 /content/yolov5/testImages/Snap-03.jpeg      8 FRMonocytes  0 IAMonocytes 265.9ms
image  4/27 /content/yolov5/testImages/Snap-04.jpeg     35 FRMonocytes  0 IAMonocytes 264.6ms
image  5/27 /content/yolov5/testImages/Snap-05.jpeg     60 FRMonocytes  0 IAMonocytes 266.6ms
image  6/27 /content/yolov5/testImages/Snap-06.jpeg     10 FRMonocytes  0 IAMonocytes 279.6ms
image  7/27 /content/yolov5/testImages/Snap-07.jpeg      8 FRMonocytes  1 IAMonocyte  274.4ms
image  8/27 /content/yolov5/testImages/Snap-08.jpeg     22 FRMonocytes  0 IAMonocytes 273.0ms
image  9/27 /content/yolov5/testImages/Snap-09.jpeg      6 FRMonocytes  0 IAMonocytes 272.8ms
image 10/27 /content/yolov5/testImages/Snap-10.jpeg      6 FRMonocytes  1 IAMonocyte  272.2ms
image 11/27 /content/yolov5/testImages/Snap-11.jpeg      1 FRMonocyte   0 IAMonocytes 262.9ms
image 12/27 /content/yolov5/testImages/Snap-1193.jpeg   16 FRMonocytes  1 IAMonocyte  262.2ms
image 13/27 /content/yolov5/testImages/Snap-1194.jpeg   16 FRMonocytes  1 IAMonocyte  278.9ms
image 14/27 /content/yolov5/testImages/Snap-1195.jpeg    3 FRMonocytes  1 IAMonocyte  264.5ms
image 15/27 /content/yolov5/testImages/Snap-12.jpeg     10 FRMonocytes  0 IAMonocytes 284.1ms
image 16/27 /content/yolov5/testImages/Snap-1208.jpeg   15 FRMonocytes  1 IAMonocyte  260.0ms
image 17/27 /content/yolov5/testImages/Snap-1209.jpeg   28 FRMonocytes  0 IAMonocytes 266.9ms
image 18/27 /content/yolov5/testImages/Snap-1210.jpeg    3 FRMonocytes  7 IAMonocytes 263.7ms
image 19/27 /content/yolov5/testImages/Snap-1223.jpeg    2 FRMonocytes  5 IAMonocytes 301.2ms
image 20/27 /content/yolov5/testImages/Snap-1224.jpeg    3 FRMonocytes  0 IAMonocytes 274.0ms
image 21/27 /content/yolov5/testImages/Snap-1225.jpeg    2 FRMonocytes  2 IAMonocytes 289.5ms
image 22/27 /content/yolov5/testImages/Snap-1226.jpeg    7 FRMonocytes 14 IAMonocytes 268.4ms
image 23/27 /content/yolov5/testImages/Snap-236.jpeg    17 FRMonocytes  0 IAMonocytes 258.1ms
image 24/27 /content/yolov5/testImages/Snap-237.jpeg    19 FRMonocytes  0 IAMonocytes 284.2ms
image 25/27 /content/yolov5/testImages/Snap-238.jpeg    16 FRMonocytes  1 IAMonocyte, 263.5ms
image 26/27 /content/yolov5/testImages/Snap-250.jpeg    11 FRMonocytes  1 IAMonocyte, 261.7ms
image 27/27 /content/yolov5/testImages/Snap-263.jpeg     9 FRMonocytes  0 IAMonocytes 259.7ms
                                                       -----          ----
                                                        344            56

Speed: 1.1ms pre-process, 271.2ms inference, 0.9ms NMS per image at shape (1, 3, 640, 640)
Results saved to runs/detect/exp - 27 labels saved to runs/detect/exp/labels
```

Fig. 8. Results of the applied model and MMA post-processing of the detected free and ingested/adhered monocytes in a set of 27 images.

5   Conclusions and Further Research

The monocyte monolayer assay (MMA) is a test used to determine the compatibility of blood between donors and recipients before a blood transfusion takes place. Currently, this process is very labor intensive, and no viable computer-based solutions have been proposed for it to date. At least 3,200 or more monocytes have to be counted under a microscope on the small eight-chamber glass slide in order to make this determination. Overall, the development of a viable computer-based solution for MMA has the potential to greatly improve the efficiency and accuracy of blood transfusion compatibility testing.

In this work, we presented a model based on the YOLOv5s deep learning network that can detect and classify monocytes in images taken with a camera attached to a microscope. Performance score, including F1, mAP scores, precision, and recall are all above 90%. The proposed approach can quickly locate and classify the monocytes simultaneously, objectively, and efficiently. Experiments involving a real data set demonstrate that this approach can successfully identify and classify monocytes in an image and determine the Monolayer Index in post-processing. This workflow can ease and accelerate the medical laboratory technician's repetitive, cumbersome, and error-prone counting process, and therefore contributes to the accuracy of medical diagnosis systems.

Acknowledgement. Research reported in this manuscript was supported by a grant from the National Institute of General Medical Sciences of the National Institutes of Health under Award Number 3P2OGM103424-20. Additional support was provided by the Abe Sadoff Chair in Bioinformatics and Lisa Burke Bioinformatics Professorship endowed funds.

References

1. Common objects in context, http://cocodataset.org/
2. Abadi, M., et al.: Tensorflow: a system for large-scale machine learning. In: Proceedings of the 12th USENIX Conference on Operating Systems Design and Implementation, OSDI 2016, USENIX Association, USA, pp. 265–283 (2016)
3. Arndt, P.A., Garratty, G.: A retrospective analysis of the value of monocyte monolayer assay results for predicting the clinical significance of blood group alloantibodies. Transfusion **44**(9), 1273–1281 (2004)
4. Ben-Kiki, O., Evans, C., döt Net, I.: YAML ain't markup language (YAML^{TM}) version 1.2.2 (2021). http://yaml.org/spec/1.2.2/
5. Biewald, L.: Experiment tracking with weights and biases. Softw. Avail. wandb.com **2**, 233 (2020)
6. Bisong, E.: Google colaboratory. In: Building Machine Learning and Deep Learning Models on Google Cloud Platform, pp. 59–64. Apress, Berkeley, CA (2019). https://doi.org/10.1007/978-1-4842-4470-8_7
7. Boral, L.I., Hill, S.S., Apollon, C.J., Folland, A.: The type and antibody screen, revisited. Am. J. Clin. Pathol. **71**(5), 578–581 (1979). https://doi.org/10.1093/ajcp/71.5.578
8. Branch, D.R., Gallagher, M.T., Mison, A.P., Sy Siok Hian, A.L., Petz, L.D.: In vitro determination of red cell alloantibody significance using an assay of monocyte-macrophage interaction with sensitized erythrocytes. Br. J. Haematolo. **56**(1), 19–29 (1984). https://doi.org/10.1111/j.1365-2141.1984.tb01268.x
9. Buslaev, A., Iglovikov, V.I., Khvedchenya, E., Parinov, A., Druzhinin, M., Kalinin, A.A.: Albumentations: fast and flexible image augmentations. Information **11**(2), 125 (2020). https://doi.org/10.3390/info11020125
10. Dwyer, B., Nelson, J., Solawetz, J.: Roboflow (2022). http://roboflow.com
11. Girshick, R., Donahue, J., Darrell, T., Malik, J.: Rich feature hierarchies for accurate object detection and semantic segmentation. In: Proceedings of the IEEE Conference on Computer Vision and Pattern Recognition, pp. 580–587 (2014)
12. Harris, C., et al.: Array programming with NumPy. Nature **585**(7825), 357–362 (2020). https://doi.org/10.1038/s41586-020-2649-2
13. Lin, T.-Y., et al.: Microsoft COCO: common objects in context. In: Fleet, D., Pajdla, T., Schiele, B., Tuytelaars, T. (eds.) ECCV 2014. LNCS, vol. 8693, pp. 740–755. Springer, Cham (2014). https://doi.org/10.1007/978-3-319-10602-1_48
14. Marquez, L.A.P., Chakrabarty, S.: Automatic image segmentation of monocytes and index computation using deep learning. In: 2022 IEEE International Conference on Bioinformatics and Biomedicine (BIBM), pp. 2656–2659 (2022). https://doi.org/10.1109/BIBM55620.2022.9994922
15. Nance, S., Arndt, P., Garratty, G.: Predicting the clinical significance of red cell alloantibodies using a monocyte monolayer assay. Transfusion **27**(6), 449–452 (1987). https://doi.org/10.1046/j.1537-2995.1987.27688071692.x

16. Paszke, A., et al.: Pytorch: an imperative style, high-performance deep learning library. In: Wallach, H., Larochelle, H., Beygelzimer, A., d' Alché-Buc, F., Fox, E., Garnett, R. (eds.) Advances in Neural Information Processing Systems, vol. 32. Curran Associates, Inc. (2019)
17. Redmon, J., Divvala, S., Girshick, R., Farhadi, A.: You only look once: Unified, real-time object detection. In: 2016 IEEE Conference on Computer Vision and Pattern Recognition (CVPR), pp. 779–788 (2016). https://doi.org/10.1109/CVPR.2016.91

ALE: A Simulation-Based Active Learning Evaluation Framework for the Parameter-Driven Comparison of Query Strategies for NLP

Philipp Kohl[1]([⊠])(iD), Nils Freyer[1](iD), Yoka Krämer[1](iD), Henri Werth[3](iD),
Steffen Wolf[1](iD), Bodo Kraft[1], Matthias Meinecke[1](iD), and Albert Zündorf[2]

[1] FH Aachen – University of Applied Sciences, 52428 Jülich, Germany
{p.kohl,freyer,y.kraemer,s.wolf,kraft,meinecke}@fh-aachen.de
[2] University of Kassel, 34121 Kassel, Germany
zuendorf@uni-kassel.de
[3] laizee.ai, Cologne, Germany
henri@laizee.ai
https://laizee.ai/

Abstract. Supervised machine learning and deep learning require a large amount of labeled data, which data scientists obtain in a manual, and time-consuming annotation process. To mitigate this challenge, Active Learning (AL) proposes promising data points to annotators they annotate next instead of a subsequent or random sample. This method is supposed to save annotation effort while maintaining model performance.

However, practitioners face many AL strategies for different tasks and need an empirical basis to choose between them. Surveys categorize AL strategies into taxonomies without performance indications. Presentations of novel AL strategies compare the performance to a small subset of strategies. Our contribution addresses the empirical basis by introducing a reproducible active learning evaluation (ALE) framework for the comparative evaluation of AL strategies in NLP. The framework allows the implementation of AL strategies with low effort and a fair data-driven comparison through defining and tracking experiment parameters (e.g., initial dataset size, number of data points per query step, and the budget). ALE helps practitioners to make more informed decisions, and researchers can focus on developing new, effective AL strategies and deriving best practices for specific use cases. With best practices, practitioners can lower their annotation costs. We present a case study to illustrate how to use the framework.

Keywords: Active learning · Query learning · Natural language processing · Deep learning · Reproducible research

1 Introduction

Within the last decades, machine learning (ML) and, more specifically, deep learning (DL) have opened up great potential in natural language processing (NLP) [8,9].

D. Conte et al. (Eds.): DeLTA 2023, CCIS 1875, pp. 235–253, 2023.
https://doi.org/10.1007/978-3-031-39059-3_16

Supervised machine learning algorithms constitute a powerful class of algorithms for NLP tasks like text classification, named entity recognition (NER), and relation extraction [25,39]. Researchers and corporations require large amounts of annotated data to develop state-of-the-art applications. However, domain-specific labeled data is often scarce, and the annotation process is time, and cost-intensive [16]. Therefore, reducing the amount of data needed to train a model is an active field of research [16]. Practitioners and researchers agree on the need to reduce annotation costs. Estimating these expenses is a research field on its own [4,16,49].

Next to transfer learning [43] from large language models or few-shot learning [47], active learning (AL) is another approach to reduce the required data size by systematically annotating data [44,48]. AL strategies (also called *query* or *teacher*) propose data points to the annotator in the annotation process, which could increase the model's performance more than a randomly chosen data point. Consequently, the model may perform better with fewer data when annotating the proposed data points instead of considering random ones. Figure 1 shows the human-in-the-loop paradigm with pool-based annotation approaches. Instead of selecting random or sequential data points, as processed in a standard annotation process, AL strategies forward beneficial samples to the annotator. Active learning has empirically proven significant advances in specific domains, including computer vision or activity recognition [2,31]. [10,13] report up to 60% of annotation effort reduction.

The performance of an AL strategy is typically measured by the performance metric of the corresponding model when trained on the same amount of systematically proposed data compared to randomly proposed data. However, AL strategies often use heuristics that may be task, data, or model sensitive [44]. Little can be said about their performance in a comparable way. Surveys often classify AL strategies into different categories but do not provide exhaustive performance benchmarks [42,45]. Researchers developing new AL strategies compare their performance with random sampling and a few other strategies [15,53]. The research field of NLP lacks a standardized procedure to evaluate AL strategies systematically for a given task, dataset, and model.

ALE facilitates researchers and practitioners to gain more insights into task-dependent taxonomies and best practices. The goal is to enable practitioners to adapt the most valuable AL strategy based on comparable previous experiments. Therefore, this paper introduces the Active Learning Evaluation (ALE)[1] Framework. Our framework aims to close the *comparison gap* [56] between different AL strategies. ALE enables researchers in NLP to implement and evaluate their own AL strategies in a reproducible and transparent way. By offering default implementations, practitioners can directly compare their AL strategies and datasets. Using the *MLFlow* platform[2], ALE simplifies visualizing and reporting experimental results. At the same time, practitioners can customize the configurations and adapt the framework to their specific requirements.

---

[1] https://github.com/philipp-kohl/Active-Learning-Evaluation-Framework.
[2] https://mlflow.org/.

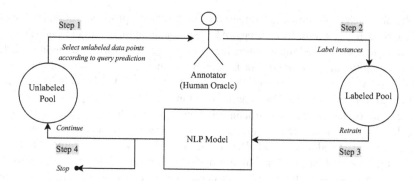

Fig. 1. Active learning cycle (following [16,45]): **1st step**: The AL strategy proposes the next batch of data points to the annotator. **2nd step**: The annotator labels the given increment. **3rd step**: The training component trains a model using the updated set of labeled data points. **4th step**: Stop the process if the model achieved the desired performance, or the unlabeled pool is empty. Otherwise, repeat the process.

2   Related Work

Active learning started with classical machine learning tasks (such as support vector machines, decision trees, etc.) [45]. With the rise of deep learning architectures, researchers transfer the knowledge of classical approaches to those using deep learning [42]. Active learning has gained significant attention across various domains, such as computer vision and natural language processing (NLP) [42,56]. In computer vision, users apply active learning to tasks like image classification, object detection, and video processing. In NLP, it has been used for text classification, entity recognition, and machine translation.

Researchers classify active learning approaches based on several taxonomies: (1) Explorative and exploitative ([7,31]) approaches: Exploration is model-agnostic and uses information about the data distribution, while exploitation uses model information to propose valuable data points. (2) [45] distinguish stream-based from pool-based approaches. Stream-based methods sample a single data point and decide whether to present it to the human oracle (annotator). In contrast, the pool-based method selects a subset of the unlabeled dataset. (3) [16] compares traditional and real-world active learning, which pays attention to the complexity of the actual annotation process. Real-world strategies do not assume a perfect oracle but take into account that humans are individuals that make mistakes. Relying on a single human annotation as the gold label can result in errors. Addressing this issue impacts the annotation time, cost, and quality.

Despite using a single AL strategy, [35] shows how to determine the most valuable method from a pre-selected subset on the fly. The authors select the best strategy by incrementally re-weighting the number of new data points each strategy proposes while performing the active learning cycle (see Fig. 1). Thus, the method finds the subset's best strategy over iterations. ALE does not address the dynamic selection of the best strategy in the annotation process, but it helps to compare the sub-selected AL strategies in isolation to select a promising strategy for the annotation process. But understanding [35]'s approach as a single AL strategy, ALE can compare this method to others.

Researchers developed many active learning strategies in various domains [42,45]. However, annotation tools for NLP tasks seem to prefer the basic method of uncertainty sampling as AL strategy: INCEpTION [24] or Prodigy [36]. Other tools like doccano [38] plan to implement active learning in the future.

The taxonomies create a significant challenge for practitioners seeking to adopt active learning for their specific use case. Developing best practices for similar use cases is necessary to address this challenge. We created an evaluation framework for AL strategies in a consistent manner. This framework allows practitioners to compare different AL strategies using the same set of parameters, thereby facilitating the selection of the most appropriate strategy for their specific use case. Surveys on active learning propose various methodologies, but they are limited in their ability to provide a comprehensive performance comparison of different AL strategies [42,45,48]. [56] provides a performance comparison for 19 AL strategies.

Similar to our objective, [54] implemented *libact*, an active learning comparison and application framework for classical machine learning methods with scikit-learn [40]. *libact* offers the feature to automatically select the best-performing strategy by the *Active Learning By Learning (ALBL)* algorithm [19] over iterations in the annotation process. While *libact* focuses on classical machine learning with scitkit-learn, ALE enables users to apply their preferred ML/DL tools, such as scikit-learn, TensorFlow, PyTorch, or SpaCy. We want to inspect the NLP domain and choose the deep learning framework *SpaCy*[3] as the default implementation.

[20,56] developed a CLI tool (*DeepAL*) to compare different pool-based AL strategies with PyTorch[4] models in the image vision domain. ALE focuses on the NLP domain and offers a sophisticated experiment configuration and tracking management. This leads to a reproducible and transparent experiment environment. Furthermore, we use parallelization and dockerization to address the long computation times and enable cloud infrastructure usage. Additionally, we facilitate experiments on the cold start phase [55] with the same strategies as for active learning.

3   Ethical Considerations

Despite the ethical considerations regarding NLP [6,29,51], more specific ethical implications are worth evaluating for active learning, as it may introduce or reinforce statistical biases to the data [11]. Especially regarding safety or life path-relevant decision-making, introducing biases to the training data may cause ethically significant harms [50]. Thus, when applying active learning strategies, it is urgent to monitor and control biases in the resulting data[5]. In contrast, efforts have been made to diminish biases in already biased datasets using AL strategies that optimize for certain fairness metrics [3]. Further research in that direction would be desirable and is generally supported by the ALE framework by adding additional metrics such as statistical parity [3].

---

[3] https://spacy.io/.

[4] https://pytorch.org/.

[5] A similar concern was made within the regulatory framework proposal on artificial intelligence by the European Union, classifying the respective systems as *high risk*.

4   ALE Framework

We want to facilitate researchers and practitioners to draw data-based decisions for selecting an appropriate AL strategy for their use case. To assess the AL strategy's performance, we simulate the human-in-the-loop as a perfect annotating oracle using the gold label from the used dataset (see Fig. 2). After each query step, we train a model and evaluate its performance on the test set. These metrics serve as a performance indicator of the AL strategy. The random selection of data points represents the baseline strategy. We average the performance across several random seeds to address the model's and strategy's stability [33,41].

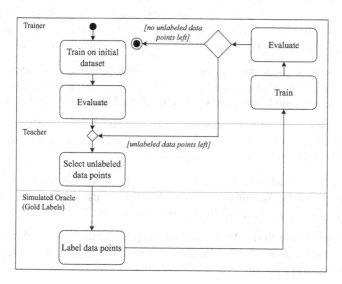

Fig. 2. The *simulator* manages the AL simulation cycle: The trainer starts the initial training on the first increment of data points proposed by a teacher component (omitted in this figure) and evaluates it. The shown teacher proposes unlabeled data points to the simulated oracle, which simulates the human annotator by adding the according gold label. The updated dataset is used for another training and evaluation cycle. Once there are no unlabeled data points left, the process stops.

In the following sections, we describe the features and implementation of the ALE framework in detail:

Configuration System. To address varying use cases, we have created a sophisticated configuration system. The user controls the parameters for the active learning simulation: initial dataset size, number of proposed data points by query, comparison metric, budget, and dataset.[6]

Experiment Documentation and Tracking. The framework documents each experiment with the configuration and results (e.g., metrics and trained models) of all substeps in MLFlow. The tracking facilitates own subsequent analysis to reveal regularities.

---

[6] Full configuration documentation can be found on GitHub.

Resume Experiments. Active learning simulation computation is time-consuming (hours to days), depending on the configuration (especially dataset size and step size). The training processes determine the majority of the consumption part. Thus, the framework can continue experiments on pauses or errors.

Reproducible. In conjunction with the experiment configuration, including fixed seeds and persisting the git commit hash enables reproducible research.

Average with Different Seeds. To avoid beneficial or harmful seeds, the user can specify more than one seed. The framework processes different simulation runs with these seeds and averages the results. Thus, we achieve a less random biased result [33,41]. If the results show high variances, the model or the query strategy might not be stable. We avoid simulating the active learning process for each seed sequentially. Instead, we offer a configurable amount of concurrent threads to simulate the process.

Arbitrary Usage of Datasets. The user can employ their preferred datasets. No additional code must be written as long as the dataset aligns with the conventions. Otherwise, the user has to implement a new converter class. See details in Subsect. 5.2.

Arbitrary Usage of ML/DL Framework. The framework uses SpaCy as the default deep learning framework. The user can change the framework by implementing the trainer (Subsect. 6.2) and corpus abstraction layer.

Easy to Test Own Strategies. The users can implement their own configurable strategies by adding a config file, implementing a single class, and registering the class with the *TeacherRegistry*. See the usage in the case study in Sect. 7.

Experiment with Cold Start. Analog to AL strategies, the user can test different strategies for the initial training dataset. Initial training strategies and AL strategies can be used interchangeably.

Containerized. We provide all involved services as docker services. In consequence, the environment can be bootstrapped rapidly. Furthermore, we provide a docker image for the framework facilitating the run of experiments on cloud infrastructure.

Parallel Computation. In conjunction with the containerization, the user can run experiments in parallel, and the framework reports the result to a central MLFlow instance.

5   Architecture

The architecture of our python framework focuses on flexibility and ease of adoption by leveraging open-source tools and allowing developers to integrate their own workflows and machine learning libraries. We use MLFlow to track configurations, pipeline steps, artifacts, and metrics. Dockerization allows users to execute the framework remotely. Figure 3 shows a brief overview of our architecture.

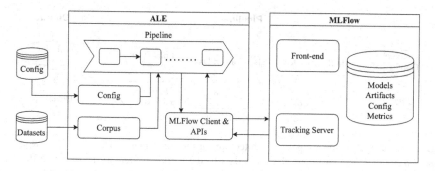

Fig. 3. The architecture consists of two components: *ALE* itself and *MLFlow*. We provide both components as dockerized applications. Every pipeline step will log its configuration, metrics, and artifacts to the tracking server. The data scientist uses MLFlow to inspect and compare experiments via the MLFlow front-end.

5.1 Configuration

The configuration system is one of the core components of the framework. It allows setting up different environments for testing AL strategies . The hydra framework [52] achieves this by dynamically creating a hierarchical configuration. This configuration is composed of different files and command line parameters. To set up a reliable end-to-end pipeline to test different strategies, we provide the following configuration modules: *converter*, *data*, *experiment*, *teacher*, *trainer*, and *MLFlow*. See GitHub for detailed explanations.

We provide a simple way to create reproducible experiments: We leverage the python API of MLFlow and combine it with hydra to make use of MLFlow's tracking capabilities and hydra's powerful configuration system. The configuration is stored in .*yaml* files. When running the framework, the configuration files are processed by hydra, stored as python objects and accessible to the pipeline components (Subsect. 5.3). The configuration object is passed down and logged to the MLFlow tracking server. The logged configuration enables the framework to detect already processed pipeline steps and load them from the tracking server, providing a simple yet powerful caching mechanism.

5.2 Dataset

NLP tasks like text classification and span labeling are based on big amounts of labeled data. To focus on the comparison of different AL strategies and not on converting data, we adopt the *convention over configuration* principle [5] to *convention over coding*. If the users follow these conventions, they do not have to write code, but at most, a short configuration.

We introduce the following conventions: (1) The raw data should be in JSONL format. (2) The converted format is SpaCy DocBin. (3) We add ids to each raw entry and use it for global identification throughout the framework.

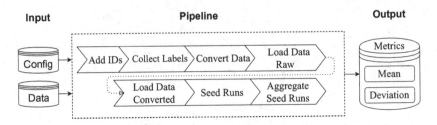

Fig. 4. ALE uses the shown pipeline architecture for the experiments. The *Load Data Raw* and *Load Data Converted* will copy the input data to the working directory and rename the files according to the convention. The copied data is logged to the tracking server. Tracking all the input and metadata allows running subsequent runs remotely, fetching data from the tracking server as needed. Figure 5 shows details for *Seed Runs* and *Aggregate Seed Runs*.

At the moment, the framework handles text classification and span labeling tasks. In the case of span labeling tasks, each data point contains a list of the labeled spans: each span is represented by the start character offset, end character offset, and the according label (`"labels":[[0,6,"LOC"]]`). For text classification tasks, the label is simply the label name (`"label":"neg"`)[7]. For our case study, we used the IMDb corpus [32] and the dataset originating from the Text RETrieval Conference (TREC) [18]. Subsection 7.1 shows details.

5.3 Pipeline

The framework provides a customizable yet fixed setup. We decided to build a step-by-step pipeline with customizable components. The *MLFlowPipeline* class acts as the executor. It collects pipeline components and initializes the pipeline storage and configuration. The executor will run every child pipeline component after another. We use MLFlow to track our experiments. Each pipeline step reports parameters and results to their MLFlow *run*. Figure 4 shows the pipeline steps. Order of execution is essential as the different components depend on each other. For example, the *load data converted* step needs the data from the *convert data* step. Before the framework executes a pipeline step, it will check if a run with matching parameters and revision was already executed. To achieve this functionality, the executor and every component log their run parameters to MLFlow. If a run matches and the match is marked as successful, the framework fetches the result of the run. Otherwise, if the matched run is marked as failed, the run will try to resume the failed run.

6   Core Composition

Four components build the core unit of the framework: *simulator*, *teacher*, *trainer*, and *corpus* (see Fig. 6). These form a logical unit responsible for selecting data proposed to the simulated oracle using gold labels as annotations (see Fig. 2). The *registry* classes

---
[7] In multi-label classification tasks the label is a list of labels.

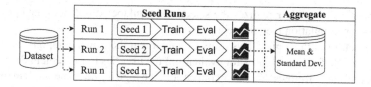

Fig. 5. The last two steps of our pipeline contain the actual core logic. The *Seed Runs* component step starts n-different simulation runs, each with its own distinct seed. A run iteratively proposes new data based on the strategy and trains and evaluates the model. This will lead to n-different results for one strategy. To get one comparable result, the *Aggregate Seed Runs* step will fetch the resulting metrics and calculate the mean and deviation over all runs.

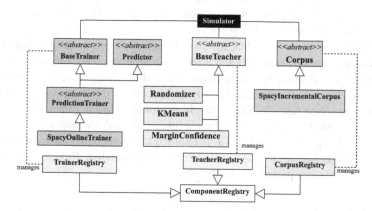

Fig. 6. The *simulator* (black) holds instances of the *trainer* (blue), *teacher* (beige), and *corpus* (grey). The configuration specifies which implementations the simulator fetches from the registries. The registry knows all implementations of each component. We distinguish between *BaseTrainer* and *Predictor* to hide the training API for the teachers. The teacher can only perform predictions but no training. (Color figure online)

collect all implementations of the abstract classes *teacher*, *trainer*, and *corpus* according to an adaption of the behavioral registry pattern[8] used in Python.

The **simulator** class[9] takes care of the high-level management of the data proposal by setting up the instances of corpus, trainer, and teacher using the classes provided by the configuration. Hereby, the simulator runs active learning cycle simulation (see Fig. 1) for every seed as a run on MLFlow. A **corpus** instance tracks the annotation progress: e.g., which documents the simulator has already annotated and which documents need annotations. Furthermore, it provides a trainable corpus based on the annotated data. Thus, the corpus and trainer represent a unit. The **trainer** uses a ML/DL framework to train and evaluate a model. We provide a default implementation *Spacy-OnlineTrainer*, with SpaCy (see Subsect. 6.2). We allow the teachers to make predictions with the trained model. Thus, the trainer class implements the abstract methods of

---

[8] https://github.com/faif/python-patterns/blob/master/patterns/behavioral/registry.py.
[9] The simulator is called *AleBartender* and *AleBartenderPerSeed* in the code.

the *Predictor*. The **teacher** realizes the selection of unlabeled potential data points for the annotation simulation (see Subsect. 6.1).

The structure of this unit allows users to implement their own components, such as AL strategies the framework uses for evaluation: the user has to implement the abstract class. If the strategy follows the folder convention, the registry finds the implementation, and the user can evaluate the new strategy via the hydra configuration. The same applies to the other components.

6.1 Teacher

The basic functionality of the teacher component is to propose new data points to the *Simulator* (see Fig. 6). Therefore, a teacher needs to implement the *BaseTeacher's* abstract method presented in Listing 1.1.

Additionally, if needed, the teacher may use metrics from training by implementing the BaseTeacher's *after_train()* and *after_initial_train()* methods to perform further computations after the corresponding training step, depending on the implemented AL strategy[10]. AL teacher strategies are broadly divided into two areas, namely exploration and exploitation (see Sect. 2). While purely exploration-based strategies are model-independent, exploitation-based strategies exploit the model's state of training and, thus, are model dependent [7,31]. ALE is required to handle both classes of AL strategies and hybrid techniques that use exploitation and exploration. Therefore, the teacher takes the *predictor* as a parameter (see Fig. 6).

```
1 @abstractmethod
2 def propose(self,
3                 potential_ids: List[int],
4                 actual_step_size: int,
5                 actual_budget: int) -> List[int]:
6     # See text for documentation
7     pass
```

Listing 1.1. BaseTeacher's abstract propose *method* to be implemented. The *propose* method takes a list of potential (unlabeled) data point ids, the actual step size, and the budget of the propose step. Step size and budget are limited to the remaining length of the not annotated data points. The method returns a list of data point ids according to the corresponding AL strategy.

6.2 Trainer

The trainer component takes the proposed data in each iteration and trains a model. The trainer first evaluates the model on the development dataset to pass information to the teacher. Afterward, it evaluates the model on the test set for performance comparison. Thus, the framework provides feedback on the quality of the newly proposed batch based on the objective metric, which serves as a benchmark for comparing the performance of different AL strategies.

---

[10] For instance, multi-armed bandit strategies use the gain in model performance after each training to estimate the reward over clusters of data to draw from [1,26].

ALE offers an abstraction layer that enables users to use an arbitrary machine learning or deep learning framework, such as PyTorch, TensorFlow, or PyTorch Lightning. Therefore, the user must implement the abstract methods specified in the *Prediction-Trainer* class: *train, evaluate, store/restore* for the resume process, and *predict* for exploitation strategies.

We focus on the NLP domain, and thus we select SpaCy as the baseline framework due to its sophisticated default implementations and practicality for production use cases [37]. During each iteration of the active learning process, the framework generates a new batch of data to be annotated and calls the trainer with the updated corpus.

The trainer initiates the SpaCy training pipeline in the constructor and reuses it in each subsequent training iteration. We also tested training from scratch after each *propose step*. However, this approach takes time to create the SpaCy pipeline[11]. Thus, we switched to online/continual learning [17,34]. Continual learning has other challenges, like the *catastrophic forgetting problem* [22,23]. The framework addresses this issue by training not only on the new proposed data but also on all previously proposed data. Thus, the trainer starts with a checkpoint from the previous iterations and fine-tunes the weights with all data available up to that iteration[12].

7   Case Study

To demonstrate the implementation of different datasets and AL strategies in ALE, we implemented an exploitation and an exploration based AL strategy. Both strategies are pool-based and applied to text classification tasks. We evaluated their performance compared to a randomizer and visualized the results in ALE on the IMDb sentiment dataset [32] and the TREC dataset [18,30]. As we stick to the convention of using a SpaCy model, we may simply use the implemented pipeline components. Therefore, we use the *SpacyIncrementalCorpus* implementation of ALE's *Corpus* class for both corpora and the implemented *SpacyOnlineTrainer*.

7.1   Experiment Setup

We compared the strategies on two text classification tasks: a binary sentiment classification task with the IMDb reviews corpus and a multiclass text classification task with the TREC coarse dataset.

The *IMDb Reviews Corpus* is a dataset for binary sentiment classification [32]. As a benchmark dataset, it provides $50,000$ documents, split into a training set of size $25,000$ and a test set of size $25,000$, which we randomly split into a test and dev set of sizes $12,500$ each.

Originating from the Text RETrieval Conference (TREC) is a multiclass question classification dataset[18]. It contains $5,452$ annotated questions for training and 500 annotated questions for testing. For our case study, we further split the training set

---

[11] We empirically tested different configurations and noticed computation time reduction up to ten percent.

[12] We compared the average of five simulation runs with and without the online approach. The latter achieves faster convergence. The final scores are nearly equal.

to train and dev set, s.t. we obtained a total of 546 annotated documents (10%) for evaluation and 4906 documents for training. We trained our model on the coarse labels for our case study, consisting of five classes.

The two presented datasets differ in complexity. The sentiment analysis task involves larger documents and a comparatively simple binary classification task. In contrast, the question classification task consists of short documents (single sentences/questions) and a comparatively difficult multiclass classification task with five classes. For our case study, we use SpaCy's bag-of-words *n-gram model* for sentiment classification[13] for runtime efficiency and as we considered it sufficient for the merely demonstrational purposes of the case study. However, on the question classification task, we did not obtain usable results with the simple model and, thus, used SpaCy's *transformer model* for question classification[14].

7.2   Implemented Active Learning Strategies

We implemented and compared an exploration- and an exploitation-based AL strategy for the case study, as the two classes of AL strategies are considered to form a trade-off. While exploration-based methods are computationally less expensive and generally yield a more representative sample of the data such that they increase the external validity of the model, exploitation-based methods have shown significant theoretical and practical performance improvements for many domains [7,57]. For comparison, we implemented a baseline randomizer strategy. The randomizer draws a random sample of step size N data points for each propose step. Therefore, we needed to implement the *propose* method of the *BaseTeacher*.

Exploration-Based Teacher (K-Means). The implemented exploration-based AL strategy is based on the cluster hypothesis [27]. The cluster hypothesis claims that documents within each cluster are more likely to belong to the same class, i.e., to have the same prospective label. The teacher initially clusters the datasets using TFIDF-vectors and the k-means algorithm [46], where k equals the number of labels. The further the document is away from the center of its respective cluster, the closer it is assumed to be to the decision boundary [14]. Thus, given a step size N, the teacher returns the data points farthest from the respective centers in each propose step. To do so, we need to implement the *propose* method of the teacher only. The propose method returns the farthest N unseen documents from the centers according to the euclidean distance[15]. There is no need to implement *after_train* or the *after_initial_train* methods of the BaseTeacher (see Subsect. 6.1), as the presented AL strategy does not take any feedback from the training.

Exploitation-Based Teacher (Margin-Confidence). For exploitation, we implemented a margin-confidence AL strategy. Margin-based AL strategies assume that, given a data point, smaller margins between the possible prediction classes of the model mean less

---

[13] For details see https://spacy.io/api/architectures#TextCatBOW.

[14] For details see https://spacy.io/api/architectures#transformers.

[15] We follow [14] using the euclidean distance and TFIDF-vectors. Other approaches such as cosine similarity and doc2vec are interesting alternatives.

certain predictions by the model [21]. For instance, we may assume a sentiment classifier to be more confident if the prediction *negative* has a considerably higher confidence value than *positive*, compared to both possible predictions having almost equal confidence values. Thus, for each propose step, the teacher needs the current state of the trained model to make predictions on the remaining unlabeled data. To reduce the computational costs, the implemented teacher only predicts a random sample of size M, which denotes the *budget* of the AL strategy. For the binary case, the margin confidence for a given data point is given by the margin between the scores of both labels. In the multiclass case, we adopted the best versus second best label, as stated in the margin-confidence approach by [21]. The margin between both classes with the highest predicted scores constitutes the margin confidence of the model for a given data point. Then, the teacher returns the step size N data points with the smallest margin confidences. As for the exploration-based teacher, there is no need to implement the BaseTeacher's *after_train* or the *after_initial_train* methods.

7.3   Experiment Configuration

The margin-confidence teacher uses a budget of 5000 datapoints. The experiment is configured for the case study to use a GPU, a step size of $N_1 = 1000$, and a second step size of $N_2 = 0.2 * M$, where M is the dataset size. The initial training ratio is 0.05 for all experiments conducted. The tracked metric is the macro F1 score [12]. In principle, any other performance metric could be used at this point if correctly implemented by the respective trainer. Especially if the *after_train* methods are implemented, the teacher receives the full metrics set implemented by the trainer and, thus, may use different metrics for proposing than the tracking metrics. Finally, both AL strategies and the randomizer were evaluated on 5 random seeds (42, 4711, 768, 4656, 32213).

Experiments were conducted using a private infrastructure, which has a carbon efficiency of 0.488 kgCO$_2$eq/kWh. A cumulative of 50 h of computation was performed on hardware of type RTX 8000 (TDP of 260 W). Total emissions are estimated to be 5.62 kgCO$_2$eq of which 0 percents were directly offset. Estimations were conducted using the MachineLearning Impact calculator presented in [28].

7.4   Results

We evaluated both AL strategies compared to the baseline randomizer on all 5 seeds, as given in Subsect. 7.3.

IMDb Sentiment Classification. On the IMDb reviews dataset, the margin-confidence strategy performed superior to both the randomizer and the k-means strategy for N_1 and N_2. For step size N_1, the k-means strategy performed inferior to the randomizer. The margin-confidence strategy reaches the threshold of $f1 = 0.85$ on average after proposing 5250 data points only. The randomizer reaches that threshold after proposing 9250, and the k-means teacher after proposing 12250 data points (see Fig. 7a). Similarly, for step size N_2, we observed a slightly inferior performance of the k-means strategy compared to the randomizer. The randomizer reaches the threshold of $f1 = 0.85$ after proposing 11250 data points, while the k-means strategy reaches the threshold after

11750 data points. Again, the margin-confidence strategy outperformed both the randomizer and the k-means strategy by a significant margin, reaching the threshold after proposing 4250 data points (see Fig. 7b).

TREC Question Classification. On the coarse question classification task of the TREC dataset, we observe a minimum threshold of ≈ 1000 proposed data points for the transformer model to learn from the data. With this insight, the initial data ratio may be adapted. For N_1, both the k-means and the margin-confidence strategy performed inferior to the randomizer. All three AL strategies exceeded the threshold of $f1 = 0.85$ after proposing 4245 data points. However, the average $f1$ score of the randomizer was higher than the scores of the k-means and the margin-based strategy for every propose step. Yet, the randomizer showed the highest seed dependent deviation (see Fig. 7c). Therefore, the k-means strategy could, given its low seed-dependent deviation and comparatively good performance, be a more stable strategy for annotating data points for this task.

For N_2, we also do not observe significant advances in the AL strategies compared to the randomizer. The randomizer performs superior to the k-means strategy and the margin-based strategy. More specifically, while both the randomizer and the margin-confidence strategy reach the threshold of $f1 = 0.85$ after proposing 1545 data points, and the k-means strategy after proposing 1845 data points (see Fig. 7d), the randomizer's scores are ahead of the magin-confidence strategy's scores for most propose steps.

7.5 Discussion

This case study does not represent a scientific investigation on the performance of the presented strategies but demonstrates the purposes and the usability of the framework presented in this paper. The k-means strategy performs poorly in our case study and should be investigated further. A possible source of its performance may be the vectorization or the within-cluster homogeneity. If a single cluster has larger within-cluster distances, it may be proposed disproportionately often, thus introducing biases to the proposed dataset. Furthermore, the strategies were tested for different datasets and different models. Comparing their performance on multiple datasets using the same model could reveal further insights. Both strategies did not perform significantly better than the randomizer for the TREC coarse task, neither for N_1 nor N_2. Further experiments with, first, more random seeds to increase generalizability and, second, other classifiers may gain further insights on this matter. Moreover, the initial data ratio should be adopted. Lastly, we observed a task-dependent effect of step sizes on the model's overall performance. Given the lower step size N_2, the trained models exceeded the given threshold earlier and reached significantly better results after proposing significantly fewer data points (e.g., $f1 > 0.9$ after ≈ 1545 data points for N_1, after ≈ 4245 data points for N_2). The chosen model architecture and the online training convention may cause this effect. Still, the step size can be identified as one of the relevant parameters to investigate when evaluating AL strategies using ALE. On the IMDb sentiment analysis task, we observed the margin-confidence strategy after training on fewer data points to perform better than the randomizer after all data points. A possible reason may be, again,

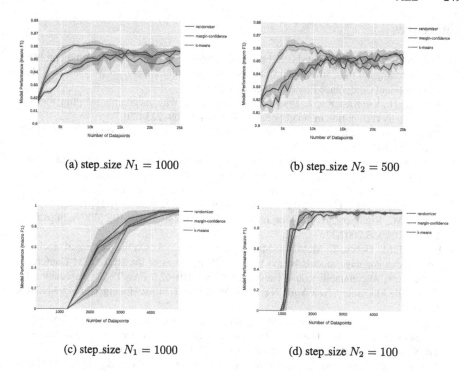

(a) step_size $N_1 = 1000$ (b) step_size $N_2 = 500$

(c) step_size $N_1 = 1000$ (d) step_size $N_2 = 100$

Fig. 7. Mean macro F1-scores of the models for each propose step on the IMDb dataset (a, b) and the TREC Coarse dataset (c, d) with deviations (min-max).

be the warm start of the model's training, being initialized with the previous model's weights and, thus, running into a different local optimum. A scientific evaluation of the effect would be desirable.

8   Conclusion and Future Work

This study presents a configurable and flexible framework for the fair comparative evaluation of AL strategies (*queries*). The framework facilitates the implementation and testing of new AL strategies and provides a standardized environment for comparing their performance. It tracks and documents all parameters and results in a central MLFlow instance. Due to the long-running active learning simulation computations, we provide containerization for using cloud infrastructure and parallelize independent computations. The demonstration of two AL strategies and the instructions for implementing additional strategies shows the usage and may motivate other researchers to contribute their query strategies.

We plan to conduct a survey for developing best practices in the field, enabling researchers and practitioners to select appropriate strategies for their specific NLP task reliably. Additionally, an expansion to the area of relation extraction tasks would allow us to reduce annotation effort in a highly time-consuming domain referring to the annotation process in the NLP domain.

References

1. Agrawal, S., Goyal, N.: Analysis of Thompson sampling for the multi-armed bandit problem. In: Proceedings of the 25th Annual Conference on Learning Theory, pp. 39.1–39.26. JMLR Workshop and Conference Proceedings, June 2012
2. Alemdar, H., Van Kasteren, T., Ersoy, C.: Active learning with uncertainty sampling for large scale activity recognition in smart homes. IOS Press **9**, 209–223 (2017)
3. Anahideh, H., Asudeh, A., Thirumuruganathan, S.: Fair Active Learning, March 2021. arXiv:2001.01796 [cs, stat]
4. Arora, S., Nyberg, E., Rosé, C.P.: Estimating annotation cost for active learning in a multi-annotator environment. In: Proceedings of the NAACL HLT 2009 Workshop on Active Learning for Natural Language Processing - HLT '09, p. 18. Association for Computational Linguistics, Boulder, Colorado (2009). https://doi.org/10.3115/1564131.1564136
5. Bächle, M., Kirchberg, P.: Ruby on rails. IEEE Softw. **24**(6), 105–108 (2007). https://doi.org/10.1109/MS.2007.176
6. Bender, E.M., Gebru, T., McMillan-Major, A., Shmitchell, S.: On the dangers of stochastic parrots: can language models be too big?. In: Proceedings of the 2021 ACM Conference on Fairness, Accountability, and Transparency, pp. 610–623. FAccT '21, Association for Computing Machinery, New York, NY, USA, March 2021. https://doi.org/10.1145/3442188.3445922
7. Bondu, A., Lemaire, V., Boullé, M.: Exploration vs. exploitation in active learning: a Bayesian approach. In: The 2010 International Joint Conference on Neural Networks (IJCNN), pp. 1–7, July 2010. https://doi.org/10.1109/IJCNN.2010.5596815
8. Brown, T.B., et al.: Language models are few-shot learners. In: Proceedings of the 34th International Conference on Neural Information Processing Systems, pp. 1877–1901. NIPS'20, Curran Associates Inc., Red Hook, NY, USA, December 2020
9. Devlin, J., Chang, M.W., Lee, K., Toutanova, K.: BERT: pre-training of deep bidirectional transformers for language understanding. In: Proceedings of the 2019 Conference of the North American Chapter of the Association for Computational Linguistics: Human Language Technologies, Volume 1 (Long and Short Papers), pp. 4171–4186. Association for Computational Linguistics, Minneapolis, Minnesota, June 2019. https://doi.org/10.18653/v1/N19-1423
10. Du, B., Qi, Q., Zheng, H., Huang, Y., Ding, X.: Breast cancer histopathological image classification via deep active learning and confidence boosting. In: Kůrková, V., Manolopoulos, Y., Hammer, B., Iliadis, L., Maglogiannis, I. (eds.) ICANN 2018. LNCS, vol. 11140, pp. 109–116. Springer, Cham (2018). https://doi.org/10.1007/978-3-030-01421-6_11
11. Farquhar, S., Gal, Y., Rainforth, T.: On statistical bias in active learning: how and when to fix it, May 2021
12. Fawcett, T.: An introduction to ROC analysis. Pattern Recognit. Lett. **27**(8), 861–874 (2006). https://doi.org/10.1016/j.patrec.2005.10.010
13. Feng, D., Wei, X., Rosenbaum, L., Maki, A., Dietmayer, K.: Deep active learning for efficient training of a LiDAR 3D object detector. In: 2019 IEEE Intelligent Vehicles Symposium (IV), pp. 667–674, June 2019. https://doi.org/10.1109/IVS.2019.8814236
14. Gan, J., Li, A., Lei, Q.L., Ren, H., Yang, Y.: K-means based on active learning for support vector machine. In: 2017 IEEE/ACIS 16th International Conference on Computer and Information Science (ICIS), pp. 727–731, May 2017. https://doi.org/10.1109/ICIS.2017.7960089
15. Gu, Y., Jin, Z., Chiu, S.C.: Combining active learning and semi-supervised learning using local and global consistency. In: Loo, C.K., Yap, K.S., Wong, K.W., Teoh, A., Huang, K. (eds.) ICONIP 2014. LNCS, vol. 8834, pp. 215–222. Springer, Cham (2014). https://doi.org/10.1007/978-3-319-12637-1_27

16. Herde, M., Huseljic, D., Sick, B., Calma, A.: A survey on cost types, interaction schemes, and annotator performance models in selection algorithms for active learning in classification. IEEE Access **9**, 166970–166989 (2021). https://doi.org/10.1109/ACCESS.2021.3135514

17. Hoi, S.C.H., Sahoo, D., Lu, J., Zhao, P.: Online learning: a comprehensive survey, October 2018. https://doi.org/10.48550/arXiv.1802.02871

18. Hovy, E., Gerber, L., Hermjakob, U., Lin, C.Y., Ravichandran, D.: Toward semantics-based answer pinpointing. In: Proceedings of the First International Conference on Human Language Technology Research (2001)

19. Hsu, W.N., Lin, H.T.: Active learning by learning. In: Proceedings of the AAAI Conference on Artificial Intelligence, vol. 29, no. 1, February 2015. https://doi.org/10.1609/aaai.v29i1.9597

20. Huang, K.H.: DeepAL: deep active learning in python, November 2021. https://doi.org/10.48550/arXiv.2111.15258

21. Joshi, A.J., Porikli, F., Papanikolopoulos, N.: Multi-class active learning for image classification. In: 2009 IEEE Conference on Computer Vision and Pattern Recognition, pp. 2372–2379, June 2009. https://doi.org/10.1109/CVPR.2009.5206627

22. Kaushik, P., Gain, A., Kortylewski, A., Yuille, A.: Understanding catastrophic forgetting and remembering in continual learning with optimal relevance mapping, February 2021. https://doi.org/10.48550/arXiv.2102.11343

23. Kirkpatrick, J., et al.: Overcoming catastrophic forgetting in neural networks. Proc. Natl. Acad. Sci. **114**(13), 3521–3526 (2017). https://doi.org/10.1073/pnas.1611835114

24. Klie, J.C., Bugert, M., Boullosa, B., Eckart de Castilho, R., Gurevych, I.: The inception platform: machine-assisted and knowledge-oriented interactive annotation. In: Proceedings of the 27th International Conference on Computational Linguistics: System Demonstrations, pp. 5–9. Association for Computational Linguistics, Santa Fe, New Mexico, August 2018

25. Kowsari, K., Jafari Meimandi, K., Heidarysafa, M., Mendu, S., Barnes, L., Brown, D.: Text classification algorithms: a survey. Information **10**(4), 150 (2019). https://doi.org/10.3390/info10040150

26. Kuleshov, V., Precup, D.: Algorithms for multi-armed bandit problems, February 2014

27. Kurland, Oren: The cluster hypothesis in information retrieval. In: de Rijke, M., et al. (eds.) ECIR 2014. LNCS, vol. 8416, pp. 823–826. Springer, Cham (2014). https://doi.org/10.1007/978-3-319-06028-6_105

28. Lacoste, A., Luccioni, A., Schmidt, V., Dandres, T.: Quantifying the carbon emissions of machine learning, November 2019. https://doi.org/10.48550/arXiv.1910.09700

29. Leidner, J.L., Plachouras, V.: Ethical by design: ethics best practices for natural language processing. In: Proceedings of the First ACL Workshop on Ethics in Natural Language Processing, pp. 30–40. Association for Computational Linguistics, Valencia, Spain (2017). https://doi.org/10.18653/v1/W17-1604

30. Li, X., Roth, D.: Learning question classifiers. In: COLING 2002: The 19th International Conference on Computational Linguistics (2002)

31. Loy, C.C., Hospedales, T.M., Xiang, T., Gong, S.: Stream-based joint exploration-exploitation active learning. In: 2012 IEEE Conference on Computer Vision and Pattern Recognition, pp. 1560–1567, June 2012. https://doi.org/10.1109/CVPR.2012.6247847

32. Maas, A.L., Daly, R.E., Pham, P.T., Huang, D., Ng, A.Y., Potts, C.: Learning Word Vectors for Sentiment Analysis. In: Proceedings of the 49th Annual Meeting of the Association for Computational Linguistics: Human Language Technologies, pp. 142–150. Association for Computational Linguistics, Portland, Oregon, USA (2011). https://aclanthology.org/P11-1015

33. Madhyastha, P., Jain, R.: On model stability as a function of random seed. In: Proceedings of the 23rd Conference on Computational Natural Language Learning (CoNLL), pp. 929–

939. Association for Computational Linguistics, Hong Kong, China (2019). https://doi.org/10.18653/v1/K19-1087

34. Masana, M., Liu, X., Twardowski, B., Menta, M., Bagdanov, A.D., van de Weijer, J.: Class-incremental learning: survey and performance evaluation on image classification, October 2022. https://doi.org/10.48550/arXiv.2010.15277

35. Mendonça, V., Sardinha, A., Coheur, L., Santos, A.L.: Query strategies, assemble! Active learning with expert advice for low-resource natural language processing. In: 2020 IEEE International Conference on Fuzzy Systems (FUZZ-IEEE), pp. 1–8, July 2020. https://doi.org/10.1109/FUZZ48607.2020.9177707

36. Montani, I., Honnibal, M.: Prodigy: a modern and scriptable annotation tool for creating training data for machine learning models. Prodigy Explosion. https://prodi.gy/

37. Montani, I., et al.: Explosion/spaCy. Zenodo, March 2023. https://doi.org/10.5281/zenodo.7715077

38. Nakayama, H., Kubo, T., Kamura, J., Taniguchi, Y., Liang, X.: Doccano: text annotation tool for human (2018). https://github.com/doccano/doccano

39. Nasar, Z., Jaffry, S.W., Malik, M.K.: Named entity recognition and relation extraction: state-of-the-art. ACM Comput. Surv. **54**(1), 1–39 (2022). https://doi.org/10.1145/3445965

40. Pedregosa, F., et al.: Scikit-learn: machine learning in Python. J. Mach. Learn. Res. **12**(85), 2825–2830 (2011)

41. Pham, H.V., et al.: Problems and opportunities in training deep learning software systems: an analysis of variance. In: Proceedings of the 35th IEEE/ACM International Conference on Automated Software Engineering, pp. 771–783. ASE '20, Association for Computing Machinery, New York, NY, USA, January 2021. https://doi.org/10.1145/3324884.3416545

42. Ren, P., et al.: A survey of deep active learning. ACM Comput. Surv. **54**(9), 1–40 (2022). https://doi.org/10.1145/3472291

43. Ruder, S., Peters, M.E., Swayamdipta, S., Wolf, T.: Transfer Learning in Natural Language Processing. In: Proceedings of the 2019 Conference of the North American Chapter of the Association for Computational Linguistics: Tutorials, pp. 15–18. Association for Computational Linguistics, Minneapolis, Minnesota (2019). https://doi.org/10.18653/v1/N19-5004, https://aclanthology.org/N19-5004

44. Schröder, C., Niekler, A.: A survey of active learning for text classification using deep neural networks, August 2020. https://doi.org/10.48550/arXiv.2008.07267

45. Settles, B.: Active Learning. Synthesis Lectures on Artificial Intelligence and Machine Learning, Springer International Publishing, Cham (2012). https://doi.org/10.1007/978-3-031-01560-1

46. Shah, N., Mahajan, S.: Document clustering: a detailed review. Int. J. Appl. Inf. Syst. **4**(5), 30–38 (2012). https://doi.org/10.5120/ijais12-450691

47. Song, Y., Wang, T., Mondal, S.K., Sahoo, J.P.: A comprehensive survey of few-shot learning: evolution, applications, challenges, and opportunities, May 2022. https://doi.org/10.48550/arXiv.2205.06743

48. Sun, L.L., Wang, X.Z.: A survey on active learning strategy. In: 2010 International Conference on Machine Learning and Cybernetics, vol. 1, pp. 161–166, July 2010. https://doi.org/10.1109/ICMLC.2010.5581075

49. Tomanek, K., Wermter, J., Hahn, U.: An Approach to Text Corpus Construction which Cuts Annotation Costs and Maintains Reusability of Annotated Data. In: Proceedings of the 2007 Joint Conference on Empirical Methods in Natural Language Processing and Computational Natural Language Learning (EMNLP-CoNLL), pp. 486–495. Association for Computational Linguistics, Prague, Czech Republic (2007). https://aclanthology.org/D07-1051

50. Vallor, S.: Technology and the Virtues: A Philosophical Guide to a Future Worth Wanting. Oxford University Press, Oxford (2016)

51. Weidinger, L., et al.: Ethical and social risks of harm from Language Models (2021). https://arxiv.org/abs/2112.04359

52. Yadan, O.: Hydra - A framework for elegantly configuring complex applications (2019). https://github.com/facebookresearch/hydra

53. Yan, X., et al.: A clustering-based active learning method to query informative and representative samples. Appl. Intell. **52**(11), 13250–13267 (2022). https://doi.org/10.1007/s10489-021-03139-y

54. Yang, Y.Y., Lee, S.C., Chung, Y.A., Wu, T.E., Chen, S.A., Lin, H.T.: Libact: pool-based active learning in python, October 2017. https://doi.org/10.48550/arXiv.1710.00379

55. Yuan, M., Lin, H.T., Boyd-Graber, J.: Cold-start active learning through self-supervised language modeling. In: Proceedings of the 2020 Conference on Empirical Methods in Natural Language Processing (EMNLP), pp. 7935–7948. Association for Computational Linguistics, Online, November 2020. https://doi.org/10.18653/v1/2020.emnlp-main.637

56. Zhan, X., Wang, Q., Huang, K.H., Xiong, H., Dou, D., Chan, A.B.: A comparative survey of deep active learning, July 2022 (2022). https://arxiv.org/abs/2203.13450

57. Zhang, Y., Xu, W., Callan, J.: Exploration and exploitation in adaptive filtering based on Bayesian active learning. In: Proceedings of the Twentieth International Conference on International Conference on Machine Learning, pp. 896–903. ICML'03, AAAI Press, Washington, DC, USA, August 2003

Exploring ASR Models in Low-Resource Languages: Use-Case the Macedonian Language

Konstantin Bogdanoski[1]([✉]) [iD], Kostadin Mishev[1,2] [iD], Monika Simjanoska[1,2] [iD], and Dimitar Trajanov[1] [iD]

[1] Faculty of Computer Science and Engineering, Ss. Cyril and Methodius University, Rugjer Boshkovikj 16, Skopje, North Macedonia
konstantin.b@live.com, {kostadin.mishev,monika.simjanoska, dimitar.trajanov}@finki.ukim.mk
[2] iReason LLC, 3rd Macedonian Brigade 37, Skopje, North Macedonia
https://finki.ukim.mk/, https://ireason.mk

Abstract. We explore the use of Wav2Vec 2.0, NeMo, and ESPNet models trained on a dataset in Macedonian language for the development of Automatic Speech Recognition (ASR) models for low-resource languages. The study aims to evaluate the performance of recent state-of-the-art models for speech recognition in low-resource languages, such as Macedonian, where there are limited resources available for training or fine-tuning. The paper presents a methodology used for data collection and preprocessing, as well as the details of the three architectures used in the study. The study evaluates the performance of each model using WER and CER metrics and provides a comparative analysis of the results. The findings of the research showed that Wav2Vec 2.0 outperformed the other models for the Macedonian language with a WER of 0.21, and CER of 0.09, however, NeMo and ESPNet models are still good candidates for creating ASR tools for low-resource languages such as Macedonian. The research presented provides insights into the effectiveness of different models for ASR in low-resource languages and highlights the potentials for using these models to develop ASR tools for other languages in the future. These findings have significant implications for the development of ASR tools for other low-resource languages in the future, and can potentially improve accessibility to speech recognition technology for individuals and communities who speak these languages.

Keywords: Machine learning · Deep learning · Automatic speech recognition · Inference metric · ASR pipeline · Speech-to-text · Wav2Vec2 · ESPNet · NeMo · Natural language processing · Macedonian language · Low-resource languages · Word error rate · WER · Character error rate · CER

1 Introduction

Automatic Speech Recognition (ASR) [26] has become a crucial technology in modern society for improving human-machine interaction and enriching communication methods between individuals with disabilities. However, the development of ASR systems for low-resource languages has remained a challenging task due to the lack of training

D. Conte et al. (Eds.): DeLTA 2023, CCIS 1875, pp. 254–268, 2023.
https://doi.org/10.1007/978-3-031-39059-3_17

data and language-specific resources. To address this issue, recent studies have explored the use of self-supervised pretraining methods such as Wav2Vec 2.0, which have shown promising results in improving the performance of ASR systems for low-resource languages. Moreover, the availability of open-source toolkits such as NeMo and ESPNet has made it easier to train and evaluate ASR models for various languages. In this paper, we present a study that investigates the effectiveness of three ASR architectures, Wav2Vec 2.0, NeMo, and ESPNet for the purpose of creating an ASR model in Macedonian language which is considered a low-resource language with limited language-specific resources. We explore the potential of these architectures for the development of ASR tools and assess their performance using Word Error Rate (WER) and Character Error Rate (CER) metrics. We begin by describing the details of the three architectures used in the study.

- Wav2Vec 2.0 is a self-supervised pretraining method that learns representations from raw audio data without requiring phonetic transcriptions.
- ESPNet is an open-source toolkit that focuses on low-latency and low-resource ASR systems.
- NeMo is an open-source toolkit that provides prebuilt components for building end-to-end ASR systems.

We believe that this study provides valuable insights into the effectiveness of different ASR models for low-resource languages and contributes to the development of ASR tools for languages with limited resources. The rest of the paper is organized as follows. Section 2 presents the related work in the field. The description of the dataset and the architectures is provided in Sect. 3. The results from the experiments are presented in Sect. 4. In the final Sect. 5 we conclude the work and highlight future directions for research.

2 Related Work

Developing speech-to-text translation models for Slavic languages has always been a significant challenge. Unlike other languages, words of Slavic origin can undergo changes in form depending on grammatical, morphological, and contextual relationships with other words. As a result, this increases the vocabulary size and requires more processing time, ultimately leading to higher error rates in word recognition.

Previous attempts to generate such a model sounded far from natural language and exhibited a significant number of errors. In 2004, scientists from Belarus, Russia, Poland, and Germany developed an STT (Speech-To-Text) system for Slavic languages, including Macedonian [7]. To solve the aforementioned problem, they used a speech representation at the morpheme level, significantly reducing the size of the dictionary. Their work includes creating linguistic and acoustic resources for speech-to-text conversion in Russian and creating an STT system for continuous Russian speech based on morphemes [20].

In 2010, several Czech students explained in detail the problems of speech recognition in Slavic languages [17] and offered solutions for them. Greater attention is particularly given to the rich morphology, free word order, and strict grammar [6]. Until then, all languages had used an N-gram language model based on probability as the standard model for speech recognition. However, because of the rich morphology, this model is not suitable for Slavic languages, so different solutions, including smooth n-grams, bi-grams with multiple words, and class-based bi-grams, have been proposed. Smooth n-grams serve as an optimization of the n-gram model. The Witten-Bell formula is seen as the best technique for the smoothing process, while also implementing a decoder [14, 15]. Bi-grams and multi-word types represent a model where bi-grams use a sequence of multiple words, which contributes to reducing the WER error by 10 percent [17]. Class-based bi-grams solve the problem of strict grammar by defining a limited number of classes based on linguistic and morphological categories.

A few years later, in 2016, automatic speech recognition (ASR) was developed only for South Slavic languages [16], which are close to the Macedonian language. An accurate linguistic and acoustic model was fully developed through web-based resources. The model uses an N-gram LM Model, but with the addition of frequently combined compound words in the vocabulary, which helped improve its performance.

On the other hand, a related study was conducted whose main goal was to convert speech to text using the Android platform. As this is a very current topic for development by engineers in recent years, there is also a need to develop such a system for Android users. The system requires access to the microphone, and then converts speech signals into text, which is appropriately saved in files. The idea is driven by the fact that this solution could help people with disabilities, especially blind people who cannot send SMS messages. By using this system, their speech is converted into text, which is then appropriately sent to the telephone number. A Markov model (Hidden Markov Model - HMM) was used, which proved to be a flexible and successful model for solving such problems [19].

An interesting and important research project was conducted using the LUT (Listen-Understand-Translate) system, which is a unified framework for translating audio signals into text and giving semantic meaning to sentences. The system was designed by scientists inspired by neuro-science, human perception, and cognitive systems. The experiments were conducted using English, German, and French languages [5].

The challenges of creating high-quality speech-to-text translation models, particularly for languages with limited resources were addressed in [11], where Mi alongside his researchers proposed a method for modifying target-side data for languages with low data quantities. Specifically, the method generates a large number of paraphrases for target-side data using a generation model that incorporates properties of statistical machine translation (SMT) and recurrent neural networks (RNN). A filtering model is then implemented to select the source that offers the highest success rate, according to speech-target paraphrases of the possible candidates. The experimental results showed that the proposed method achieved significant and consistent improvements compared to several basic models on the PPDB dataset for English, Arabic, German, Latvian, Estonian, Slovenian, and Swedish languages.

To introduce the results of paraphrase generation in low-resource language translation, two strategies are proposed:

– Audio-text paired combinations
– Multi-reference training

Experimental results have shown that speech-to-text models trained on new audio-textual datasets, which combine paraphrase generation, generally result in significant improvement over the baseline, especially in languages with few resources [11].

In 2019, Google researchers successfully introduced a synthesizer model that is multi-speaker and multilingual. This model is based on the Tacotron model [23], which can actually produce high-quality results for generating sound from text in multiple languages. Importantly, the model can transfer knowledge from one language to another. For example, Macedonian language can be synthesized using a voice from a user speaking English, without the need for any bilingual datasets or parallel examples. The same applies to languages that are not similar, i.e., are drastically distant in their speech, such as English and Mandarin [27].

When it comes to languages with a small amount of data, in 2019, researchers from the University of Beijing developed a model that functions with speech recognition (speech-to-text), whose basis involves transferring knowledge gained from examples, for the Tujia language, a Sino-Tibetan language spoken natively by the Tujia people in Hunan Province, China, which is at risk of extinction. In this study, convolutional neural networks and bidirectional long short-term memory neural networks were used to extract cross-lingual acoustic properties and train weights from shared hidden layers for the Tujia language and the Chinese phonetic corpus. According to the experiment, the recognition error rate of the proposed model is 46.19%, which is 2.11% lower than the model that only uses a dataset composed of the Tujia language. Thus, the researchers in Beijing demonstrated that this approach is feasible and effective [25].

3   Methodology

In this section, we describe the architectures used in our study, including Wav2Vec2, NeMo, and ESPNet, and how they work to solve ASR tasks. We also provide an insight into the dataset which we used to train aforementioned models.

3.1   Wav2Vec2

Wav2Vec2 [1] is a speech recognition model developed by Facebook AI Research (FAIR) that has shown impressive results on a range of speech-related tasks. Wav2Vec2 is based on a self-supervised learning approach that uses large amounts of unlabelled speech data to learn speech representations that can be used for downstream tasks such as automatic speech recognition.

The self-supervised learning approach used in Wav2Vec2 involves pre-training the model on large amounts of unlabelled speech data. The model is trained to predict masked speech representations using a contrastive loss function. Specifically, the model learns to predict the masked speech representations from the unmasked speech representations, while simultaneously maximizing the similarity between the predicted masked representations and the actual masked representations.

After pre-training, the model is fine-tuned on a smaller labelled dataset for the specific downstream task. This fine-tuning process adapts the pre-trained representations to the specific task at hand, leading to better performance on the downstream task.

Wav2Vec2 has shown impressive results on a range of speech-related tasks, including automatic speech recognition, speaker recognition, and speech classification. In particular, Wav2Vec2 has shown strong performance on low-resource languages, where labelled data is scarce. This makes Wav2Vec2 a promising model for speech recognition tasks using low-resource Macedonian language data.

The pre-trained version of Wav2Vec2, which we used is - *facebook/wav2vec2-base*.

3.2 NeMo

NeMo (Neural Modules) is an open-source toolkit [10] developed by NVIDIA that enables researchers and developers to easily build and train state-of-the-art neural network models for a variety of natural language processing (NLP) tasks, including ASR [8]. NeMo provides a modular and flexible approach to building and training models, allowing users to easily experiment with different architectures, data processing pipelines, and optimization strategies.

One of the key features of NeMo is its pre-built collection of neural modules, which are reusable building blocks that can be combined to create complex models for various NLP tasks. These modules include popular neural network layers and architectures, as well as specialized modules for speech recognition, such as acoustic models and language models [22]. NeMo also supports a variety of data formats and processing techniques, including raw waveform data and various spectrogram representations.

NeMo's ASR modules are designed to handle low-resource scenarios and can be trained on small datasets with limited amounts of labeled data. One such module is the QuartzNet architecture [9], which is a family of end-to-end ASR models that use 1D convolutional neural networks (CNNs) to process raw audio waveforms. QuartzNet models are highly efficient and can achieve state-of-the-art performance on a range of ASR benchmarks, even when trained on small datasets.

Another key aspect of NeMo is its integration with other popular deep learning frameworks, such as PyTorch [18] and TensorFlow [4], which allows users to easily incorporate NeMo modules into their existing workflows. NeMo also provides a variety of training and optimization tools, including distributed training, mixed-precision training, and automatic mixed-precision optimization, which can help improve training speed and reduce memory usage.

Overall, NeMo offers a powerful and flexible toolkit for building and training ASR models, particularly in low-resource scenarios. Its modular approach and pre-built collection of neural modules make it easy to experiment with different architectures and data processing pipelines, while its integration with popular deep learning frameworks and optimization tools make it an attractive choice for researchers and developers working on NLP tasks.

3.3 ESPNet

ESPNet (End-to-End Speech Processing toolkit) is an open-source toolkit [24] for automatic speech recognition (ASR) developed by the Center for Speech and Language Technologies (CSLT) at Tsinghua University. ESPNet provides an end-to-end pipeline for ASR, which means that the entire process, from audio input to text output, is handled within a single neural network model.

ESPNet's architecture is based on a combination of convolutional neural networks (CNNs) and recurrent neural networks (RNNs). The input audio waveform is first transformed into a sequence of Mel filterbank features, which are then fed into a CNN encoder. The CNN encoder outputs a sequence of higher-level representations that are then fed into a RNN decoder, which generates the final output text.

One of the key features of ESPNet is its support for low-resource languages. ESPNet includes several techniques to help improve ASR performance when there is limited training data available, such as knowledge distillation, transfer learning, and unsupervised pre-training.

ESPNet has achieved state-of-the-art performance on several low-resource ASR benchmarks, including the GlobalPhone dataset [21], which includes Macedonian speech data. In a recent study, ESPNet achieved a word error rate (WER) of 36.7% on Macedonian speech data, outperforming other popular ASR toolkits such as Kaldi and DeepSpeech.

ESPNet's end-to-end architecture and support for low-resource languages make it a promising toolkit for low-resource ASR tasks, including those involving Macedonian speech data.

Wav2Vec 2.0, ESPNet, and NeMo are all valuable architectures for ASR tasks, each with its unique strengths and weaknesses. Wav2Vec 2.0 is well-suited for low-resource languages due to its ability to leverage pre-training and fine-tuning on small amounts of labeled data. ESPNet is designed specifically for low-resource languages and has shown promising results. NeMo provides a flexible and extensible framework for ASR research and includes pre-built models that have achieved state-of-the-art results on several benchmarks.

Ultimately, the choice of model will depend on the specific task and available resources. Through this research, we aim to show which architecture would offer the best foundation for future tasks and analysis.

3.4 Dataset

Datasets are crucial for training and evaluating the performance of automatic speech recognition (ASR) models. These models require large amounts of high-quality audio recordings and their corresponding transcriptions to learn how to recognize and transcribe speech accurately. Without a sufficient amount of training data, the models may not be able to generalize well to new speech signals and can result in poor performance. Additionally, for low-resource languages, datasets play an even more critical role as they provide the foundation for developing and evaluating ASR models. Therefore, the availability and quality of datasets are essential for the development and improvement of ASR systems, and they are an important consideration when comparing different ASR models [3].

The dataset used in this study is a collection of audio data in the Macedonian language, which was collected with the help of volunteers who manually transcribed online audio resources from:

- Mass media (news) - obtained from macedonian media;
- Audio available on the internet - transcription from YouTube;
- Everyday sayings and phrases, and
- Key historical occurrences and figures,

but also provided their own voice for a plethora of audio files. To ensure the quality of the transcriptions, they were provided with guidelines on the transcription process, including punctuation, formatting, and common pronunciation conventions.

The total size of the dataset is 10 h of audio data. The diversity of the audio files was an important consideration, with different dialects, geo-locations, sex, and age-groups being taken into account. In Table 1 there is detailed statistics regarding the dataset, including geo-location, sex, length of characters (minimum length, maximum length, average length) and age range.

Table 1. Dataset statistics

Statistic	Value
Total number of voices	10
Male voices	6
Female voices	4
Total audio time	10 h
Total male audio time	05:50:00
Total female audio time	04:10:00
Longest sequence of letters	517 chars
Shortest sequence of letters	2 chars
Average length of sequence	57 chars
Cities where audio originates from	Skopje Gostivar Veles Shtip Prilep
Age range	23–70 yrs

The transcriptions are stored in plain text format, where each line contains the transcription of a single audio file. The format of the transcription includes the start and end of each spoken segment in the audio file and the corresponding text in the Macedonian language. Additionally, any non-speech segments, such as silences or background noise, are marked with specific tags to inform the models.

The dataset is being preprocessed, meaning we removed all special characters, removed excess whitespaces, and made all letters to be lowercase. We did this to standardize the dataset, so the models will have an easier job to learn from it.

The vocabulary used is the alphabet of the Macedonian language, known as *kirilica*(Cyrillic alphabet, pronounced as ki-ri-li-tsa). It comprises of 31 letters, of which 5 are vowels and 26 are consonants.

Complimenting the letters are the special tokens needed for speech recognition:

- PAD - Padding token, value which is used to make all inputs have the same length;
- UNK - Unknown token, value which is given to a letter, not present in the vocabulary;
- Delimiter - " ", white space value which is used to separate words in the sentence.

4   Results and Analysis

This section presents the evaluation results of the three models, Wav2Vec 2.0, ESP-Net, and NeMo, on the low-resource Macedonian dataset. In this section, we present the performance of each model on the test set, providing a quantitative analysis of the recognition accuracy of each model, and comparing them with each other. Moreover, we analyze the results to identify strengths and weaknesses of the models, as well as potential areas of improvement for future work.

The hardware of the machine used to train the models is shown in Table 2. The Python version which we used is 3.7.

Table 2. Hardware configuration

Device	Value
CPU	Core i7 4790 @ 3.60 GHz
GPU	Nvidia GTX 1060 6 GB
RAM	16 GB

For performance evaluation, we kept an eye on the following metrics:

- Training loss;
- Validation loss;
- Word Error Rate (WER);
- Character Error Rate (CER), and
- Inference output.

After completing a certain segment of training (updating the model parameters), the validation process is done as follows:

- Choose random values from the dataset used for testing (in our case, we use 20% of the data for validation, 80% for training).
- Feed the selected values to the model.
- Check the error rate with the output from the model.
- Update the model parameters for greater accuracy in the next iteration.

The WER metric is used to calculate the difference between the recognized text from the speech and the actual text spoken in the data. The error rate is calculated by dividing the total number of errors (insertions - I, deletions - D, and substitutions of letters - S) by the total number of words/characters in the actual text spoken - N. Mathematically, it can be represented with the equation shown in (1).

$$WER = (S + D + I)/N \qquad (1)$$

Aside from WER, we also stated that we used CER - which represents the Character Error Rate, to evaluate the performance of our models. CER calculates the percentage of incorrect characters in the recognized text compared to the actual transcript. It is a more granular metric than WER and can help us identify specific areas where the model may be struggling. Like WER, CER is also calculated by dividing the total number of errors (insertions, deletions, and substitutions) by the total number of characters in the reference transcript. The equation shown in (2) solves CER.

$$CER = (S + D + I)/N = (S + D + I)/(S + D + C) \qquad (2)$$

In continuation, we will focus on Wav2Vec 2.0, to explain as much as possible about our results and approach. The total length of the training for Wav2Vec2 model took around 40 h of nonstop training. Validation was done on 2000 steps and the total number of training steps was 263040, and in Fig. 1, it can be noticed how the model gains knowledge, and the rate of errors is gradually trending downwards - the model is getting better.

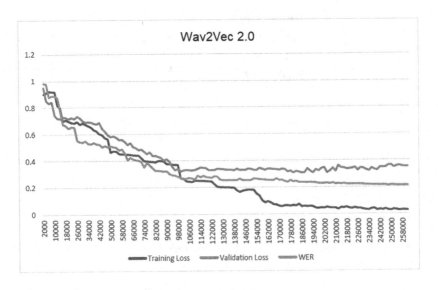

Fig. 1. Wav2Vec 2.0 Training steps

In the last 60,000 training steps, the curves started to smooth out, thus showing that the model learned as much as possible from the provided dataset.

After training the model, we tested the model with unknown audio data not used in the training set. The audio represented the following sentence: "Zdravo jas sum Konstantin Bogdanoski" (translated to: "Hi, I am Konstantin Bogdanoski"). The predicted output from the audio (In Cyrillic) is shown in Fig. 2.

здраво јас сум констаонтин богданоски

Fig. 2. Prediction

In our test, out of a total of 36 characters, the model made a mistake with two, interpreting two characters as one - "ta" with "o", which amounts to a 5% error of the entire phrase. With an increase in the dataset and diversification of the speakers from which it is composed, these results will improve.

4.1 Cross-Model Results

In this sub-segment, we will take a look at the results across all three pretrained models as shown in Table 3. Even though, Wav2Vec 2.0 distniguished as best model, the other two models were also very close to Wav2Vec2's performance. There are several factors that could contribute to the observed results:

- Model architecture: The three models use different architectures, which could affect their performance. For example, Wav2Vec2 is a self-supervised pre-training model that uses a convolutional neural network, while NeMo and ESPNet are end-to-end models that use a combination of convolutional and recurrent neural networks.
- Hyperparameters: The models have been trained with the same hyperparameters, such as learning rate or batch size, but may perform differently and affect performance.

Overall, the results suggest that Wav2Vec2 outperformed NeMo and ESPNet on both WER and CER metrics.

Table 3. Models and results

Model	Train. loss	Valid. loss	WER	CER
Wav2Vec2	0.02	**0.32**	**0.21**	**0.09**
NeMo	**0.015**	0.35	0.23	0.12
ESPNet	0.022	**0.32**	0.24	0.1

Table 4 presents detailed statistics of CER, including the number of substitutions, insertions, deletions and hits of the characters. This is important because it allows us to identify specific areas where the models are struggling and can be improved upon. For example, if we notice a high number of substitutions, it may indicate that the model is having difficulty distinguishing between certain phonemes in the language. By analyzing these statistics, we can fine-tune the model and improve its overall accuracy for speech recognition tasks in Macedonian. Additionally, these statistics can serve as a benchmark for future research on ASR in Macedonian, and can aid in the development of other models and techniques for speech recognition in this language.

On our dataset, the CER value for the models Wav2Vec2, NeMo and ESPNet, came to 0.09, 0.12 and 0.1, respectively. These values are optimistic, since the closer CER is to 0, the greater accuracy the model provides. This could be because of the fact that each letter in the Macedonian language, corresponds to a certain phoneme.

Table 4. CER statistics

Model	Subs.	Dels.	Ins.	Hits
Wav2Vec2	1290	1364	9003	126863
NeMo	1853	2623	12737	138967
ESPNet	1797	2686	13234	172588

Aside from the performance metrics, we also took notice of the inference output of the machine, which is crucial because it measures the real-world performance of a model in a production environment. It evaluates how well the model can handle new, unseen data and provide accurate predictions or results, as shown in Table 5. With the inference metric, we notice that the NeMo model could provide a result faster than the other models. Being made by NVidia, NeMo is utilizing the hardware optimally, compared to the other models, even though they are not that far behind.

Table 5. Inference metrics (seconds)

Model	Wav2Vec2		ESPNet		NeMo	
Time	CPU	GPU	CPU	GPU	CPU	GPU
Total	1122.04	38.4	1150.55	40.3	1134	34.24
Min	0.07	0.0105	0.06	0.00934	0.07	0.009
Max	3.74	0.0113	4	0.0205	3.88	0.0134
Avg	0.5118	0.0175	0.621	0.0183	0.586	0.0169
Comp.	29.22x	0.034x	28.54x	0.035x	33.12x	0.03x

Table 6 and Table 7 present the results for the inference metrics at GPU and CPU, respectively. This information can be useful for selecting the appropriate hardware for deploying the ASR system in real-world scenarios, as well as for optimizing the system's performance. Additionally, it can help identify any potential bottlenecks in the system that may need to be addressed for better efficiency.

Table 6. GPU Inference metrics

Time	Model		
	Wav2Vec2	ESPNet	NeMo
Total	38.4	40.3	37.24
Min	10.5 ms	9.34 ms	9 ms
Max	11.3 ms	20.5 ms	13.4 ms
Avg	17.5 ms	18.3 ms	16.9 ms

Table 7. CPU Inference metrics

Time	Model		
	Wav2Vec2	ESPNet	NeMo
Total	1122.04	1150.55	1134
Min	0.07	0.06	0.07
Max	3.74	4	3.88
Avg	0.5118	0.621	0.586

In Table 5, the row named **Comp.**, compares the performance between using CPU and GPU. For example, when using Wav2Vec2, using the CPU takes 29.22x longer than it takes the GPU to reach a conclusion (Total time only). Throughout all models, we notice a vast improvement in using the GPU as the heavylifter in tasks such as automatic speech recognition.

In this case, a GPU with CUDA cores is faster than CPU in inference metrics because GPUs have many more cores than CPUs, this enables them to perform many more calculations simultaneously. GPUs are designed to perform parallel computations on large amounts of data, whereas CPUs are optimized for sequential processing of smaller amounts of data. This makes GPUs much more efficient at performing the calculations required for deep learning models, which require a large number of matrix multiplications and other operations on large datasets.

Additionally, CUDA is a parallel computing platform and programming model developed by NVIDIA that allows developers to write code that can be executed in parallel on NVIDIA GPUs. This allows the GPU to perform many more calculations simultaneously, further accelerating the inference process.

5 Conclusion

After conducting experiments with various models for speech recognition on our Macedonian audio dataset, we can conclude that Wav2Vec 2.0 provided the best performance results in terms of accuracy and efficiency. It outperformed the other models (NeMo and ESPNet) and also demonstrated solid inference performance. Wav2Vec 2.0, scored a training loss of 0.02, validation loss of 0.32, WER of 0.21, and CER of 0.09. Therefore, we recommend the use of Wav2Vec 2.0 for speech recognition tasks on Macedonian audio files.

In a special use case we could even use Blanket Clusterer's features to have a visual representation of what the models have learned with our approach [2].

To conclude the research, in the future these models offer the foundation to create a Voice Chat Bot, which can be built with AWS Lambdas, in which we use at least 3 Lambdas:

- The first Lambda will receive the audio from the user, and generate text from it (our research proved this is possible even with low-resource languages - such as the Macedonian language)
- The second Lambda will perform NLP tasks, such as sentiment analysis, and provide textual output
- The last Lambda will create an audio output, generated from the text received from the previous Lambda in the pipeline - an option is to use MAKEDONKA [12]

Thanks to AWS's features, this will provide a cost-effective way of using the technology, although performance might be affected due to the architecture of AWS's ready-to-use platform (Fig. 3).

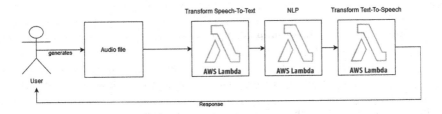

Fig. 3. Pipeline for AWS's Lambda use

Furthermore, with this kind of approach, opportunities to build voice bots, which will handle basic tasks such as Customer Support, will increase significantly. The ability to accurately transcribe and analyze speech in real-time will also have implications in fields such as healthcare, education, and entertainment. Additionally, as the technology continues to improve, it is likely that we will see a wider range of languages and dialects supported, making speech recognition and analysis more accessible to people around the world.

The successful application of Wav2Vec 2.0 in this research opens up exciting possibilities for future research in the area of ASR with the Macedonian language. With further exploration and experimentation, it is possible to improve the performance of the model by increasing the size and diversity of the training dataset. Additionally, future research could investigate the application of the Wav2Vec 2.0 model in other areas, such as natural language processing, sentiment analysis and future implementations and adaptations could also incorporate Abstract Meaning Representation (AMR) [13].

References

1. Baevski, A., Zhou, Y., Mohamed, A., Auli, M.: Wav2vec 2.0: a framework for self-supervised learning of speech representations. Adv. Neural Inf. Process. Syst. **33**, 12449–12460 (2020)
2. Bogdanoski, K., Mishev, K., Trajanov, D.: Blanket clusterer: a tool for automating the clustering in unsupervised learning (2022)
3. Dekker, R.: The importance of having data-sets (2006)
4. Developers, T.: Tensorflow. Zenodo (2021)
5. Dong, Q., et al.: Listen, understand and translate: triple supervision decouples end-to-end speech-to-text translation. In: Proceedings of the AAAI Conference on Artificial Intelligence, vol. 35, pp. 12749–12759 (2021)
6. Hajič, J.: Disambiguation of rich inflection: computational morphology of Czech. Karolinum (2004)
7. Hoffmann, R., Shpilewsky, E., Lobanov, B.M., Ronzhin, A.L.: Development of multi-voice and multi-language text-to-speech (TTS) and speech-to-text (STT) conversion system (languages: Belorussian, Polish, Russian). In: 9th Conference Speech and Computer (2004)
8. Hrinchuk, O., et al.: Nvidia nemo offline speech translation systems for IWSLT 2022. In: Proceedings of the 19th International Conference on Spoken Language Translation (IWSLT 2022), pp. 225–231 (2022)
9. Kriman, S., et al.: QuartzNet: deep automatic speech recognition with 1D time-channel separable convolutions. In: ICASSP 2020–2020 IEEE International Conference on Acoustics, Speech and Signal Processing (ICASSP), pp. 6124–6128. IEEE (2020)
10. Kuchaiev, O., et al.: NeMo: a toolkit for building AI applications using neural modules. arXiv preprint arXiv:1909.09577 (2019)
11. Mi, C., Xie, L., Zhang, Y.: Improving data augmentation for low resource speech-to-text translation with diverse paraphrasing. Neural Netw. **148**, 194–205 (2022)
12. Mishev, K., Karovska Ristovska, A., Trajanov, D., Eftimov, T., Simjanoska, M.: MAKE-DONKA: applied deep learning model for text-to-speech synthesis in Macedonian language. Appl. Sci. **10**(19), 6882 (2020)
13. Mitreska, M., Pavlov, T., Mishev, K., Simjanoska, M.: xAMR: Cross-lingual AMR end-to-end pipeline (2022)
14. Nouza, J.: Strategies for developing a real-time continuous speech recognition system for Czech language. In: Sojka, P., Kopeček, I., Pala, K. (eds.) TSD 2002. LNCS (LNAI), vol. 2448, pp. 189–196. Springer, Heidelberg (2002). https://doi.org/10.1007/3-540-46154-X_26
15. Nouza, J., Drabkova, J.: Combining lexical and morphological knowledge in language model for inflectional (Czech) language. In: Seventh International Conference on Spoken Language Processing (2002)
16. Nouza, J., Safarik, R., Cerva, P.: ASR for south Slavic languages developed in almost automated way. In: INTERSPEECH, pp. 3868–3872 (2016)

17. Nouza, J., Zdansky, J., Cerva, P., Silovsky, J.: Challenges in speech processing of Slavic languages (case studies in speech recognition of Czech and Slovak). In: Esposito, A., Campbell, N., Vogel, C., Hussain, A., Nijholt, A. (eds.) Development of Multimodal Interfaces: Active Listening and Synchrony. LNCS, vol. 5967, pp. 225–241. Springer, Heidelberg (2010). https://doi.org/10.1007/978-3-642-12397-9_19

18. Pytorch, A.D.I.: Pytorch (2018)

19. Reddy, B.R., Mahender, E.: Speech to text conversion using android platform. Int. J. Eng. Res. Appl. (IJERA) **3**(1), 253–258 (2013)

20. Ronzhin, A.L., Karpov, A.A.: Implementation of morphemic analysis for Russian speech recognition. In: 9th Conference Speech and Computer (2004)

21. Schultz, T.: GlobalPhone: a multilingual speech and text database developed at Karlsruhe university. In: Seventh International Conference on Spoken Language Processing (2002)

22. Tamburini, F.: Playing with nemo for building an automatic speech recogniser for Italian. In: CLiC-it (2021)

23. Wang, Y., et al.: Tacotron: towards end-to-end speech synthesis. arXiv preprint arXiv:1703.10135 (2017)

24. Watanabe, S., et al.: ESPNet: end-to-end speech processing toolkit. arXiv preprint arXiv:1804.00015 (2018)

25. Yu, C., Chen, Y., Li, Y., Kang, M., Xu, S., Liu, X.: Cross-language end-to-end speech recognition research based on transfer learning for the low-resource Tujia language. Symmetry **11**(2), 179 (2019)

26. Yu, D., Deng, L.: Automatic Speech Recognition, vol. 1. Springer, Heidelbergt (2016). https://doi.org/10.1007/978-1-4471-5779-3

27. Zhang, Y., et al.: Learning to speak fluently in a foreign language: multilingual speech synthesis and cross-language voice cloning. arXiv preprint arXiv:1907.04448 (2019)

Facilitating Enterprise Model Classification via Embedding Symbolic Knowledge into Neural Network Models

Alexander Smirnov⬢, Nikolay Shilov(✉)⬢, and Andrew Ponomarev⬢

SPC RAS, 39, 14 Line, 199178 St. Petersburg, Russian Federation
{smir,nick,ponomarev}@iias.spb.su

Abstract. In many real life applications, the volume of available data is insufficient for training deep neural networks. One of the approaches to overcome this obstacle is to introduce symbolic knowledge to assist machine-learning models based on neural networks. In this paper, the problem of enterprise model classification by neural networks is considered to study the potential of the approach mentioned above. A number of experiments are conducted to analyze what level of accuracy can be achieved, how much training data is required and how long the training process takes, when the neural network-based model is trained without symbolic knowledge vs. when different architectures of embedding symbolic knowledge into neural networks are used.

Keywords: Neuro-symbolic artificial intelligence · Deep neural networks · Machine learning · Enterprise model classification

1 Introduction

Neural network-based machine learning approaches have become widespread during the last decades. Artificial neural networks are actively used to solve a wide range of information processing tasks. However, the practical application of deep neural networks requires significant amounts of training data, which limits their application in tasks where the collection of such data is complicated or even impossible.

In contrast to neural network-based (sub-symbolic) knowledge, existing symbolic knowledge can be easily adapted to new problem domains without the need to train models on large amounts of data. Thus, the synthesis of neural networks and symbolic paradigms seems very promising. Such a synthesis is called neural-symbolic artificial intelligence [1].

Neuro-symbolic artificial intelligence refers to a very wide range of methods. This particular paper considers the problem of improving the process of neural network-based models on a limited amount of training data using symbolic knowledge.

The idea of neural network-based enterprise modeller assistance was considered in [2] and later developed in [3, 4]. It was shown that successful application of neural machine learning models in enterprise modelling requires knowledge of the modelling context. Hence, identification of the model class (e.g., process model, concept model, product-service model) is important. As a result, the problem of enterprise model classification is considered in the paper since currently there are no available datasets of enterprise models, however, neural networks are penetrating into the enterprise modelling domain, what makes the training process efficiency increase demanded.

Thus, the research question to be answered in the paper is "If symbolic knowledge can facilitate the neural network training so that better results can be achieved on the same (limited) volume of training data for the problem of enterprise model classification?".

The paper is structured as follows. The next section describes the state-of-the-art in the area of integration of symbolic and sub-symbolic knowledge. Then, the data used and the research methodology are presented. These are followed by the experimentation results and their discussion. Section 5 concludes the paper and outlines future work directions.

2 State of the Art Review

In this section, different existing approaches to embedding symbolic knowledge into neural networks are considered together as well as architectures for building symbolic knowledge-aware neural networks.

2.1 Approaches to Embedding Symbolic Knowledge into Neural Networks

Combining neural networks with symbolic knowledge can be done in different ways. In [5] four different integration approaches called *neural approximative reasoning, neural unification, introspection,* and *integrated knowledge acquisition* have been identified.

The term *neural approximative reasoning* refers to methods that use neural networks to generate approximate inferences in intelligent systems [6]. One of the most obvious ways for this type of integration is to implement an approximative neural network for existing rules.

The *neural unification* approach involves simulation of the mechanism used for automatic theorem proving, namely performing a sequence of logical statements leading to a confirmation or refutation of the original statement [7]. Knowledge coding in neural networks is done using selected network elements and special types of links to describe the elements of the assertion. The reasoning is carried out on the basis of minimization of the index of *energy* at updating the state of a neural network. Training of such networks is usually performed using a meta-interpreter that generates training patterns, for example, by encoding successful proofs in the Prolog language. A neural network trained on such examples of evidence is able to generalize a control strategy for selecting arguments when proving assertions.

Introspection refers to methods and techniques by which the AI observes its own behavior and improves its own performance [8]. This approach can be implemented using neural networks that observe the sequence of steps performed by the AI when making logical inferences. This approach is often referred to as *control knowledge*. When the observed behavior of the AI is coded appropriately, the neural network can learn to avoid incorrect pathways and come to its conclusions faster.

The *integrated knowledge acquisition* approach [9] is based on the following assumptions: (a) AI depends mainly on a human expert, who formulates knowledge in the form of symbolic statements (rules); (b) it is almost impossible for an expert to describe her/his knowledge in the form of rules (in particular, it is very difficult to describe knowledge acquired by experience). Therefore, for example, AI may not be able to identify a diagnosis as an experienced physician. Thus, the main challenge is how to extract knowledge from a limited set of examples (small data) for AI use. Machine learning models extend the capabilities of classical AI in the area of logical inference by the ability to generalize and process incomplete data. That is, it is possible to use neural network learning algorithms with a teacher to extract patterns from examples, and then a symbolic rule generator can translate those patterns into rules implemented, for example, in Prolog. On the other hand, having explicit symbolic knowledge (rules) might reduce the amount of training data for identifying implicit patterns.

The above approaches are mostly focused on different macro scenarios of using machine learning methods in intelligent systems. One can see that e.g., *neural approximative reasoning* is mainly focused on the use of machine learning methods for approximation of existing knowledge using neural networks; *neural unification* – on the implementation of methods of logical inference and theorem proving; *introspection* – on neural network training observing other AI models; and *integrated knowledge acquisition* – on neural network training observing expert knowledge. Thus, to answer the research question set in this paper, the *neural approximative reasoning* and *integrated knowledge acquisition* are the most appropriate approaches to embed symbolic knowledge into neural network-based classifier.

2.2 Architectures Aimed at Integration of Symbolic and Neural Knowledge

Wermter and Sun [10] suggest the following classification of architectures aimed at integration of symbolic and neural knowledge. The architectures are schematically shown in Fig. 1 and described below.

Unified architecture: symbolic knowledge is directly encoded in a neural network:

- *Localist connectionist architecture*: separate fragments of the neural network are aimed at encoding symbolic knowledge [11–13].
- *Distributed neural architecture*: symbolic and neural network knowledge is encoded by non-overlapping fragments of the neural network [14, 15].

Transformation architecture (similar to the unified architecture, but includes mechanisms for translation (transformation) of sub-symbolic knowledge representations into symbolic and/or vice versa). Usually, such architecture is implemented through mechanisms of extraction of symbolic knowledge (e.g., rules) from a trained neural network [16].

Hybrid modular architecture: symbolic and neural network knowledge are encoded in separate modules (symbolic knowledge module and neural network knowledge module):

- *Loosely coupled architecture*: information can be transferred from one module to another in one direction only. As a rule, in models with such architecture symbolic

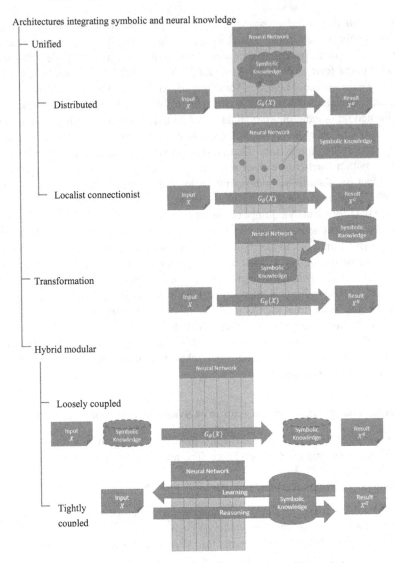

Fig. 1. Architectures integrating symbolic and neural knowledge

knowledge is used either for pre-processing and/or data augmentation before its transmission to the neural network, or for post-processing of the neural network output [17–19].

- *Tightly coupled architecture*: information is exchanged through common data structures in either direction [20–22].

- *Fully integrated architecture*: modules are interconnected through several channels or even based on their overlapping fragments [23].

Unlike approaches described in Sect. 2.1, the architectures above are not aimed at particular usage scenarios and their choice is primarily determined by the specific problem being solved. Thus, if symbolic knowledge is described by separate modules (e.g., ontologies), which are dynamic (evolving), it is reasonable to use *hybrid modular architecture*, separating symbolic and neural network knowledge, because in this case symbolic knowledge can be adjusted to some extent, without affecting the neural network knowledge. Otherwise (when working with static knowledge), *unified* and *transformational architectures* look more attractive because of their flexibility and diversity of ways to incorporate symbolic knowledge into neural network models. In this paper the *loosely coupled hybrid modular architecture* is considered since it is preferable to keep the symbolic knowledge separated with possibilities of its modification and extension.

3 Research Approach

3.1 Problem and Dataset

As it was mentioned, the case study is the problem of enterprise model classification is considered. The model class is identified based on the types of model nodes and their quantity. The dataset was collected based on student assignment works on the enterprise modelling university course validated and corrected by professors in the area of Business Informatics. It consists of 112 models, so it is too small for neural network training in a regular way. The models belong to the following 8 unbalanced classes:

- Business Process Model (43 models),
- Actors and Resources Model (12 models),
- 4EM General Model (7 models),
- Concepts Model (10 models),
- Technical Components and Requirements Model (10 models),
- Product-Service-Model (4 models),
- Goal and Goal & Business Rule Model (13 models),
- Business Rule and Business Rule & Process Model (13 models).

The average number of nodes per model is 27.3. The nodes of the models are of 36 different types, however, only the following 20 meaningful node types are considered: *Rule, Goal, Organizational Unit, Process, Resource,* IS Technical Component, *IS Requirement, Unspecific/Product/Service, Feature, Concept, Attribute, Information Set, External Process, Problem, Cause, Role, Constraint, Component, Opportunity, Individual.*

To evaluate the classification models, the 5-fold cross-validation procedure has been carried out. In accordance with this procedure, the data set is divided into 5 subsets of approximately equal size, and 5 experiments are performed so that each of the subsets is used once as the test set and the other subsets are put together to form the training set.

3.2 Methodology

Since the classification is based only on the presence of nodes of certain types in the model and does not take into account the graph topology, the neural network architecture with three fully connected layers has been selected (Fig. 2). The first two layers are followed by the rectified linear unit (ReLU) activation function:

$$ReLU(x) = (x)^+ = \max(0, x) \tag{1}$$

The size of the input vector is 20 (the number of node types), the intermediate vectors have sizes of 64 and 32. The output vector has size 8 that corresponds to the number of enterprise model classes. The model class is defined as the position of the highest output value in the vector of 8 numbers.

The pre-processing consists of two operations. First, the number of contained in a model nodes of each of 20 types is counted. Then, this vector is normalized (divided by the biggest number of nodes), so it contains only numbers between 0 and 1.

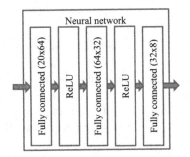

Fig. 2. Used neural network architecture.

Application of Neural Network Only. The first experiment applies only the neural network to the model classification without using any symbolic knowledge (Fig. 3). The cross entropy is used as the loss function:

$$CELoss = -\sum_{c \in C} log \frac{exp(x_y)}{\sum_{c \in C} exp(x_c)} \cdot y_c \tag{2}$$

where
    C is the set of classes,
    x_i is a prediction result for class i,
    y_i is a target class.

In this and the other two experiments the training is performed using the learning rate of 10^{-3} (selected as after performing several experiments with different learning rate values) and Adam optimizer [24] since it achieves better results in most cases, has faster computation time, and requires fewer parameters for tuning. The early stopping mechanism is used to finish the training when the accuracy on the test set does not improve for 20 epochs in a row.

Application of Semantic Loss Function. The second experiment applies the semantic loss function [20] to the neural network output along with the Cross Entropy loss function (Fig. 4). The semantic loss function makes it possible to apply logical constraints to the neural network output vectors and use this knowledge to improve the training process. The applied rules included the possibility for a model to be of a certain class if it contains a node of certain type (e.g., "if the model contains a node of type *Rule*, it can belong only to one of the following three classes: *4EM General Model, Goal Model and Goal & Business Rule Model*, or *Business Rule Model and Business Rule & Process Model*"). Such rules have been defined for all 20 types of nodes (20 rules in total). The other training parameters remain unchanged.

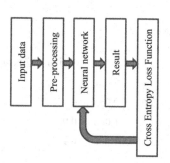

Fig. 3. Experiment with neural network only.

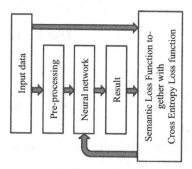

Fig. 4. Experiment with semantic loss function.

Application of Symbolic Pre-processing. The third experiment assumes including additional inputs, which are results of applying rules to the original input data (Fig. 5). The rules are the same as described in the section above, and the additional 8 inputs reflect the possible model classes (1 if the corresponding class is possible and 0 otherwise). As a result the first layer of the neural network has size 28 instead of original 20. All other training parameters remain unchanged.

4   Results and Discussion

The experimentation was performed using a server with Intel Core i9--10900X S2066 3.7G processor, 64 Gb RAM, and MSI RTX3090 GPU. JupyterLab 2.0.1 environment was used with Python 3.6.9 and PyTorch 1.7.1. The results are presented in Tables 1–4.

Tables 1 and 2 show the accuracy and number of errors on test sets for each of 5 folds as well as the average and total values. These numbers indicate how well the built machine learning models can generalize data from samples and classify enterprise models, which were not used during training. One can see that the architecture using the semantic loss function was not very successful for the task at hand. In fact, it performed even a little worse than the pure neural network, though the difference is not significant.

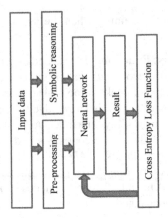

Fig. 5. Experiment with symbolic pre-processing.

Table 1. Accuracy on the test sets.

Experiment	Fold					Weighted Average
	1	2	3	4	5	
Neural network	0.957	0.957	0.909	0.864	0.955	0.929
Semantic loss	0.957	0.913	0.909	0.818	1.000	0.920
Symbolic pre-processing	0.957	1.000	0.955	0.955	1.000	0.973

Table 2. Number of errors on the test sets.

Experiment	Fold					Total
	1	2	3	4	5	
Neural network	1	1	2	3	1	8
Semantic loss	1	2	2	4	0	9
Symbolic pre-processing	1	0	1	1	0	3

Table 3. Accuracy on the training set.

Experiment	Fold					Weighted Average
	1	2	3	4	5	
Neural network	0.991	0.973	0.964	0.946	0.991	0.973
Semantic loss	0.991	0.973	0.973	0.938	0.991	0.973
Symbolic pre-processing	0.991	1.000	0.991	0.991	1.000	0.995

Table 4. Number of errors on the training set.

Experiment	Fold					Total
	1	2	3	4	5	
Neural network	1	3	4	6	1	15
Semantic loss	1	3	3	7	1	15
Symbolic pre-processing	1	0	1	1	0	3

At the same time, application of the symbolic pre-processing gave a significant improvement over the pure neural network: the number of errors decreased in 2.67 times. This can be explained by the fact that in this case added inputs strongly correlate with the expected result what gave the positive effect on the machine learning model efficiency.

Tables 3 and 4 show the accuracy and number of errors not only for the test sets, but also for the training sets. They are aimed at demonstration of the capability of machine learning models not only to generalize but also to learn the training samples. One can see that application of the semantic loss function does not decrease this capability, and the model with symbolic pre-processing is capable of learning the training samples perfectly.

5 Conclusion

The research question stated in the paper is "If symbolic knowledge can facilitate the neural network training so that better results can be achieved on the same (limited) volume of training data for the problem of enterprise model classification?".

In order to answer this question, a state of the art in the corresponding areas has been analysed and a number of experiments performed. The use case for the experiments was the classification of enterprise models based on the types of contained nodes.

Three neural network-based architectures have been analysed: pure neural network, application of the semantic loss function, and symbolic data pre-processing. The experimental results showed that symbolic knowledge indeed can significantly improve the capabilities of the neural networks to classify enterprise models using only very limited training data (the used dataset consists of only 112 samples). This improvement was achieved by the machine learning model with symbolic data pre-processing. At the same time, introduction of the semantic loss function did not produce any improvement.

The experimentation with the use case from only one problem domain can be mentioned as a limitation of the presented research. So, it is possible to speak about the efficiency of the symbolic pre-processing and inefficiency of the semantic loss only in the context of the considered use case.

Planned future work will extend the research in two main dimensions. On the one hand, it is planned to experiment with several use cases and bigger datasets (so that the validation set could be used for model hyper-parameter fine-tuning and early stopping and the test set for measuring the models' performance), and on the other hand, more architectures integrating symbolic and sub-symbolic knowledge will be considered.

Acknowledgements. The research is funded by the Russian Science Foundation (project 22-11-00214).

References

1. d'Avila Garcez, A., Lamb, L.C.: Neurosymbolic AI: The 3rd Wave (2020)
2. Borozanov, V., Hacks, S., Silva, N.: Using machine learning techniques for evaluating the similarity of enterprise architecture models. In: Giorgini, P., Weber, B. (eds.) CAiSE 2019. LNCS, vol. 11483, pp. 563–578. Springer, Cham (2019). https://doi.org/10.1007/978-3-030-21290-2_35
3. Shilov, N., Othman, W., Fellmann, M., Sandkuhl, K.: Machine learning-based enterprise modeling assistance: approach and potentials. In: Serral, E., Stirna, J., Ralyté, J., Grabis, J. (eds.) PoEM 2021. LNBIP, vol. 432, pp. 19–33. Springer, Cham (2021). https://doi.org/10.1007/978-3-030-91279-6_2
4. Shilov, N., Othman, W., Fellmann, M., Sandkuhl, K.: Machine learning for enterprise modeling assistance: an investigation of the potential and proof of concept. Softw. Syst. Model. **22**, 619–646 (2023). https://doi.org/10.1007/s10270-022-01077-y
5. Ultsch, A.: The integration of neural networks with symbolic knowledge processing. In: New Approaches in Classification and Data Analysis, pp 445–454 (1994)
6. Guest, O., Martin, A.E.: On logical inference over brains, behaviour, and artificial neural networks. Comput. Brain Behav. **6**, 213–227 (2023). https://doi.org/10.1007/s42113-022-00166-x
7. Picco, G., Lam, H.T., Sbodio, M.L., Garcia, V.L.: Neural unification for logic reasoning over natural language (2021)
8. Prabhushankar, M., AlRegib, G.: Introspective learning : a two-stage approach for inference in neural networks (2022)
9. Mishra, N., Samuel, J.M.: Towards integrating data mining with knowledge-based system for diagnosis of human eye diseases. In: Handbook of Research on Disease Prediction Through Data Analytics and Machine Learning. IGI Global, pp 470–485 (2021)
10. Wermter, S., Sun, R.: An overview of hybrid neural systems. In: Wermter, S., Sun, R. (eds.) Hybrid Neural Systems 1998. LNCS (LNAI), vol. 1778, pp. 1–13. Springer, Heidelberg (2000). https://doi.org/10.1007/10719871_1
11. Pitz, D.W., Shavlik, J.W.: Dynamically adding symbolically meaningful nodes to knowledge-based neural networks. Knowl. Based Syst. **8**, 301–311 (1995). https://doi.org/10.1016/0950-7051(96)81915-0
12. Arabshahi, F., Singh, S., Anandkumar, A.: Combining symbolic expressions and black-box function evaluations in neural programs (2018)
13. Xie, Y., Xu, Z., Kankanhalli, M.S., et al.: Embedding symbolic knowledge into deep networks. In: Advances in Neural Information Processing Systems (2019)
14. Hu, Z., Ma, X., Liu, Z., et al.: Harnessing deep neural networks with logic rules (2016)
15. Prem, E., Mackinger, M., Dorffner, G., Porenta, G., Sochor, H.: Concept support as a method for programming neural networks with symbolic knowledge. In: Jürgen Ohlbach, H. (ed.) GWAI 1992. LNCS, vol. 671, pp. 166–175. Springer, Heidelberg (1993). https://doi.org/10.1007/BFb0019002
16. Shavlik, J.W.: Combining symbolic and neural learning. Mach. Learn. **14**, 321–331 (1994). https://doi.org/10.1007/BF00993982
17. Li, Y., Ouyang, S., Zhang, Y.: Combining deep learning and ontology reasoning for remote sensing image semantic segmentation. Knowl. Based Syst. **243**, 108469 (2022). https://doi.org/10.1016/j.knosys.2022.108469

18. Dash, T., Srinivasan, A., Vig, L.: Incorporating symbolic domain knowledge into graph neural networks. Mach. Learn. **110**(7), 1609–1636 (2021). https://doi.org/10.1007/s10994-021-059 66-z
19. Breen, C., Khan, L., Ponnusamy, A.: Image classification using neural networks and ontologies. In: Proceedings. 13th International Workshop on Database and Expert Systems Applications, pp. 98–102. IEEE (2002)
20. Xu, J., Zhang, Z., Friedman, T., et al.: A semantic loss function for deep learning with symbolic knowledge. Proc. Mach. Learn. Res. **80**, 5502–5511 (2018)
21. Yang, Z., Ishay, A., Lee, J.: NeurASP: embracing neural networks into answer set programming. In: Proceedings of the Twenty-Ninth International Joint Conference on Artificial Intelligence. International Joint Conferences on Artificial Intelligence Organization, California, pp. 1755–1762 (2020)
22. d'Avila, G.A.S., Gabbay, D.M., Ray, O., Woods, J.: Abductive reasoning in neural-symbolic systems. Topoi **26**, 37–49 (2007). https://doi.org/10.1007/s11245-006-9005-5
23. Lai, P., Phan, N., Hu, H., et al.: Ontology-based interpretable machine learning for textual data. In: 2020 International Joint Conference on Neural Networks (IJCNN), pp. 1–10. IEEE (2020)
24. Kingma, D.P., Ba, J.: Adam: a method for stochastic optimization (2014)

Explainable Abnormal Time Series Subsequence Detection Using Random Convolutional Kernels

Abdallah Amine Melakhsou[1,2(✉)] and Mireille Batton-Hubert[1]

[1] Mines Saint-Etienne, Univ. Clermont Auvergne, CNRS, UMR 6158 LIMOS,
Institut Henri Fayol, 42023 Saint-Etienne, France
{amine.melakhsou,batton}@emse.fr
[2] elm.leblanc, 6 Rue des Écoles, 29410 Saint-Thégonnec Loc-Eguiner, France

Abstract. To identify anomalous subsequences in time series, it is a common practice to convert them into a set of features prior to the use of an anomaly detector. Feature extraction can be accomplished by manually designing the features or by automatically learning them using a neural network. However, for the former, significant domain expertise is required to design features that are effective in accurately detecting anomalies, while in the latter, it might be complex to learn useful features when dealing with unsupervised or one-class classification problems such as anomaly detection, where there are no labels available to guide the feature extraction process. In this paper, we propose an alternative approach for feature extraction that overcomes the limitations of the two previously mentioned approaches. The proposed method involves calculating the similarities between subsequences and a set of randomly generated convolutional kernels and the use of the One-Class SVM algorithm. We tested our approach on voltage signals acquired during circular welding processes in hot water tank manufacturing, the results indicate that the approach achieves higher accuracy in detecting welding defects in comparison to commonly used methods. Furthermore, we introduce in this work an approach for explaining the anomalies detected by making use of the random convolutional kernels, which addresses an important gap in time series anomaly detection.

Keywords: Time series anomaly · Random convolutional kernels · Explainability

1 Introduction

Identifying anomalies in time series has gained a lot of attention in recent years due to its vast range of applications, such as fraud detection and fault identification in manufacturing processes. Time series anomalies are classified into three types: point anomalies, subsequence anomalies, and full time series anomalies. Defects in industrial processes frequently manifest as irregular subsequences in the time series of the process variables, which is the case of arc welding where defects, such as burn-through and misalignment, often occur in a small part of the weld seam and exhibit themselves as abnormal subsequences in the signals of welding variables such as the voltage or current.

In the problem of abnormal subsequence detection, converting the raw data into a set of features is a commonly employed approach intended to obtain a representation allowing effective anomaly detection. Features for time series can be classified into two categories: engineered features and deep features. Engineered features are meaningful characteristics of the time series that are hand-crafted prior to the use of an anomaly detector. They can be extracted from the time domain, the frequency domain and the time-frequency domain. Despite their advantage of being interpretable, manually designing the optimal features that allow accurate anomaly detection can be a time-consuming task, and it requires significant human expertise. To bypass this shortcoming, deep learning can be used in order to automatically extract features. However, this might be challenging for unsupervised and one-class classification problems like anomaly detection, where there are no labels available to guide the feature extraction process. Moreover, advanced deep learning models typically require a significant amount of data and training can have an important computation complexity. Additionally, the features that are automatically extracted through deep learning are often difficult to interpret.

In addition to the above shortcomings, most of the methods proposed for anomaly detection in the literature focus on detection accuracy while ignoring the aspect of explaining the predictions of the anomaly detector [12]. Specifically, only little attention is given to explaining models based on temporal data [19].

To overcome the above shortcomings and in an attempt to fill these research gaps, we present in this paper an approach for explainable abnormal time series subsequence detection where the features are similarities to random convolutional kernels. These features are then used as inputs for a One-Class Support Vector Machine (OC-SVM) model that learns the hypersurface containing the normal data. In the deployment stage, subsequences that do not belong to the learned region are detected as abnormal. When an abnormal subsequence is detected, we would like to have information about why it is anomalous. For this purpose, we propose in this work an explainability approach that makes use of the random kernels. Overall, the proposed approach is computationally efficient, requires no prior knowledge about the data or the application domain, and can detect anomalous subsequences with high accuracy while providing explanations for the detected anomalies.

This paper is organized as follows: We review related works in Sect. 2. In Sect. 3 we present the proposed methods. We give a brief description of the circular welding process used in the manufacturing of hot water tanks in Sect. 4 and in Sect. 5 we present the results of the experiments, and we finish by conclusions and future works.

2   Related Work

There is a plethora of methods proposed for abnormal time series subsequences detection. These methods can be classified based on two main factors: the level of supervision of the machine learning model, which can be supervised, semi-supervised, or unsupervised, and the type of representation of the subsequences, which can be raw or transformed. When working with raw data, distance-based approaches are one of the most used methods in the literature. Famous distance-based techniques include the k Nearest Neighbours (k-NN) algorithm and its variants. [10] used the Local Outlier Factor (LOF) with the Dynamic Time Warping (DTW) dissimilarity in order to detect abnormal subsequences in energy consumption time series. [7] use the average of the distances of the subsequence to its k nearest neighbours as the anomaly score. Clustering is another technique that can deal with the raw data for anomaly detection. [2] proposed a method that consists of summarizing the subsequences of a time series into only one subsequence that best captures the normal behaviour. This is done by clustering the subsequences and choosing the centroid of the cluster that represents the normality, depending on the criteria of frequency and coverage. The anomaly score of an upcoming subsequence is defined here as the distance between this latter and the normal subsequence. To detect abnormal subsequences in multivariate time series, [11] propose an extension of the Fuzzy C-Means where they use a weighted Euclidean distance as a measure of dissimilarity between multivariate subsequences. Abnormal subsequences are then detected based on the reconstruction error of the subsequences using the centroid of the cluster.

Forecasting-based approaches are another class of methods that model the training data using a forecasting model. The anomaly score in this case is the forecasting error. classical methods such as Autoregressive Moving Average (ARMA) and Autoregressive Integrated Moving Average (ARIMA) were used for this purpose in numerous works as in [15, 17] and [9]). Recent works leverage deep learning for forecasting and anomaly detection in time series. Advanced neural networks such as Long Short Term Memory (LSTM) and Convolutional Neural Networks (CNN) were proposed for this purpose in [3, 14] and [16].

In some situations, it is preferable to work with a transformation of the subsequences as input to an anomaly detector rather than the raw data. This can be achieved by feature extraction, which can be classified into two types: engineered features and deep features. Engineered features are meaningful characteristics of the time series subsequences that are hand-crafted by a domain expert. [5] proposed a method based on the extraction of 6 statistics from subsequences and the use of OC-SVM. The so-called meta-features consist of kurtosis, the coefficient of variation, oscillation, regularity, square waves and the variation of the trends. When a new subsequence arrives, its 6 statistics are extracted, after which the OC-SVM predicts whether they belong to the learned normal regions. Similarly, [20] Extracted 15 statistical characteristics from the vibration signal of rolling-element bearing and used the Isolation Forest algorithm to detect defects. Despite the ease of implementation of engineered feature-based approaches and the interpretability of the extracted features, they have the limitation of requiring a high level of human knowledge for the extraction of the proper features that allow accurate anomaly detection [8]. Additionally, the extracted features are specific to the given problem and might not be applicable to other problems.

Rather than manually selecting features, some studies use deep neural networks to automatically extract features. These extracted features are commonly referred to as deep features. In [21], an autoencoder is utilized to extract features from the latent space, which are then used with OC-SVM to detect abnormal time series. Other studies, such as the one in [13], replace the kernel function of the Support Vector Data Description (SVDD), which is another formulation of OC-SVM, with a neural network. This enables simultaneous optimization of the loss functions of the SVDD and the reconstruction function of the autoencoder, which ensures that the features learned by the autoencoder are contained within a minimum volume hypersphere. However, selecting the appropriate neural network architecture and parameters can be challenging, particularly when only data from the normal class is available.

3 Methods

3.1 Random Convolutional Kernel Transform with OC-SVM

Random convolutional kernel transform (ROCKET) is a method for classifying time series data proposed by [4]. It involves extracting features from the convolutions of the time series with a high number of randomly generated convolutional kernels, which are then used as input to a linear classifier to classify the time series. The approach achieved state-of-the-art accuracy while maintaining a faster run-time than existing methods.

The feature extraction part of the approach is based on the convolution, which is a sliding dot product, between the input time series x and a random convolutional kernel w. Each dot product is defined as follows:

$$z_i = x_i * w = \left(\sum_{j=0}^{l_{kernel}-1} x_{i+(j \times d)} \times w_j \right) + b \tag{1}$$

where:

- l_{kernel} is the length of the kernel selected randomly from $\{7, 9, 11\}$ with uniform probability.
- w are the weights of the kernel sampled randomly from a normal distribution $w \sim N(0, 1)$ and are afterwards centered i.e. $w = (w - \bar{w})$
- b is the bias sampled from a uniform distribution, $b \sim U(\text{-}1, 1)$
- d is the dilation sampled on an exponential scale $d = 2^a$, $a \sim U(0, A)$ where $A = log_2 \frac{l_{input}-1}{l_{kernel}-1}$. This ensures that the kernel's length including dilation is at most equal to the length of the input time series.
- Padding: for each random kernel, a decision is made at random if the padding will be used, if yes, zeroes are appended at the start and the end of the time series.
- k is the number of kernels. The authors state that the higher the number, the higher the classification accuracy. They recommend using a default value of k = 10 000.
- Stride is always equal to 1.

Two features are then extracted from each convolution, which are:

- $ppv(Z) = \frac{1}{n} \sum_{i=0}^{n-1} [z_i > 0]$: The proportion of positive values. Where Z is the output of the convolution and n is the length of Z.
- $mv = max(Z)$: The maximum value of the output of the convolution. It gives the maximum similarity between the random kernel and the shapes in the time series.

The $2 \times k$ features are then used to train a linear classifier in the original paper, since in a high dimensional space the classes can be linearly discriminated.

The approach had gained a lot of interest in the time series classification domain. However, this kind of transformation did not receive much attention for time series anomaly detection, particularly for abnormal time series subsequences detection. In this paper, we are interested in using features extracted from convolutions with random filters to derive an explainable approach for the detection of abnormal time series subsequences. We first modify the feature extraction process as follows:

- l_{kernel} is selected randomly from $\{3, 5, 7, 9, 11\}$ with uniform probability. The added lengths are intended to capture and explain abnormal trends of the subsequences.
- Only the mv is extracted from the convolution. This is in order to derive an explainability approach that we shall present in the next section.
- We do not use padding when performing the convolution. This is because we want the maximum value (mv) to reflect the similarity of a random kernel with a pattern that is fully contained in the subsequence, in order to achieve a more accurate explainability.

The extracted max values from the convolutions with the random kernels will afterwards be used with the One-class SVM (OC-SVM) algorithm in order to learn the normal patterns of the subsequences. OC-SVM is a well-known semi-supervised algorithm that learns the boundaries of the hypersurface containing the normal data class proposed by [18] and that is formulated by the following optimization problem:

$$\min \frac{1}{2} \|w\|^2 - \rho + \frac{1}{\nu N} \sum_{i=1}^{N} \xi_i \tag{2}$$

$$s.c : \langle w, x_i \rangle \geq \rho - \xi_i$$
$$\xi_i \geq 0 \tag{3}$$

With:

- w: the vector normal to the hyperplane.
- ρ: the distance between the origin and the hyperplane.
- $\nu \in]0; 1]$: the proportion of outliers in the training data.
- ξ: slack variables used to account for outliers.

The algorithm seeks to find a hyperplane that maximizes the distance ρ between the origin of the space and the normal training data. The algorithm is a soft-margin problem that depends on the percentage of outliers in the training data, denoted by ν. This parameter serves as both an upper limit on the proportion of outliers and a lower limit on the proportion of support vectors.

To enable the OC-SVM to learn nonlinear decision boundaries, a nonlinear kernel (not to confuse with the random convolutional kernels) is used in the dual problem. This kernel computes indirectly the similarities between the data in a possibly infinite-dimensional space, without performing the projection of the data into that space. The dual problem is given as follows:

$$\min \sum_{i,j}^{N} \alpha_i \alpha_j k(x_i, x_j) \tag{4}$$

$$s.t : 0 \leq \alpha_i \leq \frac{1}{\nu N}$$

$$\sum_i^N \alpha_i = 1 \tag{5}$$

With α_i the Lagrange multipliers, k the kernel function and N the number of training data. We use in this work the Gaussian kernel defined as $k(x_i, x_j) = \exp\left(-\gamma|x_i - x_j|^2\right)$ where γ is a smoothing parameter.

To predict if an instance x' is normal, the OC-SVM uses a decision function that determines whether it is inside the learned decision boundaries. The decision function is defined as follows:

$$F(x') = (w.\phi(x') - \rho) = \left(\sum_i^N \alpha_i k(x', x_i) - \rho\right) \tag{6}$$

If the sign of $F(x')$ is positive, x is in the normal data region, otherwise, it is outside this region and is then abnormal. The anomaly score is the value of the decision function $F(x')$, the smaller the value, the more anomalous x' is.

The complete method we propose for abnormal subsequences detection is depicted in Fig. 1. In summary, the method is intended to learn the normal patterns of the subsequences by using the maximum values of the convolutions between normal subsequences and a number of random convolutional kernels. Once these features are extracted, they are used as inputs to an OC-SVM model. To predict if new subsequences are normal, first, their features are extracted using the random kernels generated in the training stage and are then analyzed by the OC-SVM model that returns an anomaly score of each of the subsequences.

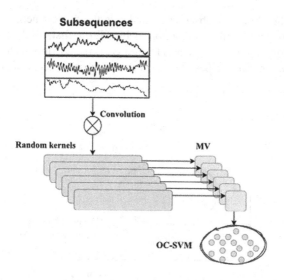

Fig. 1. The proposed approach for abnormal subsequence detection.

3.2   Explainability Using Random Convolutional Kernels

Explaining the prediction of machine learning models is becoming an important research topic in recent years. However, little attention is given to explaining models based on temporal data [19]. Hence, there is a need for research works addressing this subject. Explainability in anomaly detection consists in explaining why an anomaly is considered anomalous, and this by anomaly description and outlying property detection [12]. There are multiple types of strategies to achieve this. According to [19], they can be classified into explainability by example, feature importance and causality mechanism. In our case, we adopt the feature importance strategy where a value is attributed to each feature indicating its importance in the obtained prediction. There are two types of methods for feature importance attribution, model-agnostic and model specific [1]. Model-agnostic approaches, such as Local Interpretable Model-agnostic Explanations (LIME) and SHapley Additive exPlanations (SHAP) are designed to explain any model while model-specific approaches take into account the model structure to derive explainability, and they have the advantage of providing a more reliable explanation [12].

We propose in this paper a simple model-specific explanation approach that makes use of the random kernels. First, the approach involves estimating the importance of each random kernel in the obtained prediction. For this, we define the importance of the features as a constrained point-wise difference between the abnormal observation and its nearest neighbour from the normal training data, as Fig. 2 shows. Formally, the features' importance (FI) is estimated as follows:

$$FI_i(x) = \begin{cases} x_i - s_i & \text{if } (x_i - s_i) > 0 \\ 0 & \text{Otherwise} \end{cases} \tag{7}$$

Where x is the abnormal observation and s is the nearest neighbour of x in the training data. To understand the assumptions behind Eq. 7, first, recall that the maximum value of the convolution that we use as a feature gives the maximum similarity between the random kernel and the shapes contained in the subsequence. Suppose that we use

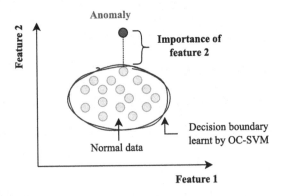

Fig. 2. An illustration of the proposed approach to estimate feature importance.

5 random and that the abnormal subsequence has a feature vector $x = [0, 0, 2, 8, 9]$, which indicates that it has a relatively high similarity with the 4th and the 5th random kernels. Suppose now that the nearest neighbour of x from the training data is given by $s = [5, 5, 2, 1, 1]$, indicating that a typical normal subsequence that x deviates the least from is characterized by high similarities with the first two random kernels while having low similarities with the last two. In order to give an informative characterization and explanation of why x is anomalous, one should answer the following question: *which shapes are present in the subsequences that made it abnormal?* which we attempt to answer by using Eq. 7, that gives the following feature importance vector $FI = [0, 0, 0, 7, 8]$, highlighting that the high similarities with the 4th and 5th random kernels and is what characterizes the abnormal subsequence, while ignoring the information that the dissimilarity with the first two kernels possibly also made the subsequence abnormal as this might not be informative for the users as they might not be aware of all the normal shapes of the subsequences.

After estimating the contribution of a random kernel to the prediction based on the defined approach, we need to find where in the subsequence the similarity with the most important random kernels is the highest, which is the second part of the explanation that we call the localization L, which indicates the localization of the pattern that makes the subsequence abnormal. For this purpose, we need to find the argument of the maximum value of the convolution $L_i = argmax(Z_i(x))$. The final explanation for the abnormal subsequence x is then a tuple $\{FI_i(x), L_i(x)\}$ with $i = [1, 2, ...k]$ (where k is the number of random kernels), arranged by $FI_i(x)$ in a decreasing order.

To test our proposed approaches for anomaly detection and explainability, we consider voltage signals obtained from the circular welding processes in the hot water tanks industry that we describe in the next section.

4   The Circular Welding Process of Hot Water Tanks

The hot water tank is a multipart product that undergoes multiple welding processes of different types. Detecting welding defects in this process by classical quality inspection techniques can be a complex task, as it requires significant expertise in the welding domain. Furthermore, inspecting every weld in the manufacturing process can be time-consuming, which leads to a decrease in efficiency. Thus, there is a need for automatic welding defect detection systems in this industry. We are interested here in the circular welding process of hot water tanks that is accomplished by a semi-automatic machine that uses Gas Tungsten Arc Welding (GTAW) (Fig. 3), which simultaneously welds the two caps with the cylinder. The tank rotates at a constant speed during the welding process, while the two torches remain stationary. The wire is automatically fed into the pool at a constant speed and the welding is done with a fixed current level. To monitor the process, two voltage signals are collected in real-time, one for each side of the tank, at a sampling rate 25 Hz. An example of a voltage signal from a normal welding cycle is shown in Fig. 4. The normal voltage signal is non-periodic, with a global trend that reflects the overall shape of the cap and the cylinder, as voltage correlates to the arc length. The trend can be different from one signal to another, and the signal can be non-stationary without presenting any anomalies.

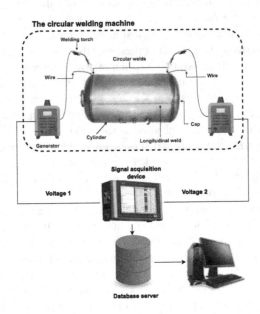

Fig. 3. The setup of the circular welding process.

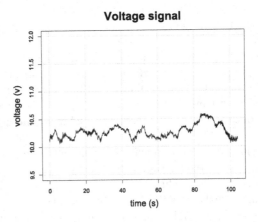

Fig. 4. A voltage signal of defect-free welding.

5   Results and Discussion

We first detail the experimental methodology followed in this study. At training, k random kernels are generated along with their attributes discussed earlier i.e. length, weights and dilation. The maximum values of the convolutions of the normal subsequences with the generated random kernels are extracted and then used as input to an OC-SVM model trained with the RBF kernel. The model parameter ν is fixed at 10^{-5} as we are only training from normal data, while the RBF kernel parameter γ is varied from 10^{-5} to 10^{-2} with a step of 10^{-1} and we retain the model achieving the best detection performance.

In the deployment stage, we first transform the upcoming subsequences using the generated convolutional kernels in the training stage. We then predict if they are normal or abnormal by feeding the extracted features from the convolutions into the trained OC-SVM model. If the anomaly score is lower than a threshold τ, the subsequence is declared abnormal. The threshold is estimated as follows:

$$\tau = \bar{F}(x^{train}) - 3 * \sigma_{F(x^{train})} \tag{8}$$

where $\bar{F}(x^{train})$ is the mean of the anomaly scores obtained from the training data and $\sigma_{F(x^{train})}$ is the standard deviation of these anomaly scores.

When a subsequence is abnormal, we follow the explainability approach presented earlier in order to give indications of why it is predicted abnormal. We retain the three most important kernels for the explanation.

We consider in the present experimentation a dataset composed of 6000 normal subsequences and 200 abnormal subsequences of length 200 obtained from the voltage signal of the circular welding. For the evaluation, we use the k-fold technique with $k = 5$. In this process, the 6000 normal subsequences are divided into 5 subsets. For each iteration, the model is trained using 4 subsets of the normal subsequences, while the 5th subset is combined with the 200 abnormal subsequences to form the test data. We repeat the training and testing 10 times for each of the five subsets, and we report here the F1-score averaged over the 10 runs along with the averaged standard deviation in order to quantify the variability of the detection due to the randomness of the kernel transform. Regarding the number of random kernels, we experiment with 100, 500, 1000 and 5000 kernels in this study.

We consider three other approaches for the comparative study. The first is based on a convolutional autoencoder that is trained to reconstruct the normal subsequences. The anomaly score of new subsequences is then the reconstruction error. The encoder consists of two convolutional layers with 32 followed by 16 kernels respectively and a kernel size of 3. The decoder consists of two transpose convolution layers with the same number of kernels and the same kernel sizes. We employ a dropout rate of 0.2 as a regularization in order to optimize the autoencoder, and we train the autoencoder with 50 epochs and a batch size of 4. In the second approach, we extract statistical features from the subsequences and employ the Isolation Forest algorithm. The extracted features are designed based on our prior knowledge of the data and the anomalies, namely skewness, maximum level shift, kurtosis, etc. The last approach is based on the k-NN algorithm, where the anomaly score is considered as the Euclidean distance between the upcoming

Table 1. F1-scores of the models.

Approach	Mean F1-score	Time (s)
Autoencoder	0.939	0.21
1-NN	0.937	6.11
Features & IForest	0.924	8.94
ROCKET & OC-SVM #kernels=100	$0.954_{\pm 0.0106}$	0.118
ROCKET & OC-SVM #kernels=500	$0.957_{\pm 0.01}$	0.36
ROCKET & OC-SVM #kernels=1000	$0.96_{\pm 0.009}$	1.00
ROCKET & OC-SVM #kernels=5000	$0.953_{\pm 0.0088}$	5.02

subsequence and its nearest neighbour in the training set. All the approaches are trained and tested on the same data and evaluated with the 5-folds technique. Moreover, the anomaly threshold of each method is estimated using the 3 standard deviations, as in Eq. 8.

Table 1 gives the results of the approaches that consist of the F1-scores and the inference time of each technique. We can notice that with only 100 random kernels, the proposed approach outperforms the other known methods both in terms of accuracy and inference time. This shows its effectiveness for the detection of abnormal time series subsequences. We also notice from the results that generally, as the number of random kernels increases, the F1-score increases while the variability decreases. However, this

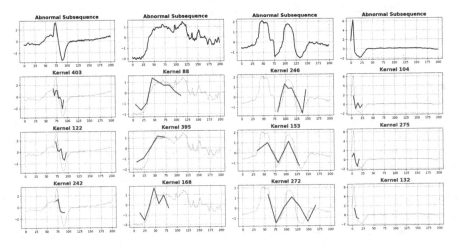

Fig. 5. Examples of abnormal subsequences along with the three most important kernels shown in blue. (Color figure online)

improvement comes at the cost of longer inference times. To overcome this problem, one can leverage feature selection and reduction techniques in order to reduce the number of features while preserving high accuracy. This will be a subject of future studies. In summary, we can conclude from the obtained results that ROCKET with OC-SVM can be accurate for the problem of anomaly detection in time series.

We are interested now in explaining the detected anomalies using the random convolutional kernels. For this, we retain the model using 500 random kernels, and we employ our approach for explainability to identify the three most important random kernels. To provide the explanation, we plot the abnormal subsequences along with the most important random kernels at their associated localization. Since the convolutions are performed with dilations, the weights of a random kernel are spaced by the amount of dilations used in the convolution in order to reflect the detected pattern in the convolution. For example, if the weights of a kernel are [0, 1, 1] and the convolution is performed using dilation of 5, then the weights will have the timestamps $t1 = L$, $t2 = t1 + 5$, $t3 = t2 + 5$, where L is the argument of the maximum value in the convolution. Figure 5 shows four abnormal subsequences and the three most important kernels for each one shown in blue. We can notice that the random kernels give an accurate indication of the abnormal shapes in the subsequences at their exact location.

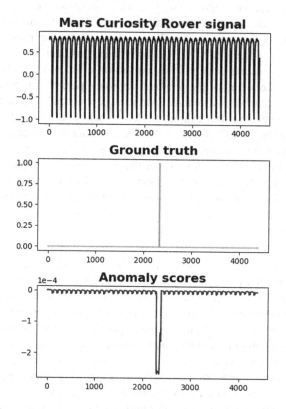

Fig. 6. Signal of Mars Curiosity Rover along with the ground truth labels and anomaly scores.

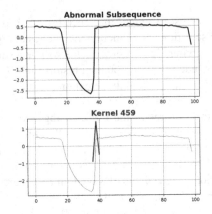

Fig. 7. The most important random kernel that explains the anomaly.

In addition to providing interpretability, the transform obtained by the random kernels can be used to cluster the abnormal subsequences, which can help in categorizing the anomalies of the welding process.

In order to evaluate the generalizability of our approach to other types of time series, we tested it on a signal acquired from Mars Curiosity Rover [6] shown in Fig. 6, which has a subtle anomaly that can be difficult to detect visually. We trained a model using 500 random kernels with the first 10% points of the signal, and we tested on the whole signal. The anomaly scores obtained that are shown in Fig. 6 demonstrate that despite the complexity of the anomaly, the proposed approach perfectly detects it, which suggests that it can be suitable as a general approach for abnormal subsequence detection.

In situations where the data exhibit subtle anomalies, as is the case of this signal, explainability is of high importance in order to understand the anomalies. We tested our approach to explain the detected anomaly in the signal, we obtained the most important kernel shown in Fig. 7, which indicates that the anomaly is explained by the more rapid rate of increase. This finding further emphasizes the effectiveness of the proposed approach in both detecting and explaining abnormal subsequences in time series data.

6　Conclusions

In this paper, we proposed a novel approach for an explainable abnormal time series subsequences detection based on feature extraction from convolutions of the subsequences with random convolutional kernels. Unlike anomaly detection methods that rely on engineered features, our approach does not require any prior knowledge of the application domain. Additionally, it does not require learning the convolutional kernels, which makes the features faster to obtain than when using deep neural networks. Results of the experiments showed that the approach has a higher accuracy and is faster both in training and testing than most of the established approaches. Furthermore, we showed that the approach we proposed for explaining anomalies gives a valuable explanation that can be useful for users to gain insights into how the predictions are made and about the abnormal shapes in the subsequences.

In future works, we would consider testing the approach on other data to further investigate it in terms of accuracy and inference time. Moreover, we would like to investigate the use of feature selection techniques and study their advantages and drawbacks in terms of accuracy and explainability. Furthermore, we would study the use of other anomaly detectors with the random kernel transform. Finally, it would be interesting to study the extension of the proposed approach to multivariate signals.

References

1. Angelov, P.P., Soares, E.A., Jiang, R., Arnold, N.I., Atkinson, P.M.: Explainable artificial intelligence: an analytical review. Wiley Interdisc. Rev.: Data Min. Knowl. Discov. **11**(5), e1424 (2021)
2. Boniol, P., Linardi, M., Roncallo, F., Palpanas, T.: SAD: an unsupervised system for subsequence anomaly detection. In: 2020 IEEE 36th International Conference on Data Engineering (ICDE), pp. 1778–1781. IEEE (2020)
3. Chauhan, S., Vig, L.: Anomaly detection in ECG time signals via deep long short-term memory networks. In: 2015 IEEE International Conference on Data Science and Advanced Analytics (DSAA), pp. 1–7. IEEE (2015)
4. Dempster, A., Petitjean, F., Webb, G.I.: Rocket: exceptionally fast and accurate time series classification using random convolutional kernels. Data Min. Knowl. Disc. **34**(5), 1454–1495 (2020)
5. Hu, M., et al.: Detecting anomalies in time series data via a meta-feature based approach. IEEE Access **6**, 27760–27776 (2018)
6. Hundman, K., Constantinou, V., Laporte, C., Colwell, I., Soderstrom, T.: Detecting spacecraft anomalies using LSTMs and nonparametric dynamic thresholding. In: Proceedings of the 24th ACM SIGKDD International Conference on Knowledge Discovery & Data Mining, pp. 387–395 (2018)
7. Ishimtsev, V., Bernstein, A., Burnaev, E., Nazarov, I.: Conformal k-NN anomaly detector for univariate data streams. In: Conformal and Probabilistic Prediction and Applications, pp. 213–227. PMLR (2017)
8. Janssens, O., et al.: Convolutional neural network based fault detection for rotating machinery. J. Sound Vib. **377**, 331–345 (2016)
9. Kromkowski, P., Li, S., Zhao, W., Abraham, B., Osborne, A., Brown, D.E.: Evaluating statistical models for network traffic anomaly detection. In: 2019 Systems and Information Engineering Design Symposium (SIEDS), pp. 1–6. IEEE (2019)
10. Lei, L., Wu, B., Fang, X., Chen, L., Wu, H., Liu, W.: A dynamic anomaly detection method of building energy consumption based on data mining technology. Energy **263**, 125575 (2023)
11. Li, J., Izakian, H., Pedrycz, W., Jamal, I.: Clustering-based anomaly detection in multivariate time series data. Appl. Soft Comput. **100**, 106919 (2021)
12. Li, Z., Zhu, Y., van Leeuwen, M.: A survey on explainable anomaly detection. arXiv preprint arXiv:2210.06959 (2022)
13. Liu, C., Gryllias, K.: A deep support vector data description method for anomaly detection in helicopters. In: PHM Society European Conference, vol. 6, p. 9 (2021)
14. Malhotra, P., Vig, L., Shroff, G., Agarwal, P., et al.: Long short term memory networks for anomaly detection in time series. In: ESANN, vol. 2015, p. 89 (2015)
15. Moayedi, H.Z., Masnadi-Shirazi, M.: Arima model for network traffic prediction and anomaly detection. In: 2008 International Symposium on Information Technology, vol. 4, pp. 1–6. IEEE (2008)

16. Munir, M., Siddiqui, S.A., Dengel, A., Ahmed, S.: DeepAnT: a deep learning approach for unsupervised anomaly detection in time series. IEEE Access **7**, 1991–2005 (2018)
17. Pincombe, B.: Anomaly detection in time series of graphs using ARMA processes. Asor Bull. **24**(4), 2 (2005)
18. Scholkopf, B., Williamson, R., Smola, A., Shawe-Taylor, J., Platt, J., et al.: Support vector method for novelty detection. In: Advances in Neural Information Processing Systems, vol. 12, no. 3, pp. 582–588 (2000)
19. Tripathy, S.M., Chouhan, A., Dix, M., Kotriwala, A., Klöpper, B., Prabhune, A.: Explaining anomalies in industrial multivariate time-series data with the help of explainable AI. In: 2022 IEEE International Conference on Big Data and Smart Computing (BigComp), pp. 226–233. IEEE (2022)
20. Wang, H., Li, Q., Liu, Y., Yang, S.: Anomaly data detection of rolling element bearings vibration signal based on parameter optimization isolation forest. Machines **10**(6), 459 (2022)
21. Wang, Z., Cha, Y.J.: Unsupervised deep learning approach using a deep auto-encoder with a one-class support vector machine to detect damage. Struct. Health Monit. **20**(1), 406–425 (2021)

TaxoSBERT: Unsupervised Taxonomy Expansion Through Expressive Semantic Similarity

Daniele Margiotta[✉], Danilo Croce[iD], and Roberto Basili[iD]

University of Rome, Tor Vergata, Italy
margiotta@revealsrl.it, {croce,basili}@info.uniroma2.it

Abstract. Knowledge graphs are crucial resources for a large set of document management tasks, such as text retrieval and classification as well as natural language inference. Standard examples are large-scale lexical semantic graphs, such as WordNet, useful for text tagging or sentence disambiguation purposes.

The dynamics of lexical taxonomies is a critical problem as they need to be maintained to follow the language evolution across time. Taxonomy expansion, in this sense, becomes a critical semantic task, as it allows for an extension of existing resources with new properties but also to create new entries, i.e. taxonomy concepts, when necessary. Previous work on this topic suggests the use of neural learning methods able to make use of the underlying taxonomy graph as a source of training evidence. This can be done by graph-based learning, where nets are trained to encode the underlying knowledge graph and to predict appropriate inferences.

This paper presents TaxoSBERT as a simple and effective way to model the taxonomy expansion problem as a retrieval task. It combines a robust semantic similarity measure and taxonomy-driven re-rank strategies. This method is unsupervised, the adopted similarity measures are trained on (large-scale) resources out of a target taxonomy and are extremely efficient. The experimental evaluation with respect to two taxonomies shows surprising results, improving far more complex state-of-the-art methods.

Keywords: Knowledge injection in neural learning · Taxonomy expansion · Unsupervised sentence embeddings

1 Introduction

Knowledge graphs are crucial resources for a large set of document management tasks, such as text retrieval, classification, or natural language inferences. Standard examples are large-scale lexical semantic graphs, such as WordNet, useful for text tagging or sentence disambiguation purposes.

Taxonomies are used in most cases for managing concepts in an organized manner [5, 13, 15]. In particular, to capture the relation "is-a" between couples of concepts, taxonomies are built as tree structures or directed acyclic graphs (DAGs). However, taxonomies are dynamic as they need to be maintained in order to follow the language

© The Author(s), under exclusive license to Springer Nature Switzerland AG 2023
D. Conte et al. (Eds.): DeLTA 2023, CCIS 1875, pp. 295–307, 2023.
https://doi.org/10.1007/978-3-031-39059-3_20

evolution across time. One maintenance task is taxonomy enrichment, which is the task of adding new nodes to an existing hierarchy. In the wider area of document and knowledge management tasks, it is a complex semantic task: it allows to extend of existing resources through the creation of new entries, i.e. concepts, that have to be harmonized with the overall taxonomic organization.

An example of this application can be found in Amazon, which exploits taxonomic descriptions of its products for simplifying the access of customers and Web surfers as well as to extract and label information from large corpora.

The construction and update of taxonomies is a costly activity that requires highly specialized human intervention usually on a large scale. This process is time and effort-consuming either in the design stage or in the population process.

Recent works [4, 10, 16, 18] try to solve the problem with the method of automatic expansion and enrichment of existing taxonomies.

Let us consider WordNet [6] the well-known large-scale lexical database of English. Nouns, verbs, adjectives, and adverbs are grouped into sets of cognitive units, (namely, synonimy sets in synsets), each expressing a distinct concept. As an example the expression *"center field"* forms, together with *"centerfield"* and *"center"* the specific concept that can be defined as *"the piece of ground in the outfield directly ahead of the catcher"*[1]. Synsets are interlinked by utilizing conceptual-semantic and lexical relations. One of the most important relations in Wordnet is the *hyperonymy* which denotes the super-types of each synset. As an example the hypernym of the above concept is the synset *"tract"*, *"piece of land"*, *"piece of ground"*, *"parcel of land"*, *"parcel"* that can be defined as *"an extended area of land"*. Since each synset is involved in a hyperonymy relation, in a resource like WordNet that counts more than 170, 000 synsets, adding a new concept to such a large-scale resource is not trivial. Recent works like [18] model the problem of adding a new concept to resources like WordNet as the process of retrieving one or more nodes involved in the hyperonymy relation. In particular, these approaches adopt supervised classification strategies: given individual concept pairs, the classifier is expected to predict whether or not the hypernym relation is valid. The cost of this assignment strictly depends on the size of the taxonomy itself and it can be extremely high for large-scale resources like WordNet. In any case, in WordNet, each concept has a definition and a set of lemmas (that is a sort of name or label of the concept) expressed in natural language, and it is one of the main features used by classifiers to infer inter-conceptual relations.

In this work, we investigate the adoption of semantic similarity measures to define a robust yet efficient method for taxonomy expansion. We base our work on recently defined neural text encoders such as Sentence-BERT (or SBERT) [8]: these are based on the encoder from a Transformer Architecture [12] that, like BERT [2], take in input a text and produces an embedding reflecting its meaning. In particular, given two sentences like *"the boy is pacing in the street"* and *"a young guy is talking a walk among the buildings"*, such an architecture produces vectors laying in the same subspaces, i.e., they will be nearer in the space with respect to a sentence like *"the guy is playing piano"* which refers to a completely different action and scenario. Architectures like SBERT, trained on millions of text pairs, have been shown to give rise to extremely

---

[1] See WordNet for sense 1 of the noun *"center field"*.

robust similarity measures between text, only modeled as the cosine similarity between their geometric counterparts. It must be said that in a taxonomy expansion, the straight application of this method may introduce noise. One of the sentences mostly similar to the definition of *"center field"* is *"the piece of ground in the outfield on the catcher's left"*, which defines the concept *"left field, leftfield, left"*, which does not generalize our target concept, but they share the same generalization. Unfortunately, the definition of the real hypernyms risks being less informative in the modeling process (i.e. ranking) of the most similar concepts. In fact, these encoders are not trained to reflect generalization (or specialization) relations but only semantic (vector) representations that determine semantic analogies, whose metric is simply the cosine metrics, as the cosine measure. To overcome this issue, it is crucial to use the knowledge underlying the taxonomy to reward concepts generalizing similar nodes.

Overall, we propose TaxoSBERT that models the taxonomy expansion problem as a retrieval task[2]: given a query concept and the set of nodes from the taxonomy, we foster to *i)* rank all nodes according to a robust semantic similarity measure and to *ii)* re-order this list to rank up all concepts that share a taxonomic relationship with the ones already in the top positions.

This method has several advantages. First of all, it is unsupervised, as methods like SBERT are trained on (large-scale) resources out of a target taxonomy. Second, it is extremely efficient: it does not require classifying a concept against the generalizations[3] of all candidates, but it is only needed to encode the query concept once, in order to measure the cosine similarity between the already encoded candidate concepts. Finally, the experimental evaluation with respect to two taxonomies shows surprising results, improving far more complex state-of-the-art methods. Without loss of generality, the proposed method can be applied to any taxonomy: it only requires each concept to be described by a short text and we think this represents a strong and efficient baseline in Taxonomy Expansion tasks.

In the rest of this paper, Sect. 2 presents the related work. The proposed TaxoSBERT is presented in Sect. 3. The experimental evaluation is presented and discussed in Sect. 4 while Sect. 5 derives some conclusions.

2   Related Works

As explored by previous work, a Taxonomy Expansion task can be solved by exploiting several strategies. One of the earliest works, such as [11], proposes the Bilinear Model, a probabilistic method that defines a joint distribution over the relations of a taxonomy where the relations are a function that connects two concepts like a link based on some features. Given a set of different relations, the model predicts the probability of relations between two concepts is true based on a Bayesian Clustered Tensor Factorization (BCTF) model, which combines good interpretability with excellent predictive power.

---

[2] The source code is publicly available at https://github.com/crux82/TaxoSBERT.

[3] Consider that, if using state-of-the-art Transfomer-based architectures, such as the BERT-based ones, the classification of a text pair requires encoding it and it is a computationally expensive task.

More recently, with the advent of neural methodologies graph-based methods were proposed, such as TaxoExpan [10]. It is a taxonomy expansion framework, which exploits positional-enhanced graph neural network (strengthened by the use of info Noise-Contrastive-Estimation loss in training, [7]) to capture relationships between query nodes and local EgoNets.

Neural nets have been also used to exploit the semantic information contained in the various taxonomies, such as the Arborist Model [4]. It learns a measure of the taxonomic relatedness over node pairs as a function of their respective feature vectors. Taxonomic relatedness is defined in terms of two learnable parameters: (i) the embeddings of nodes in the taxonomy that capture their taxonomic roles and (ii) a set of shared linear maps in the taxonomy. The parameters are trained to minimize a large margin ranking loss, with a theoretically grounded dynamic margin function. To ensure rapid convergence, Arborist employs distance-weighted negative sampling.

Finally, models that exploit together the semantics and graph structure of taxonomy are proposed in the works on TMN [18] and GeoTaxo [17]. TMN is a One-to-pair matching model which leverages auxiliary signals to augment the primal matching task combined with a novel channel-wise gating mechanism designed for regulating concept embedding and boosting the model performance. GeoTaxo is a taxonomy completion framework using both sentence-based and subgraph-based encodings of nodes under a Transformer-based encoder [12]) paradigm to perform the matching.

All of the above works increasingly improved the performance on the taxonomy completion task. We know that other encoding techniques are available and can be adopted successfully to large sets of candidate ancestors for new query concepts. One example is the transformer-based text-encoder [2], which proved to be an optimized encoder and largely applicable to generate contextualized embeddings for several phenomena (e.g. concept names or definitions), without the explicit need for pre-training. In fact, the state-of-the-art system is currently TaxoEnrich [3], a framework that exploits semantic features and taxonomic information for representing node positions as candidate ancestors for new incoming concepts. TaxoEnrich is based on four components:

1. a contextualized encoder that creates embeddings that encapsulate node (i.e. concept) meaning and taxonomic relations based on accurate pre-trained linguistic models;
2. a sequential encoder trained in the taxonomy structure, that learns representations of the candidate ancestors according to the taxonomy hierarchy;
3. a siblings encoder that extends the candidate representation aggregating information from the query and the information from the siblings of the candidate;
4. a model of the correspondence between the query and each candidate, that extends previous work by means of new candidates representation rules.

In the above approach, the transformer-based text-encoder applied to obtain the contextualized embeddings is SciBERT [1].

Furthermore, TaxoEnrich approaches the hierarchy learning on a given taxonomy $\mathcal{T}$ through linguistic patterns, the so-called Hearst Patterns [9], able to associate concepts to embeddings exclusively based on the hierarchical structure of $\mathcal{T}$. Notice that it relies on controlled, but static patterns, able to associate concept pairs through the hierarchy, but not directly applicable to other textual information in $\mathcal{T}$, e.g. concept definitions.

With this work we intend to explore the capabilities of a text-encoder pre-trained on a large number of examples in providing embeddings that are optimal in encapsulating the semantics of a text: as a consequence, there is no longer a need for defining static patterns but just natural language descriptions are sufficient. This will allow us to use a semantic similarity function among embeddings to provide a rank among a large set of candidates. Moreover, this turns out to be a method that does not require contextualized pre-training, since the models have already been pre-trained, and can therefore be applied in other contexts and situations.

3   Using Taxonomic Information for Semantic Pre-Training Neural Transformers

3.1   Problem Definition

In a taxonomy expansion task, we assume that the taxonomy $\mathcal{T} = (N, E)$ is a directed acyclic graph where N is the set of nodes and E is the set of relations (i.e. edges) that bind two nodes, expressed as $(n_i, n_j) \in N \times N$. Notice that, depending on different taxonomies, nodes $n \in N$ can be represented differently, as they exhibit different properties according to $\mathcal{T}$.

Whatever the definition of nodes $n \in N$ neural models can be trained to learn precisely which relationships are defined in E between the various node pairs in $N \times N$. The Taxonomy Completion task tries to solve a very specific task connected to this one. Given an existing taxonomy $\mathcal{T}$ and a new node $c \notin N$, find the node $n_i \in N$ that is most similar or somehow close to c, so that a new still consistent taxonomy $\mathcal{T}'$ can be generated such that $\mathcal{T}' = (N \cup \{c\}, E \cup \{(c, n_i)\})$. Thus the goal of the Taxonomy Completion task is to extend an existing taxonomy by finding the ancestors $n_i \in N$ that semantically accommodate the new c, as their direct descendant, in a consistent way.

Notice the taxonomy is a directed acyclic graph but it may express more information than the binary hierarchical connections between concept pairs in E. Nodes can also represent other information, such as concept names (e.g. through words or terms), as well concept definitions, such as glosses. In the rest of the paper, we will make use of at least the two relations expressed by E, that is $descendant\_of(n_i, n_j)$ or $ancestor\_of(n_j, n_i)$, if, for some $n_i, n_j \in N$, the pair $(n_i, n_j) \in E$ according to $\mathcal{T}$.

Given the WordNet hyponymy and hypernymy relation, we can denote the resulting taxonomy as $\mathcal{T}_{WN} = (N_{WN}, E_{WN})$[4]. As a consequence, the example sentence

the *"center field"* is a *"tract"*, that is a *"member of the geographical area"*

is true according to $\mathcal{T}_{WN}$ as the pair

$$\left((\text{``center field''}, \text{``centerfield''}, \dots), (\text{``tract''}, \text{``piece of land''}, \dots) \right) \in E_{WN}$$

---

[4] Notice that the hyponymy relation does not correspond to a perfect DAG in Wordnet, as multiple inheritances are occasionally needed for some synsets (nodes) in Wordnet. However, for the Taxonomy Enrichment task, this assumption is always satisfied, so that, in the scope of this paper, our definition is thus fully consistent.

3.2 Measuring Semantic Similarity Between Taxonomy Nodes

In this paper, we foster the idea to adopt robust semantic similarity functions to retrieve the nodes from the taxonomy $\mathcal{T}$, i.e., $n_i \in N$. In other words, the selection of candidates, as semantically plausible nodes, given a triggering concept c, requires the computation of a semantic similarity function. Instead of using existing taxonomy-driven similarity measures between nodes, it is here possible to employ neural encoders, such as BERT [2], or even sentence encoders, such as SBERT [8].

In synthesis, every node in $\mathcal{T}$ has a name or a definition. For a node $c \in N$ we derive $S(c, \mathcal{T})$ the set of labels (in WordNet the different lemmas). If we consider our running example, we can derive $S(c, \mathcal{T}) = \{$"center field", "centerfield", "center"$\}$.

These multiple lexical anchors $l_j^c \in S(c, \mathcal{T})$ of the concept c can be coupled with the textual definition def_c deriving multiple sentences in the form "The concept denoted by l_j^c is summarized by the following definition: def_c".

As an example one, of the textual description of c is, "The concept denoted by center field is summarized by the following definition: the piece of ground in the outfield directly ahead of the catcher".

When this text is made available for a generic node, a similarity measure can be defined between node pairs just by computing the cosine of the angle between the encoding of two corresponding descriptions. As the encodings are valid large dimensional vectors, the cosine measure is a valid and highly expressive similarity measure. When $|S(c, \mathcal{T})| > 1$ (i.e., a node generated multiple descriptions), a composition of potentially multiple similarity scores is applied, as discussed hereafter.

3.3 The Overall Taxonomy Expansion Algorithm

The overall algorithm can now be formalized according to the RANKING and EXPANSION steps.

Candidate Ranking. The first step is to exploit a Sentence-Encoder to obtain embeddings of the various Taxonomy concepts (both query and anchor concepts). Specifically, in our case, we are going to use the Sentence-Encoder model.

Given a new candidate node q to be added to the taxonomy $\mathcal{T}$, we aim to derive a list $rank_q$ containing all the nodes $c \in \mathcal{T}$, ordered according to the semantic similarity with respect to q.

In lines 6 and 10 we call a function that retrieves natural language phrases that describe individual concepts, i.e., $S(c, \mathcal{T})$. Then, in lines 8 and 11, the Sentence-Encoder model is applied to phrases to return semantically meaningful embeddings about concepts c. The embeddings allow us to compute a score, for each pair $\langle q, c \rangle$: it is the maximum among the scalar products between the different embeddings obtained from the taxonomy (line 15). Finally, the list of scores for individual pairs $\langle q, c \rangle$ (line 17) is kept sorted in descending order of the scores.

Taxonomic Expansion. The first rank obtained through the function *Candidate Sentence Ranking* exploits the semantic information in the Taxonomy $\mathcal{T}$, i.e. the definitions

---

Algorithm 1 Candidate RANKING (SBERT-Rank)

1: **procedure** SENTENCE_RANK(*Query q, Taxonomy T*)
2:     ▷ $rank_q$ *will contain a ranked list of members of C for the query*
3:     $rank_q = []$
4:   **for** $c \in T$ **do**
5:         ▷ *For each concept c in C generate sentences according to T*
6:         $S(c) = $ GENERATE_SENTENCES_FROM_TAX(c, T)
7:         ▷ *For each sentence $s_c \in S(c)$ get embeddings by Sentence-Encoder*
8:         $E(c) = \bigcup_{s_q \in S(q)} \{$GET_EMBEDDING$(s_c)\}$
9:         ▷ *Get embeddings also for sentences derived from q*
10:        $S(q) = $ GENERATE_SENTENCES_FROM_TAX(q, T)
11:        $E(q) = \bigcup_{s_q \in S(q)} \{$GET_EMBEDDING$(s_q)\}$
12:
13:   **for** $c \in T$ **do**
14:        ▷ *Assign a score for each pair of concepts (q, c)*
15:        $score_{q,c} = max\{\ e_q \cdot e_c \mid e_q \in E(q), e_c \in E(c)\}$
16:        ▷ *Keep the list of anchors c in a descending order according to the scores*
17:        $rank_q = $ SORTED_INSERT$((c, score_{q,c}), rank_q)$
18:   **return** $rank_q$

Algorithm 2 Taxonomic Expansion (TaxoSBERT)

1: **procedure** TAXONOMIC_RANK(*CandidateList* $rank_q$, *Threshold k, Taxonomy T*)
2:     ▷ $taxo\_rank_q$ *hosts the list of concepts reranked by the taxonomic information.*
3:     $taxo\_rank_q = []$
4:     ▷ *Populating the taxonomic rank until the threshold is reached*
5:   **for** $(c, score_{q,c}) \in rank_q$ **do**
6:        $tmp\_list = [c]$
7:        ▷ *Expand the target concept by en-queuing ancestors and descendants in T*
8:        $tmp_list = $ CONCAT$\big(tmp_list, $ANCESTOR$(c, T)\big)$
9:        **if** add_descendants **then**
10:           $tmp_list = $ CONCAT$\big($DESCENDANT$(c, T), tmp\_list\big)$
11:       $taxo_rank_q = $ CONCAT_WO_DUPLICATE$(tmp\_list, taxo\_rank_q)$
12:   **return** $taxo\_rank_q$

---

of nodes in N. A further step is to exploit hierarchical information in T: each concept has ancestors and descendants that, respectively, generalize and specialize it.

Once we have obtained the Candidate Ranking list associated with a query concept q, we analyze its members c in order (lines 3–6) by adding their direct *descendant* and *ancestor* to the final list set (lines 8 and 9). In practice, each node c in the original $rank_q$ is added, in the same order, to the final list $taxo\_rank_q$. If one or more ancestor of c exists according to the ANCESTOR(c, T) function, their list is inserted into the rest of the list, immediately before c: in this way, the source order in $rank_q$ is preserved but ancestors are given a higher priority than c. Notice we assume here that this better model the semantics of the hyponymy relation: semantic similarity is requested to rank first nodes that generalize q and not its peers in T. Our use of the similarity function aims at locating the area of the taxonomy where q is represented, Accordingly, as a variant of the previous perspective, one could also proceed by adding descendants (as shown in line 9). The function in line 11 ensures that a node (e.g., an ancestor) is added only if it is <u>not</u> already in `taxo_rank`$_q$.

Notice that the semantic locality principle adopted for each concept is exploited to adjust the order of evaluation of candidates. The result is that the initial ranked list of Candidate Ranking (CR) is expanded. In the rest of this work, having adopted SBERT [8] as a similarity measure, we will refer to this model as TaxoSBERT.

Complexity. From a computational perspective, the proposed algorithm is highly scalable and efficient. In fact, the encoding of each node in the taxonomy $\mathcal{T}$ can be executed once and offline, and all these vectors can be stacked in a matrix $\mathbf{M}_{\mathcal{T}}$. At runtime, the estimation of the cosine similarity of the embedding derived from c corresponds to matrix multiplication, while the cost of the sorting of `rank`$_q$ is $\mathcal{O}(\log(|\mathcal{T}|))$. Finally, the re-ranking of `rank`$_q$ is $\mathcal{O}(k)$ with $k << |\mathcal{T}|$. The complete visit of `rank`$_q$ is $\mathcal{O}(|\mathcal{T}|)$ as it corresponds to only one visit of all the nodes. Moreover, this step can be further reduced by only visiting the first $k << |\mathcal{T}|$ nodes in the `rank`$_q$, as this optimization would not affect the ranking of top candidates in `taxo_rank`$_q$.

4   Experimental Evaluation

In this section, we experimentally evaluate our unsupervised method in a taxonomy expansion task. We applied it to WordNet 3.0, the large lexical network of the English language [6]. For the sake of comparison, we adopted the WordNet-derived datasets presented in [18]. In particular, we targeted two taxonomies: the Noun hierarchy, including $83,073$ nodes, and the Verb one, made of $13,936$ nodes.

Moreover, we adopted the same evaluation strategy from [3]. For each taxonomy, we randomly sampled a test set made of $1,000$ nodes. For each test node, the objective is to provide the correct hypernym among nodes the nodes remaining after sampling. Given the complexity of this task (for Nouns, only one correct node among $82,073$ nodes), in [3] this task is simplified so that the method is expected to return the list of candidate nodes, ordered according to the confidence of being the correct hypernym. We adopted the following evaluation metrics:

- **Mean Rank (MR):** the rank position of the correct hypernym c of an input query q, averaged across all test examples q.
- **Recall@K (R@K):** measures the number of correctly ranked hypernyms c in the top K positions, divided by the total number of correct positions of all query concepts q.
- **Precision@K (P@K):** measures the total number of correctly ranked hypernyms c ranked among the top K positions for each query q, divided by the total number of such queries q multiplied by K.

Results are averaged by applying three different samples of the test sets.

Table 1. Experimental results of models calculated on the test set of NOUNs from WordNet.

Model	MR
TaxoExpan [10]	970.86±50.99
TMN [18]	827.37±24.31
TaxoEnrich [3]	227.84±12.25
SBERT-Rank	748.19±213.27
TaxoSBERT$_a$	**220.76±153.91**

As sentence encoder, we adopted the Sentence Bert model[5] made available at HuggingFace [14], trained or more than 1.1 Billion text pairs from more than 30 datasets (obviously Wordnet is not included).

We compared the results with respect to TaxoExpan [10], TMN [18] and TaxoEnrich [3] achieved the best results on both hierarchies using supervised methods.

Results with respect to the Noun hierarchy are presented in Table 1. The first rows show the competitive supervised methods. The last two rows present SBERT-Rank, i.e. a method only adopting the rank provided from SBERT, as defined in Algorithm 1. In practice, for each test node, SBERT-Rank provides a list ordered according to the cosine similarity measure between the embeddings derived from the test cases and all the candidate nodes. The row TaxoSBERT$_a$ reports the results of our proposed method, obtained by applying the re-rank strategy in the Algorithm 2. In particular, the symbol a means that only ancestors are considered when the re-rank step is applied.

Table 2. Different TaxoSBERT variants applied to the Noun hierarchy

Model	MR	R@1	R@5	R@10	P@1	P@5	P@10
TaxoSBERT$_d$	921.85	0.07	0.22	0.30	0.07	0.04	0.03
TaxoSBERT$_{a+d}$	266.74	0.07	0.47	0.60	0.07	0.09	0.06
TaxoSBERT$_a$	220.76	0.24	0.53	0.66	0.24	0.11	0.07

Results clearly show that the direct application of the SBERT to the taxonomy expansion task is not effective. SBERT-Rank ranks the expected hyperonym on average only at the position 748. The application of TaxoSBERT$_a$ greatly improves these results with a rank of about 220 which also outperforms all the other supervised methods. It confirms our original intuition: SBERT is effective in retrieving nodes sharing strong semantic connections among nodes, but it is not directly beneficial in a Taxonomy Expansion task. This limitation is avoided by TaxoSBERT by expanding the set of retrieved nodes.

---

[5] https://huggingface.co/sentence-transformers/all-mpnet-base-v2.

Table 3. Experimental results of models calculated on the test set of VERBs from WordNet.

Model	MR
TaxoExpan [10]	853.31±18.30
TMN [18]	832.54±29.59
TaxoEnrich [3]	320.06±14.15
SBERT-Rank	545.13±69.36
TaxoSBERT$_a$	**129.08±20.16**

The effect of selecting only ancestors or descendants in the re-rank strategy of TaxoSBERT is shown in Table 2. In particular, a re-rank strategy applied only to descendants (the model TaxoSBERT$_d$) is not effective for the targeted task. It obtains a Mean Rank of about 921, not being even competitive with respect to the average score of SBERT-Rank, i.e., 720. This seems to confirm that the semantic similarity measure correctly relates a candidate node with the area of the taxonomy expressing the same concept(s). However, in a task whose objective is to retrieve a generalizing node, specializing in the area of attention (retrieving the descendants) is not helpful. On the contrary, the retrieval of higher-level nodes is more tied to the tasks and positively boosts the quality of the overall process.

Table 4. Experimental results of models calculated on the test set of VERBs from WordNet.

Model	MR	R@1	R@5	R@10	P@1	P@5	P@10
TaxoSBERT$_d$	791.73	0.05	0.14	0.20	0.05	0.03	0.02
TaxoSBERT$_{a+d}$	180.42	0.13	0.30	0.41	0.13	0.06	0.04
TaxoSBERT$_a$	129.08	0.13	0.33	0.46	0.13	0.07	0.05

The analysis of the Recall shows the capability of TaxoSBERT in a Taxonomy Expansion task when modeled as a classification task. The R@1 of 0.24 shows that TaxoSBERT$_a$ allows the retrieval of the correct node (at the first position) for the 24% of test nodes on average, among more than 80,000 candidate nodes. This percentage raises whenever more candidates are considered, e.g., an R@5 of 0.53 suggests that more than half of test examples are associated with a correct hypernym among the first five nodes proposed by the system.

The above results are confirmed for the Verb Hierarchy as shown in Table 3. TaxoSBERT achieves an MR of about 129, whereas TaxoEnrich achieves 320, again improving the score obtained by SBERT-Rank that achieves 545 and it is not competitive. The analysis of the different variants in Table 4 still confirms the previous findings. The best R@1 outcomes from TaxoSBERT, i.e., 0.13, suggest that the 13% of suggested nodes at the first position are correct, raising to an R@5 of 0.33 and R@10 of 0.46.

Table 5 shows the error analysis, i.e., cases in which the first ranked node is not the expected one. In particular, we report four cases for both Taxonomies (column τ), the

Table 5. Error Analysis

τ	Input Synset	Gold Standard	TaxoSBERT	Rank
N	"mnemonic" (a device (such as a rhyme or acronym) used to aid recall)	"device", "gimmick", "twist" (any clever maneuver)	"method" (a way of doing something, especially a systematic way; implies an orderly logical arrangement . . .	511
	"sobriety", "temperance" (abstaining from excess)	"abstinence" (act or practice of refraining from indulging an appetite)	"natural virtue" ((scholasticism) one of the four virtues (prudence, justice, fortitude, and temperance) . . .	6
	"lockdown" (the act of confining prisoners to their cells (usually to regain control during a riot))	"imprisonment", "internment" (the act of confining someone in a prison (or as if in a prison))	"protection" (the activity of protecting someone or something)	8
	"Geophilomorpha", "order Geophilomorpha" (small elongate centipedes living in soil and under . . .	"animal order" (the order of animals)	"arthropod family" (any of the arthropods)	19
V	"hyperventilate" (produce hyperventilation in)	"treat", "care for" (provide treatment for)	"breathe", "take a breath", "respire", "suspire" (draw air into, and expel out of, the lungs)	38
	"hypophysectomize", "hypophysectomise" (remove the pituitary glands)	"remove", "take", "take away", "withdraw" (remove something concrete, as by lifting, pushing, . . .	"cut out" (delete or remove)	3
	"unbrace" (remove a brace or braces from)	"weaken" (lessen the strength of)	"relax", "unbend" (make less taut)	410
	"grow" (become larger, greater, or bigger; expand or gain)	"increase" (become bigger or greater in amount)	"enlarge" (become larger or bigger)	6

candidate synsets, the expected hypernyms, the one proposed by the system, and the rank assigned to the expected hypernyms in the produced list.

In many cases, TaxoSBERT is not able to perfectly retrieve the correct node, even though the hypernyms are strictly related, but not assigned in WordNet to the same taxonomy branch.

As an example, the concept *"sobriety"*, *"temperance" (abstaining from excess)* would be connected to *"natural virtue" ((scholasticism) one of the four virtues (prudence, justice, fortitude, and temperance) . . .)* instead of *"abstinence" (act or practice of refraining from indulging an appetite)*. In this case, the proposed generalization does not match the gold standard (from WordNet), but a nevertheless closely related concept, i.e., *"sobriety"*, *"dryness" (moderation in or abstinence from alcohol or other drugs)*.

5   Conclusion

This work presents TaxoSBERT a method for taxonomy expansion, that models the taxonomy expansion problem as a retrieval task, according to a robust semantic similarity measure and taxonomy-driven re-rank strategies. This method is unsupervised since the adopted similarity measures are trained on (large-scale) resources out of a target taxonomy and it is extremely efficient. The experimental evaluation applied to two taxonomies shows surprising results, improving far more complex state-of-the-art supervised methods.

As a future step, we have plans to extend TaxoSBERT's capabilities. It will be applied and evaluated in comparison to other taxonomies, expanding beyond the use of WordNet. Additionally, we aim to go beyond focusing solely on direct ancestors or descendants of a node. We intend to propagate this information based on the taxonomy structure and utilize these embeddings as pre-training for supervised methods.

Acknowledgements. We would like to thank the Istituto di Analisi dei Sistemi ed Informatica - Antonio Ruberti (IASI) for supporting the experimentations through access to dedicated computing resources. We acknowledge financial support from the PNRR MUR project PE0000013-FAIR.

References

1. Beltagy, I., Lo, K., Cohan, A.: SciBERT: a pretrained language model for scientific text (2019)
2. Devlin, J., Chang, M.W., Lee, K., Toutanova, K.: BERT: pre-training of deep bidirectional transformers for language understanding (2018). https://doi.org/10.48550/ARXIV.1810.04805, https://arxiv.org/abs/1810.04805
3. Jiang, M., Song, X., Zhang, J., Han, J.: TaxoEnrich: self-supervised taxonomy completion via structure-semantic representations. In: Proceedings of the ACM Web Conference 2022. ACM (2022). https://doi.org/10.1145/3485447.3511935
4. Manzoor, E., Li, R., Shrouty, D., Leskovec, J.: Expanding taxonomies with implicit edge semantics. In: Proceedings of The Web Conference 2020, pp. 2044–2054. WWW '20, Association for Computing Machinery, New York, NY, USA (2020). https://doi.org/10.1145/3366423.3380271
5. Mao, Y., et al.: Octet: online catalog taxonomy enrichment with self-supervision. In: Proceedings of the 26th ACM SIGKDD International Conference on Knowledge Discovery & Data Mining, pp. 2247–2257. KDD 2020, Association for Computing Machinery, New York, NY, USA (2020). https://doi.org/10.1145/3394486.3403274
6. Miller, G.A.: WordNet: a lexical database for English. Commun. ACM **38**(1), 39–41 (1995). https://doi.org/10.1145/219717.219748
7. van den Oord, A., Li, Y., Vinyals, O.: Representation learning with contrastive predictive coding (2019)
8. Reimers, N., Gurevych, I.: Sentence-BERT: sentence embeddings using Siamese BERT-networks. In: Proceedings of the 2019 Conference on Empirical Methods in Natural Language Processing. Association for Computational Linguistics (2019), https://arxiv.org/abs/1908.10084
9. Roller, S., Kiela, D., Nickel, M.: Hearst patterns revisited: automatic hypernym detection from large text corpora. In: Proceedings of the 56th Annual Meeting of the Association for Computational Linguistics (Volume 2: Short Papers), pp. 358–363. Association for Computational Linguistics, Melbourne, Australia (2018). https://doi.org/10.18653/v1/P18-2057, https://aclanthology.org/P18-2057
10. Shen, J., Shen, Z., Xiong, C., Wang, C., Wang, K., Han, J.: TaxoExpan: self-supervised taxonomy expansion with position-enhanced graph neural network. In: Proceedings of The Web Conference 2020, pp. 486–497. WWW 2020, Association for Computing Machinery, New York, NY, USA (2020). https://doi.org/10.1145/3366423.3380132
11. Sutskever, I., Salakhutdinov, R., Tenenbaum, J.B.: Modelling relational data using Bayesian clustered tensor factorization. In: Proceedings of the 22nd International Conference on Neural Information Processing Systems, pp. 1821–1828. NIPS 2009, Curran Associates Inc., Red Hook, NY, USA (2009)
12. Vaswani, A., et al.: Attention is all you need. In: Advances In Neural Information Processing Systems, vol. 30 (2017)
13. Vrandečić, D.: Wikidata: a new platform for collaborative data collection. In: Proceedings of the 21st International Conference on World Wide Web, pp. 1063–1064. WWW 2012 Companion, Association for Computing Machinery, New York, NY, USA (2012). https://doi.org/10.1145/2187980.2188242
14. Wolf, T., et al.: Transformers: state-of-the-art natural language processing. In: Proceedings of the 2020 Conference on Empirical Methods in Natural Language Processing: System Demonstrations, pp. 38–45. Association for Computational Linguistics, Online (2020). https://www.aclweb.org/anthology/2020.emnlp-demos.6

15. Yang, C., Zhang, J., Han, J.: Co-embedding network nodes and hierarchical labels with taxonomy based generative adversarial networks. In: 2020 IEEE International Conference on Data Mining (ICDM), pp. 721–730 (2020). https://doi.org/10.1109/ICDM50108.2020.00081
16. Yu, Y., Li, Y., Shen, J., Feng, H., Sun, J., Zhang, C.: STEAM: self-supervised taxonomy expansion with mini-paths. In: Proceedings of the 26th ACM SIGKDD International Conference on Knowledge Discovery & Data Mining. ACM (2020). https://doi.org/10.1145/3394486.3403145
17. Zeng, Q., Lin, J., Yu, W., Cleland-Huang, J., Jiang, M.: Enhancing taxonomy completion with concept generation via fusing relational representations. In: Proceedings of the 27th ACM SIGKDD Conference on Knowledge Discovery. ACM (2021). https://doi.org/10.1145/3447548.3467308
18. Zhang, J., et al.: Taxonomy completion via triplet matching network. In: Thirty-Fifth AAAI Conference on Artificial Intelligence, AAAI 2021, Thirty-Third Conference on Innovative Applications of Artificial Intelligence, IAAI 2021, The Eleventh Symposium on Educational Advances in Artificial Intelligence, EAAI 2021, Virtual Event, 2–9 February (2021), pp. 4662–4670 (2021). https://ojs.aaai.org/index.php/AAAI/article/view/16596

Towards Equitable AI in HR: Designing a Fair, Reliable, and Transparent Human Resource Management Application

Michael Danner[1] , Bakir Hadžić[2] , Thomas Weber[2(✉)], Xinjuan Zhu[3(✉)] , and Matthias Rätsch[2(✉)]

[1] CVSSP, University of Surrey, Guildford, UK
m.danner@surrey.ac.uk
[2] ViSiR, Reutlingen University, Reutlingen, Germany
{bakir.hadzic,thomas.weber,
matthias.raetsch}@reutlingen-university.de
[3] Xi'an Polytechnic University, Xi'an, Shaanxi, China
zhuxinjuan@xpu.edu.cn

Abstract. The aim of this work is the development of artificial intelligence (AI) application to support the recruiting process that elevates the domain of human resource management by advancing its capabilities and effectiveness. This affects recruiting processes and includes solutions for active sourcing, i.e. active recruitment, pre-sorting, evaluating structured video interviews and discovering internal training potential. This work highlights four novel approaches to ethical machine learning. The first is precise machine learning for ethically relevant properties in image recognition, which focuses on accurately detecting and analysing these properties. The second is the detection of bias in training data, allowing for the identification and removal of distortions that could skew results. The third is minimising bias, which involves actively working to reduce bias in machine learning models. Finally, an unsupervised architecture is introduced that can learn fair results even without ground truth data. Together, these approaches represent important steps forward in creating ethical and unbiased machine learning systems.

1 Research Problem

So far, AI algorithms often have the weakness that their decisions replicate or even amplify existing distortions in data. Recently, *Amazon* tried to use artificial intelligence in recruiting, which resulted in women were not recommended for recruitment because they were underrepresented in the available data [15]. Cathy O'Neil already described further problems of this kind in 2016 in her book "Weapons of Math Destruction" [25].

Our system starts here and addresses ethical problems in the field of personal selection from the ground up. The main focus of the research project and the resulting technical system will be on ensuring fairness, protecting privacy and providing explainability

---

M. Danner and B. Hadžić—Both authors contributed equally.

D. Conte et al. (Eds.): DeLTA 2023, CCIS 1875, pp. 308–325, 2023.
https://doi.org/10.1007/978-3-031-39059-3_21

and transparency. Several aspects of our approach deal with ethical aspects and questions of social acceptance and the ethical acceptability of artificial intelligence. Proven methods from the human resources area are raised to a new level using innovative artificial intelligence processes. This enables companies to recruit and retain highly qualified specialists and thus significantly strengthen the regional economy. The project follows goals that have already been formulated in state programs that aim to strengthen the region economically and scientifically [31].

At the same time, recruiters and human resources employees often struggle with time constraints. The market and the demand for automated processes for routine tasks (but also for more complex tasks) is therefore immense. Implementing a low-cost algorithm that requires only a minimum of additional hardware and software is urgently needed (Fig. 1).

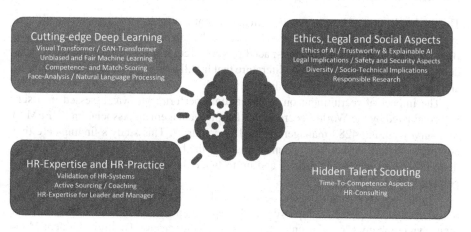

Fig. 1. Integration of AI into the recruitment process.

2 Outline of Objectives

In this work, an AI system for the application process will be developed, tested, and evaluated in its entirety. Such a trained architecture can be utilised in a multidimensional manner. Automating routine tasks will develop and acquire new competencies according to Gulliford and Dixon [14], it improves employee satisfaction, presents intelligent and comprehensible decisions throughout the entire hiring process, and efficiently utilises the available HR budget. The outcome will be modular in design, making it attractive to a wide range of businesses and other institutions. As a result, our models can be easily adapted to various businesses and are therefore universally applicable across the entire economic space.

In summary, our project focuses on four main dimensions as follows:

1. Recruitment stages: our work provides active sourcing with the support of state-of-the-art AI networks, assess the professional competence matrix and match score, and conducts long-term coaching that includes the discovery of hidden talents within the company and within recruitment pools.

2. Education level: the project's focus lies on young professionals who have just completed their academic degrees, up to the management level.
3. Professional fields: our focus is the most important professional fields with significant skills shortages. These are the automotive and medical technology industry, and the healthcare sector.
4. Plug-In: The prototype developed is a plug-in for human resource software with a universal interface. In this way, our work can be integrated into a wide range of human resource systems.

3   State of the Art

3.1   Human Resource Management and Recruiting

The modernisation of human resource management (HRM) has undergone a significant evolution as the current state of AI has proven to be effective in many time-consuming HRM processes. However, this fast-paced progress requires HR professionals and academics to dive deeper into existing literature that highlights AI-enhanced HR capabilities and areas of growth within the HR discipline [37].

The impact of recruitment on organisational performance was assessed in a survey conducted by the World Federation of People Management Association (WFPMA) in a survey among 4288 managers from 102 countries. This study's findings are that companies within 20% of the organisations with the best recruiting system, have up to three times higher revenue growth and up to twice the average profit margin of other companies [32].

Companies are focusing on and developing their own recruiting long-term strategies to recruit the best possible employees and keep the talented ones they already have in their own company [26]. Currently, the biggest challenges for Traditional Talent Management are the following:

- Demographic Change (Retirement versus lifelong learning)
- Technology advances (Artificial Intelligence)
- The growing shortage of skilled workers (War for Talents)
- Flexibilisation of the labour market with shorter employment tenures (Gig-Economy)
- Shift in recruitment communication to digital (social media and metaverse)
- Remote working environment (virtual organisations)
- Different value systems (cultural and generational differences)

The goal of our approach is to tackle these challenges and develop solutions that are fair, automatic, efficient, and intelligent. Existing HR guidelines (e.g., EN ISO-Norms) [22] should be extended and the ethical framework improved.

3.2   Artificial Intelligence and Applications Used for Recruiting

Currently, on the corporate market of German-speaking countries (DACH region), there is a great number of software that is used in the HRM sector. In the DACH region

(Germany, Austria and Switzerland) the HR software market size amounted to EUR 1.3 billion in 2020, while in 2021 it continued to grow to an amount of EUR 1.8 billion [39].

Large companies such as SAP Success Factors provide the largest share of the total HR software market size, around 60%. Nowadays it is impossible to imagine the HR sector of medium-sized and large companies without. Recruiting software accounts for about 17% of the global HR software market size and has a great potential for growth in the coming years [39]. The global recruitment software market size was valued at USD 1.7 billion in 2017 but is projected to reach the amount of 3 billion by the end of 2025, reflecting a CAGR of 7.4% [18]. See Fig. 2.

According to the prediction from Technavio [33], 35% of total market size growth in this area will originate from Europe.

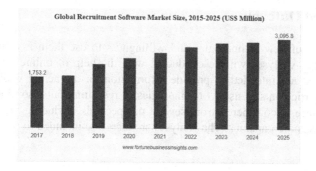

Fig. 2. Predicted market size growth [18].

There are currently various active sourcing systems on the market that can scan around one billion candidates on up to 30 social networks and 900 job portals. These systems also use machine learning, but the selected candidates are not necessarily the best possible matches and there is no transparency on why certain candidates were selected. For example, HireVue, a leading company for developing software for active sourcing started to reevaluate their algorithm after some negative feedback and criticism of their system.

3.3 Ethical and Social Aspects

The interdisciplinary research field of AI ethics is concerned with questions of comprehension, acceptance, risks, and possible mitigation strategies associated with the development and application of AI systems in society. AI-supported recruitment procedures have already been comprehensively addressed in the relevant scientific literature, as the use of AI in this field of an application implies potential risks for those affected (candidates of recruitment procedures) [5]. While AI utilisation brings many benefits to the recruiting process such as efficiency, effectiveness, and accountability, it can potentially also, even more, emphasise biased decisions by HR professionals. The dangers of "programmed discrimination" against underrepresented groups have often been highlighted in the literature [9]. AI fairness methods and metrics [2, 3] can partially overcome this.

However, the choice of suitable fairness procedures can by no means be understood as a purely technical optimisation task but requires a holistic approach [4] and a situational consideration in the concrete application [24,27]. Fairness in AI is far more than just a statistically tangible quantity [21,36], but includes justice perception of the entire socio-technical process [23]. From an ethical point of view, algorithmic fairness can be understood as a minimum standard that must be complemented by several normative criteria, such as the promotion of diversity in the company structure and practical transparency of the decision-making process for those who are affected by them [28]. Inter- and trans-disciplinary structures of technology development teams in which commercial, IT and ethics partners are in close exchange, from the beginning and throughout, are particularly adequate for overcoming these challenges [13,30].

4   Expected Outcome

The acceptance of new technologies and willingness to use them is central to their implementation. Analyses will be conducted with the help of online questionnaires sequentially and are intended to provide information on the actual willingness and readiness of participants to use AI technologies in recruiting but also globally in the HRM. Acceptance is considered from several perspectives, including the attitudes of both applicants and companies. The analyses are therefore intended to achieve the following goals:

4.1   Baseline Assessment

First, the general willingness to use and acceptance of AI systems in human resource management is to be assessed. For this purpose, a questionnaire based on the findings of Gansser and Reich [11] and Venkatesh [35] will be developed to collect demographic characteristics and various aspects of technology acceptance and willingness to use the technology. One pilot study will be conducted with a smaller sample to test the validity and reliability of the questionnaire. Based on this pilot study, a questionnaire will be adapted for its final version.

4.2   Development of Guidelines and Quality Standards

In addition, certain guidelines should be developed that an AI system should at least fulfil to guarantee a fair, ethical, and transparent way of working. To develop a fair AI system, it is necessary to include perspectives of cultural and social studies and epistemology in the research to consider the main ethical and philosophical issues that may arise within this approach. Within the framework of these guidelines, the focus will be on current forms of discrimination and the extent to which these are changing through new technologies but are still being reproduced.

4.3   Evaluation of the AI System

The design, functions and user experience of the AI systems are to be continuously evaluated and adapted by company employees throughout the entire system development period. For the evaluation of the user experience, a modified version of the System Usability Scale (SUS) [2] will be used. However, this scale is supplemented by open questions so that specific questions can be specifically adjusted to the system that is being developed.

In addition, the quality of the results of the developed system has to be constantly reviewed and compared. The AI system should therefore produce results that are comparable to the results that would be achieved by the regular HR department.

The results of the system and of the HR departments of our industry partners will be compared during system development time. Therefore both results should be scalable and comparable with each other.

4.4   Final Assessment of Perceptions and User Experience

After the development and evaluation of the AI system, the employees in HRM, persons representing minorities and different discriminated groups as well as potential users will be interviewed regarding their perceptions of the system. The survey will focus on whether the goals set (fairness, transparency, data protection) have been adequately achieved.

4.5   AI Technology

The overall goal is to identify and eliminate discriminatory tendencies in the HR context. For this purpose, state-of-the-art algorithms should be adapted so that they act without prejudice, fairly and explainable. Algorithms from the field of active sourcing, pre-selection and evaluation of interviews in the application process are considered. We use modern network architectures, such as Transformer Networks and Generative Adversarial Networks, to analyse the data in the best possible way and to document the decision-making process transparently and fairly. The discovery of internal potential should also benefit from the methods developed. The aim is to find solutions for the following use cases:

Active Sourcing. In the active sourcing procedure, potential employees are actively being searched for. Instead of the traditional recruitment process where applicants search for open positions and apply for them, with active sourcing recruiters actively search for and approach candidates to win them for their organisation.

Search for potential candidates is happening mostly through social media platforms where information on the professional skills of the users is available. Examples of those are sites like LinkedIn, Xing, Viadeo, Indeed, Glassdoor and many more. General information, education level, working experience, and skills of each user are displayed for recruiters and they can approach them if they think that their profile fits the position they have open.

But, the whole process of active sourcing is also mostly based on the decision of individual people, who are naturally affected by subjective perception. Additionally, this process takes a lot of time and effort as large data sets are to be monitored and analysed. Our development aims are to make this process automated, and faster but at the same time also more fair, transparent, and objective. With AI use, data sets of these websites could be automatically analysed and help recruiters in finding the best possible candidates for their positions.

Pre-selection. In the whole recruiting process, the most time-consuming task is the pre-selection of candidates. Each applicant's application must be fully reviewed in order to determine whether they meet the requirements for the position they are applying for.

Most of the information important for this procedure can be taken simply from the CV of the applicant. Using AI for CV parsing it should be possible to pre-sort applicants automatically and even to identify the ones that with their education and skills are the best fit for the open position [38]. AI use for CV parsing would also be very useful in the recruiting process for companies that already have a larger database of potential candidates. The solution gives companies the option to automatically determine whether their database already has someone who is qualified for the position they have open.

Video Interviews. Part of the recruiting process most prone to discrimination is an interview with the candidate. The interviewer can unintentionally discriminate against candidates based on their ethnic background, attractiveness, or other characteristics of their appearance.

It is possible to use AI to support the interview process while ensuring that it is highly structured, very transparent, and unbiased so that every participant is subjected to the same conditions. In semi-structured interviews where the interviewer is supported by AI, the system could deliver to the interviewer feedback on the nonverbal and emotional reactions of participants. For example, AI could recognise if participants are feeling uncomfortable, angry, sad, or happy while debating on some subjects. With this information, the interviewer could guide the interview process more effectively and make it more satisfying for both sides but also gather some relevant information on how participant reacts in certain situations. The main aim of our approach is to make interviews unbiased and transparent.

Detection of Internal Company Talents. Using various performance parameters, internal potential can be identified that would have remained undiscovered under traditional viewpoints. With the use of AI, employees within a corporation can be carefully supervised while their competencies are being monitored over for longer periods of time. In that way, companies can detect when an employee is making progress and is ready for taking the next step in his career so the appropriate training and development plan can be individually created to fulfil his potential and professional needs.

With the current stand of HRM practices, companies are often looking externally for new employees for higher positions, while they already have employees in their company who is ready to step out. In the same way, this approach shows the employees inside the company that their efforts and progress are valued and recognised, making them happier to be a part of the company. AI should also support decision-making here but without discrimination against vulnerable groups and with transparency for the users.

Target parameters: AI technology.

- Accurate machine learning for ethically relevant features in image recognition.
- Detection of bias in the training data.
- Minimisation of bias.
- Unsupervised network that can learn a fair outcome even without ground truth (truth values).

Target parameters: Ethics

- Development of socially accepted and ethically acceptable AI solutions for recruitment and candidate selection.
- Reflection of research and development work in the project according to the model of interdisciplinary integrated research.
- Development of specific solutions for transparency requirements, training data quality and diversity in recruiting processes.
- Creation and publication of ethical guidelines, framework, and minimum standards for candidate selection and recruiting.

5   Methodology

State-of-the-art methods of human resource management from the practice should serve as a template for the work. This applies above all to the use cases: Active sourcing, pre-selection in applicant screening selection, implementation of asynchronous video interviews and detection of internal training potential.

5.1   Integrated Research

The approach follows integrated research guidelines for the development of the project [30] in which the interdisciplinary teams are in close exchange from the beginning and throughout the entire project development process. Using methods from value-based design [10] and critical AI research [16] different ideas about the software and its use in recruiting are discussed and negotiated. Concrete instruments such as the Moral IT Deck [6] can be used to identify and address ethical risks in the use of AI human resources software.

5.2   Unbiased AI Models in HRM

To ensure the impartiality of the AI models in the HRM area, critical stop words will be used to identify and remove discriminatory tendencies at an early stage. The stop words make the system not only non-discriminatory but also explainable (Explainable AI). Stop words include words that do not correspond to our ELSA standards (i.e. words with ethnic, gender, national, age or religious affiliation). With the NLP-transformer technique, up to four times better results can be achieved than with a convolutional neural network. Therefore, we will use a network that extends BERT [20].

5.3 Machine Parsing of CVs

With our AI system, we will convert the application profiles directly into a feature vector in latent space when they are imported into the system. During this feature extraction, the profiles are translated into an abstract, higher-level language. In this way, we solve problems with different job descriptions for the same activities and can directly compare training, skills, experience and other relevant information in all languages. With this approach, our system can be used on an international scale.

Figure 3 displays an example resume alongside a sample job description from the data sets we used in our experiments. The resume, typically detailing the candidate's skills, experience, and educational background, is meticulously laid out, featuring key elements such as contact information, objective, work experience, education, and skills. Conversely, the job description outlines the requirements and responsibilities for a particular position, including the job title, duties, qualifications, and requisite skills.

5.4 Automated Structured Video Interviews

In the machine analysis of job interviews and an algorithmic evaluation of facial expressions and speech, particular attention must be devoted to personal dispositions and cultural differences. An objective foundation for unbiased and culturally invariant facial and posture analysis in video interviews is to be formed by following the Facial Action Coding System [8]. This will train an architecture that infers a latent feature vector from the action units that maps characteristics of the job competency matrix for the job-person fit. A similar procedure is to be followed for the language analysis. Instead of transcription (speech-to-text) [12], which is error-prone and highly dependent on the individual vocabulary, the so-called prosody itself is to be used as the object of analysis. Prosody is defined as the following characteristics:

1. accentuation
2. lexical tone (pitch signifies meaning)
3. intonation and sentence melody
4. quantity of all phonetic units
5. tempo, rhythm, and pauses in speech

For this purpose, the audio signal is pre-processed (e.g., Fast Fourier Transform) and then the frequency and time domain is as input to a recurrent architecture (Recurrent Neural Network, long short-term memory, Gated Recurrent Neural Network) or transformer. In this way, the totality of linguistic properties can be represented as a latent feature vector and used for a match score (Job-Person-Fit).

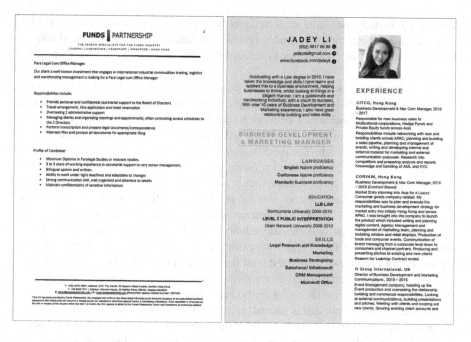

Fig. 3. Example job description and resume from our data sets.

5.5   Training of Bias-Free Architectures

There are different approaches to how AI algorithms can be made fair. For example, the available data is restructured before training so that the initial data is balanced. Alternatively, training algorithms can be modified so that discriminatory tendencies can be detected by themselves and consequently eliminated. This can be done, for example, by automatic pre-selection of the images provided for training, or by adjusted penalty terms. For loss functions of the type of Eq. 1 we define penalty terms for L_1 Eq. 2 and L_2 Eq. 3 regularisation.

$$Cost(\omega) = \sum_{i=0}^{N} y_i - \sum_{j=0}^{M} x_{ij}\omega_j{}^2 \tag{1}$$

$$P_1(\omega) = \lambda \sum_{j=0}^{M} |\omega_j| \tag{2}$$

$$P_2(\omega) = \lambda \sum_{j=0}^{M} \omega_j^2 \tag{3}$$

The last possibility is post-processing. Most networks that are compensated by post-processing are generative networks that create databases, for example. These can be sifted and sorted out in post-processing. The explainability of a neural network can be achieved by various measures such as layer-wise relevance propagation, counterfactual method, local interpretable model-agnostic explanations, or a generalised additive model. We follow two approaches to remove the bias in the data set and achieve fair predictions.

The first approach is based on data pre-processing before training, whereas the second approach is a self-learning network that learns and minimises bias. Before that, however, we need a benchmark network to determine the different labels of a person in an image. For this purpose, we already use a modified VGG-Face architecture [29] as the basis for our network. The modifications take the form of feature maps from the third and fourth convolutional blocks. Since the size of the feature maps is different from the size of the resulting feature vector, we implement an additional max-pooling layer to achieve the desired output. For the predictions of the network [7] we concatenate the soft-max layers into a single feature vector. The data set is divided into subsets, e.g., different age groups, male/female, ethnic origin, hair colour and so on. Each subset is then given a ranking factor and a weighting as a variable value. A minimum is then determined for each factor to find the least difference between the average prediction.

The second approach is based on removing the bias during training. In doing so, we develop a new unsupervised learning method to reduce the bias of the data. This consists of a deep learning network that is trained on the original learning task within the data set and then uses a specially adapted loss function to minimise the bias within the learned latent distributions-latent distributions. Each data set contains a list of weighted labels $A\ pa1, ..., anq$ and another fixed number of labels $B\ pb1, ..., bnq$ (e.g., age, occupation, hair colour and skin colour). The network evaluates all attributes of the data set during training and groups all objects into clusters according to the attributes b_i. Within each subgroup, the difference between the ethnic mean represents the bias. A non-linear operation (similar to gamma correction) is then applied to the ethnic label to calculate the differences between the labels. These differences for all clusters are the measure of the loss function, which is implemented as a categorical cross-entropy loss function and is which is minimised during training. In this way, we present a universally adaptable method to make each network fairer according to given labels.

5.6 Generating Synthetic Faces with Specific Features and Annotations

To train an AI that can measure the bias of architecture, one needs specific data for which one can vary individual characteristics in a very targeted way, e.g., degree of smiling, likeability, or trustworthiness. Deep learning models can be trained with images to determine or influence these aspects and the influence on the ranking in the suitability for a job role. In addition to image synthesis, manual annotations (ground truth) for labels are also necessary for training. Figure 4 shows an example of synthetically generated faces and an annotation tool that has been used to generate more than 3 million labels. StyleGAN and StarGAN were used to generate the images. As an innovation, transformers are also used for the synthesis of images, such as GANfomer [17, 19] and Transformer Network [34].

Fig. 4. Synthetic generated faces with specific attributes and manual annotations.

5.7 Transfer Learning

In many applications, the approach of transfer learning has already proven that the transfer of learned problems to others can considerably accelerate the training effort [40]. In addition, it can be used to tackle problems for which there was previously too little or too poor data. In our project, such a problem is the fact that for example pink-collar workers (healthcare employees) leave fewer digital fingerprints and are rarely present on classic job portals such as LinkedIn, Xing, or even Research Gate, compared to people with higher education levels or academics. We use this novel approach to transfer our findings from white-collar to pink-collar workers and other jobs with small amounts of data available.

5.8 Inclusion of the Time Domain

A significant innovation in our approach is the inclusion of the time domain in recruiting processes. Time-to-competence measure process takes into account how long it takes a candidate to acquire the skills required for the job. The Google AI blog article "Interpretable Deep Learning for Time Series Forecasting" [1] provides an overview of the novel Temporal Fusion Transformer (TFT). The authors tested the TFT against alternative methods such as DeepAR, ARIMA and long-short-term memory networks like Seq2Sep. They ran it on four data sets and tabulated their performance. The TFT outperformed the other models; both the classical methods and the neural networks of different types. Its quality losses of 50% and 90% were at least 7% lower than those of the next best model, the LSTM Seq2Seq network. Our approach could be the first AI system that can predict time-to-competence skill acquaintance in the area of HRM using TFTs.

5.9   Attractiveness of Portraits

For several years, our working group has been conducting research on the attractiveness of portraits. Companies could for example utilise this knowledge to choose appealing portraits for their advertising campaign. This topic is not without ethical controversy. Not only because a person's being is reduced to a number in the process of evaluation, but also because there are some differences in the results concerning different ethnicity. To ensure the fairness of an architecture, the underlying databases are augmented using Generative Adversarial Networks. The results of the research can be applied to any kind of discrimination.

For example, machines that would discriminate against women in the application process could be designed to be neutral. Approaches have already been developed on how to develop fair machine learning algorithms without having to perform augmentation. This can be achieved, for example, via adapted penalty terms or selective training.

5.10   Acceptance Research

To ensure that novel technology is accepted by users, it has to be adapted according to their needs and requirements. Acceptance of a newly developed system has to be systematically and thoroughly researched.

As the first step to asses and monitoring acceptance levels, a questionnaire based on relevant recent studies has to be made [35]. Psychological questionnaires measure latent constructs that cannot be captured directly and therefore need to be mapped using manifest variables. To achieve this, several diagnostic quality criteria must be fulfilled. These are in particular reliability, validity (content validity, construct validity and criterion validity) as well as objectivity, fairness and economy. Furthermore, the questionnaire must be validated concerning the test quality criteria and then, if necessary, a new, revised version of the questionnaire must be evaluated on a new sample. Then, with the help of confirmatory factor analyses (CFA) and other statistical methods, the quality criteria of the questionnaire are statistical procedures to determine the quality criteria of the questionnaire. With satisfactory validity, reliability and objectivity criteria being fulfilled, the questionnaire can be used on a large sample of participants in different stages of project development.

6   Novelties

Innovations of our previously described approach can be split into three specific subcategories: Innovations in the area of HRM, Innovations from the aspect of Artificial Intelligence and Innovations considering Ethical, Legal and Social aspects of our approach.

6.1   Novelties for HRM Applications

By reducing the workload of employees in HRM and increasing the transparency and efficiency of recruiting process current HRM trends are being raised to a new level. With our system, small and medium size companies would be provided with a much-needed tool to stay competitive with big companies with much bigger HRM departments, potential candidates databases and recruiting teams. Additionally, our approach sets strong emphasis on individuals already working in the companies and monitors their progress and career needs. Hidden talents inside the company can be identified and provided with guidance and stimulation in the most suitable way for them in their future career development.

6.2 Innovative Artificial Intelligence Approach

Our approach presents a holistic machine learning approach to the topic of HRM and recruiting. In our work, we demonstrate how the most recent advances in machine learning research can be applied to enhance HRM and recruiting procedures.

We propose the use of NLP-transformer to identify critical stop words in order to make the AI system unbiased and explainable during the preselection and CV parsing processes. For Automated and structured Video interviews instead of transcription we propose the use of prosody as the object of analysis. It's important to notice that our approach is the first AI system that can predict time-to-competence skill acquaintance in the area of HRM.

6.3 Ethical, Legal and Social Aspects

In contrast to many previous approaches, our approach deals intensively with concrete solutions for fairness, transparency and data clarity in both HRM and AI procedures. By providing explainable and unbiased AI we are offering a system that is capable to overcome discriminatory tendencies that are occurring in the traditional recruiting process.

Additionally, within our approach, we see the potential to generate a set of guidelines and instructions for the researchers and experts working on this topic in the future.

7 Sustainability Goals (Economic - Ecological - Social)

7.1 Economic Sustainability

The direct outcome of our approach with machine learning to HRM and recruiting process would be a sustainably designed labour market, as companies are getting more and more dependent on new technologies used in their practices with the constantly growing "war for talents" on the labour market. With the current trend of development in this area, small and medium size companies are going to struggle to compete in the recruiting process.

Utilising machine learning methods in their practices, small and medium size companies are getting a very needed tool to be competitive in the rapidly developing labour market. With the current trends, it seems like the problem of staff shortages is going to become even more immense in the future. Our approach offers a system that can counteract this problem and provide companies with adequate skilled professionals to meet their needs.

7.2   Ecological Sustainability

The training of the AI model in this approach is guided by the principle of using better data instead of Big Data. With this approach, the usage of energy resources will be much more efficient and a significant amount of emissions will be reduced (less computing power is needed, less emission).

7.3   Social Sustainability

By actively promoting diversity in our approach, we want to help to overcome the vicious circle that often prevents underrepresented groups from gaining access to education and social advancement. On the way to an equal society, full gender equality in the labour market is an important contribution to emancipation. This approach brings us closer to overcoming that issue. Many studies have shown that human biases in the recruiting process are often discriminating against women. Maybe the step we need to overcome that issue is the utilisation of fair, transparent and non-discriminating machine learning approaches in the recruiting process. Although many acts already regulate that people with disabilities should not be discriminated against or disadvantaged in any form on the labour market, in practice it is very difficult to determine whether in a specific case, applicants are being discriminated against based on their disability. Our model can help to clarify these and, ultimately, to overcome them or at least address them properly. AI ethic research included in this approach has the potential to generate new knowledge for professionals from the HRM sector but also for researchers and developers in the AI and IT industry. This wholesome approach to this problem has the potential to yield a comprehensive and broad framework and set of guidelines for the future AI system in the HRM sector.

8   Conclusion

This paper presents a holistic approach to the development of a fair, automated and intelligent HRM and recruiting process. Automating processes like preselection, active sourcing or video interviews, does not only save time and promotes higher efficiency of the process but also offers very powerful and successful tools for mitigating biases and unfairness which are unfortunately one of the biggest challenges in the human-guided systems. Despite being highly dependent on their training data sets, which are often based on many biases and discrimination, machine learning algorithms also offer plenty of possibilities on how to reduce or totally eliminate them. Therefore, within our paper, we outlined key advancements and techniques in the area of AI that can be used to automate the HRM and recruiting sectors in an unbiased way.

Besides that, challenges and areas where we believe there is an opportunity for improvement in future research on this subject are presented as well. In the area of HRM and recruiting there is a huge potential for machine learning breakthroughs to be utilised in order to develop more sophisticated and advanced systems. We expect that in the future, this area will receive much more attention not only from the research community but also from the industry as it's evident how AI can improve current HRM procedures. In conclusion, the comprehensive approach laid out in this paper provides a solid foundation for creating AI systems that prioritise ethics, equity, and explainability in addition to technological excellence. By adopting this strategy, system developers may establish an atmosphere for HRM and recruitment which is more inclusive, transparent and unbiased, ultimately assisting both employers and job seekers.

Acknowledgements. This work is partially supported by a grant of the BMWi ZIM program, no. KK5007201LB0.

References

1. Arik: Interpretable deep learning for time series forecasting (2021). https://ai.googleblog.com/2021/12/interpretable-deep-learning-for-time.html
2. Bangor, A., Kortum, P.T., Miller, J.T.: An empirical evaluation of the system usability scale. Int. J. Hum.-Comput. Interact. **24**(6), 574–594 (2008)
3. Bellamy, R.K., et al.: AI fairness 360: an extensible toolkit for detecting, understanding, and mitigating unwanted algorithmic bias. arXiv preprint arXiv:1810.01943 (2018)
4. Binns, R.: Fairness in machine learning: lessons from political philosophy. In: Conference on Fairness, Accountability and Transparency, pp. 149–159. PMLR (2018)
5. COM, E: Laying down harmonised rules on artificial intelligence (artificial intelligence act) and amending certain union legislative acts. Proposal for a regulation of the European parliament and of the council (2021)
6. Urquhart, L.D., Craigon, P.J.: The moral-it deck: a tool for ethics by design. J. Responsible Innov. **8**(1), 94–126 (2021)
7. Danner, M., Hadžić, B., Radloff, R., Su, X., Peng, L., Rätsch, M.: Overcome ethnic discrimination with unbiased machine learning for facial data sets. In: 18th International Joint Conference on Computer Vision, Imaging and Computer Graphics Theory and Applications - VISAPP, Lisbon, Portugal, pp. 464–471 (2023)
8. Ekman, P., Friesen, W.V.: Facial action coding system. Environ. Psychol. Nonverbal Behav. (1978)
9. Fischer, M.T., Hirsbrunner, S.D., Jentner, W., Miller, M., Keim, D.A., Helm, P.: Promoting ethical awareness in communication analysis: investigating potentials and limits of visual analytics for intelligence applications. In: 2022 ACM Conference on Fairness, Accountability, and Transparency, pp. 877–889 (2022)
10. Friedman, B.: Value-sensitive design. Interactions **3**(6), 16–23 (1996)
11. Gansser, O.A., Reich, C.S.: A new acceptance model for artificial intelligence with extensions to UTAUT2: an empirical study in three segments of application. Technol. Soc. **65**, 101535 (2021)
12. Goldsmith, J.: Dealing with prosody in a text-to-speech system. Int. J. Speech Technol. **3**, 51–63 (1999)
13. Gressel, C., Orlowski, A.: Integrierte technikentwicklung: Herausforderungen, umsetzungsweisen und zukunftsimpulse. TATuP-Zeitschrift für Technikfolgenabschätzung in Theorie und Praxis **28**(2), 71–72 (2019)
14. Gulliford, F., Dixon, A.P.: AI: the HR revolution. Strateg. HR Rev. **18**(2), 52–55 (2019)
15. Hamilton, I.A.: Amazon built an AI tool to hire people but had to shut it down because it was discriminating against women. Insider (2018). https://www.businessinsider.com/amazon-built-ai-to-hire-peoplediscriminated-against-women-2018-10. Accessed 23 May 2022
16. Hirsbrunner, S.D., Tebbe, M., Müller-Birn, C.: From critical technical practice to reflexive data science. Convergence, p. 13548565221132243 (2022)

17. Hudson, D.A., Zitnick, L.: Generative adversarial transformers. In: International Conference on Machine Learning, pp. 4487–4499. PMLR (2021)
18. Insights, F.B.: Recruitment software market size, share and industry analysis by component: regional forecast 2018–2025 (2019). https://www.fortunebusinessinsights.com/industry-reports/recruitment-software-market-100081
19. Jiang, Y., Chang, S., Wang, Z.: TransGAN: two pure transformers can make one strong GAN, and that can scale up. In: Advances in Neural Information Processing Systems, vol. 34, pp. 14745–14758 (2021)
20. Kenton, J.D.M.W.C., Toutanova, L.K.: BERT: pre-training of deep bidirectional transformers for language understanding. In: Proceedings of naacL-HLT, vol. 1, p. 2 (2019)
21. Makhlouf, K., Zhioua, S., Palamidessi, C.: On the applicability of machine learning fairness notions. ACM SIGKDD Explorations Newsl. 23(1), 14–23 (2021)
22. Resource Management, IH: Human resource management - guidelines on recruitment (2016). https://www.iso.org/standard/64149.html
23. Marcinkowski, F., Starke, C.: Wann ist künstliche intelligenz (un-) fair? Politik in der digitalen Gesellschaft—Band, p. 269 (2019)
24. Mehrabi, N., Morstatter, F., Saxena, N., Lerman, K., Galstyan, A.: A survey on bias and fairness in machine learning. ACM Comput. Surv. (CSUR) 54(6), 1–35 (2021)
25. O'neil, C.: Weapons of math destruction: how big data increases inequality and threatens democracy. Crown (2017)
26. Reilly, P.: The impact of artificial intelligence on the HR function. Which way now for HR and organisational changes, pp. 41–58 (2018)
27. Ruf, B., Detyniecki, M.: Towards the right kind of fairness in AI. arXiv preprint arXiv:2102.08453 (2021)
28. Sánchez-Monedero, J., Dencik, L., Edwards, L.: What does it mean to 'solve' the problem of discrimination in hiring? Social, technical and legal perspectives from the UK on automated hiring systems. In: Proceedings of the 2020 Conference on Fairness, Accountability, and Transparency, pp. 458–468 (2020)
29. Simonyan, K., Zisserman, A.: Two-stream convolutional networks for action recognition in videos. In: Advances in Neural Information Processing Systems, vol. 27 (2014)
30. Spindler, M., Booz, S., Gieseler, H., Runschke, S., Wydra, S., Zinsmaier, J.: How to achieve integration? Methodological concepts and challenges for the integration of ethical, legal, social and economic aspects into technological development. Das geteilte Ganze: Horizonte Integrierter Forschung für künftige Mensch-Technik-Verhältnisse, pp. 213–239 (2020)
31. Staatsministerium Baden-Württemberg: Land fördert projekte zur künstlichen intelligenz (2021). https://www.baden-wuerttemberg.de/de/service/presse/pressemitteilung/pid/land-foerdert-projekte-zur-kuenstlichen-intelligenz/
32. Strack, R., Caye, J., von der Linden, C., Quirós, H., Haen, P.: Realizing the value of people management: from capability to profitability. BCG. perspectives (2012)
33. Technavio: AI market in recruitment industry by component and geography - forecast and analysis 2022–2026 (2022). https://www.technavio.com/report/ai-market-industry-in-recruitment-industry-analysis
34. Vaswani, A., et al.: Attention is all you need. In: Advances in Neural Information Processing Systems, vol. 30 (2017)
35. Venkatesh, V., Bala, H.: Technology acceptance model 3 and a research agenda on interventions. Decis. Sci. 39(2), 273–315 (2008)
36. Verma, S., Rubin, J.: Fairness definitions explained. In: Proceedings of the International Workshop on Software Fairness, pp. 1–7 (2018)
37. Votto, A.M., Valecha, R., Najafirad, P., Rao, H.R.: Artificial intelligence in tactical human resource management: a systematic literature review. Int. J. Inf. Manag. Data Insights 1(2), 100047 (2021)

38. Waltz, C.: Deep learning model for unbiased artificial intelligence in human resource management. Master's thesis, Reutlingen University (2022)
39. Witte, W.W.: Die 25 umsatzstärksten hr-softwareanbieter 2021 (2022). https://www.hr-konjunktur.de/newsleser/die-25-umsatzstaerksten-hr-software-anbieter-2021.html
40. Zhu, Z., Lin, K., Jain, A.K., Zhou, J.: Transfer learning in deep reinforcement learning: a survey (2022)

An Explainable Approach for Early Parkinson Disease Detection Using Deep Learning

Lerina Aversano[1]([✉]) [ID], Mario L. Bernardi[1] [ID], Marta Cimitile[2] [ID],
Martina Iammarino[1] [ID], Antonella Madau[1] [ID], and Chiara Verdone[1] [ID]

[1] Department of Engineering, University of Sannio, Benevento, Italy
{aversano,bernardi,iammarino,chiverdone}@unisannio.it,
a.madau@studenti.unisannio.it
[2] Unitelma Sapienza University, Rome, Italy
marta.cimitile@unitelmasapienza.it

Abstract. Parkinson's disease (PD) is a progressive disorder that affects the nervous system and all the parts of the body controlled by it. It is the second most diffused neurodegenerative disorder, showing increasing trends in the last years and requiring new tools and procedures for diagnosis and assessment. In order to be used in medical clinics, the PD detection approaches require high effectiveness in disease detection and good capability to drive the experts in the comprehension and checking of the prediction's reasons. According to this, this paper proposes an explainable Deep Learning approach for the detection of PD from single photon emission computed tomography (SPECT) images. The approach consists of a combination of a CNN prediction model and a Gradient weighted Class Activation Mapping (Grad-CAM) interpretable technique. The validation is performed on a known dataset belonging to Parkinson's Progression Markers Initiative (PPMI). For this dataset, SPECT images of 974 patients are used showing good accuracy in the classification of healthy and PD patients and a good capability to explain the obtained prediction.

Keywords: Deep learning · Parkinson's disease detection · Explainable deep learning

1 Introduction

PD is a very diffused long-term degenerative condition of the central nervous system characterized by the gradual loss of neurons in a region of the brain called substantia nigra [11]. People affected by PD can show motor symptoms (i.e., stiffness or rest tremor) and non-motor symptoms (i.e., sleep-wake cycle disturbances, cognitive impairment, mood, and affective disorders, autonomic dysfunction, sensory symptoms, and pain). These symptoms can develop, gradually, making the early diagnosis not easy to perform since PD is characterized by a not definitive diagnostic test and cure. However, the early prediction of PD is very important since the late diagnosis can typically cause great loose neurons in the brain that can make any treatment difficult and a cure impossible [17]. Even if several approaches explore the motor symptoms [5,6],

© The Author(s), under exclusive license to Springer Nature Switzerland AG 2023
D. Conte et al. (Eds.): DeLTA 2023, CCIS 1875, pp. 326–339, 2023.
https://doi.org/10.1007/978-3-031-39059-3_22

the recent literature [25] shows good results in early diagnosis using the single photon emission computed tomography (SPECT) [8]. However, SPECT-based diagnostic tools are becoming the most used in European clinics [20] showing good performance for diagnosis in patients with first signs of PD [8,10,27]. The SPECT images are yet not easy to interpret since they require experts with the capability to evaluate and correlate the values of several parameters and clinical data. This inspection is usually made manually with a high risk of human error. This issue drives, several recent studies that propose Deep Learning (DL) [19] for detecting patterns from the SPECT images useful to recognize patients affected by PD with high accuracy and encouraging results [13,21]. However, these classification approaches have a low acceptance in medical real context [22] since they lack of explanation about the obtained predictions and this limits their utility to drive decisions. This paper answers the necessity of an explainable PD classification approach introducing a new approach for the analysis of the SPECT images and it is composed of two main components: a CNN classifier and a prediction interpreter. The CNN allows us to find patterns in the SPECT images useful to discriminate PD patients from healthy people. The prediction interpreter allows us to identify in the analyzed image, the regions of interest useful to understand the prediction results. Our idea is that this approach can better support the experts in checking and evaluating the results by indicating the patterns at the base of the prediction. The adopted interpretation technique is based on the Gradient weighted Class Activation Mapping (Grad-CAM) This technique [9,16,30] allows us to use gradients that flow through the CNN to highlight the regions of interest in the analyzed images. The proposed approach is evaluated on a real dataset showing good accuracy in the classification and a good capability in the localization of the region of interest of the SPECT images. The paper is organized into seven sections. In Sect. 2 the discussion of the background is reported. Section 3 reports the related work. Section 4 describes how the empirical validation is performed, and the obtained results are reported in Sect. 5. Section 6 discusses the threats to validity of the presented research and finally, some conclusive remarks are reported.

2   Background

2.1   Convolutional Neural Networks in Computer Vision

The approach proposed in this paper is based on a Convolutional Neural Networks (CNN) component. A CNN is a class of DL algorithms becoming dominant in computer vision tasks [32]. It typically includes three types of layers: convolution, pooling, and fully connected. The convolution and pooling layers allow the feature extraction, while the fully connected layer, relates the features extracted into the final output. The convolution layer is composed of several mathematical operations, including a type of linear operation called convolution. It aims to extract features from the input image by converting the identified patterns into numbers. Then, a nonlinearity is applied to the result through the so-called activation functions. Pooling layers perform the reduction of the spatial size of the features belonging to the previous convolutional layer (reduction of the number of input pixels that need to be processed). There are two types of pooling layers: max pooling and average pooling. The first selects the maximum value from each patch of the feature map, while the second calculates the average value of each patch [28]. Finally, the fully connected layer takes input from the previous layer and computes, for each considered class, the score. The output is an array having a size equal to the number of considered classes.

2.2 Grad-CAM Model

In the proposed approach, a Grad-CAM [26] technique is used to perform the interpretation of the classification results. Grad-CAM is a generalization of the CAM (Class Activation Mapping) [34] and it allows the generation of location maps for many CNN-based networks. Using the CAM method, the location map is generated, where the global average aggregated convolutional feature maps are directly fed into a softmax layer.

The formalization and a detailed description of Grad-CAM are reported in [26]. For a class c, Grad-CAM computes the gradient of the score y^c (before the softmax), with respect to the activations of the characteristic map A^k of a convolutional layer ($\frac{\delta y^c}{\delta A^k}$) with the aim of obtaining the class-discriminative localization map Grad-CAM $L^c_{Grad-CAM} \in \mathbb{R}^{u \times v}$ of width u and height v for any class c. To determine the neuron significance weights α^c_k, these gradients back propagated are global-average-pooled over the width and height dimensions, respectively indexed by i and j. Then the significance weights are given by the Eq. 1.

$$\alpha^c_k = \frac{1}{Z} \sum_i \sum_j \frac{\delta y^c}{\delta A^k_{ij}} \qquad (1)$$

Up until the last convolution layer to which the gradients are being propagated, the actual computation of α^c_k when backpropagating gradients with respect to activations consists of successive matrix products of the weight matrices and the gradient with respect to activation functions. Thus, this weight α^c_k encapsulates the "importance" of feature map k for a target class c and represents a partial linearization of the deep network downstream from A. After calculating the gradient score for class c, we obtain the class-discriminative localization map by performing a weighted combination of forward activation maps sent to a ReLU to focus attention on the features that have a positive influence on the interest class (Eq. 2).

$$L^c_{Grad-CAM} = ReLU \left(\sum_k \alpha^c_k A^k \right) \qquad (2)$$

The equation for α^c_k is the same as w^c_k utilized by CAM, up to a proportionality constant ($\frac{1}{Z}$) that is normalized out before display.

3 Related Work

Artificial intelligence (AI) algorithms are more always used in the clinical domain for the early diagnosis of several diseases [2–4,7,15,18]. Referring to PD, a lot of studies show encouraging results in PD diagnostics deriving from the adoption of AI from the analysis of PD's motor symptoms [5,6].

Differently from these studies, here the no-motor symptoms are explored. In particular, the proposed approach is focused on the analysis of SPECT images. This is in line with the study described in [31] where a CNN for the classification of SPECT scans in multi-site or multi-camera settings is reported. The authors perform experimentation on the same dataset used in this paper, achieving an accuracy of about 96%.

The PPMI database is also used to validate the approach proposed in [1]. The goal of this study is quite different since it proposes a CNN model to predict clinical motor function evaluation scores (Parkinson's disease rating scale) directly from longitudinal DaTScan SPECT images and non-imaging clinical measures. An ensemble classification method for the PD diagnosis is proposed in [12]. It uses multiple heterogeneous data derived from clinical tests and SPECT images. The classification accuracy obtained by this performance-weighted ensemble model on the PPMI database is almost 96%.

In [23], authors evaluated the capability of a features model, generated from SPECT brain images, to identify dopaminergic degeneration state to perform PD diagnosis. Several classifiers are also evaluated (including the support vector machines (SVM), k-nearest neighbors, and logistic regression) reaching an accuracy equal to 97.9%, obtained when all the features are used simultaneously. The described studies are generally characterized by high accuracy in PD detection but they lack any explanations useful to drive the experts to check and use the classification results [22]. However, for medical professionals to fully embrace the use of the described techniques, they need to know how prediction is obtained or what data is used to make the diagnosis in order to justify why the algorithm generated a specific prediction.

4 Approach

The explainable PD detection approach allows performing a binary classification to distinguish between people affected by PD and healthy people. The people are classified on the base of their SPECT images.

An overview of the proposed approach is reported in Fig. 1. The figure shows three main components: (a) the image processor, (b) the classifier, and (c) the Grad-CAM interpreter. In the following subsections, these components are described

4.1 The Image Processor

This component achieves the filtering and selection of the SPECT images starting from an initial image dataset. However, the image selection and normalization are performed according to the prediction goals. The main idea is that a SPECT image can be split into a given number of slices and only a few parts of these slices are of interest to our prediction. More precisely, since the region of greater interest for PD detection is the putamen and caudate [24], the slices that better represent these regions are selected. In particular, for each initial SPEC image, the processor selected a k-neighborhood of the slice with the highest pixel intensity (it is the slice that better represents the putamen and caudate regions according to the previous studies). In this study, different values of k (3, 5, 7, 9, 11, 13, and 15) are explored in the way to find the best k-neighborhood

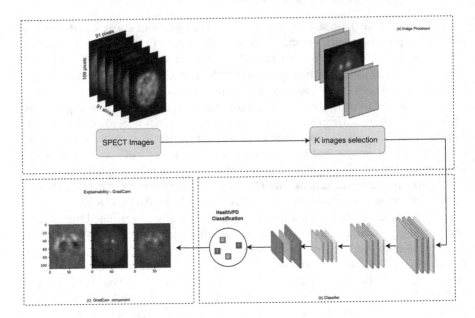

Fig. 1. The explainable PD detection approach.

according to the classification accuracy. The so obtained dataset is composed, of the k-neighborhood slices of each initial slice uniformed in size, converted in a standard format (jpeg), and labeled as healthy or PD.

4.2 Data Classification

The dataset obtained after the image processing is split into the training, test, and validation dataset to perform the CNN model training and assessment. Figure 1-(b) reports the architecture of the adopted CNN classifier. It is a Conv2D network made up of three different levels each including several layers. The figure highlights different types of layers (represented with different colors) repeated in the three levels: convolutional layer (yellow), batch normalization layer (blue), dropout layer (orange), and max pooling 2D layer (green). At the end of the three levels, we find two dense layers (violet) between which there is a dropout layer.

The proposed CNN model differs from the known transfer learning models because it can process tensors of arbitrary dimensions. This makes the CNN useful to map the multi-channels that compose an image according to its different input layers. However, in the CNN model, we include as input several tensors having different sizes respectively equal to 3, 5, 7, 9, 11, 13, 15. The multi-channel structure is followed by considering a sequence of k concatenated slices (k-neighborhood) merged into a single k-channel (RGB) image. Considering, for example, k equal to five, we consider five slices: the better slice, its two preceding slices, and its two subsequent slices. Finally, the trained CNN model achieves the image classification by labeling them as "healthy" or "PD".

4.3   Grad-CAM Interpreter

The CNN model and the predictions are sent as input to the Grad-CAM component. This component generates for the predicted images a map to localize the regions that better influence the predictions. Figure 1-(c) shows an example of Grad-CAM image processing. The input image (the output of the classification process) is represented at the center while the Grad-CAM map is reported on the right side, the other image represents the overlapping between these two images. In the Grad-CAM map image, the area influencing the classification of the corresponding input class is colored green. Considering the input image classified as "PD", the green area corresponds to the area that best discriminates patients with PD.

5   Experimental Description

In the following sub-sections respectively the adopted dataset and the experiment settings are described.

5.1   Dataset

The performance of the explainable PD detection approach is evaluated on a real dataset belonging to Parkinson's Progression Markers Initiative (PPMI)[1]. The dataset contains both clinical data and images of controls performed on Parkinson's patients and healthy people. The data collection comes from various organizations around the world through a continuous updating process. From this dataset, 843 SPECT scans of 388 PD patients and 281 scans of 278 healthy controls are extracted. For each patient and healthy person, at least one SPECT image and a maximum of 5 images are considered.

The considered images were selected randomly from the initial dataset and the adopted criteria consist to maintain a 1 to 3 ratio between healthy controls and PD patients (the initial dataset contains a greater number of images for PD patients and the healthy controls). In the obtained filtered dataset, each SPECT scan image is composed of 91 slices made up of 91×109 pixels encoded with 16-bit greyscale.

After an analysis of the initial dataset, it was observed that slice 41 better represents the region of interest (the putamen and caudate regions); this slice contains the highest striatal signal-to-background ratio according to the visual inspection performed by radiologists [33]. For this reason, according to the proposed approach (described in Section IV), a k-neighborhood of slice 41 was considered in this study and several sub-datasets from the filtered dataset are obtained for each different value of k (i.e., 3, 5, 7, 9, 11, 13, 15).

5.2   Experiment Setting

The experiments aim to evaluate the performance of the proposed approach and its explainability.

---

[1] www.ppmi-info.org.

Table 1. Classifier performance evaluated on the best hyperparameters configurations.

k	neuron_map	n_epoch	batch_size	optimizer	dropout_rate	Accuracy	Precision	Recall	F1-score
1	32	10	32	adam	0.15	0.9438	0.927	0.9057	0.9157
3	64	10	64	adam	0.15	0.9556	0.9301	0.9424	0.936
5	128	20	16	adam	0.15	0.9497	0.932	0.9192	0.9254
7	64	10	32	adam	0.15	0.9645	0.9444	0.953	0.9486
9	64	10	16	adam	0.2	0.9556	0.9334	0.9376	0.9354
11	32	10	16	adam	0.15	0.9467	0.9337	0.9076	0.9198
13	64	10	16	Nadam	0.2	0.929	0.9298	0.8573	0.887
15	32	20	32	adam	0.15	0.9408	0.9009	0.9378	0.9174

According to the first issue, we aim to evaluate the efficiency of the proposed classification process, on the considered sub-datasets. The evaluation is performed by using the following metrics: Accuracy, Precision, Recall, and F1, which can be calculated from the Confusion Matrix.

The classification is performed by splitting each considered sub-datasets into train, validation, and test datasets. The 30% of the dataset is used for the test, while from the remaining part, the 70% is used for the classifier training and the 30% is used for the validation.

This study also includes a hyperparameter optimization step using Talos[2] technology. Talos allows you to test all possible combinations of hyperparameters in one experiment. The best model is selected and saved in order to make the prediction later.

According to this, the following hyperparameters and hyperparameters values are considered:

– neurons: the dimensionality of the output space, the number of output filters in the convolution. The evaluated values are 128, 64, and 32;
– n_epochs: number of epochs to train the model. The evaluated number of epochs are 10 and 20;
– batch_size: number of samples per gradient update. The considered values are 16, 32, and 64;
– optimizer: algorithms used to decrease loss by tuning various parameters and weights, hence minimizing the loss function, providing better accuracy of model faster;
– loss: the function used to evaluate a candidate solution. The adopted loss function is the categorical cross-entropy;
– dropout_rate: the probability of cutting a node out of an iteration. Two different values of the dropout rate are considered: 0,15 and 0,2;
– last_activation: the activation function of the last layer. The used function is softmax.

The training phase of each classifier was repeated 10 times and the one obtaining the higher average F1-Score is selected. With the same F1-score, the model with the highest average Accuracy was chosen.

---

[2] https://autonomio.github.io/talos.

Finally, the experiment also aims to evaluate the ability of the proposed approach to provide correct interpretations of the predictions obtained from Grad-CAM through a manual evaluation by a team of experts.

The experimentation is carried out by using the following workstations:

- AMD Ryzen Threadripper 3960X 24-Core, with 128 GB of RAM and two GPUs NVIDIA RTX 3090 (with 24 GB of RAM);
- Intel Core i9 9940X (14 cores), with 64 GB of RAM and four GPU NVIDIA Tesla T4 (with 16 Gb of RAM).

6   Discussion of Results

This section shows and discusses the obtained results.

6.1   Classifier Performance

The Table 1 lists the metrics results for the CNN classifier. Each row of the table shows the results of a particular neighborhood; the first column specifies the k-neighborhood of the slice, the second to the fifth columns report the hyperparameters that characterize the best model; and the following columns show the results of the metrics used to validate the model.

We can observe a performance trend similar to a Gaussian as in particular the F1-score metric starts with a value equal to 0.91 for $k = 1$ and grows up to 0.94 for $k = 7$ and then decreases for greater neighborhoods.

Therefore, the neighborhood of slices that has the best performance is the one with k equal to 7; the found hyperparameters that characterize the best model are 64 as *neuron_map*, 10 epochs, 32 as *batch_size*, *adam* optimizer and *dropout_rate* equal to 0.15.

Figure 2 shows the trend of the classification metrics for all the neighborhoods of slices considered. In particular, it confirms that the neighborhood of the slice with k=7 is the one with the best performance for all the metrics considered. While considering the metrics individually, taking into account the fact that they do not undergo excessive variations overall, we can still say that the metric that varies more than the others is the recall which reaches the minimum for neighborhood 13 equal to 0.95 against the maximum reached for neighborhood 7 equal to 0.85.

Table 2. k = 1

	HC	PD
HC	62	12
PD	7	257

Table 3. k = 3

	HC	PD
HC	68	6
PD	9	255

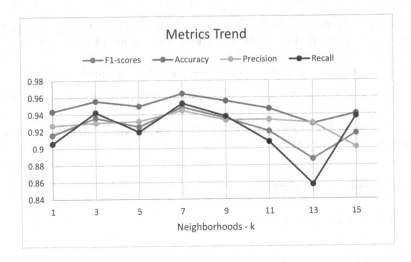

Fig. 2. Classification metrics trend

Table 4. k = 5

	HC	PD
HC	64	10
PD	7	257

Table 5. k = 7

	HC	PD
HC	69	5
PD	7	257

In Tables 2, 3, 4, 5, 6, 7, 8, 9, we report the confusion matrices for all the neighborhoods of slices considered. In particular, Table 5 shows the confusion matrix for the best-performing neighborhood ($k = 7$), as we can see in the main diagonal that out of a total of 263 PD patients, 255 are correctly classified for our model, while only 5 patients are classified as false positives and 8 as false negatives.

The results of this work are also compared with those present in [31]. The two studies have in common the use of the CNN network and the original PPMI dataset, however the preliminary use of the dataset changes. In particular, Wenzel et al. use the putaminal specific binding ratio obtained from the ROIs of the images together with the images themselves to train the network and classify, while our approach is based on using a neighborhood of slices 41 as if they were a single image per patient. The results obtained are however comparable in terms of accuracy of around 96%.

While in [23] various algorithms were used for the classification including SVM, KNN, and Logistic Regression in order to evaluate the disease starting from a dataset obtained by extracting the characteristics from SPECT images of the brain, obtaining an accuracy of 97.9%.

Although these works are comparable to ours in terms of the results obtained regarding the metrics, however, they lack any useful explainability to guide the experts to make decisions in a real environment.

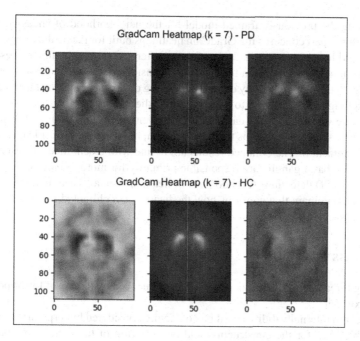

Fig. 3. An example of GradCam heatmap when k=7 for the PD patient (top) and healthy control (down).

Table 6. k = 9

	HC	PD
HC	67	7
PD	8	256

Table 7. k = 11

	HC	PD
HC	62	12
PD	6	258

Table 8. k = 13

	HC	PD
HC	54	20
PD	4	260

Table 9. k = 15

	HC	PD
HC	69	5
PD	15	249

6.2   Prediction Explainability

Despite the increasing knowledge about the concept of explainability, a consensus about the validation and assessment methods and metrics is still missing [29]. So in this study, we report a qualitative evaluation of the capability of the proposed approach to perform prediction explainability based on the analysis of the obtained interpretation maps (heatmaps) and the feedback of an expert.

Figure 3 shows an example of heatmaps generated by the proposed explainable PD detection approach. The upper part concerns the GradCam heatmaps built for a PD patient using the previously trained model for the neighborhood of slices equal to 7, while the lower part concerns the GradCam heatmaps built for the healthy control. Both the upper and lower parts are made up of three different images: the first represents the heatmap generated by GradCam for each class considered, the second is a slice 41 of a single scan of a patient or healthy control depending on the case, and the third represents the overlapping of the heatmap to the considered slice.

The figure shows that in the case of the PD patient, the heatmap highlights precisely the part of the scan that concerns the basal ganglia, highlighting the critical role of this area in PD detection. The obtained results are in line with recent medical research that shows that the basal ganglia are responsible primarily for motor control [14] playing a critical role in PD detection. A random set of heatmaps are also evaluated by an expert in the medical domain that confirms how the heatmaps highlighted area are crucial for the diagnosis.

7 Threats to Validity

Regarding threats to validity, this study can be influenced by threats of two types: internal and external.

A threat to internal validity could be not having considered hyperparameters among the possible ones for the construction and optimization of the model, this fact could certainly influence the results. However, to mitigate this threat, several experiments were carried out in order to consider the most recurring parameters in the construction of the model. As for the threats to external validity, one of them could concern the poor generalizability of the obtained results. This is due to the fact that the dataset used, although recently updated, is relatively related to the data that should be used to obtain highly generalizable estimates. The motivation behind the small number of scans within the dataset concerns a choice due to the high imbalance between patients and healthy controls within the original dataset, in fact it was decided to maintain a 1:3 ratio between healthy controls and sick. However, it is clear that the findings obtained in this study may be consolidated or disproved through new phases of future studies. Finally, the validation of the capability of the proposed approach to give interpretable results requires further experiments with the greater participation of domain experts.

8 Conclusion

This paper proposes an approach for the detection of PD patients based on a CNN classifier that considers as input the k-neighborhood slices of the analyzed SPECT images. A comparison among different values of k is proposed. Moreover, the proposed approach also includes a Grad-CAM component able to highlight the areas of the considered images that more influenced the prediction. The empirical validation is performed on a dataset derived from the PPMI dataset and composed of images of 834 PD patients and 281 healthy controls. The obtained results show good performance of the classifier that in the best hyperparameter configuration has an F1-score of 0.94. The study of the optimal value of k for the considered k-neighborhood slices also shows a gaussian trend for the F1-score with the peak when k = 7. Finally, referring to the explainability of the proposed approach, the obtained results are manually validated by an expert. The obtained masks highlight that the area that better discriminates PD patients from the healthy control is the brain's basal ganglia. This result is in line with the more recent medical literature confirming the critical role of this area on motor control. Further study will be conducted to extend the obtained results. In particular, a more extended validation of the Grad-CAM component can be proposed with a greater involvement of medical staff.

References

1. Adams, M.P., Rahmim, A., Tang, J.: Improved motor outcome prediction in Parkinson's disease applying deep learning to DaTscan SPECT images. Comput. Biol. Med. **132**, 104312 (2021)
2. Aversano, L., et al.: Thyroid disease treatment prediction with machine learning approaches. Procedia Comput. Sci. **192**, 1031–1040 (2021). https://doi.org/10.1016/j. procs.2021.08.106, https://www.sciencedirect.com/science/article/pii/S1877050921015945. knowledge-Based and Intelligent Information and Engineering Systems: International Conference KES2021
3. Aversano, L., Bernardi, M.L., Cimitile, M., Iammarino, M., Montano, D., Verdone, C.: Using machine learning for early prediction of heart disease. In: 2022 IEEE International Conference on Evolving and Adaptive Intelligent Systems (EAIS), pp. 1–8. IEEE (2022)
4. Aversano, L., Bernardi, M.L., Cimitile, M., Iammarino, M., Verdone, C.: An enhanced UNet variant for effective lung cancer detection. In: 2022 International Joint Conference on Neural Networks (IJCNN), pp. 1–8. IEEE (2022)
5. Aversano, L., Bernardi, M.L., Cimitile, M., Pecori, R.: Early detection of Parkinson disease using deep neural networks on gait dynamics. In: 2020 International Joint Conference on Neural Networks (IJCNN), pp. 1–8. IEEE (2020)
6. Aversano, L., Bernardi, M.L., Cimitile, M., Pecori, R.: Fuzzy neural networks to detect Parkinson disease. In: 2020 IEEE International Conference on Fuzzy Systems (FUZZ-IEEE), pp. 1–8. IEEE (2020)
7. Aversano, L., Bernardi, M.L., Cimitile, M., Pecori, R.: Deep neural networks ensemble to detect COVID-19 from CT scans. Pattern Recogn. **120**, 108135 (2021)
8. Badoud, S., Van De Ville, D., Nicastro, N., Garibotto, V., Burkhard, P.R., Haller, S.: Discriminating among degenerative Parkinsonisms using advanced 123i-ioflupane SPECT analyses. NeuroImage: Clin. **12**, 234–240 (2016)
9. Banerjee, P., Banerjee, S., Barnwal, R.P.: Explaining deep-learning models using gradient-based localization for reliable tea-leaves classifications. In: 2022 IEEE Fourth International Conference on Advances in Electronics, Computers and Communications (ICAECC), pp. 1–6 (2022). https://doi.org/10.1109/ICAECC54045.2022.9716699
10. Benamer, H.T., et al.: Accurate differentiation of parkinsonism and essential tremor using visual assessment of [123i]-FP-CIT SPECT imaging: the [123i]-FP-CIT study group. Mov. Disord. Official J. Mov. Disord. Soc. **15**, 503–510(2000)
11. Brown, E., et al.: Parkinson's progression markers initiative (PPMI) online expands biomarker research in Parkinson's disease (PD). Neurology, **98** (2022)
12. Castillo-Barnes, D., Ramírez, J., Segovia, F., Martínez-Murcia, F.J., Salas-Gonzalez, D., Górriz, J.M.: Robust ensemble classification methodology for i123-ioflupane SPECT images and multiple heterogeneous biomarkers in the diagnosis of Parkinson's disease. Front. Neuroinform. **12**, 53 (2018)

13. Choi, H., Ha, S., Im, H.J., Paek, S.H., Lee, D.S.: Refining diagnosis of Parkinson's disease with deep learning-based interpretation of dopamine transporter imaging. NeuroImage: Clin. **16**, 586–594 (2017). https://doi.org/10.1016/j.nicl.2017.09.010, https://www.sciencedirect.com/science/article/pii/S2213158217302243

14. Du, G., Zhuang, P., Hallett, M., Zhang, Y.Q., Li, J.Y., Li, Y.J.: Properties of oscillatory neuronal activity in the basal ganglia and thalamus in patients with Parkinson's disease. Transl. Neurodegener. **7**(1), 17 (2018). https://doi.org/10.1186/s40035-018-0123-y

15. Siva Shankar, G., Manikandan, K.: Diagnosis of diabetes diseases using optimized fuzzy rule set by grey wolf optimization. Pattern Recogn. Lett. **125**, 432–438 (2019). https://doi.org/10.1016/j.patrec.2019.06.005, http://www.sciencedirect.com/science/article/pii/S0167865519301734

16. Górski, L., Ramakrishna, S., Nowosielski, J.M.: Towards grad-cam based explainability in a legal text processing pipeline. CoRR abs/2012.09603 (2020). https://arxiv.org/abs/2012.09603

17. Iarkov, A., Barreto, G.E., Grizzell, J.A., Echeverria, V.: Strategies for the treatment of Parkinson's disease: beyond dopamine. Front. Aging Neurosci. **12**, 4 (2020)

18. Karayilan, T., Kilic, O.: Prediction of heart disease using neural network. In: 2017 International Conference on Computer Science and Engineering (UBMK), pp. 719–723 (2017). https://doi.org/10.1109/UBMK.2017.8093512

19. Khachnaoui, H., Mabrouk, R., Khlifa, N.: Machine learning and deep learning for clinical data and PET/SPECT imaging in Parkinson's disease: a review. IET Image Process. **14**(16), 4013–4026 (2020). https://doi.org/10.1049/iet-ipr.2020.1048, https://ietresearch.onlinelibrary.wiley.com/doi/abs/10.1049/iet-ipr.2020.1048

20. Lundervold, A.S., Lundervold, A.: An overview of deep learning in medical imaging focusing on mri. Zeitschrift für Medizinische Physik **29**(2), 102–127 (2019). https://doi.org/10.1016/j.zemedi.2018.11.002, https://www.sciencedirect.com/science/article/pii/S0939388918301181, special Issue: Deep Learning in Medical Physics

21. Nazari, M., et al.: Data-driven identification of diagnostically useful extrastriatal signal in dopamine transporter SPECT using explainable AI. Sci. Rep. **11**(1), 22932 (2021). https://doi.org/10.1038/s41598-021-02385-x, https://doi.org/10.1038/s41598-021-02385-x

22. Nazari, M., et al.: Explainable AI to improve acceptance of convolutional neural networks for automatic classification of dopamine transporter SPECT in the diagnosis of clinically uncertain Parkinsonian syndromes. Eur. J. Nucl. Med. Mol. Imaging **49**(4), 1176–1186 (2022). https://doi.org/10.1007/s00259-021-05569-9

23. Oliveira, F.P., Faria, D.B., Costa, D.C., Castelo-Branco, M., Tavares, J.M.R.: Extraction, selection and comparison of features for an effective automated computer-aided diagnosis of Parkinson's disease based on [123i] FP-CIT SPECT images. Eur. J. Nucl. Med. Mol. Imaging **45**(6), 1052–1062 (2018)

24. Ortiz, A., Munilla, J., Martínez-Ibañez, M., Górriz, J.M., Ramírez, J., Salas-Gonzalez, D.: Parkinson's disease detection using isosurfaces-based features and convolutional neural networks. Front. Neuroinform. **13**, 48 (2019). https://doi.org/10.3389/fninf.2019.00048, https://www.frontiersin.org/articles/10.3389/fninf.2019.00048

25. Poewe, W., et al.: Parkinson disease. Nat. Rev. Dis. Primers **3**(1), 17013 (2017). https://doi.org/10.1038/nrdp.2017.13

26. Selvaraju, R., Cogswell, M., Das, A., Vedantam, R., Parikh, D., Batra, D.: Grad-cam: visual explanations from deep networks via gradient-based localization. 2016. arXiv preprint arXiv:1610.02391 (2016)

27. Staffen, W., Mair, A., Unterrainer, J., Trinka, E., Ladurner, G.: Measuring the progression of idiopathic Parkinson's disease with [123i] *beta*-CIT SPECT. J. Neural Transm. **107**(5), 543–552 (2000)

28. Valizadeh, M., Wolff, S.J.: Convolutional neural network applications in additive manufacturing: a review. Adv. Ind. Manuf. Eng. **4**, 100072 (2022). https://doi.org/10.1016/j.aime. 2022.100072, https://www.sciencedirect.com/science/article/pii/S2666912922000046
29. Vilone, G., Longo, L.: Notions of explainability and evaluation approaches for explainable artificial intelligence. Inf. Fusion **76**, 89–106 (2021). https://doi.org/10.1016/j.inffus.2021. 05.009, https://www.sciencedirect.com/science/article/pii/S1566253521001093
30. Vuppala, S.K., Behera, M., Jack, H., Bussa, N.: Explainable deep learning methods for medical imaging applications. In: 2020 IEEE 5th International Conference on Computing Communication and Automation (ICCCA), pp. 334–339 (2020). https://doi.org/10.1109/ ICCCA49541.2020.9250820
31. Wenzel, M., et al.: Automatic classification of dopamine transporter SPECT: deep convolutional neural networks can be trained to be robust with respect to variable image characteristics. Eur. J. Nucl. Med. Mol. Imaging **46**(13), 2800–2811 (2019)
32. Yamashita, R., Nishio, M., Do, R.K.G., Togashi, K.: Convolutional neural networks: an overview and application in radiology. Insights Imaging **9**(4), 611–629 (2018). https://doi. org/10.1007/s13244-018-0639-9
33. Zhang, Y.C., Kagen, A.C.: Machine learning interface for medical image analysis. J. Digit. Imaging **30**(5), 615–621 (2017)
34. Zhou, B., Khosla, A., Lapedriza, A., Oliva, A., Torralba, A.: Learning deep features for discriminative localization. In: IEEE Conference on Computer Vision and Pattern Recognition. IEEE Computer Society (2016). https://doi.org/10.1109/CVPR.2016.319, https://doi. ieeecomputersociety.org/10.1109/CVPR.2016.319

UMLDesigner: An Automatic UML Diagram Design Tool

Vinasetan Ratheil Houndji$^{(\boxtimes)}$ and Genereux Akotenou

Institut de Formation et de Recherche en Informatique, Université d'Abomey-Calavi,
Abomey-Calavi, Benin
ratheil.houndji@uac.bj

Abstract. Software design encompasses a range of activities, starting with the request for the computerization of a process and progressing through various phases. Among these phases, modeling is particularly crucial and can sometimes be challenging. This paper presents an approach to automatically analyze software specifications and generate the corresponding UML class diagram. We use Natural Language Processing tools, namely Stanza and NLTK, and a rule-based method for data extraction. We have developed UMLDesigner, a tool that includes a text-to-diagram editor enabling users to create UML diagrams based on textual descriptions of management rules. Our model has been trained on French language data specific to the UML class diagram domain. To facilitate end-user adoption, we have containerized the model into an API and developed a web application that interfaces with the API to process text and generate diagram images. The conducted experiments demonstrate the effectiveness of UMLDesigner.

Keywords: Modeling · Class diagram · UML · NLP

1 Introduction

Unified Modeling Language (UML) [2,4] is today an essential communication tool widely used in development projects worldwide to create visual models of object-oriented software systems, including the components of the system and the relationships between them. A lot of software exists for the design of UML diagrams. Unfortunately, most available tools for creating UML diagrams are designed for manual use, which can be time-consuming and labor-intensive, particularly for complex projects. This manual approach also introduces the possibility of bias, as the analysis depends on the designer's understanding of constraints and their experience with UML.

The main objective of this work is to set up an autonomous processing system based on NLP (Natural Language Processing) that can design UML class diagrams from a descriptive text given as input, thereby reducing the time and effort required for manual UML modeling. We proceed as follows:

1. Users have to input a textual description of UML specifications;
2. Our model uses NLP to process the input text, removing some useless words and making reformulations to have needed and full-of-sense sentences;

D. Conte et al. (Eds.): DeLTA 2023, CCIS 1875, pp. 340–350, 2023.
https://doi.org/10.1007/978-3-031-39059-3_23

3. These sentences are used through a filtering layer where we implement some UML pattern to identify elements like classes, attributes, methods, relation types, and multiplicities.

Based on our model, we have developed UMLDesigner. This user-friendly web application allows one to: 1) edit the generated diagrams, 2) collaborate with others, and 3) generate code in classical programming languages such as Java and Python.

2 Overview of the State of the Art

Several works exist in the automation field for the design of UML systems. In this section, we provide an overview of the most relevant and related studies in this field to our knowledge:

- UCDA [7] (Use Case driven development assistant tool for class model generation) is a versatile tool that can serve requirement analysis and model generation purposes. It uses a well-established requirement analysis technique to gather and document requirements in a textual format. The software designer then applies the Object Model Creation Process (OMCP) to analyze the requirements and identify objects, including their attributes, associations, and behaviors. The object model is then refined using generalization and object, and classes are determined based on domain knowledge.
- LIDA [5] (LInguistic Assistant for Domain Analysis) helps analysts develop object-oriented domain models using a subset of UML. To create such models, the requirements analyst must often analyze large volumes of text. LIDA facilitates this analysis by compiling a list of the words and multi-word terms in a document and providing a graphical interface for the user to mark them as corresponding to model elements. LIDA does not offer an end to analysis, but rather a starting point, requiring significant user interaction.
- Gleek [3] is a text-to-diagram tool that turns descriptions (in its syntax) into diagrams. To our knowledge, this is the closest solution to ours. However, Gleek is based on a codified language that must be mastered beforehand. This solution is based on diagrammatic reasoning.

This paper proposes an approach for generating a UML class diagram from a descriptive text in natural language.

3 Methodology

Our model is built to recognize and process the UML class diagram. It is one of the most used structure diagrams in UML that describes how the system is divided while showing the relationships between these different components.

3.1 Overview

For the extraction of class diagrams, our approach is based on two hypotheses:

- each sentence is syntactically correct;
- each element identified as a class cannot be so if no textual entry describes at least one attribute for it.

Based on these hypotheses, we have divided our system into four modules that enable the extraction of class diagrams. These modules are text acquisition, syntactic analysis, semantic analysis, and data extraction.

Syntactic and semantic analysis is based on the natural language processing tool Stanza [6]. It is a Python natural language analysis package to facilitate text processing. Stanza contains tools that can be used in a pipeline to convert a string containing human language text into lists of sentences and words, to generate base forms of those words, their parts of speech, and morphological features, to give a syntactic structure dependency parse, and to recognize named entities. The toolkit is designed to support over 70 languages. Stanza is built with highly accurate neural network components on the PyTorch library.

In the last layer of our model, we choose an expert or rules-based approach for data extraction. We build a list of patterns to identify different class relations types.

3.2 Filtering Algorithm

This section describes our algorithm for designing the class diagrams. This algorithm, mainly based on natural language processing and pattern mining process, can take as input UML management rules in plain text format and output UML class diagrams in XML string format. This prototype only processes text in French and can extract the name of the classes, the class attributes, the methods, the relations between classes, their type, and their multiplicities. The design of the diagram is achieved in 4 modules: Management rules acquisition, Syntactic analysis, Semantic analysis, and finally, the diagram extraction module as illustrated in Fig. 1.

- **Layer 1: Management rules acquisition.** A textual document is given as an input document. To get this input, we created a web interface allowing users to enter or paste text for processing easily.
- **Layer 2: Syntactic Analysis.** Here, the input corpus of text is processed to assign labels to each text word. This is achieved through the tokenization of the input text, which enables the splitting of sentences into chunks of tokens. After that, we find for each token the Part of Speech to get each word's role in the grammar of a sentence. To optimize and generalize our model, we reduce each word to its base form, the lemma, while maintaining its meaning and context. We achieve this processing using Lemmatization. Finally, we find each word's dependencies with other words in the same sentences. As an output of this layer, each corpus word contains an ID of type integer, the original word, the lemmatization form of the word, the word Part of Speech, the word head number in the sentence, and the dependency relation of the word regarding the other word in its sentence. We illustrate below this data structure defining each word or token in the output:

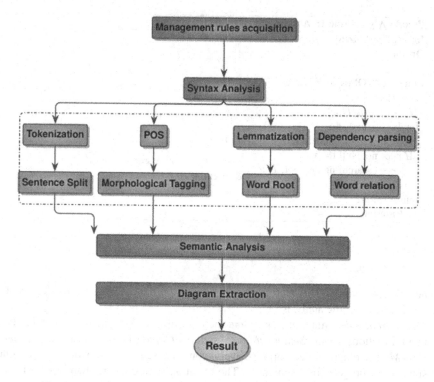

Fig. 1. Pseudo code to break complex sentence to subject-verb-object form

```
"word": {
    "id": Integer,
    "text": String,
    "lemma": String,
    "pos": String,
    "head": Integer,
    "deprel": String
}
```

- **Layer 3: Semantic Analysis.** The objective is to understand the meaning of each sentence. To achieve this, phrases are chunked into subject-verb-object structures, while dependencies that surround various sentence words are analyzed. We use the data structure in the previous step to filter the texts as much as possible and keep only essential information full of meaning. Figure 2 illustrates the pseudo-algorithm used in this layer.

We look for the root verb for each sentence and call our search engine. The *getRootNodeInSentence* function identifies the root verb in each sentence then the *searchEngine* function searches for the words based on their dependencies with

```
Pseudo-Algorithme 1:  Atomic Sentence Maker
Input: Sentence₁ ... Sentence_N
Output: List

function ATOMICSENTENCEMAKER(Doc)
 result ← []
 N ← length(Doc)
 for k ← 1 to N do
  root ← getRootNodeInSentence(Doc_k)
  If root not Null then                          & & end for
   svo ← searchEngine(root, Doc_k)  & & end if &
   result.add(svo) &
  return result
 end function
```

Fig. 2. Chunk of Multiplicity predefined list

neighboring words. This step is possible using the data structure provided by the previous step of the analysis.

The *searchEngine* function represents the core of our algorithm and contains two main functions *atomicSentenceMaker* and *umlObjectClassifier*. The *atomicSentenceMaker* function decomposes every complex sentence into one or multiple subject-verb-object form sentences. The resulting sentences are then input into the *umlObjectClassifier* function, which attempts to identify the essential components of each sentence, such as the subject, verb, object, sentence form (active or passive), and any linked articles or determinants for the subject and object.

– **Layer 4: Pattern-based Diagram Extraction.** This last module is based on defined patterns to extract our diagram. It thus allows the identification of classes, attributes, relations, methods, and multiplicities. The following rules define our UML entity extraction patterns:

  - Class identification: UML *class diagram* identification follows these rules. The subjects of sentences are systematically classified as classes. If they are also nouns associated with another noun of the type "noun+noun" then the first is identified as a class, and the second will be classified as an attribute. If sentences involve two noun phrases and a verb of the type "noun phrase+verb+noun phrase", the first noun phrase is identified as a class, and the second is temporarily classified as an attribute. After analyzing all the sentences, if the second nominal group is not classified as an attribute, it becomes a class and the verb is classified as the title of the relation associating these two classes.

  - Relationship identification: if a sentence in the text matches the pattern "group nominal + verb + group nominal" and both nominal groups are classified as classes then the verb in the sentence is classified as a relationship description text. This means that the verb describes a relationship between the two classes represented by the nominal groups. The verb category is used to identify the relationship being described. The verb category refers to the semantic category or type of the verb, such as "uses", "contain", "has a", "belongs to", etc. The type of relationship between the classes is identified based on the verb category.

- **Multiplicity identification**: when an entity is identified as a class, all the determinant groups linked to the class are filtered by exploiting the data of the dependencies between the words. Then with a dictionary, we assign the correct multiplicity as illustrated in Fig. 3:

key word	multiplicity
un	1,1
plusieurs	0.*
au moins	1,*
au plus	0,1
un et un seul	1,1
chaque	1,1
...	...

Fig. 3. Filtering model layer

- Attribute identification: To identify attributes in a sentence, two specific patterns were used: a noun followed by another noun. This pattern refers to a sentence structure where two nouns are placed one after the other. If a sentence follows this pattern and the first noun is known as being a class, then the second noun in the sequence is considered an attribute. A group nominal followed by a verb and another group nominal: This pattern refers to a sentence structure where a group of words that acts as a single noun (known as a group nominal) is followed by a verb and then another group nominal. For example, "Each user has an email address." If a sentence follows this pattern, then the second group nominal in the sequence is considered an attribute if it is not already identified as a class.

- Methods identification: To identify methods in the text, a specific pattern was recognized in the verbs. This pattern consists of a noun followed by multiple consecutive verbs. For example, "Each user can connect, buy, read, and delete a book". If a sentence follows this pattern, where multiple consecutive verbs follow a noun, then all of the verbs in the sequence are classified as class methods. This means that any verbs that match this pattern are identified as methods, which are actions or procedures that the class can perform.

4   UMLDesigner

4.1   Web Application and API

To allow the usage of our model by end users, we have packaged our algorithm into a Python API using the FastAPI framework. We have established endpoints that receive text input, which is subsequently processed by the controller implementing our model. The outcome of this processing is returned as an XML string. We have developed a web application, UMLDesigner, which offers interactive diagram-building interfaces. It has two main areas: a text input area and a diagram representation area. When you make an API call, the result that comes back in XML format is processed by a module we created in the client application. This module is designed to create a visual diagram based on XML responses. The diagram is then displayed in the diagram representation area illustrated in Fig. 4.

To turn XML data into a diagram, we use jsUml2 [1] library, a web-based visualization toolkit of automatically generated data models. UMLDesigner also offers the advantage of being able to modify the generated diagrams as needed. Using our platform, the user can invite collaborators on diagram design. It is also possible to generate Python and Java code from the diagrams.

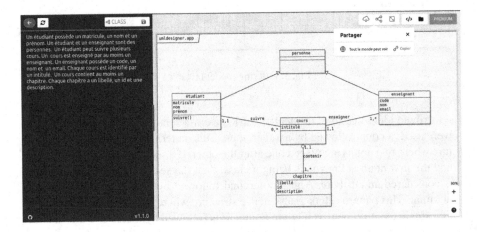

Fig. 4. UMLDesigner web application editor

4.2 Use Case

As stated in Sect. 1, the model is exclusively trained for the French language, focusing solely on class diagrams. In order to use our tool, it is necessary to establish a connection using valid credentials and subsequently create or open an existing project or diagram. Following this, the text to be processed can be modified as required. For instance, consider the example French text entered in Fig. 5.

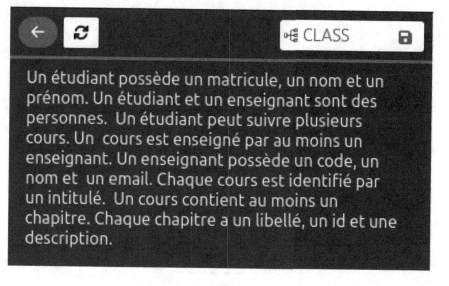

Un étudiant possède un matricule, un nom et un prénom. Un étudiant et un enseignant sont des personnes. Un étudiant peut suivre plusieurs cours. Un cours est enseigné par au moins un enseignant. Un enseignant possède un code, un nom et un email. Chaque cours est identifié par un intitulé. Un cours contient au moins un chapitre. Chaque chapitre a un libellé, un id et une description.

Fig. 5. Case study input text

To get a look at the diagram, we click on the sync icon and then we have the following result in Fig. 6

5 Results and Discussion

In UML, two different diagrams can be valid and syntactically correct even if they come from the same rules. The evaluation of the diagrams is also complicated because their interpretation is linked to the bias of the person who interprets them. It is, therefore, difficult to have a basis for comparison. We then proceeded as follows to evaluate our model:

1. we retrieved via Google Forms and web scraping a total of 20 test cases of class diagram management rules from external people with UML prerequisites;
2. we then process these management rules with our model; and
3. the images obtained joined with the management rules are submitted to the assessment of an evaluation committee. This latter has evaluated the model based on consistency as well as syntax.

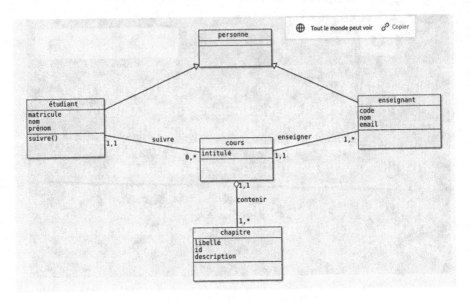

Fig. 6. Case study output diagram

Table 1 shows an overview of the results obtained. It shows the expected number of each concept (class, attribute, method, multiplicity, class association, association, aggregation, and generalization), the number of concepts found by the tool, and the number of true positives, false positives, true negatives, and false negatives of each concept.

Table 1. Overview of the results obtained after the evaluation of our model

Concept	Expected	Found	True Positive	False Positive	True Negative	False Negative
Class	84	81	81	0	0	3
Attribute	144	134	134	0	0	10
Methods	29	30	29	1	0	0
Multiplicity	84	70	70	0	0	14
Class association	4	4	3	1	0	0
Association	47	45	45	0	0	2
Aggregation	10	10	10	0	0	0
Generalization	30	30	30	0	0	0

To analyze the performance of UMLDesigner, we calculate *precision, recall,* and $F1 - score$. *precision* is the number of true positives divided by the sum of true positives and false positives, *recall* is the number of true positives divided by the sum of true positives and false negatives, and $F1 - score$ measures the balance between precision and recall.

$$precision = \frac{TruePositives}{TruePositives + FalsePositives} \qquad (1)$$

$$recall = \frac{TruePositives}{TruePositives + FalsePositives} \qquad (2)$$

$$F_1 = 2 \cdot \frac{precision \cdot recall}{precision + recall} \qquad (3)$$

Table 2 shows their respective values for each concept.

Table 2. Precision, recall, and F1 score of the model for each concept.

Concept	Precision	Recall	F1 Score
Class	100%	96.43%	98.18%
Attribute	100%	93.06%	96.43%
Methods	96.67%	100%	98.28%
Multiplicity	100%	83.33%	90.91%
Association	75%	75%	75%
Association	100%	95.74%	97.83%
Aggregation	100%	100%	100%
Generalization	100%	100%	100%

The precision values indicate that UMLDesigner is highly reliable for most of the concepts, with precision values of 100%. However, the precision for the class associations is at 75%. This suggests that the tool may need further improvement in accurately detecting class associations.

6 Perspectives

Overall, the model performs quite well. However, there are some limitations that need to be addressed. For example, the model's restricted knowledge base makes it challenging to identify the multiplicities of diagrams when encountering new or unsupported vocabulary. Another area for improvement is the model's inability to handle complex sentences with multiple clauses and sub-text, unlike simple sentences with a subject-verb-object-complement structure, which the system can handle. The model assumes that the user provides syntactically correct sentences, and sentences with spelling mistakes may cause difficulties for the model. Additionally, the filtering algorithm only recognizes classes if they have at least one attribute. We plan to address these issues in future work to enhance the accuracy and performance of the tool.

Furthermore, as our research represents an initial stage of investigation in automating the generation of UML diagrams, the focus of our work is on producing UML class diagrams from French text descriptions of entities and properties. While this specific application provides valuable insights into automating the diagram creation process, we acknowledge that a more generic methodology would be desirable. Such a methodology could be applied to other types of UML diagrams and also extend to different languages. In the current landscape of Chat-GPT and generative tools, it would be interesting to compare the output of our approach with that of a dedicated generative tool. Exploring these avenues for future research would contribute to the broader applicability and versatility of automated diagram generation techniques.

7 Conclusion

This paper proposes a rules-based approach using Natural Language Processing to turn French text into a UML class diagram image automatically. Our work was carried out in three phases: the development of our data recognition and extraction algorithm, the implementation of our API exposing the functionality of the above model, and the creation of an interactive web platform UMLDesigner that interacts with our API to make the solution usable. As main functionalities, one can create class diagrams with text elements, edit generated diagrams, collaborate on projects, and generate Java and Python code sketches from the created diagrams. Our model can well extract the class names, the multiplicities, the methods, the attributes, and the relationships between the classes. For further works, we will improve the performances of UMLDesigner on class diagrams and address some other UML diagrams.

References

1. Dujlovic, I., Obradovic, N., Kelec, A., Brdjanin, D., Banjac, G., Banjac, D.: An approach to web-based visualization of automatically generated data models. In: IEEE EUROCON 2019–18th International Conference on Smart Technologies, pp. 1–6. IEEE (2019)
2. Fowler, M.: UML distilled: a brief guide to the standard object modeling language. addison-wesley professional (2003)
3. Gleek: Gleek (2022). https://www.gleek.io/
4. Larman, C.: Applying UML and Patterns: An Introduction to Object-Oriented Analysis and Design and Iterative Development. Prentice Hall, Hoboken (2004)
5. Overmyer, S.P., Benoit, L., Owen, R.: Conceptual modeling through linguistic analysis using LIDA. In: Proceedings of the 23rd International Conference on Software Engineering, pp. 401–410. IEEE (2001)
6. Qi, P., Durrett, G., Wang, C.D.: Stanza: A python natural language processing toolkit for many human languages (2020). https://github.com/stanfordnlp/stanza
7. Subramaniam, K., Liu, D., Far, B.H., Eberlein, A.: UCDA: use case driven development assistant tool for class model generation. University of Calgary, Department of Electrical and Computer Engineering (2021)

Graph Neural Networks for Circuit Diagram Pattern Generation

Jaikrishna Manojkumar Patil[1(✉)], Johannes Bayer[2], and Andreas Dengel[2]

[1] RPTU, Kaiserslautern, Germany
jaikrishnapatil2407@gmail.com
[2] DFKI, Kaiserslautern, Germany
https://rptu.de/, https://www.dfki.de/web

Abstract. Graph neural networks (GNNs) have found majority of applications in multiple domains including physical science, molecular biology, etc. However, there is still scope of research in the application of GNNs in electrical domain. This research work tries to generate the circuit graphs that analyzes the GNNs. The end goal is to create a mechanism to iteratively predict the graph structure and thus complete a incomplete circuit diagram. The research work firstly tries to find out how well GNN can be used for predicting the missing node label or the missing node geometric features in the subset of the graph. Then, application of GNNs to anomaly detection problem is investigated. Next, GNN architecture is used to accurately predict the node label and approximate geometric features of the missing node in the circuit graph. Furthermore, a Graph Autoencoder (GAE) model is created and used for pruning the wrong edges in the circuit graph.

The GNN model created for the purpose of anomaly detection problem gave around 90% accuracy. The GNN model used for missing node feature estimation gave around 89% accuracy to predict the correct label for missing node and it also performed effectively in approximating the geometric features of the missing node correctly. The link prediction model is able to classify the correct edges 92% of the times. Finally, a mechanism is provided that iteratively predicts graph structure using the anomaly detection model, node feature prediction model and link prediction model in a cycle.

Keywords: Circuit diagrams · Graph neural network · Graph structure prediction

1 Introduction

Graphical data is generated in variety of domains like physics, chemistry, genetic engineering, electronics, image processing, etc. The introduction of Graph neural networks [8] gave a more sophisticated way to create end to end training process with input as the graph data and output as required for the problem at hand. The first very important part of any machine learning research is the dataset that is used for analysis. Therefore, the first step in our research work is to create a printed circuit diagram dataset. We already had access to the handwritten circuit diagram dataset as provided in [9]. The open source software, KiCad schematics is used to digitize this handwritten circuit

ⓒ The Author(s), under exclusive license to Springer Nature Switzerland AG 2023
D. Conte et al. (Eds.): DeLTA 2023, CCIS 1875, pp. 351–369, 2023.
https://doi.org/10.1007/978-3-031-39059-3_24

diagrams to printed circuit diagrams. The detailed explanation is mentioned in Sect. 3. Each of the kicad circuit schematics are then transformed into networkx [2] graphs to make it compatible to work with them and visualize them. These networkx graphs are then further transformed into PyTorch geometric Data instances to finally use them for training a Graph neural network using PyTorch geometric framework.

The end goal of this research is to provide a mechanism to use Graph neural network models to complete a incomplete circuit diagram to generate a sound circuit diagram in the end. This major problem statement is divided into multiple problem statements to reach this goal one step at a time. The research work starts with answering 4 research questions and providing 3 individual trained GNN models which are then used for the implementation of final pattern generation mechanism. Firstly, we try to answer *Can Graph neural networks be used to classify anomaly and non-anomaly nodes in the sample incomplete graph?* Next, we check if *Given a circuit diagram with one featureless node, is it possible to predict node label and estimate geometric features of that node based on remaining node feature values?* Furthermore, *Given a circuit diagram with extra or wrong edges/wires between nodes, is it possible to prune the unnecessary edges/wires from the input circuit diagram?* Finally, we bring it all together by asking *Given a incomplete circuit diagram, how to create a mechanism that will heal the diagram to give a sound circuit diagram in the end?*

2   Related Work

Chip floor planning is a method of designing the physical layout of a computer chip. [6] provides deep reinforcement learning approach of chip floor planning. In this approach, authors have used edge level graph convolution neural network model to learn the sound representations of chip. The netlist is represented as graph in this approach. For this edge-GNN, the input graph has almost similar input node features that we have used in our approach too, i.e. x and y coordinates, node type, width and height of node, etc. Then, iteratively it performs following update to node and edge representations:

– Edge embedding is updated by using a fully connected network to concatenation of two nodes of that particular edge.

The edge level GNN model was main pillar of the RL-based approach to chip floorplanning.

One of the major problems in Electronic Design Automation (EDA) workflow is the placement of application specific Integrated Circuits. [5] proposed a graph learning based framework called PL-GNN that will act as a guidance mechanism for placement in any design. Here, the node embedding is learn using GraphSAGE [3] GNN layers. This node representation learning is later followed by other steps to finally give automated guidance for placements. In this work, researchers have used PyTorch geometric library for carrying out experiments which is also used in our research work as well.

3   Dataset Preparation

The dataset creation process for this research work starts with using Hand drawn circuit diagrams dataset provided in [9]. Each of the unique hand drawn circuit diagrams is manually digitalized using an open source software, KiCad Schematic. This forms a Printed Circuit diagram dataset. Now, after creating a printed circuit diagram dataset, next step was to find a way to represent these kicad schematic circuit diagrams as graphs. For this reason, a python script(KiCad Converter) is used to convert each of the KiCad circuit diagrams into a networkx graph. Additionally, KiCad Converter can also be used to convert a networkx graph to a KiCad schematic file, which the can be visualized using a KiCad Schematic software. Finally, another python script (GNN Converter) is used to transform each of the networkx graphs into a PyTorch-geometric Data instances. GNN Converter also has the ability to decode a PyTorch-geometric [1] Data instance into a networkx graph. Furthermore, another python script (Png Converter) is often used to visualize networkx graphs as a resemblance of electronic circuit diagrams instead of just nodes and edges representation.

Additionally, in order to incorporate port/connector information in a PyTorch geometric Data instance, a new graph transformation method is added. This method takes as input a networkx graph generated by Kicad converter and transforms it into another networkx graph where ports are added as nodes in the graph.

3.1 Printed Circuit Diagram Dataset

An open-source software, KiCad is used for manually developing the digital versions of circuit diagrams from [9]. A segmented version of sample circuit diagram from hand-written circuit diagram dataset [9] and its corresponding KiCad schematic is shown in Fig. 1a and Fig. 1b respectively.

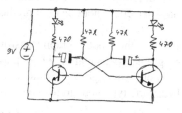 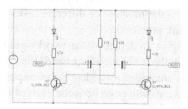

(a) Segmented Hand-Drawn (from [9])   (b) Digitized Manually in KiCad

Fig. 1. Sample Circuit Diagram

The printed circuit diagram dataset consists of 240 circuit diagrams. There are in total 40 types of nodes or components used to build these circuit diagrams. These kicad-schematics are further loaded as networkx [2] graphs to make this data accessible for multiple machine learning problems such as using graph neural network architectures for anomaly detection, graph denoising, node label prediction, etc.

3.2 KiCad Converter: KiCad Schematic to a Networkx Graph

Kicad converter is a script written in Python programming language to convert a KiCad schematic file to a networkx graph. For a sample kicad schematic, firstly a kicad schematic file is parsed and information of all the components or symbols and their connection in the schematic file is extracted. Then, a networkx [2] graph object is created and subsequently nodes are added into the graph object. Each node represents a symbol or component in the circuit diagram. Each node in the networkx graph has some unique information stored in it.

3.3   Graph Transformation (Ports as Nodes Representation)

Another way of representing circuit diagrams as networkx graphs is provided in the code reposiotry. Every circuit diagram component has one or more than more ports or connectors. These connectors are the means of connection of one component to another component. In the kicad converter, there was no use of ports while forming edges between two nodes. In this representation, however port is considered as a node. Now, in kicad converter for a Wire in circuit diagram, it added one edge in the networkx graph between the two electronic circuit diagram components. In this representation, for a single wire, three edges are added.

- First edge: between source node to source port.
- Second edge: between source node and target port.
- Third edge: between target port and target node.

The major reason for adding this graph transformation method after the Kicad converter was to provide a way to embed port information i.e. (x, y coordinate) of nodes into the input graph. This is mainly useful when GNN converter explained in Sect. 3.4 is used to create input PyTorch geometric data instance. This graph transformation method provides a way to incorporate port information in the input tensor of each Graph data while training the models for different tasks like Anomaly detection, node feature prediction and link prediction.

Figure 2a depicts the networkx graph that is given as the input to graph transformation method. On the other hand, Fig. 2b shows the graph representation where ports of nodes are added as nodes in the output networkx graph. Although not visible to naked eye, there is an edge between each of the "detached_port" type node to the corresponding node.

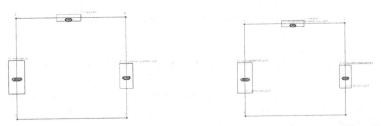

(a) Circuit graph: Before graph transformation   (b) Circuit graph: After graph transformation

Fig. 2. Sample networkx circuit graph

3.4   GNN Converter: Networkx Graph to PyTorch Geometric Instance and Vice Versa

GNN converter contains two important methods, Encoder and Decoder described as follows.

x co-ordinate of node (range: 0 - 1)	y co-ordinate of node (range: 0 - 1)	width of node (range: 0 - 1)	height of node (range: 0 - 1)	rotation of node (range: 0 - 1)	One hot encoded vector representing node label (size: total number of node labels = 41)

Fig. 3. General node feature vector.

Fig. 4. Sample node feature vector.

Encoder: Convert Networkx Graph to PyTorch Geometric Data Instance. To carry out the experiments for this research, PyTorch geometric library [1] in python is used to create Graph neural network architectures or models. In order to work with this framework, it becomes mandatory to represent each graph in the dataset as a PyTorch geometric Data instance. The encoder method is thus created as a part of GNN converter script. Encoder method of GNN converter script converts a networkx graph [2] to a PyTorch geometric [1] Data instance.

Sample PyTorch geometric Data instance is given as:
```
Data(edge_index=[2, 33], x=[28, 5], y=[28, 41],
num_nodes=28)
```
In this sample Data instance, the edge_index tensor means that there are 33 edges in the given graph. The first tensor of length 33 represents the source nodes of the edges and the second tensor of same length represents the target nodes for corresponding source nodes. Tensor x is of shape (28, 5) which means that for 28 nodes in the graph, each has a tensor $[x, y, width, height, rotation]$ assigned to it. Tensor y is the one hot encoded tensor of shape (28, 40) with 40 unique classes over the dataset. Each tensor in the y tensor represents one hot encoding of that node in the graph. The general representation of an encoded node feature vector is depicted in Fig. 3 along with a sample node feature vector shown in Fig. 4.

Note that this encoding also includes the data transformations on position attributes to get them in the range of 0 to 1. Two types of data transformations were tried in different experiments. Each of these data transformations are described in detail in Sect. 3.4.

Decoder: Convert to Networkx Graph from PyTorch Geometric Data Instance.
The decode method in the GNN converter script will convert the Data instance into a
networkx graph. Inverse one hot encoding is used to get the label for the node. Additionally note that this decoding also includes the inverse transformations on position
attributes to get them in their original ranges. The equations for inverse transformations
for each of the geometric attributes are provided in Sect. 3.4. Finally, the `edge_index`
tensor from PyTorch geometric Data instance can be used to add the edges in the
decoded networkx graph.

Types of Transformations

1. *Min max normalization.* In this type of data transformation, the transformation for
   each of the geometric attribute values of nodes is transformed in such a way that
   the values are between 0 and 1. Later, to get the graph in original coordinate space
   inverse transformation is carried out.
2. *Log transformation.*
   In this type of data transformation, the transformation for each of the geometric
   attribute values (v) of nodes is calculated as follows:

$$v = \ln v \tag{1}$$

Now, after performing log transformation, we perform min max normalization again
on this to get the values between 0 and 1. Now, to get the graph in original coordinate
space inverse transformation is carried out using following equation:

$$v = e^v \tag{2}$$

4   Methodology

Given a incomplete circuit diagram, the question is how to create a mechanism that will
heal the diagram to give a sound circuit diagram in the end? The goal here is to provide
a mechanism that is able generate a sound circuit diagram out of the incomplete circuit
diagram.

4.1   Mechanism

The mechanism uses 3 core Graph neural network models. These 3 models are
anomaly detection model Sect. 5.1, node feature prediction model Sect. 5.2 and link
prediction model Sect. 5.3. Each iteration of the mechanism consists of following steps:

1. Detect the anomalies using the anomaly detection model in the incomplete circuit
   diagram.
2. For each detected anomaly node, add a new empty node to it. Then predict the features of the new node using node feature prediction model.
3. Each new node is also marked as anomaly node by default.

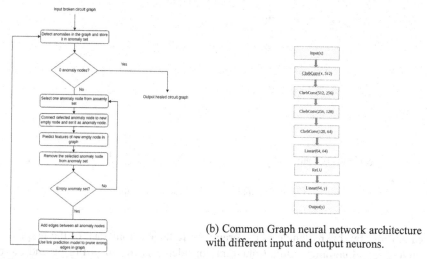

(b) Common Graph neural network architecture with different input and output neurons.

(a) Iterative circuit graph structure prediction process.

Fig. 5. Algorithm for graph structure prediction and core GNN architecture

4. Add edges between all the anomaly nodes in the updated graph.
5. Use link prediction model to predict the correct edges from the newly added edges and prune other edges from the graph.

The mechanism halts only when there are no more anomaly nodes detected in the graph. The flowchart representing the process is depicted in Fig. 5a.

5   Experimental Setup

The printed circuit diagram dataset described in Sect. 3 is used for working on all problem statements. Inorder to use the circuit diagrams in the dataset as set of graph structures, the first step is to use KiCad converter Sect. 3.2. This allows to generate networkx graph for each of the KiCad schematic. However, inorder to encode port information into the input graph, before using GNN converter, the networkx graph is passed through graph transformation method mentioned in Sect. 3.3. PyTorch-Geomteric library is used to create and train graph neural network architectures. Hence, the generated networkx graphs are converted to PyTorch-geometric Data instances using GNN converter Sect. 3.4. This is done to make the graph data compatible to be used with PyTorch-geometric library.

5.1   Anomaly Detection

After the general process is completed, in order to satisfy the requirements of the problem statement at hand, the random rectangular region is considered and nodes for the

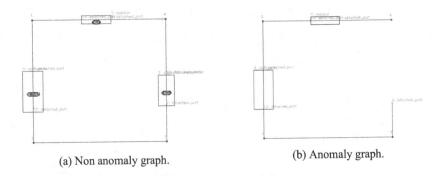

(a) Non anomaly graph.   (b) Anomaly graph.

Fig. 6. Anomaly and Non-anomaly graph

area covered by this rectangle in the graph are removed from each of the graphs. The sample original complete circuit diagram considered is shown in Fig. 6a and the one with random rectangular region removed from original graph is depicted in Fig. 6b. The nodes that are not affected by this process are set as non-anomaly nodes and the nodes that were neighbors of the removed nodes are set as anomaly nodes. We have represented non-anomaly nodes using Blue color, while the red nodes represent the anomaly nodes. In this specific instance, it can be seen that the incomplete circuit diagram has 2 less nodes than the complete circuit diagram, i.e. capacitor and one of the ports of capacitors. Now, given such incomplete circuit diagram, the goal is to classify the nodes as anomaly and non-anomaly nodes.

In order to provide input and output to a graph neural network model, two additional tensors namely, input and output are added to each of the Data instances externally. The input tensor for each Data instance is of shape (number of nodes, 46).

The 45 features are categorized as:

- 5 Geometric features, i.e. x, y, $width$, $height$, $rotation$.
- 41 features representing one hot encoding of a particular node label.

On the other hand, the output tensor is of shape (number of nodes, 1) representing the health of a node, 0 represents non-healthy or anomaly node and 1 represents healthy or non-anomaly node. Now, the sample PyTorch geometric Data instance is given as:

```
Data(edge_index=[2, 33], x=[28, 5], y=[28, 40],
num_nodes=28, input=[28, 46], output=[28, 1])
```

Data Augmentation. Data augmentation is one of the most important part of training any neural network model if less data is available for training or even to provide variability to the data at hand. In the case of anomaly detection model, data augmentation is carried out in two ways. Firstly, for each unique and complete graph in the dataset, selecting one random rectangular region to remove will generate one input graph for anomaly detection training. However, to add more data in training set and facilitate our model to learn swiftly we have selected for each unique graph, $N = 10$ random regions to be removed from it one by one. This results in $N = 10$ unique incomplete input graphs for a single original graph in the dataset. Additionally, each input graph can be noised M times to give M different input graphs. Here, "shaking" basically means randomly adding or subtracting the x, y coordinate values of each nodes in the given circuit diagram. It was seen that adding such data helped the model to generalize the results well which was very necessary when this particular anomaly detection model is used in our pattern generation mechanism. When both these data augmentations were combined, for each unique complete graph in dataset, M*N incomplete input graphs were added to training set.

GNN Architecture. The graph neural network architecture that suited best for this particular problem statement is depicted in Fig. 5b. The input to the graph neural network model is a tensor of size 46. This is given as input to first Chebyshev layer followed by 3 more Chebyshev graph neural network layers [4]. This is followed by 2 post processing linear layers, finally giving the output of 1 neuron. The output gives the state of node in a incomplete circuit diagram.

Table 1. Hyperparameters for training GNN model for anomaly problem.

Optimizer	Learning rate	Loss function	Number of epochs	Batch size
Adam	0.001	Binary Cross Entropy loss	150	10

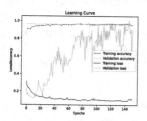

Fig. 7. Learning curve: Training for anomaly detection problem.

Training. The input to the Graph neural architecture is a graph with 46 input node features. First 5 node features indicate the spatial features of the node, i.e. x co-ordinate, y co-ordinate, width of node, height of node, rotation of node. The remaining 41 features indicate the type of node or symbol. The values of these features is 1 if node belongs to that class else it is 0. The output of each node in the graph is 0(non-anomaly node) or 1(anomaly node). The model is trained with hyper parameters given in Table 1. The learning curve Fig. 7 depicts the training and validation accuracy and Binary Cross Entropy loss for all the epochs.

5.2   Missing Node Feature Estimation

Once the general process is completed, inorder to satisfy the requirements of the problem statement at hand, for a sample circuit graph, a random node was chosen and its complete information is removed from the graphs Data instance. One way to do this which is used in this research work is to set the node features to a tensor of zeros of shape (1, 46) for that selected random node.

In order to provide input and output to a graph neural network model, two additional tensors namely, input and output are added to each of the Data instances externally. The input tensor for each Data instance is of shape (number of nodes, 47). The 47 features are categorized as:

- 5 Geometric features, i.e. x, y, $width$, $height$, $rotation$. Note: if blind bit for a node is 1, it is a zero tensor of shape (1, 5) instead of the true transformations of geometric features.
- 41 features representing one hot encoding of a particular node.
  Note: if blind bit for a node is 1, it is a zero tensor of shape (1, 41) instead of the true one hot encoding of node.
- 1 blind bit, In a specific graph only one node will has this bit set to 1.
  (blind bit = 1 means the node has missing node information and model should be trained on this.
  (blind bit = 0 means the node has all the node information available.

The output tensor on the other hand is of shape (number of nodes, 46) representing the geometric features as first 5 entries and remaining 41 as one hot encoding of each node in the graph.

Now, the sample PyTorch geometric Data instance is given as:

Data(edge_index=[2, 33], x=[28, 5], y=[28, 41], num_nodes=28, input=[28, 47], output=[28, 46]).

Data Augmentation. Similar to anomaly detection problem, here the training data only has 240 graphs as input and model was trained for this 240 random nodes, one from each graph. Additionally, since majority of nodes in the graph are "junctions", majority of random nodes chosen for each graph were junctions. This caused bias in the training process and the results were not that good. To solve this bias problem, a new data augmentation method is used. Now, each of the nodes in a graph gets a chance to be a node with missing label information. So instead of a single sample graph with N number of nodes, it augments it to N graphs, each with one missing node label information in it. This improved the training process significantly.

Furthermore, this model with above data augmentation worked well when input graph has only 1 missing node in the graph. However, in our final mechanism of pattern generation in the initial stages there are more than 1 missing node. So, this model was not generalizing well for our final problem statement. To improve the results, additional data augmentation method was used. Instead of removing a random node from complete circuit diagram, now the circuit graph is first broken using random rectangular mask as done in Sect. 5.1. Now, in this incomplete circuit diagram random node information was removed. This training data made the node feature prediction model more robust to unknown input graphs provided in patter generation framework.

Lastly, as mentioned in Sect. 5.1, graph is "noised" in every iteration to make model robust to noise.

GNN Architecture. The graph neural network architecture that suited best for this particular problem statement is depicted in Figure Fig. 5b. The input to the graph neural network model is a tensor of size 46. This is given as input to first Chebyshev layer followed by 3 more Chebyshev graph neural network [4] layers. This is followed by 2 post processing linear layers, finally giving the output of 45 neurons. This will give the estimated geometric features and predicted one hot encoding for missing node label.

Table 2. Hyper parameters for training GNN model for node estimation problem.

Optimizer	Learning rate	Loss function	Number of epochs	Batch size
Adam	0.001	2*MSE loss + Cross Entropy loss	100	25

Training. During training and validation, the input tensor of a Data instance as mentioned in previous section is passed as an input to the GNN model and the output tensor. Now, for one sample input graph the output will be some tensor of size 45 representing the node features. For training, a loss function is calculated by comparing this predicted output and true output given in the "output" tensor of Data instance. Firstly, we use the first 5 feature values of both true and estimated tensor and calculate mean squared loss for each feature separately and add twice the factor of each of them. Then calculate Cross entropy loss using the latter 40 entries in tensor. Now add both of these calculated losses to give the total loss that is used for training the model The hyper parameters chosen for training this GNN model are mentioned in Table 2.

The learning curve shown in Fig. 8a depicts the training and validation accuracy and total loss (2*MSE loss + Cross Entropy loss) for all the epochs. It is observed that the model achieved highest validation accuracy of around 78%. Additionally, the learning curve shown in Fig. 8b depicts this individual training losses i.e. for x, y, $width$, $height$, $rotation$ and lastly the classes for all the epochs.

5.3 Link Prediction

In order to satisfy the requirements of the problem statement at hand, for a sample circuit graph, some random nodes were selected and connected all these nodes in the circuit graph. Each PyTorch geometric instance now also has true link labels tensor that carries true class of each edge in the graph. Here, graph autoencoder(GAE) [7] is used as a GNN model to predict correct edges in the input graph. In order to provide input and output to a graph neural network model, two additional tensors namely, input and link_labels are added to each of the Data instances externally. The input tensor

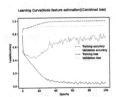

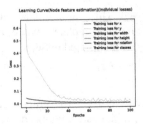

(a) Learning curve for node estimation problem: Combined loss/accuracy.

(b) Learning curve for node estimation problem: Individual loss/accuracy.

Fig. 8. Learning curves for Node estimation problem.

for each Data instance is of shape (number of nodes, 46). The 46 features are categorized as:

- 5 Geometric features, i.e. x, y, $width$, $height$, $rotation$. Note: if blind bit for a node is 1, it is a zero tensor of shape (1, 5) instead of the true transformations of geometric features.
- 41 features representing one hot encoding of a particular node.

The link_label tensor on the other hand is of shape (1, Number of edges in the graph). Each entry of this tensor is either 0 or 1 representing the correct presence/absence of the edge at that index in edge_index tensor.

Now, the sample PyTorch geometric Data instance is given as:

Data(edge_index=[2, 33], x=[28, 5], y=[28, 41], num_nodes=28, input=[28, 46], link_lables=[1, 33])

Data Augmentation. Here, data was augmented by selecting N random incomplete graphs for a single complete graph in the datset and add it to training set. This gives the model to learn more instances that it will face during the last step of final pattern generation mechanism.

Additionally, as mentioned in Sect. 5.1, graph is " noised" in every iteration to make model robust to noise. This is very helpful as it trains the model to work on instance with slightly wrong node feature predictions while generating pattern in our final mechanism.

GNN Architecture. Graph Auto encoder is widely used architecture for link prediction in lot of recent research works. Hence, we tried it and gave us best results for link prediction in our case. Graph Auto encoder model comprises of Encoder and Decoder model. The Encoder part works same as the GNN architecture depicted in Fig. 5b. The encoder will learn perfect node embedding for each node in the graph. Now, the decoder performs dot product between node embeddings in particular edge. The decoder will assign finally some value to every edge in the graph. If this value is greater than 0, then edge is predicted as correct else the edge is set as incorrect and can be pruned from the graph.

5.4 Training

Table 3. Hyper parameters for training GNN model for link prediction problem.

Optimizer	Learning rate	Loss function	Number of epochs	Batch size
Adam	0.001	2BCEWithLogits loss	100	15

During training and validation, the input tensor of a Data instance as mentioned in previous section is passed as an input to the Encode method of Graph Auto encoder model. For each epoch, this will generate a learned node embedding for every node in the graph. Then, these node embeddings are passed to decode method to calculate dot product of each node embedding for corresponding edge in the graph. This will give a logit for each edge in the graph. For training, a loss function is calculated by comparing this predicted logit value and true output given in the "link_labels" tensor of Data instance. The hyperparameters chosen for training this GAE model are mentioned in Table 3.

The learning curve shown in Fig. 9 depicts the training and validation accuracy and loss for all the epochs from 1 to 100. It is observed that the model achieved highest validation accuracy of around 90% and training accuracy of 94%.

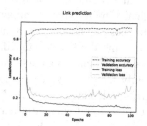

Fig. 9. Learning curve: Training and validation for link prediction problem.

6 Results

6.1 Anomaly Detection

A confusion matrix depicted in Fig. 10a is generated to evaluate the generated model. It is found that the saved model gave around 90% accuracy for the test graphs. Out of 78 anomaly nodes, it correctly predicts 52 anomaly nodes. On the other hand, out of total 425 non-anomaly nodes, it correctly predicts 407 non-anomaly nodes.

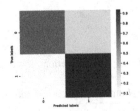
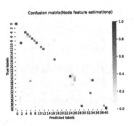

(a) Confusion matrix: Evaluation of anomaly de- (b) Confusion matrix: Evaluation of node esti-
tection problem.   mation problem.

Fig. 10. Confusion matrix for first 2 problems.

6.2 Missing Node Feature Estimation

It is found that the saved model gave 89 % accuracy for the test graphs. From the
confusion matrix, it can be decided which labels or classes does the model predict very
correctly and the ones that it predicts very badly. The components that the model is able
to classify much better than other ones are junction, voltage.dc, voltage.ac, inductor,
xor, detached_ports, etc. However, there are some labels that the model is predicting
very wrong. The labels that can be included in this category can be gnd, gnd_ref, or, etc.
Now to evaluate how this graph neural network model works for the both classes and
geometric features, one way is to use our best friends, GNN decoder and png converter
to see the missing node features in a circuit diagram form.

Figure 2b depicts the original sample circuit diagram we have been considering
almost throughout this report to visualize node features and its predicted labels for one
of the missing nodes of this graph. Inorder to prove that data augmentation has worked
well in the node feature prediction problem, let us first "noise" the geometric attributes
of graph. Now, we have randomly removed one of the node features from this noised
graph and set the tensor to 0. This is depicted in Fig. 11. Now, the node feature estimation model is used to predict the label and geometric features of this missing node in
the graph. The prediction is depicted in Fig. 12. It can be clearly seen that the model
is robust to the additional noise in the graph and predicts correct label and geometric
features (Fig. 13).

6.3 Link Prediction

It is found that the saved link prediction model gave 92 % accuracy for the test graphs.

6.4 Sample Pattern Generation Process

Pattern generation in the input incomplete circuit diagram depicted in Fig. 14a is carried
out as follows:

1. Detect anomaly nodes using the anomaly detection model in the input graph.
   Figure 14b depicts the input graph embedding the anomaly nodes represented as
   red colored nodes and non-anomaly nodes as blue colored nodes.

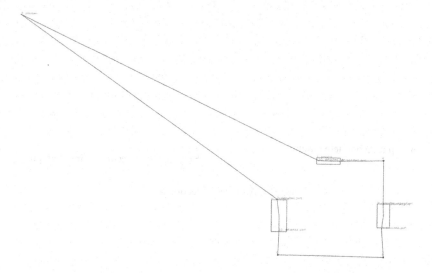

Fig. 11. Noised graph with junction tensor set to 0.

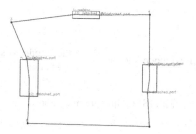

Fig. 12. Noised graph with correctly predicted junction class and correctly approximating the geometric features

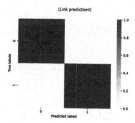

Fig. 13. Confusion matrix: Evaluation of link prediction problem.

2. Add empty node to one anomaly node and predict the features of newly added node using the node feature prediction model. Figure 15a depicts newly added node attached to a junction as a port. Set this new node as anomaly node.

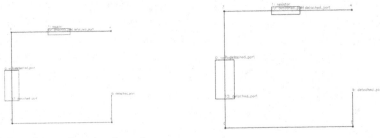

(a) Input incomplete circuit graph.

(b) Anomaly detection in input graph.

Fig. 14. Anomaly detection step.

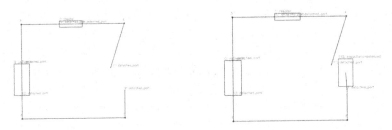

(a) Added node to first anomaly node and pre-dicted features.

(b) Added node to second anomaly node and predicted features.

Fig. 15. Adding anomaly nodes.

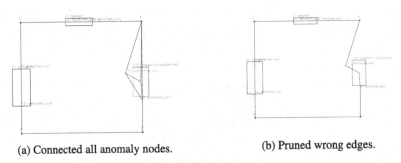

(a) Connected all anomaly nodes.

(b) Pruned wrong edges.

Fig. 16. Link prediction step.

3. Add empty node to other anomaly node and predict the features of newly added node using the node feature prediction model. Figure 15b depicts newly added node attached to a port as a unpolarized capacitor. Set this new node as anomaly node.

4. Connect all the anomaly nodes to each other in the updated graph. Figure 16a depicts a graph where all 4 anomaly nodes have edges between them.

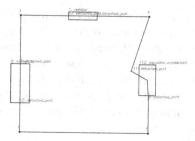

Fig. 17. Anomaly detection in graph with newly added nodes and edges. (Blue nodes are non-anomaly nodes) (Color figure online)

5. Now, we want to prune the wrong edges from the graph using link prediction model. Figure 16b depicts an updated graph with pruned edges. It can be seen that only edge between capacitor and port survived and rest were pruned by the model.
6. Again, repeat the iteration by starting with anomaly detection model to predict anomalies in the updated graph. Figure 17 depicts graph with no anomaly nodes as all the nodes are blue colored. Hence, the mechanism is stopped and it outputs the current graph as the completed graph.

7   Conclusion and Future Work

7.1   Conclusion

Firstly, this research work uses the graph structure of electronic circuit diagrams and make use of graph neural network layers to create a model to learn this graph structure followed by a multi layer perceptron. The major goal of this research work is to investigate if given a incomplete circuit diagram, is there any way to learn the incomplete part? For this investigation 4 different research questions were formed. Firstly, *Is it possible to find anomaly nodes in the incomplete or damaged circuit diagram?* To answer this question a GNN model was trained with sample incomplete circuit graphs to classify the anomaly and non-anomaly nodes correctly. The results provided in Sect. 5.1 showed that 90% of the times model correctly classifies anomaly and non-anomaly nodes in the incomplete circuit graph. This proves that the answer to this third research is Yes. Then, we move to the case where we already know that the circuit diagram is incomplete such that only one electronic component is missing in the diagram and we also know the anomaly node(s) where it is incomplete (using Sect. 5.1). Then consider that the edges are added between the anomaly nodes and a new blank node with no information at all. This situation leads to our research question, *Given a circuit diagram with one feature-less node, is it possible to predict node label and estimate geometric features of that node based on remaining node feature values?* Sect. 6.2 provides a detailed description on how a Graph neural network model was able to convert this featureless node to a node with correctly predicted label and almost correctly estimated geometric features. Then we move on to the last step required to complete our mechanism that we are trying to achieve, i.e. Link prediction. Given some sample circuit diagram with extra wrong edges in the graph, we are trying to prune these edges to clean the circuit diagram. This leads to our second last research question, *Given a circuit diagram with extra or wrong edges/wires between nodes, is it possible to prune the unnecessary edges/wires from the input circuit diagram?* Sect. 5.3 gives in detail description of how Graph Auto encoders is used to train a model to prune unnecessary edges from the circuit graph. It was found that model was able to learn swiftly and provided 92% accuracy on test set. It is also found that data augmentation played a very important role in anomaly detection, node feature prediction and link prediction tasks.

Finally, we also provide a mechanism to heal or complete a incomplete circuit diagram step by step in Sect. 4. We also explained the mechanism with the help of a small example and hence we provided answer to our last research question postulated in the start, i.e. *Given a incomplete circuit diagram, how to create a mechanism that will heal the diagram to give a sound circuit diagram in the end?*

7.2 Future Scope

Since the existing training set is limited in size, follow-up experiments can be performed on larger training sets with more data augmentation strategies. This will make the mechanism robust to more instances where it can go wrong currently. Some new data augmentation methods in each of the individual models should improve the results. Furthermore, currently each node of graph includes its geometric features, node type and geometric features of port. However, some more information can be encoded in the node instance like value of resistance in resistor, inductance in inductor, etc. This will help the GNN model as well to learn more effectively. Additionally, all models can be trained jointly and then the overall performance of the mechanism can be evaluated for this trained model.

References

1. Fey, M., Lenssen, J.E.: Fast graph representation learning with PyTorch geometric. arXiv preprint arXiv:1903.02428 (2019)
2. Hagberg, A., Swart, P., Chult, D.S.: Exploring network structure, dynamics, and function using networkX. Technical report, Los Alamos National Lab. (LANL), Los Alamos, NM (United States) (2008)
3. Hamilton, W., Ying, Z., Leskovec, J.: Inductive representation learning on large graphs. In: Advances in Neural Information Processing Systems, vol. 30 (2017)
4. He, M., Wei, Z., Wen, J.R.: Convolutional neural networks on graphs with Chebyshev approximation, revisited. arXiv preprint arXiv:2202.03580 (2022)
5. Lu, Y.C., Pentapati, S., Lim, S.K.: VLSI placement optimization using graph neural networks. In: Proceedings of the 34th Advances in Neural Information Processing Systems (NeurIPS) Workshop on ML for Systems, Virtual, pp. 6–12 (2020)
6. Mirhoseini, A., et al.: A graph placement methodology for fast chip design. Nature **594**(7862), 207–212 (2021)
7. Pan, S., Hu, R., Long, G., Jiang, J., Yao, L., Zhang, C.: Adversarially regularized graph autoencoder for graph embedding. arXiv preprint arXiv:1802.04407 (2018)

8. Scarselli, F., Gori, M., Tsoi, A.C., Hagenbuchner, M., Monfardini, G.: The graph neural network model. IEEE Trans. Neural Netw. **20**(1), 61–80 (2009). https://doi.org/10.1109/TNN.2008.2005605

9. Thoma, F., Bayer, J., Li, Y.: A public ground-truth dataset for handwritten circuit diagram images. CoRR abs/2107.10373 (2021). https://arxiv.org/abs/2107.10373

Generative Adversarial Networks for Domain Translation in Unpaired Breast DCE-MRI Datasets

Antonio Galli[iD], Michela Gravina[iD], Stefano Marrone[(✉)][iD], and Carlo Sansone[iD]

Department of Electrical Engineering and Information Technology (DIETI),
University of Naples Federico II, Naples, Italy
{antonio.galli,michela.gravina,stefano.marrone,
carlo.sansone}@unina.it

Abstract. Generative Adversarial Networks (GAN) are more and more gaining attention in the computer vision domain thanks to their ability to generate synthetic data, in particular in the context of domain adaptation and image-to-image translation. These properties are attracting the medical community too, in order to solve some complex biomedical challenges, such as the translation between different medical imaging acquisition protocols. Indeed, as the actual acquisition protocol is strongly dependent on factors such as the operator, the aim, the centre, etc., gathering cohorts of patients all sharing the same typology of imaging is an open challenge. In this paper, we propose to face this problem by using a GAN to realise a domain translation architecture in the case of breast Dynamic Contrast-Enhanced Magnetic Resonance Imaging (DCE-MRI), considering two different acquisition protocols, in the context of automatic lesion classification. Despite this work wanting to be a first step toward artificial data generation in the medical domain, the obtained results have been analysed from both a quantitative and qualitative point of view, in order to evaluate the correctness and quality of the proposed architecture as well as its usability in a clinical scenario.

Keywords: Breast MRI · Generative adversarial networks · Image-to-image translation

1 Introduction

In recent years, Machine Learning (ML) techniques have become increasingly influential in computer vision. Among all, Convolutional Neural Networks (CNNs) are a particular set of Deep learning (DL) models characterized by multiple hidden layers able to perform multiple image transformations to create highly complex features and process them for classification and other tasks, that have proven extremely successful for image analysis and manipulation. Healthcare is one of the domains that has been benefitting a huge impact of DL, and it is identified as one of the most promising applications of AI that provides "intelligent" systems to support both patients and physicians. In particular, among all healthcare sectors, medical image analysis or medical image computing is the research field experiencing the most significant impact of AI-based solutions that particularly exploit ML and DL models.

D. Conte et al. (Eds.): DeLTA 2023, CCIS 1875, pp. 370–384, 2023.
https://doi.org/10.1007/978-3-031-39059-3_25

While many AI solutions have been proposed for medical imaging, it is often challenging to obtain a high-quality set of images that can be used for the evaluation of the methodologies. Indeed, it is not easy to operate with acquisitions coming from different institutions due to data-sharing limitations. In particular, given the strong dependence of the acquisition protocol on various factors, obtaining a huge cohort by integrating data from various medical centres with the same typology of imaging is an ongoing challenge. Recently, deep learning is proving to be a promising approach for addressing this task by automating the process of translating between different medical imaging acquisition protocols by leveraging generative models. Despite over the years several generative deep neural networks have been introduced, Generative Adversarial Networks (GANs), a special deep composite architecture in which multiple CNNs are trained together, continue to be among the most widely used models, both raw or redesigned to better address task specific constraints. In particular, CycleGANs are the new frontier for imaging tasks in many fields, including Domain Adaptation and Image to Image translation [4].

This study primarily emphasizes the application of CycleGAN within the realm of medical imaging, specifically targeting image-to-image translation tasks. In particular, we will focus on breast Dynamic Contrast-Enhanced Magnetic Resonance Imaging (DCE-MRI), considering two different acquisition protocols (depending on the specific setting set by the physician and on the scanner model and vendor), in the context of automatic lesion classification. The goal is to understand if, and to what extent, GANs can be effectively used to generate artificial, but realistic, MRI samples. Despite this work wanting to be a first step toward artificial data generation in the medical domain, the obtained results will be analysed from a quantitative as well as a qualitative point of view, to evaluate the correctness and quality of the proposed solution as well as speculate an initial idea about its potential use in a clinical scenario.

The remaining sections of the paper are structured as follows: In Sect. 2, a concise examination of the existing literature on deep learning (DL)-based image generation is presented. The proposed approach and the experimental setup are described in Sects. 3 and 4, respectively. Section 5 provides an overview of the achieved results. Finally, Sect. 6 contains the concluding remarks.

2   State of the Art

Generative Adversarial Networks (GANs) are emerging as a powerful tool for data generation [12] and image-to-image translation tasks in computer vision [3]. Their ability to generate realistic data has attracted the attention of the medical community, which are interested in applying these techniques to solve biomedical challenges [13]. One important application of GANs in the biomedical field is in breast cancer detection and diagnosis, using dynamic contrast-enhanced magnetic resonance imaging (DCE-MRI) [2]. DCE-MRI is a non-invasive imaging technique used to visualize and analyze the blood flow in breast tissue, which can help to identify cancerous areas.

However, there are two variants of the DCE-MRI acquisition protocol, and the resulting images can differ significantly in quality and content. This can create challenges for classification tasks, as the availability of data may be limited for specific protocols. In this paper, the authors propose a GAN-based domain translation architecture that can be used to translate MR-images generated through one variant of the DCE-MRI acquisition protocol to the other. This allows for more data to be available for classification tasks, which can improve the accuracy of breast cancer detection and diagnosis.

The proposed model consists of two networks: a generator network that learns to translate MR-images between the two variants of the DCE-MRI acquisition protocol, and a discriminator network that distinguishes between translated images and genuine images. The networks are trained using adversarial training, where the generator is trained to generate images that can fool the discriminator into thinking they are genuine images. The authors evaluate the performance of the proposed architecture using both quantitative metrics and qualitative visual comparisons.

Previous work has explored the use of GANs for medical image synthesis and domain adaptation tasks [14, 16]. In [20], the authors address the challenges of obtaining and utilizing multi-modal neuroimaging data for studying human cognitive activities and pathologies. To overcome limitations such as cost, acquisition time, and privacy concerns, they proposed FedMed-GAN, a benchmark that combines federated learning and medical GANs. FedMed-GAN effectively synthesizes brain images, mitigates mode collapse, and outperforms centralized methods, offering a promising solution for integrating dispersed data from different institutions. Recently, Cai et al. [1] introduced a novel approach for registering dynamic contrast-enhanced magnetic resonance imaging (DCE-MRI) by leveraging a multi-domain image-to-image translation network, motion and contrast changes are disentangled, resulting in improved registration accuracy.

However, very few studies have focused specifically on breast cancer detection using DCE-MRI data. The authors compare their results to previous state-of-the-art methods for MR-image generation and demonstrate that their approach outperforms these methods in both quantitative and qualitative assessments. The proposed architecture can be used to translate MR-images between different DCE-MRI acquisition protocols, which can help to address the challenges associated with limited data availability for specific medical classification tasks. Overall, the results of this study demonstrate the potential of GANs for improving the accuracy of breast cancer detection and diagnosis using DCE-MRI data.

3   Methodology

In this work we aim to implement an unpaired domain translation on a medical dataset, considering breast MRI slices as a case study. In particular, given two different sets of images acquired by using two different acquisition protocols, namely A and B, the aim of the proposed approach is to translate images between the two domains by leveraging a CycleGAN architecture to learn a mapping between A and B and vice versa. To put on some useful notations, from here on we will denote the distributions of data from sets A and B as $x_a \sim p(a)$ and $x_b \sim p(b)$ respectively. In accordance with the proposal made in [21], the CycleGAN comprises two mappings, namely $G_{ab} : A \to B$ and $G_{ba} : B \to A$, performed by two generators sharing the same architecture. Additionally, two

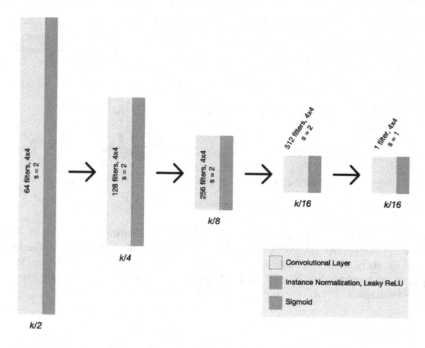

Fig. 1. The architecture of the discriminator consists of a fully convolutional network with three convolutional layers, incorporating Instance Normalization and the Leaky ReLU function. The last layer of the network utilizes a sigmoid function, and the spatial dimension of the output features is represented by the variable k.

discriminators, D_a and D_b, have been defined to i) differentiate images x_a from the translated instances $G_{ba}(x_b)$, and ii) to distinguishes between x_b and $G_{ab}(x_a)$. Each of the discriminators generates an output that represents the likelihood of the input images belonging to the authentic distribution of the desired domain.

Among all the available architectures, in this work we use the PatchGaN [6] model to implement D_a and D_b. This is motivated by the need of producing a full discrimination map, in contrast to the single scalar output produced by a CNN in the context of binary classification. The PatchGAN architecture produces a 2-D matrix of size $N \times N$, where each pixel represents a patch of size $M \times M$ from the input image. The value assigned to each pixel reflects the probability of the corresponding patch belonging to the authentic distribution of the target domain. As a result, every pixel within the matrix encompasses a receptive field of size $M \times M$. The classification probability for each patch is then determined by calculating the mean value of the corresponding patch within the matrix. Hence, the discriminator is designed as a fully convolutional network (FCN) consisting of four convolutional layers, with Instance Normalization and Leaky ReLU activation functions, and a final convolution layer utilizing a sigmoid function, as illustrated in Fig. 1.

Similarly, the generator is an FCN, sharing a design close to the one used by the discriminator, but aimed ad generating the translated image starting from the original one. In this work, we investigated two different architectures, namely the U-Net

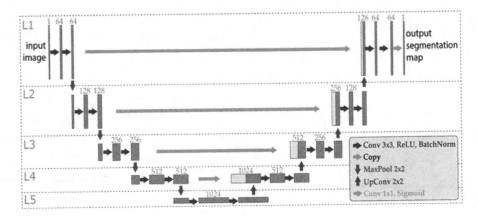

Fig. 2. The U-Net generator presents skip connections between encoder and decoder layers. The encoder side applies 2D convolution with ReLU activation, batch normalization, and max-pooling. Similarly, the decoder side uses 2D up-convolution combined with the cropped feature map from the corresponding encoder level, followed by 2D convolution, ReLU activation, and batch normalization.

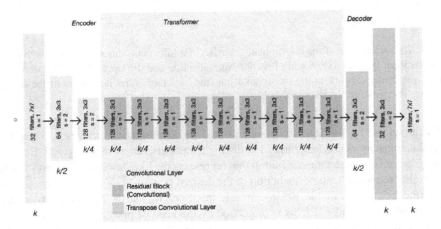

Fig. 3. The Generator, which adopts the ResNet architecture, is composed of three fundamental elements, shown as Encoder, Transformer, and Decoder. k represents the spatial dimension of the output features.

[11] and ResNet [5]. The U-Net architecture involves an FCN with skip connections between the encoder and decoder pathways. The former employs 2D convolution with ReLU activation, batch normalization, and max-pooling, while the latter employs 2D up-convolution, concatenated with the cropped feature map from the corresponding level on the encoding side, followed by 2D convolution, ReLU activation, and batch normalization. The ResNet architecture, inspired by the network proposed in [7], comprises three main components: the Encoder, the Transformer, and the Decoder. The Encoder reduces dimensionality through three convolutional layers. To facilitate domain trans-

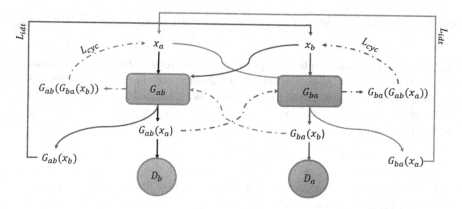

Fig. 4. CycleGAN framework implemented in this paper. The different CycleGAN image flows and cycle-consistency constraints are highlighted with different colours.

lation, the Transformer architecture incorporates nine Residual Blocks that effectively extract features. Within each Residual Block, Instance Normalization is employed in place of standard batch normalization. The utilization of Instance Normalization, initially introduced by Ulyanov et al. [17] for efficient contrast normalization, enhances the performance of the Residual Blocks. These architectural details of the considered generators based on the U-Net and on the ResNet are reported in Fig. 2 and in Fig. 3 respectively.

During the training process for the CycleGan implementation, the objective function includes three main components: adversarial losses, cycle consistency losses, and identity losses. The goal of adversarial losses is to align the generated images' distribution with the data distribution in the desired domain. Simultaneously, cycle consistency losses ensure that the learned mappings remain consistent and avoid contradictions. The identity loss ensures that the generators function as good identity mappings. To address the potential issue of vanishing gradients in GANs, a least-square loss (L_{adv}) is employed for the adversarial losses in both mapping functions. This choice of loss, proposed in [8], helps mitigate the problem and provides more stable training. More specifically, for the mapping function $G_{ab} : A \rightarrow B$ with discriminator D_b, the objective function, denoted as $L_{adv}^{A \rightarrow B}$, for $L_{adv}(G_{ab}, D_b, x_a, x_b)$ is defined as follows:

$$\begin{cases} \mathbb{E}_{x_b \sim p(b)}[(D_b(x_b) - 1)^2] + \mathbb{E}_{x_a \sim p(a)}[(D_b(G_{ab}(x_a))^2] \\ \qquad \mathbb{E}_{x_a \sim p(a)}[(D_b(G_{ab}(x_a)) - 1)^2] \end{cases} \tag{1}$$

The initial equation represents the discriminator D_b, which distinguishes between real samples x_b and the generated images $G(x_a)$. The subsequent equation pertains to the generator G_{ab}, responsible for producing images ($G_{ab}(x_a)$) resembling those from domain B. A similar adversarial loss $L_{adv}(G_{ba}, D_a, x_b, x_a)$, namely $L_{adv}^{B \rightarrow A}$, is introduced for the mapping $G_{ba} : B \rightarrow A$ and its discriminator D_a. Moreover, the cycle-consistency should be verified, ensuring that $x_a \rightarrow G_{ab}(x_a) \rightarrow G_{ba}(G_{ab}(x_a)) \approx x_a$ and $x_b \rightarrow G_{ba}(x_b) \rightarrow G_{ab}(G_{ba}(x_b)) \approx x_b$. As a consequence, the *cycle consistency loss* (L_{cyc}) is defined as follows:

$$L_{cyc}(A, B) = \mathbb{E}_{x_a \sim p(a)}[\|G_{ba}(G_{ab}(x_a)) - x_a\|_1] + \mathbb{E}_{x_b \sim p(b)}[\|G_{ab}(G_{ba}(x_b)) - x_b\|_1]$$
$$(2)$$

Finally, the utilization of identity loss (L_{idt}) serves as a regularization method, ensuring that the generator is in close proximity to an identity mapping when confronted with authentic samples from the target domain. This is demonstrated through the verification of the relationships: $x_b \to G_{ab}(x_b) \approx x_b$ and $x_a \to G_{ba}(x_a) \approx x_a$. The formulation for L_{idt} is as follows:

$$L_{idt}(A, B) = \mathbb{E}_{x_a \sim p(a)}[\|G_{ba}(x_a) - x_a\|_1] + \mathbb{E}_{x_b \sim p(b)}[\|G_{ab}(x_b) - x_b\|_1] \quad (3)$$

During the training stage, the total loss L_{tot} consists of the previously defined components, and is defined as follows:

$$L_{tot} = L_{adv}^{A \to B} + L_{adv}^{B \to A} + \lambda_1 \cdot L_{cyc} + \lambda_1 \cdot \lambda_2 \cdot L_{idt} \quad (4)$$

where λ_1 and λ_2 are the weights for L_{cyc} and L_{idt}. Figure 4 describes the overall architecture proposed in this paper, highlighting how the generators and discriminators interact.

Following the recommendation provided in [15] the Image Pool technique is employed to address the issue of the discriminator's sole focus on the most recently generated images during weight updates. This deficiency in memory not only results in divergence during training but also causes the reintroduction of artefacts by the generator that had previously been forgotten by the discriminator. Given that every generated image produced throughout the entirety of the training is considered a counterfeit to the discriminator, it is expected that the discriminator should competently classify all images generated by the generator. Building upon this observation, the training of the discriminator incorporates a historical context of previously generated images in addition to the images within the current iteration, namely the mini-batch, during training process.

4   Experimental Set-Up

In this paper, we consider two breast DCE-MRI datasets obtained from patients belonging to"Istituto Nazionale Tumori, Fondazione G. Pascale" of Naples. The first, denoted as Dataset A, consists of breast DCE-MRI scans from 39 women with an average age of 50 years (ranging from 31 to 74). Among the samples, 37 lesions were identified as malignant, while 24 were classified as benign. The imaging was performed using a 1.6 T scanner (SymphonyTim, Siemens Medical System, Erlangen, Germany) equipped with breast coils. The imaging protocol involved the acquisition of DCE T1-weighted FLASH 3D axial fat-saturated images using the following parameters: repetition time (TR)/echo time (TE) of 5.08/2.39 ms, flip angle of 15°, matrix size of 384×384, slice thickness of 1.6 mm, acquisition time of 110 s, and 128 slices covering the entire volume of the breast. The second dataset, designated as Dataset B, comprises 33 patients ranging in age from 16 to 69 years, who underwent imaging using a 1.5T scanner (Magnetom Symphony, Siemens Medical System, Erlangen, Germany) equipped with breast

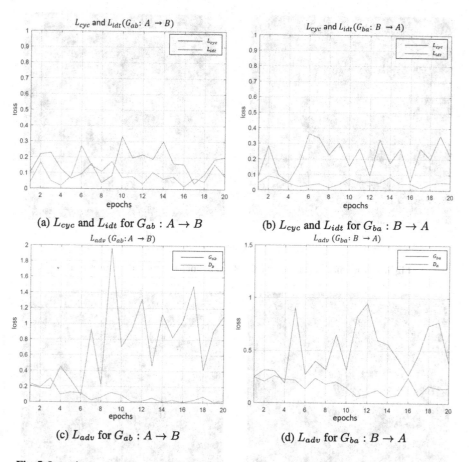

Fig. 5. Learning curves for the training of the architecture considering ResNet as Generetor.

coils. DCE FLASH 3D T1-weighted coronal images were acquired, with the following parameters: transversal relaxation time of 9.8 ms, echo time (TE) of 4.76 ms, flip angle of $25°$, field of view of 370×185 mm^2, image resolution of 256×128 pixels, slice thickness of 2mm, no gap between slices, acquisition time of 56 s, and 80 slices covering the entire breast volume.

The difference between datasets A and B consists of the involved acquisition protocol as the former considers axial volumes with *fat suppression* to suppress high signal from fat to better visualize pathology or contrast enhancement, while the latter is characterized by coronal images without *fat suppression*. Due to short relaxation times, fat has a high signal in MRI scans, which may be useful for the characterization of a lesion [10]. For both datasets, only the pre-contrast acquisition is considered, resulting in volumes of $384 \times 384 \times 128$ and $256 \times 128 \times 80$ for A and B respectively. Then, the mask reporting the information about the breast parenchyma is applied to reduce the contribution of the pixels belonging to other organs or anatomical structures. In the preprocessing step, a permutation operation is applied on the B dataset, to obtain images

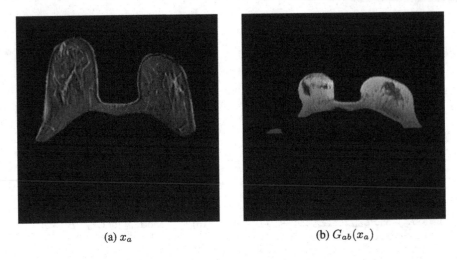

(a) x_a (b) $G_{ab}(x_a)$

Fig. 6. Some images generated with *Resnet* generator from domain A to B.

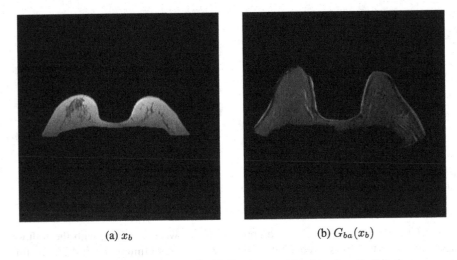

(a) x_b (b) $G_{ba}(x_b)$

Fig. 7. Some images generated with *Resnet* generator from domain B to A.

in axial projection (256 × 80 × 128). Moreover, in each volume, the first and last 10 slices are excluded and a set of 256 × 256 slices is obtained by cropping the images belonging to dataset A and using the padding operation in the B set. *It is worth noting that the sets of patients belonging to the two datasets are completely separated, making the implemented domain translation approach work with unpaired volumes.*

In Sect. 3, it was previously mentioned that the proposed CycleGan comprises two interconnected GANs. The first GAN consists of the generator G_{ab}, which performs the mapping function $G_{ab} : A \rightarrow B$, and the discriminator D_b. The second GAN includes the generator G_{ba} for the mapping function $B : Y \rightarrow A$, and the discriminator D_a. In our experiments, the domains A and B represent dataset A, which is characterized by

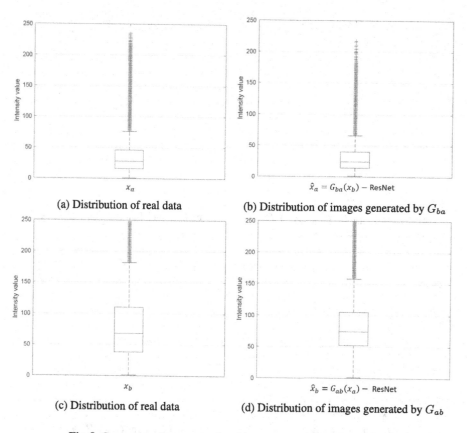

(a) Distribution of real data

(b) Distribution of images generated by G_{ba}

(c) Distribution of real data

(d) Distribution of images generated by G_{ab}

Fig. 8. Comparison between real and images by ResNet architecture.

fat suppression, and dataset B, respectively. During the experiments, the CycleGan is trained using the loss described in Eqs. 1, 2, and 3. The slices belonging to datasets A and B are divided into training, validation, and test sets based on a patient-based hold-out criterion. Specifically, five randomly selected subjects are allocated to the validation and test sets, while the remaining subjects are used as training volumes.

In each experiment, Adam solver is used for the weights optimization making the training more stable and efficient than the traditional gradient descent algorithm. The least squares function is considered for the adversarial losses, and L1-norm for cycle consistency and identity losses. In Eq. 4, the coefficients λ_1 and λ_2 are set to 10 and 0.5 respectively. The learning rate is set to 0.0002, and the instance normalization is used as suggested in [9] and [19] to support an efficient contrast normalization. In the networks, weights are initialized as a Gaussian distribution with zero-mean and standard deviation equal to 0.02. Moreover, the discriminator considers patches of size 70×70, and a history buffer with a pool size of 50 images is used, as reported by [9,15], to improve the stability of adversarial training.

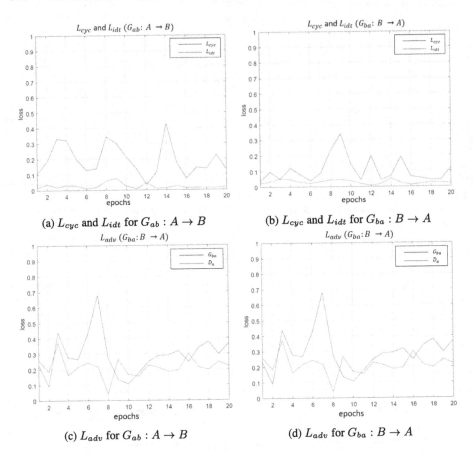

(a) L_{cyc} and L_{idt} for $G_{ab} : A \to B$ (b) L_{cyc} and L_{idt} for $G_{ba} : B \to A$

(c) L_{adv} for $G_{ab} : A \to B$ (d) L_{adv} for $G_{ba} : B \to A$

Fig. 9. Learning curves for the training of the architecture considering U-Net as Generaetor

5   Results

In this section we report the results obtained by using the proposed approach, for both the considered generators. We first show and analyse the results of the experiment using CycleGAN with a Resnet generator containing 9 residual blocks. Figures 5 depict the losses during the entire 20-epoch training process. It is evident that the validation loss starts to increase during the final epochs, suggesting that the network is prone to over-fitting. As a result, training is stopped after the 20th epoch. The following Figs. 6–7 showcase images that were generated from domain A to B and B to A respectively, illustrating accurate translation in both directions. Lastly, boxplots reported in Fig. 8 display the pixel value distributions of 30 test images. These boxplots aid in the statistical evaluation of the degree of accuracy in the domain translation achieved by this network. The distributions of pixel values have been computed for both the real and fake images and were compared to each other.

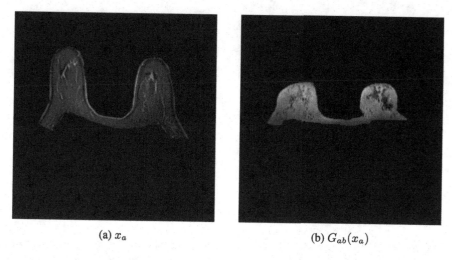

(a) x_a (b) $G_{ab}(x_a)$

Fig. 10. Some images generated with *U-net* generator from domain A to B.

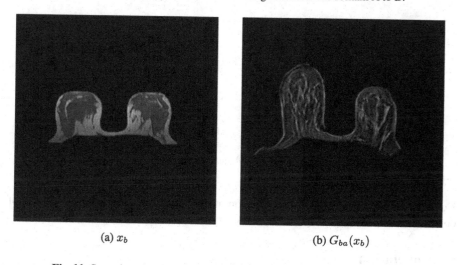

(a) x_b (b) $G_{ba}(x_b)$

Fig. 11. Some images generated with *U-net* generator from domain B to A.

Similar considerations can be made for the CycleGAN model trained by using a U-net generator having an input size of 256×256 pixels. Upon observing the training and validation images, it was noted that the translation achieved by the U-net generator was clearer and visually appealing as compared to that by the Resnet generator. The loss values over the 20-epoch training process are shown in Fig. 9. Moreover, the generated images depict that the translation is correctly executed in both directions (Figs. 10–11), preserving the shape and colour with only minor distortion in the B to A direction. Finally, the pixel value distribution of 30 test images is analyzed using boxplots in Fig. 12, which demonstrates that the domain translation is carried out accurately from a statistical standpoint. This verifies that the translated version of a sample

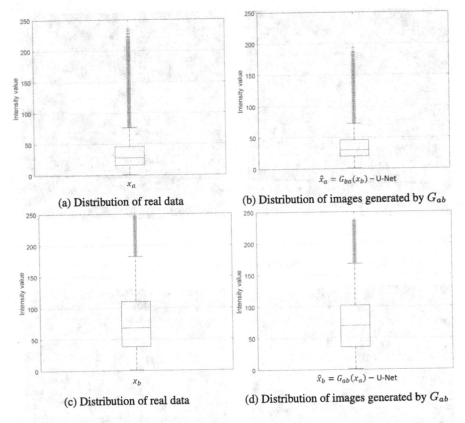

(a) Distribution of real data   (b) Distribution of images generated by G_{ab}

(c) Distribution of real data   (d) Distribution of images generated by G_{ab}

Fig. 12. Comparison between real and images by U-Net architecture

from one domain belongs to the distribution of the other domain with a certain level of confidence, further validating the success of the domain translation task.

6   Conclusion

In this study we analysed the use of Generative Adversarial Networks (GANs) for image-to-image translation in the context of multi-protocol breast DCE-MRI. After analyzing the test images and boxplots of the entire experimental set, it is evident that using U-net produces the best results in terms of both shape and colour preservation. Indeed, while Resnet yields similar results in terms of overall similarity for a non-expert eye, the translated images typically exhibit distortions. The results were evaluated qualitatively and quantitatively, however, no objective metrics have been found to be useful in assessing GAN performance during training. Indeed, the loss curve alone does not provide much information about GAN training, including CycleGAN. Therefore, a visual inspection of generated images and subjective evaluation approaches are the best means of evaluating GAN performance. There is still a need for the development of quantitative and deterministic methods for measuring GAN performance.

This study highlights the strengths and limitations of GAN networks, as previously discussed in [22] and [18]. CycleGAN is particularly effective in tasks involving colour or texture changes such as day-to-night photo translation, photo-to-painting tasks, and even collection style transfer. However, it may not be as successful in tasks that require substantial geometric changes to the image, as presented in this study. Therefore, there is a need to improve CycleGAN for translations requiring geometric changes which is an essential area for future investigation. However, an additional potential limitation of this study is the small size of the datasets, which we will try to mitigate in future works by including more patients in the study. In future works, we will further analyse this aspect, by also investigating whether it is possible to add some terms to the loss function in order to make the generator aware of biological and physiological constraints.

References

1. Cai, N., Chen, H., Li, Y., Peng, Y., Guo, L.: Registration on DCE-MRI images via multi-domain image-to-image translation. Comput. Med. Imaging Graph. **104**, 102169 (2023)
2. Desai, S.D., Giraddi, S., Verma, N., Gupta, P., Ramya, S.: Breast cancer detection using gan for limited labeled dataset. In: 2020 12th International Conference on Computational Intelligence and Communication Networks (CICN), pp. 34–39. IEEE (2020)
3. Goodfellow, I., et al.: Generative adversarial networks. Commun. ACM **63**(11), 139–144 (2020)
4. Gravina, M., et al.: Leveraging CycleGAN in lung CT Sinogram-free kernel conversion. In: Sclaroff, S., Distante, C., Leo, M., Farinella, G.M., Tombari, F. (eds.) ICIAP 2022 Part I. LNCS, vol. 13231, pp. 100–110. Springer, Cham (2022). https://doi.org/10.1007/978-3-031-06427-2_9
5. He, K., Zhang, X., Ren, S., Sun, J.: Deep residual learning for image recognition. In: Proceedings of the IEEE Conference on Computer Vision and Pattern Recognition, pp. 770–778 (2016)
6. Isola, P., Zhu, J.Y., Zhou, T., Efros, A.A.: Image-to-image translation with conditional adversarial networks. arXiv (2018). https://arxiv.org/abs/1611.07004
7. Johnson, J., Alahi, A., Fei-Fei, L.: Perceptual losses for real-time style transfer and super-resolution. In: Leibe, B., Matas, J., Sebe, N., Welling, M. (eds.) ECCV 2016. LNCS, vol. 9906, pp. 694–711. Springer, Cham (2016). https://doi.org/10.1007/978-3-319-46475-6_43
8. Mao, X., Li, Q., Xie, H., Lau, R.Y., Wang, Z., Paul Smolley, S.: Least squares generative adversarial networks. In: Proceedings of the IEEE International Conference on Computer Vision, pp. 2794–2802 (2017)
9. Modanwal, G., Vellal, A., Mazurowski, M.A.: Normalization of breast MRIs using cycle-consistent generative adversarial networks. arXiv (2019). https://arxiv.org/abs/1912.08061
10. Murphy, A., Niknejad, D.M.T.: Fat suppressed imaging. https://radiopaedia.org/articles/fat-suppressed-imaging?lang=us
11. Ronneberger, O., Fischer, P., Brox, T.: U-Net: convolutional networks for biomedical image segmentation. In: Navab, N., Hornegger, J., Wells, W.M., Frangi, A.F. (eds.) MICCAI 2015. LNCS, vol. 9351, pp. 234–241. Springer, Cham (2015). https://doi.org/10.1007/978-3-319-24574-4_28
12. Sannino, C., Gravina, M., Marrone, S., Fiameni, G., Sansone, C.: Lessonable: leveraging deep fakes in MOOC content creation. In: Sclaroff, S., Distante, C., Leo, M., Farinella, G.M., Tombari, F. (eds.) ICIAP 2022 Part I. LNCS, vol. 13231, pp. 27–37. Springer, Cham (2022)

13. Secinaro, S., Calandra, D., Secinaro, A., Muthurangu, V., Biancone, P.: The role of artificial intelligence in healthcare: a structured literature review. BMC Med. Inform. Decis. Mak. **21**, 1–23 (2021)
14. Shamsolmoali, P., Zareapoor, M., Granger, E., Zhou, H., Wang, R., Celebi, M.E., Yang, J.: Image synthesis with adversarial networks: a comprehensive survey and case studies. Inf. Fusion **72**, 126–146 (2021)
15. Shrivastava, A., Pfister, T., Tuzel, O., Susskind, J., Wang, W., Webb, R.: Learning from simulated and unsupervised images through adversarial training. In: Proceedings of the IEEE Conference on Computer Vision and Pattern Recognition, pp. 2107–2116 (2017)
16. Tavse, S., Varadarajan, V., Bachute, M., Gite, S., Kotecha, K.: A systematic literature review on applications of GAN-synthesized images for brain MRI. Future Internet **14**(12), 351 (2022)
17. Ulyanov, D., Vedaldi, A., Lempitsky, V.: Instance normalization: the missing ingredient for fast stylization. arXiv preprint arXiv:1607.08022 (2016)
18. Wolf, S.: Cyclegan: Learning to translate images (without paired training data) (2018). https://towardsdatascience.com/cyclegan-learning-to-translate-images-without-paired-training-data-5b4e93862c8d
19. Wolterink, J.M., Dinkla, A.M., Savenije, M.H., Seevinck, P.R., van den Berg, C.A., Isgum, I.: Deep MR to CT synthesis using unpaired data. arXiv (2017). https://arxiv.org/abs/1708.01155
20. Xie, G., et al.: Fedmed-gan: Federated domain translation on unsupervised cross-modality brain image synthesis (2022). Available at SSRN 4342071
21. Zhu, J.Y., Park, T., Isola, P., Efros, A.A.: Unpaired image-to-image translation using cycle-consistent adversarial networks. In: Proceedings of the IEEE International Conference on Computer Vision, pp. 2223–2232 (2017)
22. Zhu, J.Y., Park, T., Isola, P., Efros, A.A.: Unpaired image-to-image translation using cycle-consistent adversarial networks. arXiv (2020). https://arxiv.org/abs/1703.10593

A Survey on Reinforcement Learning and Deep Reinforcement Learning for Recommender Systems

Mehrdad Rezaei$^{(\boxtimes)}$ and Nasseh Tabrizi

East Carolina University, Greenville, NC, U.S.A.
rezaeim18@students.ecu.edu

Abstract. Recommender systems are quickly taking over our daily lives. By suggesting and customizing the recommended items, they play a significant part in solving the information overload issue. Traditional recommender systems used for simple prediction issues include collaborative filtering, content-based filtering, and hybrid techniques. With new techniques used in recommender systems, such as reinforcement learning algorithms, more difficult problems can be resolved. These issues can be resolved using Markov decision processes and reinforcement learning techniques. It is now possible to employ reinforcement learning techniques to address issues with the huge environment and states, thanks to recent advancements in the field. The development of traditional and reinforcement learning-based methods, their appraisal, difficulties, and suggested future research will be followed by a discussion of the reinforcement learning recommender system.

Keywords: Recommender systems · Reinforcement learning · Rewards

1 Introduction

Making the appropriate choice is difficult due to information overload, which is a result of the vast amount of information available on the Internet. When we have a long list of potential products to buy, it is feasible to experience it in our regular online shopping. The difficulty of making a choice from the list will increase as it gets longer. In order to help consumers, identify items of interest, recommender systems (RSs) use algorithms and software tools to predict their preferences or ratings. By anticipating user ratings and past preferences, the creation of RSs aids consumers in finding the thing they are looking for. A wide variety of RSs applications, including e-learning [1], e-commerce [2], healthcare [3], and news [4], are currently used by major corporations, including Netflix, Amazon, Facebook, and Google. To address the recommender system issue, other approaches, including content-based filtering, collaborative filtering, and hybrid methods, are also suggested. A certain amount of success in the area of making appropriate suggestions was attained through the advent of matrix factorization. According to [5], the aforementioned approaches still have issues with a cold start, scalability, serendipity, appropriate computational expense, and suggestion quality. Because of the potential in complex relationships between users, objects, and their accurate performance in the recommendation, deep learning (DL) has lately acquired acceptance in the RSs application field. DL models have the following characteristics: they are hard to understand, expensive to compute, and data-hungry [6]. The interaction between users and items, which can be handled better with Reinforcement Learning (RL) and its training agent in the environment that is a semi-supervised machine learning sector, is where past RSs methods are not helpful [6]. Deep Reinforcement Learning (DRL), a blend of classic RL techniques and DL can be used to identify the most crucial points in RL. This made it possible to apply RL to challenges in areas like robotics [7], industry automation [8], self-driving automobiles [9–11], finance [12], and healthcare [13] that have expansive state and action spaces.

Because it has the potential to reward learning without any training data, which is a unique criterion, RL is a perfect match for the recommendation problem. Many businesses now exploit the ability of RL to suggest better products to their clients, such as the YouTube video recommender system [14]. Both in the business world and in academics, RL is being used more and more in RSs. We were inspired to create this work in the area of reinforcement for RSs by how important this topic is. The main objective of our paper is to demonstrate the development of using RL in RSs to illustrate the trend that has evolved in recent years. The sample chart shown in Fig. 1 depicts the number of papers published from 2010 through January 2023.

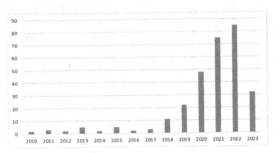

Fig. 1. Number of papers published from 2010 to 2023

2 Methodologies

We have made the decision to investigate the issues and difficulties brought on by RSs using the RL algorithm. The introduction of the applications for RL-based and non-RLbased RSs is a crucial point that is highlighted in the discussion of approaches and algorithms to address these challenges as the second goal. These issues are covered in this essay:

- First, we divide the available algorithms into RL- and DRL-based methods. After that, the categories are separated into the various RL algorithms used in the studies.
- We initially give a brief yet thorough description of each method in order to convey to the reader the key idea and contribution of the work.

- The directions for future research are offered. Finally, we provide some insights regarding ongoing research in the Reinforcement Learning Recommender System (RLRS) field to tie together our survey study.

We used the following search terms to find all relevant articles at first: recommender systems, recommender engine, recommendation, content filtering, collaborative filtering, reinforcement learning algorithm, deep reinforcement learning algorithm, and reinforcement learning for recommender systems. We then applied these terms to various databases, including IEEE Xplore, SpringerLink, ScienceDirect, ACM digital library, and Lynda.com. We also looked into well-known RSs conferences like RecSys, SIGIR, and KDD.

After compiling the articles, we searched through the publications to discover any that were pertinent to our work [1, 15]. As a result, the studies that use RL for technologies other than RSs, like conversational/dialogue management systems [16, 17], were chosen. Subsequent paragraphs, however, are indented.

3 Reinforcement Learning and Deep Reinforcement Learning

A machine learning method that studies different problems and solutions to maximize a reward through interaction between agents and their environment is called RL. Three characteristics that discriminate an RL problem [18] are: i) closed-loop problem ii) there is no need for a trainer to teach the learner, but it trains what to do to the learner with the trial-and-error method according to the policy, iii) the short and long terms results can be influenced by the actions. The crucial part of modeling the RL problem is the agent's interface and environment, as shown in Fig. 2.

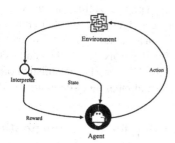

Fig. 2. The interface of RL

An agent is a decision-maker or learner; everything outside of the agent is called the environment. Information and representation that is seen outside of the agent (the environment) at time step t is called state, and the agent makes an action according to the current state. Based on the action taken, the environment is given a numerical reward and goes to a new state. The RL problems are formulated commonly as a Markov Decision Process (MDP) with the form of (S, A, R, P, γ), where S represents all possible states, A indicates actions that are available in all states, R presents reward function, P shows the probability of the transition, and finally, γ is the discount factor. The agent's aim in the RL problem is the best policy $\pi(a|s)$ to make an action that is a member of A in state $s \in S$ to maximize the cumulative reward. An RL system includes four principal parts [18]: i) Policy: It is presented by π generally, which indicates the probability of doing an action. The RL algorithm may be categorized into on-policy and off-policy techniques depending on the policy. In the first case, RL approaches are used to evaluate or improve the policy that is being used to make judgments. They enhance or assess a policy that is not the same as the one used to create the data in the latter. ii) Numeral rewards: regarding the selected actions, the environment gives a numeral reward in order to send an announcement to the agent about the action that is selected. Iii) Value function: the purpose of the value function is to indicate how good or bad the action is in the long run. iv) Model: it indicates the conduct of the environment. There are two types of algorithms that are utilized to address RL challenges: tabular and approximate. In the tabular method, tables are used to represent value functions, and an accurate policy is found because the size of spaces (action and state) is not big. Monte Carlo (MC), Temporal Difference (TD), and Dynamic Programming (DP) are popular tabular methods. The MC methods need only an instance of rewards, states, and actions that will be provided by the environment. Monte Carlo Tree Search (MCTS) is the most important algorithm of MC methods. The DP methods use an excellent model of the environment and value function in order to find good policies. Policy and value iteration can be good examples of DP methods. The TD method is a blend of the MC sampling method and the DP bootstrapping method. The TD methods, like the MC methods, may learn from the agent's interactions with the world and do not require model knowledge. From this class, Q-learning [19] and SARSA are the most important ones as they are off-policy and on-policy, respectively. In the approximate method, the aim is to search for sufficient solutions regarding the computational resources constraint because state space has a massive size. To address this, previous experiences are used. Policy Gradient methods are very popular because of their ability to learn policy parametrization and action selection without the need for a value function. Actor-critic and reinforcement are more significant methods in this category. DL is a field based on an artificial Neural Network that is used as the function in RL and suggests a deep Q-network (DQN) [21]. DQN and Deterministic policy gradient (DPG) [21] are combined and used in Deep Deterministic Policy Gradient (DDPG) [22]. In RSs, Double DQN (DDQN) and Dueling Q-network are also used [23].

4   Reinforcement Learning for Recommendation

The way a user interacts with an RS is sequential. A sequential decision problem is selecting the best products to recommend to a user [24]. This suggests that RL methods can be used to model the recommendation problem as an MDP and solve it. As was already mentioned, an agent in a typical real-world scenario tries to maximize a monetary reward by interacting with the environment. The RS algorithm attempts to recommend the best products to the user while maximizing the user's pleasure in the recommendation issue, which is comparable to this. As a result, everything outside of the RL agent, such as system users and objects, can be thought of as the agent's environment. The RS algorithm can therefore act as the RL agent. Modern RSs with vast action and state spaces make it practically impossible to apply conventional tabular RL algorithms [25].

Instead, there is a rising trend in the RS community to adopt RL techniques as a result of the appearance of DRL algorithms.

5 Reinforcement Learning Algorithms

In this section, we classify and present algorithms. After examining every algorithm, we came to the conclusion that the development of DRL has drastically changed how RLRSs are studied. Consequently, we divided RLRS methods into two major groups: RL- and DRL-based algorithms. A high-level summary of the methods and the number of publications is provided in Fig. 3. We start with RL-based methods.

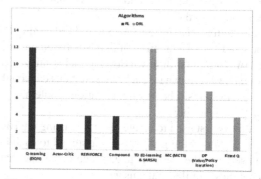

Fig. 3. Number of publications using each algorithm

5.1 RL-Based Methods

We refer to RLRSs that optimize recommendation policies using an RL algorithm but do not employ DL to estimate parameters using RL-based techniques. TD, DP, MC, and Fitted Q RL algorithms from both tabular and approximate approaches are examples of RL-based methodologies.

5.1.1 TD Methods

In the field of RS, Q-learning is a popular RL algorithm [26, 27]. WebWatcher is probably the initial RS technique to incorporate RL to raise the caliber of suggestions. The web page recommendation problem is simply treated as an RL problem, and Q-learning is used to improve the precision of their fundamental web recommendation system (RS), which uses a similarity function (based on TF-IDF) to suggest pages that are related to the user's interests. Ten years later, [28], authors expanded on this concept to provide users with custom websites. They use a sliding window to represent states using the N-gram model from the online usage mining literature [29] to overcome the state dimensionality issue.

Another TD algorithm employed by some RSs is called SARSA [30] The global and local units make up the web RS [30]. The local unit keeps track of each consumer separately, while the global unit is in charge of learning the system's overall trend, such as the most popular products. To choose the next page to recommend, the system combines the local and global models using a weighted approach. Scalability is an apparent issue with this work because it is unclear how they intend to maintain track of all users on a worldwide scale. SARSA (λ) is a modified SARSA algorithm that approximates the original SARSA algorithm's solution and was used in [30] to create a customized ontology-based web RS. The work's major objective is to combine RL techniques and epistemic logic software to suggest to a user the best website concept. In actuality, the work's contribution consists in converting epistemic data into real number arrays that may be used by approximative RL algorithms. Reference [31] employs RL for online education. Here, the RS's objective is to offer students a learning route that is customized to meet their unique needs and traits. Similar to [28], they approach the state dimensionality issue using the N-gram model.

Additionally, some studies have tested SARSA and Q-learning for policy optimization [32]. For instance, in [32]., the creation of playlists based on emotions is presented as an RL challenge. An N-gram (sliding window) model is utilized to model the states in order to manage the state space, where each state contains data on the emotion classes of the user's most recent m songs. The recommendation problem is presented in [33] as a grid world game with biclustering.

The user-item matrix is first used to create biclusters using the Bimax [34] and Bibit [35] algorithms. Then, a state in the grid world is assigned to each bicluster. Any state in the grid world can serve as the starting state, which is the one that, according to the Jaccard distance, is the most similar to the user.

5.1.2   DP Methods

DP has also been applied in [24, 36, 37] as a tubular technique. One of the earliest studies to formulate the recommendation problem as an MDP is a reference [36]. In fact, using the example of guiding a user through an airport, the study investigates the possible advantages of adopting MDP for the recommendation problem. As an early and worthwhile attempt to represent the recommendation problem as an MDP, see ref. [24]. They suggest a predictive model that can offer initial parameters for the MDP because the model parameters of an MDP-based recommender are unknown, and deploying it on the actual to learn them is too expensive. The state and transition function of this Markov chain-based prediction model is modeled based on the dataset observations. They contend that when the maximum probability is used to estimate the transition function, the simplest Markov chain exhibits the data sparsity issue. As a result, the basic version is enhanced using the three techniques of clustering, mixture modeling, and skipping. The MDP-based recommender is then launched using this prediction model. The remaining k items are used to encode state data to solve the dimensionality problem. To gauge the success of their plan, they use an online study.

5.1.3   MC Methods

Several RLRSs have adopted the ultimate tabular technique, MC [37, 39, 40]. Each song is represented as a vector of the song (spectral auditory) descriptors to address the dimensionality issue. These descriptors contain details on the song's spectral fingerprint, rhythmic elements, overall loudness, and change over time. The listener's preference for particular songs as well as his song transition behavior, are both taken into consideration by the reward function in order to speed up the learning process. The learning of listener parameters (his preferences for songs and transitions) and the planning of a song sequence make up the two main components of the DJ-MC architecture.

5.1.4 Fitted Q Methods

Some RL-based algorithms also use an approximation technique (fitted Q) for policy optimization [41–43]. In a clinical setting, RL is applied to provide lung cancer patients with different therapy options in an effort to increase patient survival [41]. They see advanced non-small cell lung cancer (NSCLC) treatment as a two-line therapy, with the RL agent's job being to advise on the optimal course of action for each line of therapy and when to start second-line therapy. The Q-function for the RL agent is improved by the use of support vector regression (SVR). Because the original SVR cannot be applied to censored data, they modify it using a -insensitive loss function [42]. [44] uses RL to find the most effective treatment options for schizophrenia patients. To address the issue of missing data, they first use multiple imputations [45], which can introduce bias and increase variance in Q-value estimations as a result of patient dropout or item missingness. They also address the second problem, which is that function approximation is challenging since clinical data is very volatile and has limited trajectories. They consequently use a fundamental linear regression model and fitted Q-iteration (FQI) [46] to train the Q-function. The motivation for the study described in [43] is that existing ad suggestion algorithms assume that all visits to a website are from new visitors and do not differentiate between a visit and a visitor. They contend that lifetime value (LTV), defined as (the total number of clicks/total number of visitors) 100, is a true choice for long-term performance, whereas Click-Through Rate (CTR) is a realistic choice for greedy performance. They use a model-free method called HCOPE, created by the same authors in [43], to solve the off-policy evaluation problem in the RS field. HCOPE uses concentration inequality to establish a lower bound on a policy's expected return.

5.2 DRL-Based Methods

This section examines DRL-based RSs that make use of DL to simulate the value function or policy. These techniques make use of the three crucial RL algorithms Q-learning, actor-critic, and REINFORCE for policy optimization. There are additional papers as well that evaluate and contrast the effectiveness of different RL algorithms for policy optimization.

5.2.1 Q-learning Methods

Possibly the first study to use DQN for a slate recommendation was Slate-MDP [47]. They offer agents that employ a sequential greedy technique to learn the value of complete slates in order to deal with the combinatorial action space created by slates (tuples) of actions. In fact, it's expected that the item slates will provide recommendations to users one at a time using a feature called sequential presentation. This presumption is followed by another that assumes one of the primordial acts will be carried out. For each slot in the slate, they additionally use an attention mechanism based on DDPG to focus the search on a small region of the action space that has the highest value. According to [48], the second assumption, which equates to the possibility of compelling a user to consume a certain food item, is not very plausible in frequent suggestion scenarios.

DQN is utilized to enhance heparin dosage recommendations [49]. They initially treat the issue as a partially observable MDP (POMDP) before employing a discriminative hidden Markov model to estimate the belief states. DQN is then used to optimize the policy. Another clinical application [50] uses a variant of DQN to improve dosage recommendations for sepsis treatment. They use both a discrete action space and a continuous state space. Due to inherent problems with the original DQN algorithm, such as the overestimation of Q values, they modify DQN as follows.

Generative adversarial networks (GANs) are used to create a user model [51], and a cascade DQN technique is then used to select the best options. In user modeling, user behavior and the reward function are simultaneously learned using a mini-max optimization technique. The best recommendation technique is then learned using DQN and the learned user model. Instead of using a single Q-network to solve the combinatorics of suggesting a list of items (k items), as is the case with other systems, k Q-networks are used in a cascaded manner to discover k optimum actions. To be more precise, the fact that follows determines the best actions: $max_{a_{1:k}} Q^*(s, a_{1:k}) = max_{a_1} (max_{a_{2:k}} Q^*(s, a_{1:k}))$.

Other efforts only include DQN or its extensions into a particular RS application without making any noticeable modifications or adaptations. In [83], the issue of proactive caching in mobile networks is described as an RL problem. The issue is divided into two issues—content recommendation and pushing— to solve the dimensionality. A based station (BS) proposes a piece of material from a group of content candidates. In the meantime, the BS has the ability to actively push content to the user's mobile device cache. As a result, two RL agents are created for pushing and recommending material to consumers. These agents are trained sequentially and collaborate to accomplish the overall objective.

It is important to note that DDQN with dueling Q networks is used for both agents. A different application study [53] formulates movie recommendation as an RL problem. The work's key contribution is to employ cross entropy to record user interest changes and prioritize experience replay in DQN. RecEnergy employs deep Q-learning in an Internet-of-things (IoT) application [47] to optimize energy use in commercial buildings. Move, Schedule Change, Personal Resources, and Coerce are just a few of the recommendations it makes to both inhabitants and the Building Management System (BMS) in order to accomplish this. Authors in [38] also discuss the impact of recommendations built on trust. The trust is utilized as the reward for the RL agent to be maximized in particular because RL is used to improve user confidence in the system.

5.2.2 Actor-Critique Methods

An actor-critic system called Wolpertinger [28] can handle extremely large action spaces (up to one million). The objective is to develop a method that is generalizable across activities and sub-linear in terms of action space. There are two parts to Wolpertinger.

Action creation is the first, and action refinement is the second. The actor creates proto-actions in continuous space in the first half, which are then translated to discrete space using the k nearest neighbor method. The critic that selects the best action with the highest Q value filters outlier actions in the second segment. Additionally, their approach is trained using DDPG. Wolpertinger can handle a recommendation task in a simulation study, but it is not specially designed for RSs.

The training of parameters also uses DDPG. One problem with this work is that when generating a list of items rather than suggesting a single item, it does not manage the combinatorics of action space. They subsequently offer page-by-page guidance [28]. They refer to proposing a collection of complementary items and putting them on display on a page when they say, "recommending a group of complementary things and putting them on a website." The actor is responsible for producing a page of content. The initial and current states are first produced by two encoders. After that, a decoder—specifically, a deconvolutional neural network—generates the actions. On the other side, the actor's action was delivered to the critic, which uses a DQN architecture, together with the actor's current state (as determined by the same approach). Once more, DDPG is used for model training. Additionally, they extend their e-commerce research to a full supply chain suggestion [54]. They use a multiagent system with shared memory that optimizes all of these situations simultaneously rather than having several scenarios in a user session, such as a welcome page and item pages (in fact, they only take into account two pages in their studies: entry and item pages). Agents (actors) work together to maximize the cumulative reward by successively interacting with the user. On the other hand, it is the duty of the global critic to exercise influence over these actors. To record user preferences in numerous contexts, the global critic uses an attention technique, with each attention only active in its respective situation.

These states are sent to the recommender, who then generates recommendations using a straightforward closest neighbor method. The user's preferences, as shown in the user's earlier comments, are used to limit the number of infractions by using a historical critic. Although this study uses multimodal data in a novel way, the actor-critic paradigm is not fully explained.

There are some additional efforts that merely optimize the recommendation policy using DDPG. Store recommendation is defined as an RL problem in [55]. First, latent Dirichlet allocation (LDA) is used to transform storage data into a continuous space. The user is then given a store recommendation via DDPG. CapsNet [56] is used for feature extraction in CapDRL [57], which is the key contribution. Particularly, user feedback (both positive and negative) is supplied into CapsNet, which transforms it into latent weight vectors and projects them into the continuous space. The recommendations are then generated using DDPG.

5.2.3 REINFORCE Methods

A conversational RS is developed by authors in [58] based on hierarchical RL [59]. The meta-controller module in the framework receives the dialogue state and foresees the goal for that state. The two types of goals that the work supports are chitchat and recommendation. A goal-specific representation module converts the dialogue state into a score vector, which is subsequently enhanced by an attention module to highlight more pertinent parts. The controller module then uses these updated scores to take the necessary actions to accomplish the given goal. The framework has two critics: an internal critic pays the controller based on the objectives, and an external critic critiques the reward for the meta controller produced by the environment.

Ref. [14] provides a helpful study in the area of RL-based movie recommendation. The adaptation of the REINFORCE approach to a neural candidate generator with a very vast action space is the main contribution of the paper. The policy gradient estimator in an online RL situation can be expressed as follows:

$$\sum_{\tau \sim \pi_\theta} \left[\sum_{t=0}^{|r|} R_t \, \Delta_\theta \, \log \pi_\theta(a_t|s_t) \right] \tag{1}$$

where Rt is the total reward, π_θ is called the parametrized policy, and $\tau = (s_0, a_0, s_1, \ldots)$. Because, unlike in conventional RL situations, online or real-time interaction between the agent and environment is infeasible and frequently only logged feedback is provided, applying the policy gradient in Eq. (1) is biased and requires rectification. The policy gradient estimator that has been off-policy-corrected is then:

$$\sum_{\tau \sim \beta} \frac{\pi_\theta(\tau)}{\beta(\tau)} \left[\sum_{t=0}^{|r|} R_t \, \Delta_\theta \, \log \pi_\theta(a_t|s_t) \right] \tag{2}$$

where β is the importance weight and $\frac{\pi_\theta(\tau)}{\beta(\tau)}$ is the $\beta(\tau)$ behavior policy because this adjustment yields a large variance for the estimator, they utilize first-order approximation, resulting in the biased estimator with the reduced variance:

$$\sum_{\tau \sim \beta} \left[\sum_{t=0}^{|r|} \frac{\pi_\theta(a_t|s_t)}{\beta(a_t|s_t)} R_t \Delta_\theta \log \pi_\theta(a_t|s_t) \right] \tag{3}$$

The work's last contribution is top-K off-policy rectification. Setting (top-K) recommendations result in an exponentially increasing action space. The off-policy adjusted estimator described in Eq. (3) is modified to the following estimator for top-K recommendation under two assumptions:

$$\sum_{\tau \sim \beta} \left[\sum_{t=0}^{|r|} \frac{\pi_\theta(a_t|s_t)}{\beta(a_t|s_t)} \frac{\partial \alpha(a_t|s_t)}{\partial \pi(a_t|s_t)} R_t \, \Delta_\theta \, \log \pi_\theta(a_t|s_t) \right] \tag{4}$$

where α is the likelihood that an item α appears in the final non-repetitive set A (top-K items).

The environment is lastly shown in [60] as a heterogeneous information network (graph) made up of people, things, and other information sources, including content, tags, reviews, friends, and so forth. Finding a path in the graph between a user and an unseen item is the objective. The article uses a multi-iteration training approach, as demonstrated below. The meta-path calculated at each iteration is stored in a meta-path base, which is comparable to a knowledge base. Initial meta-path bases are empty, and they are later filled with expert-provided meta-paths. The meta-path base is then expanded with the meta paths that the RL agent tested in each iteration. The updated meta-path base is applied to train the RL agent in the following iteration. The process is repeated until no more data can be acquired or the allotted number of iterations has been reached. The top-K recommendation is based on the nearest neighbor algorithm.

5.2.4 Compound Methods

The use of RL to suggest learning activities in a smart class by the authors [61] is rare yet intriguing. In particular, a cyber-physical-social system is built that keeps track of students' learning states by collecting multimodal data, including test results, heartbeat, and facial expression, and then presents a learning activity that is suitable for them.

A task-oriented dialogue management system is suggested and used in Ref. [62] for a variety of recommendation tasks. Two strategies are suggested for dialogue management: segmentation-based and state-based. According to context, such as demographics and purchasing history, the former segmented the user population into different categories, each with its own policy. The latter approach is built on combining agent assumptions about prior conversations, user intentions, and context. Then, a special policy is loaded with this belief vector for each user. A benchmark [55] in the field is used to evaluate the work. This benchmark comprises a range of recommendation tasks, such as recommending restaurants in Cambridge or San Francisco.

Finally, authors in EDRR [63] explain that embedding, state representation, and policy are the three elements that all RL techniques have in common. They contend that because RL approaches contain gradients with substantial variation, it is impossible to directly train the embedding module with the two other modules. Through the addition of a supervised learning (SL) unit to EDRR, they hope to solve this issue. Three different approaches are suggested to incorporate this SL signal into their method; the difference between them is whether the SL signal is used to update the state representation component.

6 Algorithm Overview

6.1 RL-Based Methods Recap

Tabular and approximation methods can be used to categorize the RL approaches in this area. Because they demand a complete grasp of the environment and have a significant computing cost, DP procedures are often not feasible among tabular alternatives. Although these algorithms have polynomial numbers of states, it is usually impossible to implement even one iteration of a policy or value iteration approach [76]. Only two RS approaches [11, 24] use DP in the RLRS literature. [24] use some approximations and a few features in their state space to make it more practical. Similar restrictions apply to the number of policy iterations that can be made, according to [64]. MC methods do not necessitate an exact grasp (or model) of the environment, in contrast to DP. However, there are a number of drawbacks to MC techniques, including the fact that they do not bootstrap. On the other hand, TD techniques have gained a lot of traction in the RS community [26]. They can be expressed using a single equation, are online, model-free, need little computation, and are appealing for this reason alone [18]. Tabular techniques typically suffer from the curse of dimensionality as the state and action spaces expand, which makes them ineffective for learning even though they may find the precise solution, i.e., the optimal value function and policy. By making the state space as small as feasible, RLRSs that employ DP and TD methods make an effort to overcome this problem. Similarly, MCTS-based methods must only keep the data from a sampling event and not the complete environment.

On the other hand, aside from the SARSA (λ) approach utilized by Mircea and Dan [65], the fitted Q technique, a flexible framework that can fit any approximation architecture to the Q-function [43], is the only type of approximate method used by RL-based RSs. Therefore, the Q-function can be approximated by any batch-mode supervised regression technique that scales well to high dimensional spaces [18]. It is possible that the computational and memory costs will rise as the number of four-tuples ((x_t, u_t, r_t, x_{t+1}), where x_t represents the system state at time t, u_t the control action executed, r_t the immediate reward, and x_{t+1} the next state of the system) increases [46]. [41, 42], and [43] mention other RLRSs that have employed this approach.

6.2 DRL-Based Methods Recap

A turning point in the development of RLRSs was the creation of DRL. Figure 1 clearly illustrates this pattern. DRL is the best choice for RSs with large state and action spaces due to its exceptional capacity for handling high-dimensional spaces. According to several studies [47, 49, 62], DQN has been the most often employed DRL algorithm by RLRSs. According to [18], DQN altered the initial Q-learning algorithm in three ways: 1) It leverages experience replay, a technique first used in [66], which uses agents' experiences over a range of time steps to modify weights throughout the training phase. 2) The current updated weights are fixed and input into a second network, whose outputs are used as Q-learning targets in order to reduce the complexity of updating weights. 3) The reward function is trimmed to be 1 for positive rewards, -1 for negative rewards, and 0 for no rewards in order to reduce the magnitude of error derivatives. All of these modifications appear to have improved DQN's stability. However, as was already said, DQN has several problems. First, when using the Q-learning approach, DQN occasionally overestimates the value of an action, which makes learning ineffective and may result in suboptimal policies [67]. Numerous RLRSs use DDQN, which was offered as a solution to this problem [50]. Second, DQN replays events at random, regardless of their importance, which makes learning time-consuming and ineffective. Only four RLRS algorithms update the original experience replay mechanism of DQN, although most DQN-based RLRSs still employ the original mechanism.

References [50] use prioritized experience replay; References [52] use stratified sampling rather than uniform sampling; References [68] utilize cross-entropy of user interest to prioritize experiences; References [50] use uniform sampling. Third, because it requires an iterative optimization process at each step, which is computationally prohibitively expensive, DQN cannot handle continuous domains. DDPG, which combines DQN and DPG, has been offered as a solution to this problem.

As opposed to action-value techniques like DQN, policy gradient methods learn a parameterized policy without the use of a value function. Policy-based systems have three advantages over action-value strategies [18]. One can get closer to determinism using policy approximate approaches, two, policy approximation may be easier than approximating value functions, and three, policy approximation methods can uncover stochastic optimal policies while value-based approaches cannot. RSs primarily employ two different approaches to policy gradients: REINFORCE and actor-critic approaches. The Monte Carlo (MC) stochastic gradient method REINFORCE modifies the policy weights directly. The REINFORCE algorithm's high variation and slow learning rate are crucial problems. These problems are caused by the MC nature of REINFORCE, which randomly selects samples. Numerous methods were employed to solve the issue of high variance in the REINFORCE-based RLRSs under study, including a baseline based on neural networks [69], first-order approximation (Satinder et al., 2000), the REINFORCE with baseline algorithm [55], and weight capping. It is unknown, nevertheless, how other REINFORCE-based RLRSs,

including [60], approach this problem. The actor-critic algorithm uses a critic rather than a baseline to address REINFORCE's challenges. It computes the value of the state-action pair supplied by the actor and offers feedback on how effective the action selected is. To be more precise, the critic is used to critique the policy created by the actor. These days, bootstrapping is a part of the policy gradient method. This lowers variance, speeds up learning, and produces a tolerable bias [18]. As already mentioned, the actor-critic algorithm is used in the well-known DRL method known as DDPG to handle continuous spaces. It is important to note that actor-critic is the second most popular RL algorithm among RLRSs [28].

7 Discussion and Future Works

RL algorithms were initially created to choose one action from a range of potential actions. However, making a list of suggested products is very frequent in the RS area, Slate, top-K, or list-wise recommendations are other names for this. The vast majority of the algorithms studied consider the problem of single-item recommendation, with a few exceptions [14, 31, 43, 47, 48]. Only references [43] thoroughly investigate this subject and modify their RL approach to address a variety of problems. When the RL agent encounters a large combinatorial action space in the future, the recommendation of a list of objects should be more thoroughly studied. However, there doesn't seem to be any reason to choose a particular RL algorithm in an RS application. Therefore, establishing a connection between the RL algorithm and the RS application is a crucial area of future research.

According to [70], an RS must be able to provide an explanation for their recommendations in addition to making them [70]. The user experience, system trust, and ability to make better decisions may all be enhanced by providing users with an explanation of the recommendations given [71–73]. Model intrinsic and model-agnostic explainable techniques can be divided into these two groups [74]. Unlike the latter, where the explanation is provided after the suggestion has been made, the former offers an explanation as part of the recommendation process. The approach we previously stated [69] may be an intrinsic explanatory approach. In contrast, RL is used to describe various recommendation algorithms as a model-neutral example [75]. The technique uses a pair of agents, one of which is in charge of coming up with explanations, while the other determines whether the user will find the explanation to be adequate. The failing RS [75] is being debugged, which is a fascinating use of explainable suggestions. In other words, using the provided explanations, we may trace the origin of systemic issues and identify the broken components. The fact that one reference [69] provides an explainable recommendation among the RLRSs evaluated in this study shows that there is a need for more research in this area. In conclusion, we believe that recommendations that can be explained are crucial for the development of future RSs and that RL can be utilized to provide better explanations.

And finally, RLRS evaluation needs to be improved. An RL agent must interact with the environment directly in order to learn what to do. This is comparable to RS online research in that the RS algorithm generates recommendations and gathers feedback from users in real time. However, the majority of the works use an offline study for evaluation, with the exception of a few methodologies presented [48]. This is particularly true given the high cost of doing online research and the significant risk associated with using an RLRS to optimize its recommendation technique for the majority of enterprises. Therefore, offline evaluation of RLRSs using accessible datasets or simulations is essential.

Future studies might concentrate on other topics. For instance, managing the global and local states of all users is one of the most important concerns. The development and implementation of RLRSs that can handle vast state spaces is a research area that is rarely investigated.

8 Conclusion

In this work, we provided a comprehensive and current survey of RLRSs. As a result of highlighting the role that DRL plays in shifting the focus of research in the RLRS sector, we divided the algorithms into two categories: RL- and DRL-based techniques. Following that, based on the RL algorithm used, such as Q-learning and actor-critic, each broad group was further broken into sub-categories. We believe that there has to be a lot more development made in the RLRS research field. Both RL and RSs are active research areas that huge businesses and industries are particularly interested in. As a result, we may expect exciting new models to appear every year. Last but not least, we think that our survey will aid future advancement in the field and understanding of key issues among scholars.

References

1. Klašnja-Milićević, A., Ivanović, M., Nanopoulos, A.: Recommender systems in e-learning environments: a survey of the state of the art and possible extensions. Artif. Intell. Rev. **44**(4), 571–604 (2015)
2. Schafer, J.B., Konstan, J., Riedl, J.: Recommender systems in e-commerce. In: ACM Conference on Electronic Commerce, pp. 158–166 (1999)
3. Sezgin, E., Ozkan, S.: A systematic literature review on health recommender systems. In: E-Health and Bioengineering Conference (EHB), pp. 1–4. IEEE (2013)

4. Karimi, M., Jannach, D., Jugovac, M.: News recommender systems survey and roads ahead. Inf. Process. Manag. **54**(6), 1203–1227 (2018)
5. Ricci, F., Rokach, L., Shapira, B.: Introduction to Recommender Systems Handbook. In: Ricci, F., Rokach, L., Shapira, B., Kantor, P. (eds.) Recommender Systems Handbook, pp. 1–35. Springer, Boston, MA (2011). https://doi.org/10.1007/978-0-387-85820-3_1
6. Zhang, S., Yao, L., Sun, A., Tay, Y.: Deep learning based recommender system: a survey and new perspectives. Comput. Surv. (CSUR) **52**(1), 1–38 (2019)
7. Jens Kober, J., Bagnell, A., Peters, J.: Reinforcement learning in robotics: a survey. J. Robot. Res. **32**(11), 1238–1274 (2013)
8. Meyes, R., et al.: Motionplanning for industrial robots using reinforcement learning. Procedia CIRP **63**, 107–112 (2017)
9. Navaei, M., Tabrizi, N.: Machine learning in software development life cycle: a comprehensive review. ENASE, pp. 344–354 (2022)
10. Sallab, A.E.L., Abdou, M., Perot, E., Yogamani, S.: Deep reinforcement learning framework autonomous driving. Electron. Imaging **2017**(19), 70–76 (2017)
11. You, C., Jianbo, L., Filev, D., Tsiotras, P.: Advanced planning for autonomous vehicles using reinforcement learning and deep inverse reinforcement learning. Robot. Auton. Syst. **114**, 118 (2019)
12. Jiang, Z., Xu, D., Liang, J.: A deep reinforcement learning framework for the financial portfolio management problem (2017). arXiv
13. Guez, A., Vincent, R.D., Avoli, M., Pineau, J.: Adaptive treatment of epilepsy via batch-mode reinforcement learning. In: AAAI, pp. 1671–1678 (2008)
14. Chen, M., Beutel, A., Covington, P., Jain, S., Belletti, F., Chi, E.H.: Top-k off policy correction for a reinforce recommender system. In: ACM International Conference on Web Search and Data Mining, pp. 456–464 (2019)
15. Smyth, B., Cotter, P.: A personalised TV listings service for the digital TV age. Knowl.-Based Syst. (2000)
16. Singh, S., Kearns, M., Litman, D., Walker, M.: Reinforcement learning for spoken dialogue systems. Neural Inf. Process. Syst. 956–962 (2000)
17. Tetreault, J., Litman, D.: Using reinforcement learning to build a better model of dialogue state. In: European Chapter of the Association for Computational Linguistics (2006)
18. Sutton, R.S., Bartom, A.G.: Introduction to Reinforcement Learning, vol. 2. MIT Press, Cambridge (2017)
19. Watkins, C.J.C.H.: Learning from delayed rewards (1989)
20. Krizhevsky, A., Sutskever, I., Hinton, G.E.: Image net classification with deep convolutional neural networks. Neural Inf. Process. Syst. 1097–1105 (2012)
21. Goldberg, D., Nichols, D., Terry, D., Oki, B.M.: Using collaborative filtering to weave an information tapestry. ACM **35**(12), 61–70 (1992)
22. Lillicrap, T.P., et al.: Continuous control with deep reinforcement learning (2015). arXiv
23. Wang, Z., Schaul, T., Hessel, M., Hasselt, H., Lanctot, M., Freitas, N.: Dueling network architectures for deep reinforcement learning. In International Conference on Machine Learning (2016)
24. Shani, G., Heckerman, D., Brafman, R.I., Boutilier, C.: An MDP based recommender system. Mach. Learn. Res. J. **6**(Sep), 1265–1295 (2005)
25. Dulac-Arnold, G., et al.:
26. Joachims, T., Freitag, D., Mitchell, T.: Webwatcher: a tour guide for the world wide web. In: IJCAI (1), pp. 770–777. Citeseer (1997)
27. Srivihok, A., Sukonmanee, P.: Ecommerce intelligent agent: personalization travel support agent using Q learning. In: 7th International Conference on Electronic Commerce, pp. 287–292 (2005)

28. Taghipour, N., Kardan, A., Ghidary, S.S.: Usage based web recommendations: a reinforcement learning approach. In: ACM Conference on Recommender Systems, pp. 113–120 (2007)
29. Mobasher, B., Cooley, R., Srivastava, J.: Automatic personalization based on web usage mining. ACM **43**(8), 142–151 (2000)
30. Thomas, P.S., Theocharous, G. Rojanavasu, P., Srinil, P., Pinngern, O.: New recommendation systemusing reinforcement learning. Spec. Issue Int. J. Comput. Internet Manag. **13**(SP 3) (2005)
31. Intayoad, W., Kamyod, C., Temdee, P.: Reinforcement learning for online learning recommendation system. In: 2018 Global Wireless Summit (GWS), pp. 167–170. IEEE (2018)
32. Chi, C.Y., Tsai, R.T.H., Lai, J.Y., Hsu, J.Y.J.: A reinforcement learning approach to emotion-based automatic playlist generation. In: 2010 International Conference on Technologies and Applications of Artificial Intelligence, pp. 60–65. IEEE (2010)
33. Choi, S., Ha, H., Hwang, U., Kim, C., Ha, J.W., Yoon, S.: Reinforcement learning based recommender system using biclustering technique (2018). arXiv preprint arXiv:1801.05532
34. Prelic, A., et al.: A systematic comparison and evaluation of biclustering methods for gene expression data. Bioinformatics **22**(9), 1122–1129 (2006)
35. Rodriguez-Baena, D.S., Perez-Pulido, A.J., Aguilar-Ruiz, J.S.: A biclustering algorithm for extracting bit-patterns from binary datasets. Bioinformatics **27**(19), 2738–2745 (2011)
36. Bohnenberger, T., Jameson, A.: When policies are better than plans: decision theoretic planning of recommendation sequences. In: International Conference on Intelligent User Interfaces, pp. 21–24 (2001)
37. Liebman, E., Saar-Tsechansky, M., Stone, P.: Dj-mc: a reinforcement earning agent for music playlist recommendation (2014). arXiv
38. Qi, F., Tong, X., Yu, L., Wang, Y.: Personalized project recommendations: using reinforcement learning. EURASIP J. Wirel. Commun. Netw. **2019**(1), 1–17 (2019). https://doi.org/10.1186/s13638-019-1619-6
39. Wang, Y.: A hybrid recommendation for music based on reinforcement learning. In: Lauw, H., Wong, R.W., Ntoulas, A., Lim, E.P., Ng, S.K., Pan, S. (eds.) Advances in Knowledge Discovery and Data Mining. PAKDD 2020. LNCS, vol. 12084, pp. 91–103. Springer, Cham (2020). https://doi.org/10.1007/978-3-030-47426-3_8
40. Zou, L., Xia, L., Ding, Z., Yin, D., Song, J., Liu, W.: Reinforcement learning to diversify Top-N recommendation. In: Li, G., Yang, J., Gama, J., Natwichai, J., Tong, Y. (eds.) Database Systems for Advanced Applications. DASFAA 2019. LNCS, vol. 11447, pp. 104–120. Springer, Cham (2019). https://doi.org/10.1007/978-3-030-18579-4_7
41. Zhao, Y., Zeng, D., Socinski, M.A., Kosorok, M.R.: Reinforcement learning strategies forclinical trials in nonsmall cell lung cancer (2011)
42. Shortreed, S.M., Laber, E., Lizotte, D.J., Scott Stroup, T., Pineau, J., Murphy, S.A.: Informing sequential clinical decision making through reinforcement learning: an empirical study. Mach. Learn. **84**(1–2), 109–136 (2011)
43. Theocharous, G., Thomas, P.S., Ghavamzadeh, M.: Personalized ad recommendation systems for lifetime value optimization with guarantees. In: Twenty-Fourth International Joint Conference on Artificial Intelligence (2015)
44. Vapnik, V.: The Nature of Statistical Learning Theory. Springer science & business media (2013)
45. Little, R.J.A., Rubin, D.B.: Statistical Analysis with Missing Data, vol. 793. John Wiley, Hoboken (2019)
46. Ernst, D., Geurts, P., Wehenkel, L.: Tree-based batch mode reinforcement learning. J. Mach. Learn. Res. **6**(Apr), 503–56 (2005)

47. Sunehag, P., Evans, R., Dulac-Arnold, G., Zwols, Y., Visentin, D., Coppin, B.: Deep reinforcement learning with attention for slate Markov decision processes with high dimensional states and actions (2015). arXiv preprint arXiv:1512.01124
48. Ie, E., et al.: Reinforcement learning for slate-based recommender systems: a tractable decomposition and practical methodology (2019). arXiv preprint arXiv:1905.12767
49. Nemati, S., Ghassemi, M.M., Clifford, G.D.: Optimal medication dosing fromsuboptimal clinical examples: a deep reinforcementlearning approach. Eng. Med. Biol. Soc. 2978–2981. IEEE (2016)
50. Raghu, A., Komorowski, M., Ahmed, I., Celi, L., Szolovits, P., Ghassemi, M.: Deep reinforcement learning for sepsis treatment (2017). arXiv preprint arXiv:1711.09602
51. Chen, X., Li, S., Li, H., Jiang, S., Qi, Y., Song, L.: Generative adversarial user model for reinforcement learning based recommendation system. In: International Conference on Machine Learning, pp. 1052–1061 (2019)
52. Chen, S.Y., Yu, Y., Da, Q., Tan, J., Huang, H.K., Tang, H.H.: Stabilizing reinforcement learning in dynamic environment with application to online recommendation. In: Proceedings of the 24th ACM SIGKDD International Conference on Knowledge Discovery & Data Mining (2018)
53. Yuyan, Z., Xiayao, S., Yong, L.: A novel movie recommendation system based on deep reinforcement learning with prioritized experience replay. In: 2019 IEEE 19th International Conference on Communication Technology (ICCT), pp. 1496–1500. IEEE (2019)
54. Zhao, X., Xia, L., Yin, D., Tang, J.: Model-based reinforcement learning for wholechain recommendations (2019). arXiv preprint arXiv:1902.03987
55. Casanueva, I., et al.: Deep reinforcement learning for recommender systems. In: 2018 International Conference on Information and Communications Technology (icoiact), pp. 226–233. IEEE (2018)
56. Hinton, G.E., Sabour, S., Frosst, N.: Matrix capsules with EM routing. In: International Conference on Learning Representations (2018)
57. Zhao, C., Hu, L.: CapDRL: a deep capsule reinforcement learning for movie recommendation. In: Nayak, A., Sharma, A. (eds.) PRICAI 2019: Trends in Artificial Intelligence. PRICAI 2019. LNCS, vol. 11672, pp. 734–739. Springer, Cham (2019). https://doi.org/10.1007/978-3-030-29894-4_59
58. Greco, C., Suglia, A., Basile, P., Semeraro, G.: Converse-et-impera: exploiting deep learning and hierarchical reinforcement learning for conversational recommender systems. In: Esposito, F., Basili, R., Ferilli, S., Lisi, F. (eds.) AI*IA 2017 Advances in Artificial Intelligence. AI*IA 2017. LNCS, vol. 10640, pp. 372–386. Springer, Cham (2017). https://doi.org/10.1007/978-3-319-70169-1_28
59. Kulkarni, T.D., Narasimhan, K., Saeedi, A., Tenenbaum, J.: Hierarchical deep reinforcement learning: integrating temporal abstraction and intrinsic motivation. Neural Inf. Process. Syst. 3675–3683 (2016)
60. Liang, H.: Drprofiling: deep reinforcement user pro ling for recommendations in heterogenous information networks. IEEE Knowl. Data Eng. (2020)
61. Liu, S., Chen, Y., Huang, H., Xiao, L., Hei, X.: Towards smart educational recommendations with reinforcement learning in classroom. In: International Conference on Teaching, Assessment, and Learning for Engineering, pp. 1079–1084. IEEE (2018)
62. Den Hengst, F., Hoogendoorn, M., Van Harmelen, F., Bosman, J.: Reinforcement learning for personalized dialogue management. In: International Conference on Web Intelligence (2019)
63. Fotopoulou, E., Zafeiropoulos, A., Feidakis, M., Metafas, D., Papavassiliou, S.: An interactive recommender system based on reinforcement learning for improving emotional competences in educational groups. In: Kumar, V., Troussas, C. (eds.) Intelligent Tutoring Systems. ITS 2020. LNCS, vol. 12149, pp. 248–258. Springer, Cham (2020). https://doi.org/10.1007/978-3-030-49663-0_29

64. Mahmood, T., Ricci, F.: Learning and adaptivity in interactive recommender systems. In: Conference on Electronic Commerce, pp. 75–84 (2007)
65. Preda, M., Popescu, D.: Personalized web recommendations: supporting epistemic information about end-users. In: The 2005 IEEE/WIC/ACM International Conference on Web Intelligence (WI'05), pp. 692–695. IEEE (2005)
66. Lin, L.-J.: Self-improving reactive agents based on reinforcement learning, planning and teaching. Mach. Learn. **8**(3–4), 293–321 (1992)
67. Thrun, S., Schwartz, A.: Issues in using function approximation for reinforcement learning. Connectionist Models Summer School Hillsdale. Lawrence Erlbaum, NJ (1993)
68. Yu, T., Shen, Y., Zhang, R., Zeng, X., Jin, H.: Vision-language recommendation via attribute augmented multimodal reinforcement learning. In: ACM International Conference on Multimedia, pp. 39–47 (2019)
69. Xian, Y., Fu, Z., Muthukrishnan, S., De Melo, G., Zhang, Y.: Reinforcement knowledge graph reasoning for explainable recommendation. In: ACM SIGIR Conference on Research and Development in Information Retrieval, pp. 285–294 (2019)
70. Zhang, Y., Chen, X.: Explainable recommendation: a survey and new perspectives (2018). arXiv:1804.11192
71. Cosley, D., Lam, S.K., Albert, I., Konstan, J.A., Riedl, J.: Is seeing believing how recommender system interfaces a ect users' opinions. In: Conference on Human Factors in Computing Systems, pp. 585–592 (2003)
72. Chen, L., Pu, P.: Trust building in recommender agents workshop on web personalization, Recommender Systems and Intelligent User Interfaces at the 2nd International Conference on E-Business, pp. 135–145. Citeseer (2005)
73. Tintarev, N., Mastho, J.: Exective explanations of recommendations: usercentered design. In: ACM Conference on Recommender Systems, pp. 153–156 (2007)
74. Lipton, Z.C.: The mythos of model interpretability. Queue **16**(3), 31–57 (2018)
75. Wang, X., Chen, Y., Yang, J., Wu, L., Wu, Z., Xie, X.: A reinforcement learning framework for explainable recommendation. In: Conference on Data Mining, pp. 587–596. IEEE (2018)
76. Barto, A.G.: Reinforcement learning and dynamic programming. In: Analysis, Design and Evaluation of Man Machine Systems, pp. 407–412. Elsevier (1995)

GAN-Powered Model&Landmark-Free Reconstruction: A Versatile Approach for High-Quality 3D Facial and Object Recovery from Single Images

Michael Danner[1] , Patrik Huber[3(✉)] , Muhammad Awais[1(✉)] ,
Matthias Rätsch[2(✉)] , and Josef Kittler[1(✉)]

[1] CVSSP, University of Surrey, Guildford, UK
{m.danner,muhammad.awais,j.kittler}@surrey.ac.uk
[2] ViSiR, Reutlingen University, Reutlingen, Germany
matthias.raetsch@reutlingen-university.de
[3] University of York, York, UK
patrik.huber@york.ac.uk

Abstract. In recent years, 3D facial reconstructions from single images have garnered significant interest. Most of the approaches are based on 3D Morphable Model (3DMM) fitting to reconstruct the 3D face shape. Concurrently, the adoption of Generative Adversarial Networks (GAN) has been gaining momentum to improve the texture of reconstructed faces. In this paper, we propose a fundamentally different approach to reconstructing the 3D head shape from a single image by harnessing the power of GAN. Our method predicts three maps of normal vectors of the head's frontal, left, and right poses. We are thus presenting a model-free method that does not require any prior knowledge of the object's geometry to be reconstructed.

The key advantage of our proposed approach is the substantial improvement in reconstruction quality compared to existing methods, particularly in the case of facial regions that are self-occluded in the input image. Our method is not limited to 3d face reconstruction. It is generic and applicable to multiple kinds of 3D objects. To illustrate the versatility of our method, we demonstrate its efficacy in reconstructing the entire human body.

By delivering a model-free method capable of generating high-quality 3D reconstructions, this paper not only advances the field of 3D facial reconstruction but also provides a foundation for future research and applications spanning multiple object types. The implications of this work have the potential to extend far beyond facial reconstruction, paving the way for innovative solutions and discoveries in various domains.

1 Research Problem

Reconstructing a 3D object from a single image is an important task in the field of computer vision and has many useful applications such as 3D animation, virtual reality, object recognition, and scene understanding [17]. 3D reconstruction from a single view

© The Author(s), under exclusive license to Springer Nature Switzerland AG 2023
D. Conte et al. (Eds.): DeLTA 2023, CCIS 1875, pp. 403–418, 2023.
https://doi.org/10.1007/978-3-031-39059-3_27

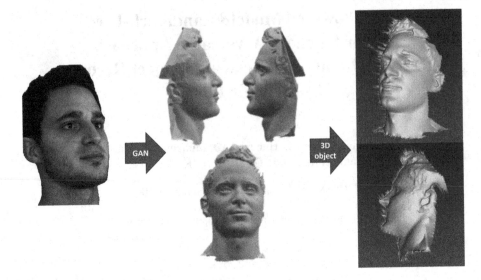

Fig. 1. Workflow: From single view image to three maps of normal vectors of the head's frontal, left and right poses to partial 3D representations.

involves recovering the 3D structure of an object from a single image of the object. This is a tremendously challenging task due to factors such as the diversity of object poses and the uncertainty of depth information. 3D Reconstruction of various objects from single or multiple images has been considered in the literature but faces are by far the most studied 3D object. In recent years, a 3D reconstruction of faces from single images using deep convolutional neural networks (DCNNs) has achieved remarkable results. Nevertheless, most of the methods of 3D facial reconstruction are based on parametric 3D morphable models (3DMM). However, 3DMM methods make specific assumptions about the geometry of the object and are more difficult to extend to generic 3D objects (Fig. 1).

Recent works investigate voxel [22], point cloud [12] and surface representations [8] for modelling the generic 3D objects as an alternative to 3DMM. Reconstructing a 3D object from a single image is an important task in the field of computer vision and has many useful applications such as 3D animation, virtual reality, object recognition, and scene understanding. 3D reconstruction from a single view involves recovering the 3D structure of an object from a single image of the object. This is a tremendous challenge due to factors such as the diversity of facial and body pose and the uncertainty of depth information.

In recent years, 3D reconstruction of an object from a single image using deep learning has achieved remarkable results. Many 3D-based models have made great strides in solving various tasks that have used 3D data directly, such as classification, segmentation of object parts and completion of 3D shapes. The availability of large amounts of data also encourages researchers to formulate and address the problem of reconstruction from a single source. Many approaches, with limitations and special assumptions on the input image, have been proposed to predict the 3D geometry. Recent work investigates voxel, point cloud and surface representations for modelling generic 3D objects. Volumetric methods were first used to infer the 3D structure of an object from a single view. However, the volumetric display suffers from a lack of information and high computations during the training process.

In the case of human faces and bodies, there are multiple 3D reconstruction approaches to recreate the poses and shapes. However, due to the economy and ambiguity of the 2D graphics features, this work is still a major challenge. The depth and scale of the 3D network forecast from a single RGB image are fuzzy, the facial expressions and emotions of faces are diverse, and the rotation of the joints of the human body is complex.

Most current methods of 3D facial reconstruction are based on parametric 3D morphable models, and an SMPL model is often used in human body reconstruction tasks. These models attempt to rely on an iterative optimisation scheme to estimate the 3D shapes in accordance with the 2D viewing angle. However, the mapping between the 3D shape and the parameters of the deformable human model is highly non-linear. The pose space is represented in the 3-D rotation, which can cause problems of non-minimal representation or discontinuity. In addition, the parametric methods usually lose accurate geometric information on the mould surface.

In this work, we are investigating the 3D reconstruction of a human face from a single 2D image. Our approach aims to not be dependent on these models, but instead to rely on the performance of current generative adversarial networks (GAN).

For this purpose, we propose a CNN model that solves the task of single-view reconstruction. The model has an encoder-generator form in which the encoder extracts useful features from the input image and three generators derive representations of normal vectors of the object shown in the 2D image. The three generators create a front view and two profile views to improve the quality of the reconstruction.

Overall, the most important contributions to this paper can be summarised as follows:

- A revolutionary new framework is proposed for the 3D reconstruction, with models and landmarks being completely dispensed with. In order to avoid the spatial loss caused by the parametric method, we generate several views of the object in normal vector representation, from which we can calculate the 3D mesh or point cloud.
- Due to the absence of dependencies on models, this approach can be transferred to any 3D object. The only requirement is the availability of sufficient training data for these objects.
- The proposed approach recreates the 3D mesh model of the human face or body from a single image. We evaluate the approach based on the mainstream data sets and compare the results with the state-of-the-art.

2   State of the Art

In this section, we review the relevant literature on the prediction and integration of normals and 3D object reconstruction methods, applicable to the human face and body.

Normal Integration. Computing the 3D shape of a surface from a set of normals is a classical problem in 3D reconstruction called normal integration, where a constant of integration must be specified. Various applications in computer vision such as shape from shading, photo-metric stereo vision and deflectometry are based on normal integration. The solution is not as simple as it might seem, even if the normals are known at every pixel of an image. Queau [18] presents most of the existing methods on normal integration and points out the problem of depth discontinuities which occur whenever there are occlusions.

A deep neural network approach to normal vector field generation has been introduced by Bouafif [4]. It predicts the map of normal vectors of a human face from a single photo and a 3D model is then generated from it. However, this work does not cope with occlusion and does not allow the depth parameters to be estimated.

3D Object Reconstruction. Inferring the 3D structure of an object from a single image is an ill-posed problem, but many attempts have been made using various approaches such as structure from motion [10] and SLAM [5]. However, ShapeFromX, where X can be shadow, texture, etc. requires prior knowledge of the nature of the input image. The most 3D data representations that are used in deep learning are volumetric data, meshes, and point clouds. 3D-GAN [25] proposed a generative adversarial network (GAN) to generate 3D objects from a probabilistic space using volumetric CNN. They mapped a low-dimensional probabilistic space to the 3D object space.

A parameterisation-based 3D reconstruction is proposed in [21] that generates geometry images which encode x; y; z surface coordinates. Three separate encoder-decoder networks were used to generate the shape images. The networks take an RGB image or a depth image as an input and learn the x; y; and z geometry images respectively.

To avoid the limitation of the volumetric representation, point clouds are used to represent 3D data. A 3D Point Cloud is a set of disordered 3D points that approximate the geometry of 3D objects.

3D Meshes are commonly used to represent 3D shapes. They are attractive for real applications as the shape details can be modelled accurately.

Human Face Reconstruction. Huber et al. [13] introduce a novel approach to fitting 3D Morphable Models using local image features, resolving a non-differentiable feature extraction operator issue through a learning-based cascaded regression method.

Sharma and Kumar [20] introduce a voxel-based 3D face reconstruction framework utilising deep sequential learning, encompassing variational autoencoder (VAE), bidirectional long short-term memory (BiLSTM), and triplet loss training techniques. Additionally, the fuzzy c-means clustering method addresses voxel sparsity from various poses, with VAE facilitating efficient dimensionality reduction and BiLSTM optimizing VAE embedding by leveraging the sequential data property of facial features.

With GANFIT, Gecer et al. [9] present a novel perspective on optimization-based 3D face reconstruction, harnessing the capabilities of modern machine learning techniques like GANs and face recognition networks as statistical texture models and energy functions respectively. Representing the first application of GANs for model fitting, the proposed approach demonstrates high-fidelity, identity-preserving 3D reconstructions, as evidenced by both qualitative and quantitative experiments.

Afzal et al. [1] introduce an automatic method for reconstructing a 3D image from a single input image, involving facial feature extraction, depth determination using multivariate Gaussian distribution, and detail recovery via the Shape from Shading technique. The proposed method demonstrates efficiency in recovering high details and aligns with the Basel face model, with the accuracy and efficiency verified through comparison with existing methods and examination of RMS error.

Human Body Reconstruction. BodyNet is an end-to-end multi-task network architecture capable of predicting 3D human body shape from a single image with intermediate tasks and volumetric regression significantly enhancing results. Varol et al. [24] present a flexible, extendable building block for future 3D body information applications, including virtual cloth-change, with a potential exploration of 3D tasks reliant on intermediate representations and 3D body shape understanding under clothing as future research directions.

Litany et al. [15] present a unique graph-convolutional approach for shape completion, exhibiting robustness to non-rigid deformations, low sample complexity during training, and versatility in reconstructing various styles of missing data. Evaluation results indicate promising prospects for real-world scan shape completion, with future work to focus on disentangling shape and pose representation, enhancing initialization for noisy real-world data, and addressing shape completion with unknown topology.

Zhang and Xiao [26] introduce a novel framework for reconstructing detailed 3D human body mesh, where 2D features are integrated into 3D space via perceptual feature transformation and flexible mesh deformation is facilitated by a coarse-to-fine graph convolution network. Acknowledging room for improvement in reconstructing invisible areas and reducing reliance on cumbersome 3D annotations, future research directions include incorporating a generative adversarial network to address ambiguities and employing weak supervision to lessen the dependency on 3D annotations.

Evaluation for 3D Mesh Registration. Cignoni, Rocchini and Scopigno [6] introduce Metro, a novel tool aimed at mitigating the limitations of many existing simplification methods by enabling the comparison of two surfaces, such as a complex triangulated mesh and its simplified counterpart, through surface sampling. Metro provides both numerical and visual results, showing error metrics and approximation error mapping respectively, offering a high degree of generality without making assumptions on the approach used to construct the simplified representation.

FAUST is a dataset introduced by Bogo et al. [3] created to evaluate 3D mesh registration techniques, containing the first high-resolution, real human meshes with ground-truth correspondences, demonstrating that real data registration poses significant challenges compared to synthetic datasets. Furthermore, we established ground-truth scan-to-scan correspondences through a new technique that utilises both shape and appearance information, yielding alignments accurate to within 2mm, thus creating a valuable resource for benchmarking and learning non-rigid shape models.

The Florence 2D/3D Hybrid Face Dataset from Bagdanov et al. [2] is designed for research in face recognition, that combines accurate 3D models and corresponding 2D videos of human faces, captured under a range of conditions. The dataset includes high-resolution 3D scans of multiple subjects, diverse video sequences at various resolutions and zoom levels, and captures subjects in controlled settings as well as less constrained indoor and outdoor environments, aimed to bridge the gap between 2D computer vision algorithms and 3D model-based methods.

The Headspace dataset [7] is a comprehensive collection of 3D human head images from 1519 subjects wearing latex caps, captured using a static 3D scanning system, with various auxiliary components including original PNG images, a FLAME head model fitting, and low-resolution pose-normalized JPEG images; the dataset, designed for non-commercial university-based research, was prepared under the guidance of the Alder Hey Craniofacial Unit and supported by QIDIS and Google Faculty Awards.

3   Methodology

Our main goal is to reconstruct a complete 3D shape of an object from a single image. Inspired by the work of Bouafif et al. [4], we introduce three important innovations to achieve a significant improvement in the quality of 3D face reconstruction.

1. As [4], we use a surface normal map estimated from the input 2D image as a stepping stone for 3D reconstruction, but importantly, we create three complementary views of the surface normal vector field to improve its estimate in self occluded parts of the face.
2. We develop a novel GAN architecture to generate the three normal vector fields using a shared CNN encoder backbone feeding three separate decoders.
3. We propose a method of 3D face reconstruction which fuses multiple normal vector fields into a single 3D shape estimate.

Our method takes a single RGB image as input and the network generates three normal vector maps as output. In the experiment of 3D head reconstruction, we recover normal vector maps for the frontal and two profile views and preserve identity from photo-realistic face images under different poses. The normal frontal and profile images are rendered from publicly available datasets using orthographic projection which is important for the reconstruction part using normal integration.

Firstly, we introduce the proposed network structure.

3.1   Network Structure

The proposed GAN architecture aims to generate, from any given facial input image, surface normal maps corresponding to the frontal and profile poses. A training data unit is composed of a rendered facial input image and three rendered ground-truth normal maps N_F, N_L, N_R. Both the input and output training images have $256 \times 256 \times 3$ pixel size and are stored in the PNG raster graphics file format. The input facial training images have been augmented by pitch and yaw rotation, different lighting conditions and different perspective settings while the output normal surface maps are always orthographic. Afterwards, the ground truth and generated normal surface maps are injected as the input of the discriminator to check if the generated maps are marked as real or fake.

The generator model is based on an encoder-decoder network. To adapt the model to our problem, the generator trains one encoder and three decoders for each output task, see Fig. 2 and 3.

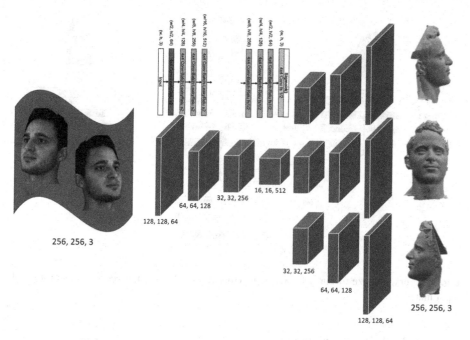

Fig. 2. Network architecture: Surface normal generator network.

3.2   Training Data

The deep learning training process we implemented for this study capitalises on a dataset comprised of approximately 700,000 rendered 3D head images. These images are utilised as the ground truth, and their surface normal representations form the crux of the training data. An example subset is shown in Fig. 4.

The training phase involved feeding the model with the rendered 3D head images. Each image was inputted into the model and propagated through the layers of the network, generating a prediction. The model's prediction was then compared to the ground truth image to compute the loss. This loss value was used to adjust the model's parameters during the backpropagation phase, with the aim of minimising the discrepancy between the predicted output and the ground truth.

After the initial training phase, the model underwent a fine-tuning process. This involved further training on a smaller learning rate to refine the model parameters and enable the model to better capture intricate details and complex patterns in the data, especially on the frontal and profile representation.

Through this rigorous deep learning training process, the model was effectively trained to create accurate surface normal representations from 3D head images, further expanding its potential applications in various fields requiring detailed and precise 3D reconstructions.

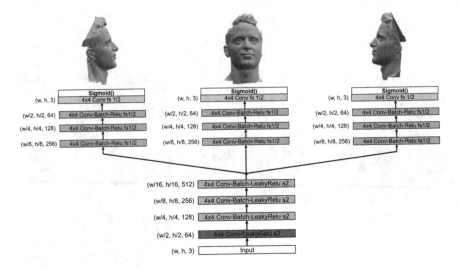

Fig. 3. Layers and size of our network architecture to train the normal vector poses of a facial input image.

3.3   3D Head Reconstruction

Our 3D reconstruction solution is based on the integration of normals. To facilitate the calculations, the network output is trained for orthographic projection by attaching a 3D frame to the camera whose origin is located at the optical centre, so that the camera and z axis coincide with the optical axis.

Normal Integration Using Quadratic Regularisation. At first, the gradients of the three output normal surface maps N_F, N_L, N_R are calculated where each normal map consists of the three normal components N_x, N_y and N_z. Having u, v as the pixels of the normal images, the gradients are calculated by:

$$g(u, v) = \begin{bmatrix} p(u, v) \\ q(u, v) \end{bmatrix} = \begin{bmatrix} -\frac{N_x(u,v)}{N_z(u,v)} \\ -\frac{N_y(u,v)}{N_z(u,v)} \end{bmatrix} \tag{1}$$

Assuming orthographic projection, a 3D-point projects orthogonally onto the image plane, so that $x(u, v) = u$ and $y(u, v) = v$. In our approach, we use quadratic regularisation to solve the integration problem. With $\nabla z = [\partial_u z, \partial_v z]^\top$ and $g = [p, q]^\top$ and $p = -\frac{n_1}{n_3}$ and $q = -\frac{n_2}{n_3}$ we have to minimise the function:

$$F_{L_2}(z) = \iint_{(u,v) \in \Omega} \| \nabla z(u, v) - g(u, v) \|^2 \, du dv \tag{2}$$

$\Omega \subset \mathbb{R}^2$ is the reconstruction domain which is used as a mask for our calculations.

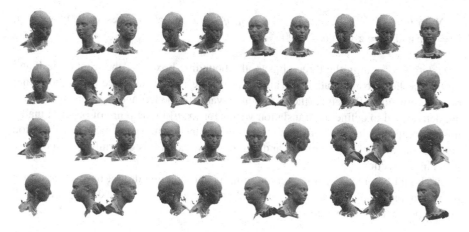

Fig. 4. Layers and size of our network architecture to train the normal vector poses of a facial input image.

Fig. 5. Normal integration. From left to right: p, q, Ω, resulting mesh

We use the algorithm described in [18] to compute the resulting mesh using quadratic regularisation with minor modifications. The mask that represents Ω is reduced by a small amount at the edge to eliminate discontinuities at the border, and the addition of Gaussian noise to the mask is skipped. Figure 5 shows the gradient matrices p and q as well as the fundamental mask and the resulting mesh.

Mesh Transformations. The meshes for frontal and side views have to be aligned in 3D space by rotation, translation vectors to move the mesh and stretching algorithms to converge the mesh outlines.

Rotation: To align the three meshes the two profile objects for left and right view have to be rotated by 90°, respectively −90°. The centre of rotation is the local origin of the coordinate system.

$$M' = M \times \begin{bmatrix} x \cdot \cos(\theta) - y \cdot \sin(\theta) \\ x \cdot \sin(\theta) + y \cos(\theta) \\ z \end{bmatrix} \tag{3}$$

With $\theta = 90°$ this results in $x' = -y$, $y' = x$ and $z' = z$ respectively $x' = y$, $y' = -x$ and $z' = z$ with $\theta = -90°$

Translation: To align the three objects, the bounding boxes for the frontal mesh and profile meshes are calculated. We aim to move the meshes in space so the bounding boxes fit together. Therefore, the minimum x-value of the left and right profile meshes are determined to define the translation vector for (x, y) to the point of origin. Finally, the minimum x-value of the frontal mesh is determined to move the mesh in x direction and the midpoint of the minimum and maximum value of y is determined to move the mesh in y direction.

It should be noted that the specific minima and maxima do not have to be in one plane and that the 3D objects can overlap slightly or not at all. These discrepancies will be rectified in the following steps.

Stretching: We develop a stretching algorithm that evaluates the factors for each slice in z direction, which is possible in this case, since the generated meshes only have vertices that have an integer value of z.

For each z plane we do the following:

Determine the point a with minimum $\min_{ry}$ of M_{right} where $x < \frac{x_{max}}{2}$. Determine the point a' with minimum $\min_{fy}$ of $M_{frontal}$ and calculate the distance d_{rf} from $\min_{fy}$ to $\min_{ry}$. Stretch M_{right} by multiplying its y component with the d_{rf} and a factor f_y.

Apply the same for M_{left} and the maximum values and points b and b'.

To stretch the frontal mesh $M_{frontal}$ calculate the distance from a to a' and the distance from b to b'. The average distance defines the factor to stretch the frontal mesh in x direction. See Fig. 6.

Shrink-Wrap Modifier. The shrink-wrap modifier allows an object to "shrink" to the surface of another object. It moves each vertex of the object being modified to the closest position on the surface of the given mesh [16]. This modifier is now used to merge the side meshes with the profile object. Finally, the three different meshes are merged into a single 3D object and an export function can save the result, in our experiments we used the STL file format.

3.4   3D Body Reconstruction

Using the example of the reconstruction of an entire human body, we show that the proposed procedure can easily be transferred to other tasks. These problems for face and body are usually based on different models and cannot be combined with one another. With this approach, it is possible for the first time to keep the entire work chain unchanged; the generative network only needs to be trained for the changed problem.

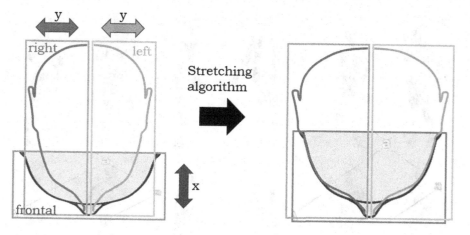

Fig. 6. Stretching algorithm.

4   Experiments

4.1   Datasets

The Headspace Dataset [7] is a set of 3D scanned images of the human head that we used in our first experiment. A series of rendered images is to be created for each person in the data set, with holes in the 3D mesh being cleaned up with an automatic fill-holes tool. The subjects are then rotated horizontally in 15-degree steps from -105 to $+105$ and additionally from -30 to $+30°$ vertically. This procedure is repeated several times with a variable perspective view and with the use of different lighting settings. As a result, we received a training data set of more than 400,000 images. As ground truth values for the GAN, the subjects are then rendered in the frontal and profile views with an orthographic perspective. To do this, the surface textures are replaced by the normal vectors as shaders for the rendering process.

For the human body 3D reconstruction task we use the MPI FAUST [3] training data set, augment the data with synthesised 3D bodies and evaluate the result on the MPI FAUST test data.

4.2   Evaluation

Evaluation of our proposed method in 3D face reconstruction on single images of the MICC dataset of 53 subjects. According to Tran [23] the estimated and ground truth shape is cropped at a radius of $95mm$ around the tip of the nose.

3D Shape Reconstruction on MICC Dataset. First, we would like to make available accurate and complete 3D models of faces to researchers who are primarily interested in the analysis of 3D meshes and textures of human faces. That is, our dataset is designed to be useful for research on pure 3D analysis techniques (Fig. 7).

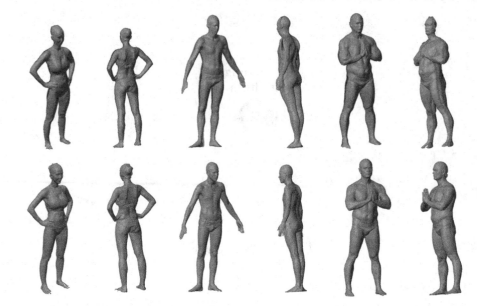

Fig. 7. Comparison of our qualitative results (top row) with the ground truth data (bottom row). A difficulty with this approach is the self-occlusion, as can be seen in the example of the hands in the last column, which cannot be solved by a frontal normal vector image without further information.

We have also designed our dataset to go beyond the scope of 3D analysis techniques, allowing researchers to investigate the possibility of reducing the gap between 2Da computer vision algorithms and those methods that work on more precise, 3D models. In particular, our dataset is thought of in the context of evaluating the use of 3D information in computer vision problems like 3D face pose estimation and 3D face recognition directly from video data or still images.

To this end, the pipeline of data acquisition is designed to provide both 3D data and 2D videos consistent with each other.

Hausdorff Distance. The Hausdorff distance is a widely used metric for quantifying the geometric difference between two 3D models in the realm of mesh processing. By providing a measure of similarity or dissimilarity between two given shapes, the Hausdorff distance plays a critical role in various applications, such as mesh registration, shape retrieval, and model comparison, see Fig. 8 (Table 1).

To compute the Hausdorff distance, a sampling approach is employed, which entails selecting a set of points on mesh X and, for each point x, identifying the closest point y on mesh Y. Mathematically, the Hausdorff distance (HD) between two point sets X and Y can be defined as:

$$HD(X, Y) = max(h(X, Y), h(Y, X)) \tag{4}$$

Table 1. 3D estimation accuracy on the MICC dataset. The top is single view methods, the bottom is multi-frame. 3DRMSE in real-world mm. Zhu [27], Tran [23], 3DMM [19], Huber et al. [14], Hassner et al. [11]

Method	3DRMSE	RMSE
Single view		
3DMM	1.75	3.64
Hassner	1.83	3.29
Tran	1.57	3.18
Multi frame		
3DMM	1.61	3.31
Zhu	1.89	3.84
Huber et al.	1.84	3.73
Tran	1.53	3.14
Model-free single view		
ours	1.88	3.82

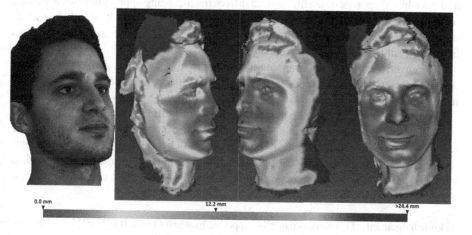

Fig. 8. Experimental outcomes for 3D facial reconstruction assessed using the Hausdorff distance metric. The Hausdorff distance values are expressed in real-world units, specifically in millimetres.

where $h(X, Y)$ is the directed Hausdorff distance from X to Y, calculated as:

$$h(X,Y) = d(x,y), max(x \in X), min(y \in Y) \tag{5}$$

Here, d(x, y) represents the Euclidean distance between points x and y. The directed Hausdorff distance from X to Y accounts for the most considerable minimum distance from any point in X to its closest counterpart in Y. By considering the maximum of the directed Hausdorff distances in both directions (from X to Y and from Y to X), the Hausdorff distance captures the most significant discrepancy between the two meshes.

It is important to note that the Hausdorff distance is sensitive to outliers and noise present in the 3D models. Therefore, in some applications, a modified or relaxed version of the Hausdorff distance, such as the partial Hausdorff distance or the average Hausdorff distance, may be more appropriate. These variants help mitigate the impact of outliers and provide a more robust measure of similarity between 3D models.

In summary, the Hausdorff distance serves as a valuable tool for computing the geometric difference between two 3D models, offering insights into their relative similarity and enabling the development of advanced mesh processing applications.

5 Conclusion

In conclusion, this conference paper presents a novel approach to 3D facial reconstruction by harnessing the power of Generative Adversarial Networks (GANs). By predicting three maps of normal vectors for the head's frontal, left, and right poses, our method offers a model-free solution that eliminates the need for prior knowledge of the object's geometry.

This innovative technique achieves a reconstruction quality that already comes close to model-based approaches. This method differs from existing approaches that are based on the adaptation of 3D morphable models. In further work, the results should be further improved and the possibility created to combine this approach with morphable models in order to achieve even higher accuracy.

Moreover, our approach transcends the realm of 3D facial reconstruction and demonstrates its adaptability for a wide range of 3D objects, as evidenced by its efficacy in reconstructing the entire human body. This versatility suggests potential applications across various fields, opening up new possibilities for future research and innovation.

Ultimately, this paper contributes a groundbreaking method to the field of 3D facial reconstruction, while simultaneously laying the groundwork for further advancements in object reconstruction across multiple domains. The potential impact of this work is vast, as it paves the way for innovative solutions and discoveries that could revolutionise not only facial reconstruction but numerous other fields as well.

Acknowledgement. This work is partially supported by a grant of the BMWi ZIM program, no. KK5007201LB0.

References

1. Afzal, H.M.R., Luo, S., Afzal, M.K., Chaudhary, G., Khari, M., Kumar, S.A.P.: 3D face reconstruction from single 2D image using distinctive features. IEEE Access **8**, 180681–180689 (2020). https://doi.org/10.1109/ACCESS.2020.3028106
2. Bagdanov, A.D., Del Bimbo, A., Masi, I.: The Florence 2D/3D hybrid face dataset. In: Proceedings of the 2011 Joint ACM Workshop on Human Gesture and Behavior Understanding, J-HGBU 2011, pp. 79–80. ACM, New York (2011). https://doi.org/10.1145/2072572. 2072597, https://doi.acm.org/10.1145/2072572.2072597
3. Bogo, F., Romero, J., Loper, M., Black, M.J.: FAUST: dataset and evaluation for 3D mesh registration. In: Proceedings of IEEE Conference on Computer Vision and Pattern Recognition (CVPR). IEEE, Piscataway (2014)

4. D Bouafif, O., Khomutenko, B., Daoudi, M.: Monocular 3D head reconstruction via prediction and integration of normal vector field. In: Farinella, G.M., Radeva, P., Braz, J. (eds.) Proceedings of the 15th International Joint Conference on Computer Vision, Imaging and Computer Graphics Theory and Applications, VISIGRAPP 2020, Volume 5: VISAPP, Valletta, Malta, 27–29 February 2020, pp. 359–369. SCITEPRESS (2020). https://doi.org/10.5220/0008961703590369

5. Bresson, G., Alsayed, Z., Yu, L., Glaser, S.: Simultaneous localization and mapping: a survey of current trends in autonomous driving. IEEE Trans. Intell. Veh. **2**(3), 194–220 (2017). https://doi.org/10.1109/TIV.2017.2749181

6. Cignoni, P., Rocchini, C., Scopigno, R.: Metro: measuring error on simplified surfaces. Comput. Graph. Forum **17**(2), 167–174 (1998). https://doi.org/10.1111/1467-8659.00236

7. Dai, H., Pears, N.E., Smith, W.A.P., Duncan, C.: Statistical modeling of craniofacial shape and texture. Int. J. Comput. Vis. **128**(2), 547–571 (2020). https://doi.org/10.1007/s11263-019-01260-7

8. Feng, L., Lai, J., Zhang, L.: 3D surface reconstruction based on one non-symmetric face image. In: Li, S.Z., Lai, J., Tan, T., Feng, G., Wang, Y. (eds.) SINOBIOMETRICS 2004. LNCS, vol. 3338, pp. 268–274. Springer, Heidelberg (2004). https://doi.org/10.1007/978-3-540-30548-4_31

9. Gecer, B., Ploumpis, S., Kotsia, I., Zafeiriou, S.: GANFIT: generative adversarial network fitting for high fidelity 3D face reconstruction. In: IEEE Conference on Computer Vision and Pattern Recognition, CVPR 2019, Long Beach, CA, USA, 16–20 June 2019, pp. 1155–1164. Computer Vision Foundation/IEEE (2019). https://doi.org/10.1109/CVPR.2019.00125

10. Häming, K., Peters, G.: The structure-from-motion reconstruction pipeline - a survey with focus on short image sequences. Kybernetika **46**(5), 926–937 (2010). https://www.kybernetika.cz/content/2010/5/926

11. Hassner, T.: Viewing real-world faces in 3d. In: IEEE International Conference on Computer Vision, ICCV 2013, Sydney, Australia, 1–8 December 2013, pp. 3607–3614. IEEE Computer Society (2013). https://doi.org/10.1109/ICCV.2013.448

12. Hg, R.I., Jasek, P., Rofidal, C., Nasrollahi, K., Moeslund, T.B., Tranchet, G.: An RGB-D database using microsoft's kinect for windows for face detection. In: Yétongnon, K., Chbeir, R., Dipanda, A., Gallo, L. (eds.) Eighth International Conference on Signal Image Technology and Internet Based Systems, SITIS 2012, Sorrento, Naples, Italy, 25–29 November 2012, pp. 42–46. IEEE Computer Society (2012). https://doi.org/10.1109/SITIS.2012.17

13. Huber, P., Feng, Z., Christmas, W.J., Kittler, J., Rätsch, M.: Fitting 3D morphable face models using local features. In: 2015 IEEE International Conference on Image Processing, ICIP 2015, Quebec City, QC, Canada, 27–30 September 2015, pp. 1195–1199. IEEE (2015). https://doi.org/10.1109/ICIP.2015.7350989

14. Huber, P., et al.: A multiresolution 3D morphable face model and fitting framework. In: Magnenat-Thalmann, N., et al. (eds.) Proceedings of the 11th Joint Conference on Computer Vision, Imaging and Computer Graphics Theory and Applications (VISIGRAPP 2016) - Volume 4: VISAPP, Rome, Italy, 27–29 February 2016, pp. 79–86. SciTePress (2016). https://doi.org/10.5220/0005669500790086

15. Litany, O., Bronstein, A.M., Bronstein, M.M., Makadia, A.: Deformable shape completion with graph convolutional autoencoders. In: 2018 IEEE Conference on Computer Vision and Pattern Recognition, CVPR 2018, Salt Lake City, UT, USA, 18–22 June 2018, pp. 1886–1895. IEEE Computer Society (2018). https://doi.org/10.1109/CVPR.2018.00202

16. van Overveld, K., Wyvill, B.: ShrinkWrap: an efficient adaptive algorithm for triangulating an ISO-surface. Vis. Comput. **20**(6), 362–379 (2004). https://doi.org/10.1007/s00371-002-0197-4

17. Peng, B., Wang, W., Dong, J., Tan, T.: Learning pose-invariant 3D object reconstruction from single-view images. Neurocomputing **423**, 407–418 (2021). https://doi.org/10.1016/j.neucom.2020.10.089

18. Quéau, Y., Durou, J., Aujol, J.: Normal integration: a survey. J. Math. Imaging Vis. **60**(4), 576–593 (2018). https://doi.org/10.1007/s10851-017-0773-x

19. Romdhani, S., Vetter, T.: Estimating 3D shape and texture using pixel intensity, edges, specular highlights, texture constraints and a prior. In: 2005 IEEE Computer Society Conference on Computer Vision and Pattern Recognition (CVPR 2005), 20–26 June 2005, San Diego, CA, USA, pp. 986–993. IEEE Computer Society (2005). https://doi.org/10.1109/CVPR.2005.145

20. Sharma, S., Kumar, V.: Voxel-based 3D face reconstruction and its application to face recognition using sequential deep learning. Multim. Tools Appl. **79**(25–26), 17303–17330 (2020). https://doi.org/10.1007/s11042-020-08688-x

21. Sinha, A., Unmesh, A., Huang, Q., Ramani, K.: SurfNet: generating 3D shape surfaces using deep residual networks. In: 2017 IEEE Conference on Computer Vision and Pattern Recognition, CVPR 2017, Honolulu, HI, USA, 21–26 July 2017, pp. 791–800. IEEE Computer Society (2017). https://doi.org/10.1109/CVPR.2017.91

22. Szeliski, R.: Rapid octree construction from image sequences. CVGIP: Image Underst. **58**(1), 23–32 (1993)

23. Tuan Tran, A., Hassner, T., Masi, I., Medioni, G.: Regressing robust and discriminative 3D morphable models with a very deep neural network. In: Proceedings of the IEEE Conference on Computer Vision and Pattern Recognition, pp. 5163–5172 (2017)

24. Varol, G., et al.: BodyNet: volumetric inference of 3D human body shapes. In: Ferrari, V., Hebert, M., Sminchisescu, C., Weiss, Y. (eds.) ECCV 2018, Part VII. LNCS, vol. 11211, pp. 20–38. Springer, Cham (2018). https://doi.org/10.1007/978-3-030-01234-2_2

25. Wu, J., Zhang, C., Xue, T., Freeman, B., Tenenbaum, J.: Learning a probabilistic latent space of object shapes via 3D generative-adversarial modeling. In: Lee, D.D., Sugiyama, M., von Luxburg, U., Guyon, I., Garnett, R. (eds.) Advances in Neural Information Processing Systems 29: Annual Conference on Neural Information Processing Systems 2016, 5–10 December 2016, Barcelona, Spain, pp. 82–90 (2016)

26. Zhang, S., Xiao, N.: Detailed 3D human body reconstruction from a single image based on mesh deformation. IEEE Access **9**, 8595–8603 (2021). https://doi.org/10.1109/ACCESS.2021.3049548

27. Zhu, X., Lei, Z., Yan, J., Yi, D., Li, S.Z.: High-fidelity pose and expression normalization for face recognition in the wild. In: Proceedings of the IEEE Conference on Computer Vision and Pattern Recognition, pp. 787–796 (2015)

GAN-Based LiDAR Intensity Simulation

Richard Marcus[1(✉)] [iD], Felix Gabel[1] [iD], Niklas Knoop[2], and Marc Stamminger[1] [iD]

[1] Chair of Visual Computing, Friedrich-Alexander-Universität Erlangen-Nürnberg,
Erlangen, Germany
{richard.marcus,felix.gabel,marc.stamminger}@fau.de
[2] e:fs TechHub GmbH, Gaimersheim, Germany
niklas.knoop@efs-techhub.com

Abstract. Realistic vehicle sensor simulation is an important element in developing autonomous driving. As physics-based implementations of visual sensors like LiDAR are complex in practice, data-based approaches promise solutions. Using pairs of camera images and LiDAR scans from real test drives, GANs can be trained to translate between them. For this process, we contribute two additions. First, we exploit the camera images, acquiring segmentation data and dense depth maps as additional input for training. Second, we test the performance of the LiDAR simulation by testing how well an object detection network generalizes between real and synthetic point clouds to enable evaluation without ground truth point clouds. Combining both, we simulate LiDAR point clouds and demonstrate their realism.

Keywords: LiDAR simulation · GAN · Autonomous driving

1 Introduction

Autonomous Driving is increasingly getting more attention in the automotive industry and research. A big part of the success depends on the quality of perception systems. LiDAR is a very important sensor for perception as it provides highly accurate distance measurements. Even though the capabilities of LiDAR sensors are increasing rapidly, they are still expensive by themselves and acquiring training data in large quantities is difficult.

Simulation environments provide a possible solution, since generating synthetic data to train perception systems is much cheaper than using real data and also allows the simulation of rare or dangerous situations. However, the quality of the simulated data is often not realistic enough to transfer the trained perception system to the real world. For the simulation of LiDAR sensors, ray tracing can be employed. But for this to generate realistic data, the simulation environment needs to have an extremely high degree of realism and the ray tracing implementation needs to account for all the effects that occur in real LiDAR sensors. Additionally, every specific LiDAR sensor has its own specifications and characteristics. Of particular interest is the reflection behavior of the

R. Marcus—Supported by the Bayerische Forschungsstiftung (Bavarian Research Foundation) AZ-1423-20.

D. Conte et al. (Eds.): DeLTA 2023, CCIS 1875, pp. 419–433, 2023.
https://doi.org/10.1007/978-3-031-39059-3_28

surfaces hit by the LiDAR rays. Materials with higher reflectivity result in a stronger signal, but can also cause light to be reflected away from the detector like a mirror. This gives important cues, comparable to a grayscale camera image. The respective measurement for this is the *LiDAR intensity*. While the strength of the detected signal also depends on the distance, sensors like the Velodyne HDL32E [23] return a distance adjusted value.

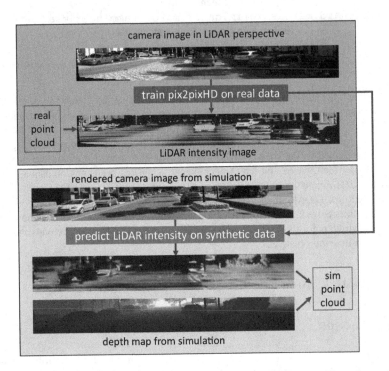

Fig. 1. Our LiDAR point cloud simulation approach consists of two steps: First, the neural network is trained on real data (blue box). Second, it is fed with synthetic data to simulate point clouds (green box). We show LiDAR intensity images using a color map for better visibility and keep depth maps in grayscale. (Color figure online)

An option to directly capture these effects is a learning-based approach that uses data from real test drives to simulate LiDAR sensors. If we can then simulate LiDAR intensities, we can use them to filter a depth map, which is easily available in simulation environments, so that we can obtain a realistic synthetic LiDAR point cloud in the end. Figure 1 summarizes this approach. Distinctive parts in the point cloud then become visible like in real LiDAR point clouds, which can improve performance in perception systems such as 3D object detection or semantic segmentation. For the image translation task, we use pix2pixHD [26], which follows the concept of General Adversarial Networks (GANs) [9]. We create a polar-grid image of the point cloud from the LiDAR sensor. To have a direct mapping, we also project the camera picture into the

LiDAR perspective so that each camera pixel corresponds to a LiDAR point. This process is described in greater detail in Sect. 3.1. As this results in two matching images, pix2pixHD can learn to translate from RGB to LiDAR intensity. While one use case for this would be to simulate LiDAR point clouds from real test drives without the presence of an actual LiDAR sensor, we focus on simulating realistic point clouds from synthetic data. If we then use the simulated data as input for training real world systems, we obtain a *Real2Sim2Real* pipeline that consists of the following steps:

- *Ground Truth:* Generate LiDAR intensity images from the reflectivity measurements in LiDAR point clouds (see Fig. 2).
- *Training Input:* Derive data from camera images (e.g., depth and segmentation maps) and project these images into perspective of the LiDAR sensor (see Fig. 3).
- *Real2Sim:* Train a neural network on real data to predict LiDAR intensity images. Using synthetic input data, the network can then simulate LiDAR point clouds (see Fig. 7).
- *Sim2Real:* Train real world perception systems (e.g., object detection) on LiDAR point clouds from simulated data.

Our contribution in this is that we provide segmentation and depth data in addition to the camera data to pix2pixHD. By doing so, we can recover information about surfaces in the scene where no returns have been measured by the LiDAR sensor, which enables performing the image translation task in the LiDAR perspective. Furthermore, we evaluate the quality of the simulated data beyond direct metrics between ground truth and generated data. Without this we would have to rely on visual comparisons as no LiDAR ground truth point clouds are available for the synthetic data. Thereby, we demonstrate that by leveraging real data, our simulation produces realistic LiDAR point clouds.

After establishing the foundations in the related work, there are two core parts in our paper that are both evaluated, corresponding to our pipeline. First, the simulation of LiDAR intensities in Sect. 3 and secondly, the generation of synthetic point clouds in Sect. 4. This is followed by a discussion of the implications for practical use cases in developing autonomous driving systems and ends with the conclusion.

2   Related Work

Simulation Environments. There are multiple capable simulation environments to test and develop advanced driving assistance and autonomous driving systems. However, open source tools like CARLA [5] or AirSim [22] only have basic sensor implementations, which are not sufficient for realistic simulation. Vista [2] on the other hand employs data-based sensors, but is less focused on optimal integration into driving simulation workflows. Commercial products, especially NVIDIA DRIVESim [17], are promising but usually employ high quality physics-based implementations. Our approach on the other hand is supposed to be integrated easily without requirements to the simulation environment and should allow simple implementation of new sensors. A further relevant approach is Pseudo-LiDAR [27,28]. Here, predicted depth maps are converted and down sampled to mimic LiDAR point clouds so that they can be used for object detection. We make use of this concept to generate point clouds from synthetic depth maps.

Image2Image Translation. For learning-based LiDAR simulation, Image2Image translation has shown promising results. This can be achieved effectively with GANs as shown by pix2pix [13]. In contrast to this, CycleGAN [29] can translate between two domains without the need for paired data. Further improvements for paired images were made by pix2pixHD [26] and SPADE [19]. Vid2Vid [25] improves the temporal consistency for a video sequences and World-Consistent Video2Video Translation [14] further proposes solutions for long-term consistency. Even though, test drive data is most of the time given as video sequences, approaches that process multiple consecutive frames simultaneously are problematic for LiDAR intensity, because the reflected light strongly depends on the angle. As this changes when the car is moving, warping between frames cannot be simply performed to improve the training. Additionally, using the camera images as input already gives good data for temporal consistency.

LiDAR Simulation. Methods for LiDAR simulation based on machine learning use similar approaches. LiDARSim [15] learns a mapping between a reconstructed dense point cloud and individual scans using a U-Net [20] based architecture. However, this approach makes it difficult to integrate the LiDAR simulation in simulation environments. There also is research regarding unpaired training, in this case using simulated unrealistic and real point clouds or point clouds from different sensors [21]. While this offers great flexibility and various use cases, there is a conceptual problem to learn the LiDAR behavior for specific materials or objects in the scene.

This behavior can be learned with paired GAN-based approaches that utilize camera images during training [10,16]. These approaches operate in the camera perspective, but the former further uses depth and segmentation images as additional input. The latter also simulates LiDAR intensity in addition to dropping rays. Our approach incorporates both concepts but performs the image translation in the LiDAR perspective, which has the distinct advantage that there is no need for upscaling the LiDAR image to the camera resolution. Aside from usual interpolation artifacts, LiDAR data has the problem that gaps can either result from the sampling pattern of the sensor or the reflection properties of the surfaces hit. Only blurring the LiDAR points avoids larger interpolation errors, but also prevents the network to discern between the two different kinds of gaps. Fully matching the resolution of the camera, on the other hand, tends to also fill reflection gaps in the LiDAR data. As our method transforms the camera images into the LiDAR perspective, it does not suffer from these problems. In addition, training can be completed faster because only the lower resolution of the LiDAR sensor is used.

Training Data. For training the network, we focus on the KITTI [7,8] dataset, as it includes various benchmarks and hand labeled data, which is helpful for training on high quality data, so the network can generalize well to the perfect synthetic samples and gives many options for evaluation. It uses a Velodyne [24] HDL-64E LiDAR sensor with 64 laser rows. Furthermore, VKITTI [3,6] creates a virtual model of five sequences of the KITTI dataset in a semiautomatic process. While the result is not a direct digital twin, as the geometry and object placement differs, it still provides a good starting point for comparative evaluations. VKITTI provides stereo color images and corresponding depth, class segmentation and instance segmentation maps. It also contains the extrinsic and intrinsic camera parameters for each frame and bounding boxes for the cars. Thus, we make this the basis for our synthetic data.

3 Simulating LiDAR Intensity

To leverage the simulation environment, a data-based sensor should make use of the available ground truth data. The generation of realistic materials is often challenging and would be even more problematic to obtain from real world sequences. Instead, we focus on readily available data: RGB renderings (camera images), depth, semantic and instance segmentation masks.

3.1 Generating the Training Data

As these are supposed to be used during inference in the simulation, the training also needs to be performed with these modalities as input. The ground truth for the training is obtained from only the LiDAR data.

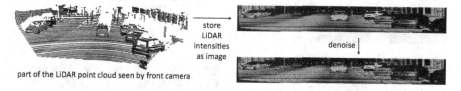

Fig. 2. Data processing and generation for LiDAR intensity simulation with pix2pixHD.

LiDAR Data Processing. To transform a LiDAR point cloud into an image without sampling gaps, we must recover the way the points should have been measured. Given an elevation angle and the azimuth (resulting from the laser arrays and rotating sensor), a LiDAR point is located at the detected distance at these angle coordinates. However, if we reverse this and try to map the individual points to pixel coordinates, we notice that this is not consistent. There can be gaps in the resulting image, which only should be the case when there is no detection at all. While the columns are quite accurate, the points are often registered at the wrong row. To obtain a dense image, we exploit that the points in the KITTI files are ordered linewise. Consequently, jumps between quadrants allow assignment of rows for each point.

As our goal is to learn a mapping between camera and LiDAR , only the overlapping parts can be used. The camera only covers about 80 degree of the 360-degree LiDAR panorama and the lowest rows of the sensor are blocked by the ego car and also not in the camera perspective. This effectively leads to a crop of 372 × 44 pixels, where each pixel gives a LiDAR intensity and lies inside the camera view. As a last step, we denoise the polar-grid LiDAR images with LIDIA, a universal learned denoiser, as shown in Fig. 2. Even though noise is part of realistic LiDAR intensity, we decided to exclude it from our experiments as good as possible to focus on how well the network can learn the ideal reflective properties of materials. However, this could be analyzed in future work, as different materials could produce specific noise patterns in the measured intensities.

Camera Data Processing. For the real input data, we need to start from the available sensors: LiDAR and RGB camera. Based on these, we can predict dense depth maps via the depth completion network PENet [12]. Depth completion means that the network uses the LiDAR point cloud as basis to predict a dense depth map. Hence, we not only use LiDAR data in the ground truth but also as input. This is no issue as we strictly want to predict reflection behavior in the form of LiDAR intensity. Perfect depth maps are already available in the simulation environment for inference, later. The predicted depth maps have a slightly lower resolution of 1216×352 pixels due to the training process, so we crop the original camera image as well.

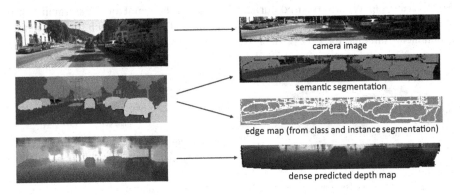

Fig. 3. Processing KITTI images for training: a projection from camera perspective to LiDAR perspective is necessary.

For semantic and instance segmentation, we rely on manually labeled data, as given for the Semantic Instance Segmentation Evaluation Benchmark [1]. However, this could also be achieved with an appropriate neural network. Instead of directly using the instance segmentation maps for training, we follow pix2pixHD and generate edge maps from the segmentation and instance data.

To project the camera data into the LiDAR perspective, we also need depth information for every pixel. Here, we can use the same predicted depth map that is one of the inputs for learning the intensity prediction. Due to the offset between camera and LiDAR sensor on the vehicle, there are parts in the image that were not seen by the LiDAR because of occlusions. We detect these areas generously by checking whether multiple pixels from the camera image would fall into the same LiDAR pixel coordinate and mask out occluded areas. These masks are then added as *don't care* labels to the segmentation masks. When projecting the camera images, we also adjust the resolution to the LiDAR crop (372 × 44) so that each LiDAR ray corresponds to one camera pixel, as shown in Fig. 3.

3.2 Results and Evaluation

We use the pix2pixHD implementation from Imaginaire [18] and train for 20 epochs with batch size 8, enable the local enhancer network, but deactivate horizontal image flipping during training. Otherwise, we keep the default configuration.

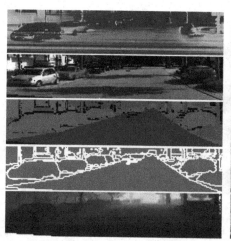 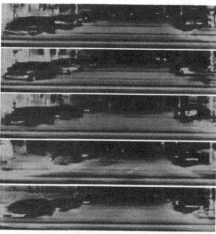

(a) First row: ground truth LiDAR intensity, below: input images. Images are cropped slightly for better visibility.

(b) First row: LiDAR intensity prediction from combined input, below: predictions from individually trained networks.

Fig. 4. Qualitative evaluation with sequence from the KITTI validation split using different input data configurations.

We train on 16 sequences (about 6000 images) of the training and validation data of the segmentation benchmark and use the five sequences (about 2000 images) that are also modeled in VKITTI for evaluation. In particular, we compare between different runs of camera images and derived images individually, as well as one variant where we use all available input data. Images are always scaled to a resolution of 512 × 64. As metric, we use the Fréchet inception distance (FID) [11], which has proven to be an effective way to compare similarity between generated and real images. The numerical results can be seen in Table 1 and the corresponding images in Fig. 4.

In general, the numbers show a clear picture of what the network learns from the data and how this generalizes to unseen data. Class segmentation and depth provide information that applies very similarly to materials and geometry across scenes and thus promote generalization capabilities. The same is true for the edge maps in theory, albeit they contain considerably less information by themselves. The RGB camera

Table 1. FID between ground truth and predictions on validation set at epoch 20.

Input Data	RGB	Semantic	Edges	Depth	Combined
FID	89.44	77.01	102.61	81.43	73.81

images behave very differently. They are full of high frequent information, which also makes it very hard for the network to apply on new data. The value of including RGB images is not diminished by this, as they provide valuable information for simulating intensities that correspond to the specific materials in the frame and also help with temporal stability.

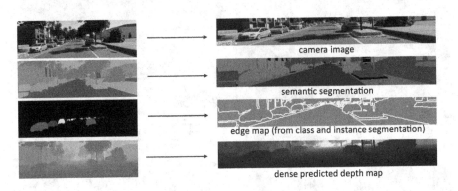

camera image

semantic segmentation

edge map (from class and instance segmentation)

dense predicted depth map

Fig. 5. Processing VKITTI images for training: instead of projection, we directly downsample the images to a resolution of 372 × 44 pixels.

Generalization on Synthetic Data. VKITTI provides the necessary data, but it has to be adjusted slightly to adhere to the training format. The depth maps need to be scaled into the same range as in the completed depth images. The color-based classes of VKITTI need to be mapped to KITTI class identifiers. The color-based instance segmentation maps of VKITTI need to be mapped to subsequent numbers. The conversion results are shown in Fig. 5. Even though the training is performed on data converted into the LiDAR perspective, the network itself learns a mapping between camera-derived data and LiDAR intensity. This means that we can directly evaluate the network on the synthetic images. As long as the aspect ratio and resolution are comparable, the generated data is plausible. Without further projection, this would simulate a LiDAR sensor that has the same position as the camera and a resolution corresponding to the training resolution. A higher resolution LiDAR could be simulated by simply sampling a different set of pixels from the source image.

We show the results of the networks trained on the different kinds of real world input images with synthetic data input from VKITTI in Fig. 6, using the corresponding KITTI validation sequence. This is the same sequence for which we have already showed the predicted intensities in Fig. 4, but there are notable differences between the real and the synthetic scene itself. Still, the predictions align reasonably well, especially considering that there is a domain gap, not only between camera and rendered image but also between perfect and predicted depth maps as well as between the segmentation maps because of different class labels. However, we cannot evaluate the performance quantitatively in the same way. We provide a solution to this by creating point clouds from the predicted intensity and analyze them for the task of training object detection.

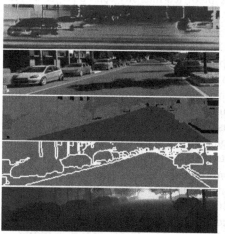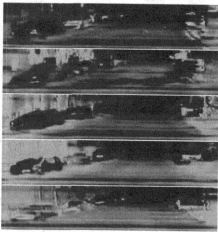

(a) First row: ground truth LiDAR intensity (KITTI), below: input images. Images are cropped slightly for better visibility.

(b) First row: LiDAR intensity prediction from combined input, below: predictions from individually trained networks.

Fig. 6. Qualitative evaluation with the VKITTI sequence corresponding to the KITTI sequence in Fig. 4 using different input data configurations.

4 Object Detection with Simulated LiDAR Point Clouds

To further evaluate the realism of the simulated point clouds, we train an object detection network on VKITTI and evaluate the performance on real KITTI data. Furthermore, we also evaluate a model trained on real KITTI data on the synthetic point clouds. In each case, the network architecture is Voxel-R-CNN [4]. The unstructured point clouds are first encoded into voxels. This means that it should be rather robust against sensor noise or aliasing from the depth projection in synthetic point clouds.

Generating Point Clouds from Synthetic Data. In order to use the VKITTI dataset analogously to the KITTI dataset, we have to generate the point clouds from the depth maps and reconstruct the calibration files. For the image files, we use the RGB images of the left VKITTI camera (Camera_0) and generate the 3D and 2d bounding box information in the correct format with the information from the supplied text files. We remove bounding boxes that exceed a maximum distance of 80 m and exclude scenes that contain no objects completely.

In contrast to the real KITTI data, we define no objects as completely visible, because VKITTI gives very accurate occlusion values, which however do not exceed a visibility of 90This has no effect on the training process itself, only during evaluating on the VKITTI sequences there will be no easy category for evaluating object detections. Instead, these samples will be part of the moderate difficulty.

For a specific LiDAR sensor position, there are three possibilities. First, the point cloud generated from the depth map and the predicted intensity can be projected into the LiDAR coordinate system. Second, equivalent to the training procedure, the camera data can be projected first and in this form be fed into the network. Last, an additional camera could be simulated at the desired LiDAR position. This should in most cases be the most accurate solution, but cannot be applied on already existing data like VKITTI. For our evaluation, we choose the first option and convert the depth maps to point clouds following Pseudo-LiDAR [27]. We make one important modification: the intensity of LiDAR points is set to 0 instead of 1. Training on real point clouds and evaluating on synthetic point clouds with intensity 1 causes the detection performance to collapse because the detector then considers every point as retro-reflective. Adding the intensity simulation enables us to assign an intensity to points or drop them completely with low intensity. We apply this conservatively and only drop points with 0 intensity (see Fig. 7). We only use the network once for each frame and upscale the low resolution intensity map to the cropped KITTI resolution of 1216×352, which assigns an intensity of 0 to the border pixels from the original KITTI and VKITTI resolution of 1242×375 pixels. This allows using the full available resolution of the depth map. After generating a point cloud from it, we sparsify the point cloud according to the desired sensor specifics following the Pseudo-LiDAR method [28].

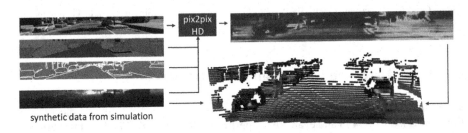

synthetic data from simulation

Fig. 7. Generating synthetic point clouds from depth maps and predicted LiDAR intensity.

Generating point clouds from VKITTI results in about 2000 samples for training. We generate three different sets of simulated VKITTI point clouds and compare them against real KITTI point clouds, following the procedure above: one with 32 lines and two with 64 lines, once without further processing and once with dropping points according to the predicted intensity.

(a) Real point cloud (blue) and point cloud from projected synthetic depth (red).

(b) Using simulated intensity in addition to the depth (green).

Fig. 8. Comparison between synthetic and real point clouds. Using the same image area as during training causes the lowest LiDAR rays to be cut off.

The reason for also evaluating a variant with only 32 lines is to gain insights whether the observed effects could also simply result from having less points overall. Visualizations of the 64 line point clouds are given in Fig. 8. This also shows the geometry discrepancy between VKITTI and KITTI. However, this is not a problem for comparing the general object detection performance, because we do not compare point clouds individually. The point cloud with simulated intensity on the right (8b) matches the real point cloud structure, having more sparsity in the distant parts and around reflective and transparent surfaces.

To analyze the results, we follow the KITTI object detection benchmark, which comes with 7481 point cloud and labels that are roughly split in half for training and evaluation. It divides the samples for evaluation into easy, moderate and hard depending on the size, occlusion and truncation in the camera image. The network outputs 3D and 2D bounding boxes, a top-down bird's eye view detection (bev) and average orientation similarity (aos), while we follow the KITTI object evaluation to calculate the detection scores. We always train Voxel-RCNN for 80 epochs for stable results and average the last 5 epochs to account for remaining fluctuations between epochs.

Evaluation on KITTI. In the first experiment, we have trained a Voxel R-CNN on KITTI point clouds as well as all types of our synthetic point clouds. All of these models are then evaluated on the real data, see Table 2. First, we notice that there is a significant gap between the models trained on real and synthetic data for generalization. The difficult part is to determine how much can be attributed to the individual scene differences, point cloud structure or data set selection. We see that simply reducing the point density to 32 lines decreases performance. In theory, an increase could have been possible if there were no sparse car samples in the synthetic for the network to learn to detect them in real data. The *IntensitySim* point clouds, on the other hand, also consist of fewer points, but the detection performance increases for almost every metric, which means that points have been dropped correctly. Only for the easy samples, the performance

Table 2. Voxel R-CNN Objected detection evaluation on KITTI point clouds, the first column specifies the data set used for training.

easy	3D@0.70	bev@0.70	3D@0.50	bev@0.50	2D	aos
VKITTI (64 Lines)	63.78	75.32	79.93	80.24	79.24	78.58
VKITTI (32 Lines)	61.4	73.61	78.28	78.62	77.3	76.62
IntensitySim (64)	63.22	80.33	85.69	86.15	83.83	82.1
KITTI	91.94	95.36	98.51	98.55	98.46	98.43
moderate	3D@0.70	bev@0.70	3D@0.50	bev@0.50	2D	aos
VKITTI (64 Lines)	50.5	62.43	68.54	68.91	66.59	65.05
VKITTI (32 Lines)	47.56	59.57	65.16	65.51	63.73	61.81
IntensitySim (64)	51.74	70.77	75.51	79.71	72.78	70.54
KITTI	82.9	91.19	94.78	95.46	94.65	94.52
hard	3D@0.70	bev@0.70	3D@0.50	bev@0.50	2D	aos
VKITTI (64 Lines)	48.7	62.07	69.01	70.02	66.84	65.05
VKITTI (32 Lines)	45.2	58.15	64.04	65.42	62.9	60.93
IntensitySim (64)	50.27	69.74	74.76	79.07	71.95	69.42
KITTI	80.45	88.94	94.49	94.59	92.34	92.14

slightly decreases compared to the regular VKITTI point cloud. This is not unexpected because the detection of these rely on a representative number of dense car samples in the training set.

Table 3. Voxel R-CNN Objected detection evaluation on VKITTI point clouds, the first column specifies the data set used for training (no samples for easy difficulty).

moderate	3D@0.70	bev@0.70	3D@0.50	bev@0.50	2D	aos
VKITTI (64 Lines)	88.78	95.79	98.94	99.25	98.59	97.89
VKITTI (32 Lines)	72.93	90.74	94.3	97.54	89.78	88.48
IntensitySim (64)	86.33	99.11	99.89	99.92	99.78	98.35
KITTI	82.9	91.19	94.78	95.46	94.65	94.52
hard	3D@0.70	bev@0.70	3D@0.50	bev@0.50	2D	aos
VKITTI (64 Lines)	77.18	87.84	92.47	93.94	90.59	89.51
VKITTI (32 Lines)	61.2	80.42	85.12	89.96	79.68	77.85
IntensitySim (64)	81.19	95.02	96.89	96.97	95.97	94.37
KITTI	80.45	88.94	94.49	94.59	92.34	92.14

Training on KITTI and Evaluation on VKITTI. The second experiment evaluates the detection performance of a model trained on real KITTI data when applied on synthetic data, in our case the VKITTI point clouds. With the perfect synthetic data, object detection is often too exact in simulation environments. On the other hand, there might be a domain gap, that prevents the network from reaching the optimal performance. The data in Table 3 shows that the latter largely is the case. This is significant as the performance increases despite having less points per sample for detection. Furthermore, it indicates that the intensity was simulated realistically. Again, simply reducing the point cloud resolution causes the network trained on higher resolution KITTI point clouds to miss many objects.

5 Discussion

When designing a system that in the end is supposed to bring autonomous driving functions to the street, we carefully have to consider the limitations of the simulation capabilities. In our case, we want the network to learn a mapping between data from camera and LiDAR . Yet, the information given in single camera frames cannot determine the reflection behavior of the materials seen by the LiDAR . Materials exist, that look exactly the same in the camera images but not for the LiDAR . Trying to test specific situations with such a sensor in the simulation is therefore a problematic approach when we want to replicate real world behavior exactly. Conversely, the data-based sensor can be used to generate diverse virtual training data.

Going back from this general observation to the data of our experiments, we consider improving the training data as very important. For the real data set, higher resolution LiDAR sensors could improve the training drastically and the data projection methods can be optimized. Here, both the predicted depth and the occlusion masking can introduce artifacts in the training data. While VKITTI offered interesting insights because of its similarity, having a dynamic simulation environment would allow greater flexibility for generating synthetic training data and thus also experiments that investigate how to optimize generalization from synthetic to real data, in particular, when it is based on real data.

6 Conclusion

We have proposed a pipeline to simulate LiDAR point clouds, including intensity, from real data and validate the realism by observing the generalization of an object detection network. In particular, we employed pix2pixHD for Image2Image translation and Voxel-R-CNN for object detection using KITTI point clouds and synthetic data from VKITTI as starting points. Converting these appropriately resulted in reasonable visual quality and FID scores, showing that using all available data as input performed best. The data from the object detection experiments strongly indicates that our LiDAR simulation approach can increase the realism of the synthetic point clouds, thus creating a valuable starting point for evaluating different data configurations.

References

1. Abu Alhaija, H., Mustikovela, S.K., Mescheder, L., Geiger, A., Rother, C.: Augmented reality meets computer vision: efficient data generation for urban driving scenes. Int. J. Comput. Vis. **126**(9), 961–972 (2018). https://doi.org/10.1007/s11263-018-1070-x

2. Amini, A., et al.: Vista 2.0: an open, data-driven simulator for multimodal sensing and policy learning for autonomous vehicles (2021)
3. Cabon, Y., Murray, N., Humenberger, M.: Virtual kitti 2 (2020)
4. Deng, J., Shi, S., Li, P., Zhou, W., Zhang, Y., Li, H.: Voxel R-CNN: towards high performance voxel-based 3d object detection. arXiv:2012.15712 (2020)
5. Dosovitskiy, A., Ros, G., Codevilla, F., Lopez, A., Koltun, V.: CARLA: An Open Urban Driving Simulator. arXiv:1711.03938 [cs] (2017)
6. Gaidon, A., Wang, Q., Cabon, Y., Vig, E.: Virtual worlds as proxy for multi-object tracking analysis (2016)
7. Geiger, A., Lenz, P., Stiller, C., Urtasun, R.: Vision meets robotics: the KITTI dataset. Int. J. Rob. Res. 32(11), 1231–1237 (2013). https://doi.org/10.1177/0278364913491297
8. Geiger, A., Lenz, P., Urtasun, R.: Are we ready for autonomous driving? The KITTI vision benchmark suite. In: 2012 IEEE Conference on Computer Vision and Pattern Recognition, pp. 3354–3361 (2012). ISSN: 1063–6919
9. Goodfellow, I.J., et al.: Generative adversarial networks (2014)
10. Guillard, B., et al.: Learning to simulate realistic lidars (2022)
11. Heusel, M., Ramsauer, H., Unterthiner, T., Nessler, B., Hochreiter, S.: GANs trained by a two time-scale update rule converge to a local Nash equilibrium (2018)
12. Hu, M., Wang, S., Li, B., Ning, S., Fan, L., Gong, X.: PENet: towards precise and efficient image guided depth completion. In: 2021 IEEE International Conference on Robotics and Automation (ICRA), pp. 13656–13662. IEEE (2021)
13. Isola, P., Zhu, J.Y., Zhou, T., Efros, A.A.: Image-to-Image Translation with Conditional Adversarial Networks. arXiv:1611.07004 [cs] (2018)
14. Mallya, A., Wang, T.-C., Sapra, K., Liu, M.-Y.: World-consistent video-to-video synthesis. In: Vedaldi, A., Bischof, H., Brox, T., Frahm, J.-M. (eds.) ECCV 2020. LNCS, vol. 12353, pp. 359–378. Springer, Cham (2020). https://doi.org/10.1007/978-3-030-58598-3_22
15. Manivasagam, S., et al.: LiDARsim: Realistic LiDAR Simulation by Leveraging the Real World. arXiv:2006.09348 [cs] (2020)
16. Marcus, R., Knoop, N., Egger, B., Stamminger, M.: A lightweight machine learning pipeline for LiDAR-simulation. In: Proceedings of the 3rd International Conference on Deep Learning Theory and Applications. SCITEPRESS - Science and Technology Publications (2022). https://doi.org/10.5220/0011309100003277
17. NVIDIA: Nvidia drive sim (2021). https://developer.nvidia.com/drive/drive-sim
18. NVLabs: Imaginaire, a pytorch library that contains optimized implementation of several image and video synthesis methods developed at NVIdia (2020). https://github.com/NVlabs/imaginaire
19. Park, T., Liu, M.Y., Wang, T.C., Zhu, J.Y.: Semantic image synthesis with spatially-adaptive normalization. In: Proceedings of the IEEE Conference on Computer Vision and Pattern Recognition (2019)
20. Ronneberger, O., Fischer, P., Brox, T.: U-Net: convolutional networks for biomedical image segmentation. In: Navab, N., Hornegger, J., Wells, W.M., Frangi, A.F. (eds.) MICCAI 2015. LNCS, vol. 9351, pp. 234–241. Springer, Cham (2015). https://doi.org/10.1007/978-3-319-24574-4_28
21. Sallab, A.E., Sobh, I., Zahran, M., Essam, N.: LiDAR sensor modeling and data augmentation with GANs for autonomous driving (2019)
22. Shah, S., Dey, D., Lovett, C., Kapoor, A.: AirSim: High-Fidelity Visual and Physical Simulation for Autonomous Vehicles. arXiv:1705.05065 [cs] (2017)
23. Velodyne: Velodyne hdl32e (2023). https://velodynelidar.com/wp-content/uploads/2019/08/63-9114-rev-a-hdl32e-hdl-32e-software-version-v2.0-manual.pdf
24. Velodyne: Velodyne (2023). https://velodynelidar.com/

25. Wang, T.C., et al.: Video-to-Video Synthesis. arXiv:1808.06601 [cs] (2018)
26. Wang, T.C., Liu, M.Y., Zhu, J.Y., Tao, A., Kautz, J., Catanzaro, B.: High-resolution image synthesis and semantic manipulation with conditional GANs. In: Proceedings of the IEEE Conference on Computer Vision and Pattern Recognition (2018)
27. Wang, Y., Chao, W.L., Garg, D., Hariharan, B., Campbell, M., Weinberger, K.Q.: Pseudo-LiDAR from Visual Depth Estimation: Bridging the Gap in 3D Object Detection for Autonomous Driving. arXiv:1812.07179 [cs] (2018)
28. You, Y., et al.: Pseudo-lidar++: accurate depth for 3d object detection in autonomous driving. In: ICLR (2020)
29. Zhu, J.Y., Park, T., Isola, P., Efros, A.A.: Unpaired Image-to-Image Translation using Cycle-Consistent Adversarial Networks. arXiv:1703.10593 [cs] (2020)

Evaluating Prototypes and Criticisms for Explaining Clustered Contributions in Digital Public Participation Processes

Lars Schütz[1,2]($\boxtimes$) (ID), Korinna Bade[1] (ID), and Andreas Nürnberger[2] (ID)

[1] Department of Computer Science and Languages, Anhalt University of Applied Sciences, Köthen (Anhalt), Germany
{lars.schuetz,korinna.bade}@hs-anhalt.de
[2] Faculty of Computer Science, Otto von Guericke University Magdeburg, Magdeburg, Germany
andreas.nuernberger@ovgu.de

Abstract. We examine the use of prototypes and criticisms for explaining clusterings in digital public participation processes of the e-participation domain. These processes enable people to participate in various life areas such as landscape planning by submitting contributions that express their opinions or ideas. Clustering groups similar contributions together. This supports citizens and public administrations, the main participants in digital public participation processes, in exploring the submitted contributions. However, explaining clusterings remains a challenge. For this purpose, we consider the use of prototypes and criticisms. Our work generalizes the idea of applying the k-medoids algorithm for computing prototypes on raw data sets. We introduce a centroid-based clusterings method that solely considers clusterings. It allows the retrieval of multiple prototypes and criticisms per cluster. We conducted a user study with 21 participants to evaluate our centroid-based clusterings method and the MMD-critic algorithm for finding prototypes and criticisms in clustered contributions. We examined whether these methods are suitable for text data. The related contributions originate from past, real-life digital public participation processes. The user study results indicate that both methods are appropriate for clustered contributions. The results also show that the centroid-based clusterings method outperforms the MMD-critic algorithm regarding accuracy, efficiency, and perceived difficulty.

Keywords: Example-based explanations · Prototypes and criticisms · Clustering · Text data

1 Introduction

Clustering data is a widely used technique across various fields such as machine learning [9], bioinformatics [14,30], information retrieval [6], and text mining [2]. It helps us to discover patterns in data by grouping together supposedly similar data instances. However, explaining a clustering, which can be conceived as

© The Author(s) 2023
D. Conte et al. (Eds.): DeLTA 2023, CCIS 1875, pp. 434–455, 2023.
https://doi.org/10.1007/978-3-031-39059-3_29

a set of clusters, can be challenging because the concept of a cluster is often fuzzy [8]. It might not be obvious to us why certain data instances belong to one particular cluster instead of another one. Furthermore, it might not be clear what the underlying concept of a cluster actually is. So instead of focusing on a cluster and its associated data instances as a whole, we could rather try to identify representative examples for this cluster. In the context of explaining raw data sets, such representative examples are typically known as prototypes and criticisms. A prototype is a data instance that best describes a data set [12], and a criticism is a data instance that provides insights into parts of the data set that prototypes do not explain well [17]. The use of criticisms in addition to prototypes is beneficial because it has been shown that criticisms can simplify human understanding and reasoning [17]. We, however, examine whether prototypes and criticisms can be directly inferred from clusterings so that the associated clusters can be explained. This is the major motivation for this work.

Digital public participation processes enable individuals to engage in different life areas such as city budgeting planning or planning public green city spaces. Their goal is to ensure that diverse opinions can be heard and considered in the design and decision-making processes [5]. This might lead to more acceptance and trust in the final planning result. Different groups participate in digital public participation processes. Two main groups are citizens and public administrations. While citizens submit contributions in order to voice their ideas or concerns, public administrations have to assess these contributions, typically by accepting or rejecting them. In the end, there is a collection of contributions to explore in digital public participation processes. However, this exploration is challenging. For instance, citizens, on the one hand, might want to identify if their opinion is already shared by others. So they are looking for similar contributions. This can be cumbersome. Public administrations, on the other hand, need to be consistent in deciding whether a contribution is accepted or rejected, i.e., when they reject one particular contribution, they should also reject all other contributions that have the same meaning. Thus, the public administrations might also look for similar contributions. This is an elaborate endeavor. In this regard, clustering can help to group together similar contributions. However, as we have already pointed out, the interpretation of clusterings is challenging. That is why we examine prototypes and criticisms for clusterings. We hypothesize that the clustering of contributions and the use of prototypes and criticisms for explaining the associated clusters can support citizens and public administrations in digital public participation processes. Thus, this work is also motivated by demands from practice.

In the remainder of this paper, we present related work first (Sect. 2). We then describe a novel centroid-based clusterings method for finding prototypes and criticisms (Sect. 3). This method is a generalization of the k-medoids [15] algorithm when used for finding prototypes in raw data sets. Our centroid-based clusterings method is able to retrieve multiple prototypes and criticisms per cluster. This method directly infers prototypes and criticisms from clusterings. We also briefly explain the key idea of the MMD-critic algorithm [17] that can be considered a general baseline for computing prototypes and criticisms (Sect. 3).

The main part of this paper is about a user study that we conducted with 21 participants to evaluate our centroid-based clusterings method and the MMD-critic algorithm for finding prototypes and criticisms in clustered contributions of digital public participation processes. In this regard, we describe the associated experiments (Sect. 4), and we present the results of the user study (Sect. 5). Finally, we conclude our research and provide potential for future work (Sect. 6).

In summary, our main contributions are (1) the introduction of the novel centroid-based clusterings method for finding prototypes and criticisms in clusterings, (2) the idea of applying the MMD-critic algorithm to clusterings, (3) the application of the previous two methods as well as other, naive methods to text data from digital public participation processes of the e-participation domain, and (4) the evaluation of all methods in a user study.

2   Related Work

This work is part of the broad field interpretable machine learning. The need for methods in this field has long been recognized [1,18], and research activities in this field continue to increase for years [7]. One specific subarea of interpretable machine learning is about example-based explanations such as prototypes and criticisms.

In psychology, the concept of prototypes refers to the idea that certain stimuli or objects serve as the best or most representative examples of a particular category. These prototypes are thought to be stored in memory and used as a reference point for making judgments about other stimuli or objects that belong to the same category [25]. The underlying prototype theory posits that people form mental representations of prototypes based on the most representative examples of a category [29]. People tend to rate category members that are similar to the prototype as more typical or representative of the category. The distance from an example to a prototype also affects the human categorization process, i.e., the nearer an example is to the prototype, the more likely it is for this example to be categorized in that prototype's category [13,21]. The related exemplar theory even considers multiple representative examples of a category to extract important properties of the category [22]. The inclusion of edge cases can provide valuable information about the boundaries and limits of a category [17].

Considering the computation of prototypes for raw data sets, there is the k-medoids [15] clustering algorithm. It assigns all data instances of a raw data set to different clusters. Each cluster then contains a medoid. This medoid can be used as a prototype for this particular cluster. However, only one prototype per cluster is output by the k-medoids algorithm. This might not be enough for explaining a cluster according to the exemplar theory. Other related clustering algorithms such as the k-means [19,20] and fuzzy c-means [4] algorithms do not guarantee to return a medoid. They provide cluster centroids that typically are not actual data instances of the data set, which however is required for prototypes and criticisms. Furthermore, there is no concept of criticisms in any of the previously mentioned clustering algorithms. We, however, consider the computation of multiple prototypes and criticisms per cluster. In contrast to previous work, we want to use prototypes and criticisms to explain individual clusters of a clustering.

There are also the MMD-critic [17] and ProtoDash [12] algorithms for computing prototypes and criticisms. We briefly explain the MMD-critic algorithm in Sect. 3. The ProtoDash algorithm is a generalization of or an extension to the MMD-critic algorithm because it additionally computes importance weights for each prototype. These methods are model-agnostic [23], i.e., in our context, they can be applied to raw data sets and to individual clusters of a clustering, although we have not yet observed that this has already been done. However, these methods would not work directly with the clusterings but perform other extensive computations on top instead. To the best of our knowledge, there is no other (model-specific) method that directly infers prototypes and criticisms from clusterings. At the same time, we also observe that these methods have been applied to image data rather than text data [17]. Additionally, the evaluation of these methods using user studies appears to be rare.

Referring briefly to digital public participation processes, there exist general applications of natural language processing and machine learning methods that aim to provide sophisticated computer-aided support for the participants. For example, citizens are clustered to connect citizens with common interests, or comprehensive contributions are summarized to provide an overview of the contents [3]. Furthermore, certain approaches exist which utilize similarity-based ranking methods to aid public administrations in assessing contributions more effectively [26]. There is also work that considers the clustering of contributions [27]. However, that particular work focuses on comparing such clusterings. We focus on providing representative examples for a clustering instead. Furthermore, there are semi-supervised approaches to automatically categorize contributions by their topics while reducing the efforts needed to label the contributions [24]. However, the interpretation of the methods used is not covered at all. There is also a broader perspective that considers the visual analytics [16] approach including machine learning methods for digital public participation processes [28], e.g., for decision-making. However, no promising machine learning algorithm is described, and there is no relation to or mention of interpretable machine learning. We, however, consider it important in that domain.

3 Prototypes and Criticisms for Clusterings

We examine two primary methods for identifying prototypes and criticisms in clusterings as a foundation for our user study. On the one hand, we introduce a centroid-based clusterings method. On the other hand, we briefly examine the MMD-critic algorithm as a baseline. Both methods operate under the assumption that we have a pre-existing clustering of data, such as one obtained through the k-means algorithm. Our goal is to present this clustering to a user by focusing on prototypes and criticisms. To achieve this, we adapt existing approaches.

3.1 Centroid-Based Clusterings Method

We adapt the idea of inferring prototypes directly from clusterings similar to directly using the medoids of a k-medoids clustering as prototypes. However, we introduce a novel generalization of this idea in three aspects. First, we take clustering algorithms into account that do not directly yield actual data instances from the data set as prototypical examples. Second, we consider the selection of more than one prototype for every cluster of a clustering. Third, we also allow the retrieval of criticisms for each cluster.

Consider a set of data instances $X = \{\mathbf{x}_1, \mathbf{x}_2, \ldots, \mathbf{x}_n\}, \mathbf{x}_i \in \mathbb{R}^m$. We use the partitioning $\{X_1, X_2, \ldots, X_k\}$ of X into k disjoint subsets $X_1, X_2, \ldots, X_k$ to denote a clustering of X. The subsets $X_1, X_2, \ldots, X_k$ represent the associated clusters. Every cluster X_i then contains at least one representative data instance which we denote by $\mathbf{p}_{i,1}, 1 \leq i \leq k$. This is then the first prototype of the i-th cluster. Using a k-medoids clustering, for example, $\mathbf{p}_{i,1}$ is the medoid of the i-th cluster. For all centroid-based clustering algorithms such as k-means, we retrieve the centroid $\boldsymbol{\mu}_i = \frac{1}{|X_i|} \sum_{\mathbf{x} \in X_i} \mathbf{x}$ for every cluster X_i through the clustering algorithm. If this centroid is equal to an actual data instance from the data set, we consider it as the first prototype for the particular cluster, i.e., $\mathbf{p}_{i,1} = \boldsymbol{\mu}_i$. Otherwise, we select the nearest data instance to $\boldsymbol{\mu}_i$ as the first prototype. This approach describes our first generalization aspect. Through this little adaptation, we can use any centroid-based clustering algorithm as a basis. The second generalization aspect is about choosing p prototypes for every cluster X_i rather than being limited to choosing only one prototype, $1 \leq p \leq \min_{1 \leq i \leq k} |X_i|$. This is motivated by the exemplar theory. For this purpose, we consider the next $p-1$ nearest data instances to $\boldsymbol{\mu}_{i,1}$ as the next prototypes $\mathbf{p}_{i,2}, \mathbf{p}_{i,3}, \ldots, \mathbf{p}_{i,p}$. In case of a tie, the selection can be made randomly. For the computation of criticisms, we adopt a similar approach as with the selection of prototypes. However, we consider the c most distant data instances to $\boldsymbol{\mu}_{i,1}$ as the criticisms $\mathbf{c}_{i,1}, \mathbf{c}_{i,2}, \ldots, \mathbf{c}_{i,c}, 1 \leq c \leq \min_{1 \leq i \leq k} |X_i|$, with $\mathbf{c}_{i,1}$ being the most distant criticism, $\mathbf{c}_{i,2}$ being the second most distance criticism etc. This approach is our third generalization aspect. Figure 1 shows an example.

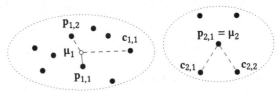

Fig. 1. Example of our centroid-based clusterings method with a different number of prototypes and criticisms per cluster. Prototype $\mathbf{p}_{i,j}$ is the j-th prototype of cluster i, criticism $\mathbf{c}_{i,j}$ is the j-th criticism of cluster i, and $\boldsymbol{\mu}_i$ is the centroid of cluster i.

3.2   MMD-Critic for Clusterings

We briefly explain the MMD-critic algorithm [17] because we consider it as a model-agnostic baseline in our user study. While this algorithm has been successfully applied to raw data sets, we exclusively apply it to clusterings instead.

The MMD-critic algorithm relies on the maximum mean discrepancy (MMD) between two distributions for identifying prototypes. In our context, we need to consider a distribution of prototypes and a distribution of data instances. The MMD represents how much these distributions differ from each other. The MMD can be estimated by the empirical MMD $\hat{f}$ using the set of n data instances X, the set of p prototypes P, and a kernel function k as stated in (1) [11].

$$\hat{f}(X, P) = \left[\frac{1}{n^2} \sum_{i,j=1}^{n} k(\mathbf{x}_i, \mathbf{x}_j) + \frac{1}{p^2} \sum_{i,j=1}^{p} k(\mathbf{p}_i, \mathbf{p}_j) - \frac{2}{np} \sum_{i,j=1}^{n,p} k(\mathbf{x}_i, \mathbf{p}_j) \right]^{\frac{1}{2}} \quad (1)$$

The MMD-critic algorithm greedily selects p prototypes. Each data instance is examined and evaluated as a potential prototype until p prototypes have been found. For this purpose, the squared empirical MMD is calculated once when the particular data instance would be added as a prototype and once when it is not considered as a prototype. The data instance for which the difference between these two values is the lowest is selected as the next prototype, because then the discrepancy between the prototypes and the data instances has been reduced the most. This approach can be applied to every cluster of a clustering.

The criticisms are selected based on the witness function $g(\mathbf{x})$ that computes how much two distributions differ at a specific data instance $\mathbf{x}$. In our context, we consider the distribution of prototypes and the distribution of data instances. Its empirical estimate $\hat{g}(\mathbf{x})$, given a set of n data instances X, a set of p prototypes P, and a kernel function k, is stated in (2) [11].

$$\hat{g}(\mathbf{x}) = \frac{1}{n} \sum_{i=1}^{n} k(\mathbf{x}, \mathbf{x_i}) - \frac{1}{p} \sum_{i=1}^{p} k(\mathbf{x}, \mathbf{p_i}) \quad (2)$$

The MMD-critic algorithm greedily identifies the criticisms. So each data instance is tested again, however, the data instance that has the highest witness function value is selected as the next criticism. This procedure is repeated until c criticisms have been found. This procedure can be applied to each cluster of a clustering.

4   Experiments

We conducted a user study to evaluate our centroid-based clusterings method and the MMD-critic algorithm for finding prototypes and criticisms in clustered contributions. We also wanted to find out whether these methods are suitable for text data in general, as previous research in this area has primarily focused on image data. The data used in the user study were contributions from past, real-life digital public participation processes. This is our specific application scenario. In the following, we provide details about the participants, data set, task, design and procedure, and research questions and measures.

4.1 Participants

After contacting students and employees at our institution, we were able to recruit a total of 21 participants (ten female, eleven male) for our user study. In the post-experiment questionnaire, the participants reported their ages within the following groups: ten participants were 21–30 years old, six participants were 31–40 years old, two participants were 41–50 years old, and three participants were 51–60 years old. The participants were also asked to rate their proficiency as computer users. None of them considered themselves inexperienced, and none considered themselves beginners. Eleven participants reported having average experience with computers, while ten participants considered themselves advanced computer users. The participants in our study have diverse educational backgrounds. Among them are two computer science professors, eight data science students, two employees with degrees in marketing, and one employee each with degrees in German studies and marketing, office administration, and IT administration. Additionally, six participants are employees with degrees in computer science.

4.2 Data Set and Task

We merged two data sets of contributions from past, real-life digital public participation processes into one larger data set to be used in the user study. Overall, the contributions consider either ideas for a regional development planning project or complaints about city noise sources. The contributions are in German. Each contribution belongs to exactly one of ten categories. The literal translations of these categories are: (1) residence, (2) city equipment, (3) pedestrian and bicycle paths, (4) miscellaneous, (5) play/sport/exercise, (6) vegetation, (7) aviation noise, (8) train noise, (9) miscellaneous noise, and (10) traffic noise. We considered these categories as the ground truth for the clustering of the contributions, i.e., every category actually represents a single cluster. This allowed us to bypass the manual as well as the computer-assisted clustering of the contributions, reducing the potential of inaccuracies. However, the participants in the

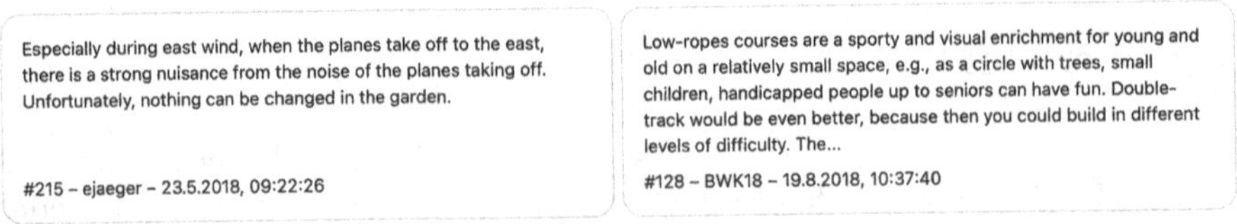

Especially during east wind, when the planes take off to the east, there is a strong nuisance from the noise of the planes taking off. Unfortunately, nothing can be changed in the garden.

#215 – ejaeger – 23.5.2018, 09:22:26

Low-ropes courses are a sporty and visual enrichment for young and old on a relatively small space, e.g., as a circle with trees, small children, handicapped people up to seniors can have fun. Double-track would be even better, because then you could build in different levels of difficulty. The...

#128 – BWK18 – 19.8.2018, 10:37:40

Fig. 2. Two sample contributions of the data set used in the user study. The literal translations of the contents are shown.

user study were unaware of the existence of these categories and their origin. In order to provide a clearer understanding of the contributions, we provide two sample contributions with their literal translations in Fig. 2.

The participants were tasked with repeatedly assigning a reference contribution to one of three groups of other contributions. For each reference contribution, they were told to find the best fitting group of similar contributions. We hypothesize that it should be easier to assign a reference contribution to a cluster when the cluster is well-explained. We explain a cluster by selecting a subset of contributions from the cluster.

We used four different methods to populate the clusters with the contributions: (1) the MMD-critic algorithm (MMD), (2) our centroid-based clusterings method (CEN), (3) a random method (RND), and (4) a raw method (RAW). The MMD, CEN, and RND methods find prototypes and criticisms in the cluster to populate the particular cluster, whereas the RND method randomly selects data instances from the cluster without any specific selection criteria for prototypes and criticisms. The RAW method simply includes all contributions of a cluster.

We exclusively considered the contents of the contributions. We preprocessed the contributions before utilizing the MMD-critic algorithm and our centroid-based clusterings method for determining the prototypes and criticisms for each cluster. This involved tokenizing the content and creating averaged word embeddings through the use of pre-trained models [10]. For the remaining two methods, we simply queried the original data set. We always choose two prototypes and two criticisms per cluster except for the raw method. This is motivated by the exemplar theory. The resulting contributions were arranged vertically for each cluster, starting with the prototypes. We have limited the contributions to a maximum of five lines for display purposes, so that reading long texts does not take up too much time for a whole experiment per participant.

4.3 Design and Procedure

The user study employed a within-subject design. We presented the four methods in a balanced Latin block design to avoid potential bias due to usage order. Every participant completed the task three times for each method, and the trials were averaged per method for the final analysis. The reference contribution and groups were varied across trials to eliminate potential memorization effects. Importantly, the participants were not informed about the number of categories or their labels, nor were they aware of the method used for each trial.

In the beginning (phase 1), we explained the purpose of the experiment and the task to the participants. The participants were informed that they will perform the task multiple times with breaks in between. The participants were encouraged to solve the task as quickly and accurately as possible.

After completing phase 1, participants received a brief, self-paced tutorial that introduced them to the basic graphical elements and layout of the user study system (phase 2). Figure 3 shows the graphical user interface of the user study system. The tutorial emphasized key elements by highlighting them and providing text descriptions via tooltips. Participants learned about the reference contribution (cf. (1) in Fig. 3), the three groups of contributions arranged in a three-column layout (cf. (2)–(4) in Fig. 3), and the button group containing three buttons (cf. (5) in Fig. 3), each corresponding to a group of contributions. Participants were reminded that their decisions were final, meaning that once a button was clicked, the task would be completed and the contribution would be assigned irrevocably.

After completing phase 2, the actual experiment began (phase 3). The participants solved the task, where the reference contribution was randomly chosen from the three selected clusters. Upon completion, they were required to take a mandatory break of at least ten seconds, which was enforced by the user study system (phase 4). During this break, participants were asked to rate the perceived difficulty of the previous task on a 5-point Likert scale, with 1 indicating "very easy" and 5 indicating "very difficult". Once they had answered this question and once the break had ended, the participants clicked a button to proceed to the next task. Phase 3 and phase 4 repeated eleven more times. In the final phase (phase 5), the participants answered a post-experiment question-

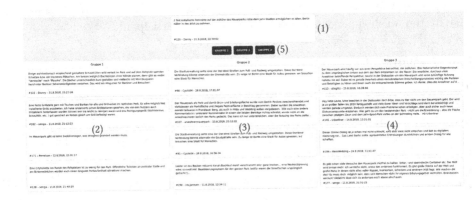

Fig. 3. Graphical user interface of the user study system consisting of the reference contribution (1), the three groups with different contributions aligned in a column layout (2–4) (each group represents a cluster by displaying selected prototypes and criticisms), and the button group (5) (one button for each group) for selecting the alleged group of the reference contribution

naire, providing information on their gender, age group, computer proficiency, and educational and job background.

4.4 Research Questions and Measures

We defined the following research questions q_1 to q_6:

- What is the accuracy of the cluster assignments (q_1)?
- How long does it take to complete the tasks (q_2)?
- What is the perceived difficulty of the tasks (q_3)?
- Is there a correlation between accuracy, time, and difficulty (q_4)?
- Do participants explore all clusters (q_5)?
- How do the methods compare with each other (considering the previous research questions) (q_6)?

We then defined the following measures to address the research questions:

- *Accuracy*: Ratio of correct assignments and the total number of assignments (per method) (for q_1 and q_6)
- *Efficiency*: Time spent to complete the task (per method) (for q_2 and q_6)
- *Difficulty*: Participant's perceived level of difficulty (per method) (for q_3 and q_6)
- *Correlation*: Spearman's correlation coefficient ρ (per method) (for q_4 and q_6)
- *Exploration*: Kernel density estimation of recorded mouse positions (per method) (for q_5 and q_6)

5 Results

This section presents the findings of the conducted user study. We refer to the research questions and measures used to evaluate the proposed methods for computing prototypes and criticisms.

5.1 Accuracy

For all methods, the 21 participants completed the repeated task with a mean accuracy of 0.746 (SD = 0.119). The results are clearly better than random guessing. However, there is still potential for improvement. Table 1 lists all accuracies per method. Our centroid-based clusterings method leads to the best results (mean = 0.794, SD = 0.197), followed by the MMD-critic algorithm (mean = 0.762, SD = 0.261). The raw method resulted in the lowest accuracy (mean = 0.698, SD = 0.256). Figure 4 provides a detailed comparison of the paired mean difference effect sizes between each method pair. Additionally, we report these effect sizes in Table 2. This table also lists the corresponding p-values computed using the Wilcoxon matched-pairs signed rank test at a significance level of $\alpha = 0.05$. The largest effect with an absolute paired mean difference of 0.096 ($p = 0.279$) can be observed between the centroid-based clusterings method and the raw method. These findings demonstrate that prototypes and criticisms accurately represent a cluster. Both the centroid-based clusterings method and the MMD-critic method are effective for textual data representations, with little differences between the two methods. Furthermore, the results suggest that a sophisticated method for finding prototypes and criticisms is preferable to randomly selecting an arbitrary subset of some data instances from a cluster.

5.2 Efficiency

We observe a mean efficiency of 2.345 (SD = 0.812) minutes needed for completing one task for all methods. This generally depends on how long the contributions are and how willing the participants are to actually read the contributions carefully. Table 3 lists the average efficiency values per method. Our centroid-based clusterings method achieved the best result (mean = 2.043, SD = 0.876), followed by the MMD-critic algorithm (mean = 2.325, SD = 1.053). The raw method achieved the worst efficiency (mean = 2.596, SD = 1.806). The random method (mean = 2.416, SD = 1.195) achieves similar values to the raw method. This indicates that prototypes and criticisms lead to more efficient results than randomly selecting a subset of data instances or selecting all data instances. Figure 5 shows the paired mean difference effect sizes for all pairs of methods. We also provide these effect sizes and the p-values computed with the Wilcoxon matched-pairs signed rank test at significance level $\alpha = 0.05$ in Table 4. We find that the centroid-based clusterings method is more efficient than the MMD-critic method due to the absolute paired mean difference of 0.282 min ($p = 0.191$). To our surprise, there is almost no effect to observe between the MMD-critic method and the random method due to the absolute paired mean difference of 0.091 ($p = 0.973$), i.e., in terms of efficiency, both methods perform the same.

Table 1. Average accuracy (mean (SD)) per method. The best average accuracy is in bold.

Method	Accuracy
MMD	0.762 (0.261)
CEN	**0.794** (0.197)
RND	0.730 (0.291)
RAW	0.698 (0.256)

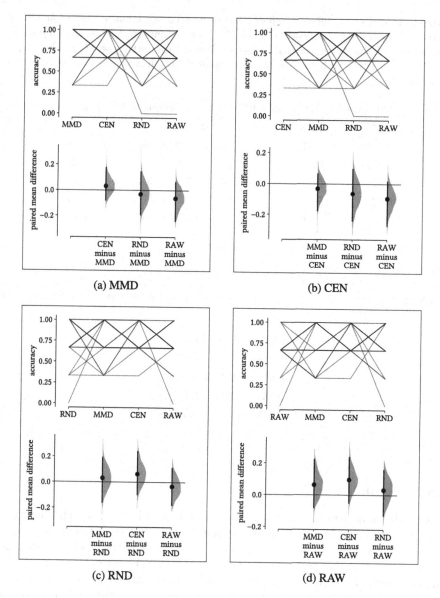

Fig. 4. Four shared-control estimation plots of paired mean differences between the methods (a) MMD, (b) CEN, (c) RND, (d) RAW and all other methods each for comparing the accuracy. For each plot, all methods are plotted on the top axes as a slopegraph, the paired mean differences are plotted on the bottom axes as bootstrap sampling distributions, the mean difference is depicted as a dot in each case, and the 95% confidence interval is indicated by the vertical error bar each. It becomes immediately apparent that the CEN method outperforms all other methods while the RAW method performs the worst.

Table 2. Paired mean difference effect sizes of the accuracies, bias-corrected and accelerated 95% bootstrap confidence intervals (CI, 1000 resamples), and associated p-values computed with the Wilcoxon matched-pairs signed rank test ($\alpha = 0.05$). The largest absolute effect size is in bold. The absolute value of the effect size equals the effect size of m_1 minus m_2.

Method m_1	Method m_2	Effect size (m_2 minus m_1) [CI]	p
MMD	CEN	0.032 [−0.079, 0.175]	0.791
MMD	RND	−0.032 [−0.190, 0.143]	0.793
MMD	RAW	−0.064 [−0.238, 0.063]	0.266
CEN	RND	−0.064 [−0.238, 0.095]	0.768
CEN	RAW	**−0.096** [−0.270, 0.016]	0.279
RND	RAW	−0.032 [−0.175, 0.111]	0.615

Table 3. Average efficiency (mean (SD)) per method. The best average efficiency is in bold.

Method	Time in min
MMD	2.325 (1.053)
CEN	**2.043** (0.876)
RND	2.416 (1.195)
RAW	2.596 (1.806)

5.3   Difficulty

The results show a mean perceived difficulty of 3.000 (SD = 0.404). This indicates that the task was generally neither easy nor difficult for the participants. Table 5 lists all perceived difficulty values per method. The centroid-based clusterings method obtained the best results (mean = 2.873, SD = 0.853). Interestingly, the raw method (mean = 2.905, SD = 1.050) achieved the second best results. The many contributions when using the raw method obviously did not trigger overburdening. Only then follows the MMD-critic method (mean = 3.079, SD = 0.802). This means that the participants found the task easier when they used the centroid-based clusterings method instead of the MMD-critic method. The highest perceived difficulty value was recorded when the random method (mean = 3.143, SD = .0764) was used. However, we have to acknowledge that all values are still close to each other. Figure 6 shows the paired mean difference effect sizes for all pairs of methods. Again, we provide these effect sizes and the p-values computed with the Wilcoxon matched-pairs signed rank test at a significance level of $\alpha = 0.05$ in Table 6. There is almost no effect in perceived difficulty

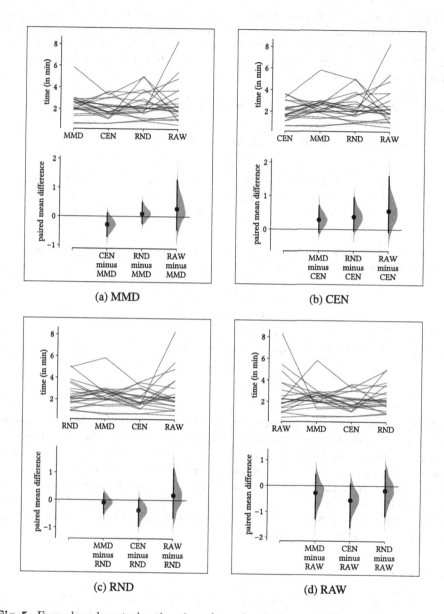

Fig. 5. Four shared-control estimation plots of paired mean differences between the methods (a) MMD, (b) CEN, (c) RND, (d) RAW and all other methods each for comparing the efficiency. For each plot, all methods are plotted on the top axes as a slopegraph, the paired mean differences are plotted on the bottom axes as bootstrap sampling distributions, the mean difference is depicted as a dot in each case, and the 95% confidence interval is indicated by the vertical error bar each. The CEN method clearly outperforms all other methods in terms of efficiency. The RAW method performs the worst.

Table 4. Paired mean difference effect sizes of the task completion time (in min), bias-corrected and accelerated 95% bootstrap confidence intervals (CI, 1000 resamples), and associated p-values computed with the Wilcoxon matched-pairs signed rank test ($\alpha = 0.05$). The largest absolute effect size is in bold. The absolute value of the effect size equals the effect size of m_1 minus m_2.

Method m_1	Method m_2	Effect size (m_2 minus m_1) [CI]	p
MMD	CEN	−0.282 [−0.697, 0.103]	0.191
MMD	RND	0.091 [−0.235, 0.502]	0.973
MMD	RAW	0.271 [−0.442, 1.276]	0.973
CEN	RND	0.373 [−0.039, 0.946]	0.168
CEN	RAW	**0.553** [−0.075, 1.599]	0.562
RND	RAW	0.180 [−0.636, 1.153]	0.785

Table 5. Average difficulty (mean (SD), based on a 5-point Likert scale) per method. The best average difficulty is in bold.

Method	Difficulty
MMD	3.079 (0.802)
CEN	**2.873** (0.853)
RND	3.143 (0.764)
RAW	2.905 (1.050)

between the centroid-based clusterings method and the raw method because of the absolute paired mean differences of 0.032 ($p = 0.747$). The largest effect, however, can be observed between our centroid-based clusterings method and the random method due to the absolute paired mean difference of 0.270 ($p = 0.601$). In contrast, the MMD-critic method and the random method only differ by a small amount of 0.064 ($p = 0.917$). Thus, the centroid-based clusterings method should be preferred when deciding between a sophisticated method for computing prototypes and criticisms.

5.4   Correlation

We are interested in the correlations between (1) accuracy–efficiency, (2) accuracy–difficulty, and (3) efficiency–difficulty although we cannot derive causal relationships. All results are shown in Fig. 7. Table 7 lists the numeric values.

Concerning (1) accuracy–efficiency, the Spearman's correlation coefficients ρ for the centroid-based method ($\rho = 0.074$) is almost 0, i.e., there is no correlation between the accuracy and efficiency for this method. However, the Spearman's correlation coefficient ρ for the MMD-critic method ($\rho = -0.394$) shows a weak monotonic decreasing relationship, i.e., as the accuracy of the cluster assignments increased, the time needed for finishing the task decreased.

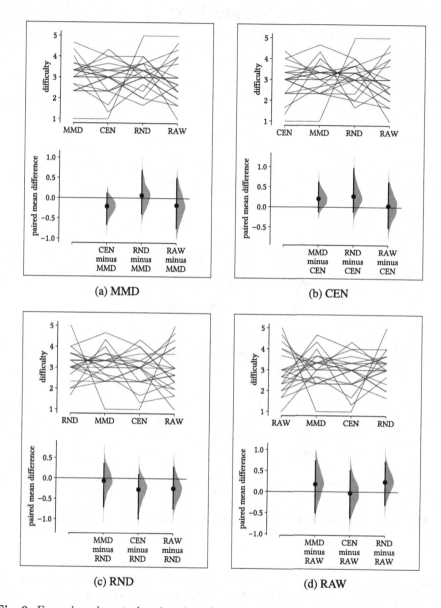

Fig. 6. Four shared-control estimation plots of paired mean differences between the methods (a) MMD, (b) CEN, (c) RND, (d) RAW and all other methods each for comparing the perceived difficulty. For each plot, all methods are plotted on the top axes as a slopegraph, the paired mean differences are plotted on the bottom axes as bootstrap sampling distributions, the mean difference is depicted as a dot in each case, and the 95% confidence interval is indicated by the vertical error bar each. The CEN method once again outperforms all other methods. However, the difference to the RAW method is very small. The RND method performs the worst.

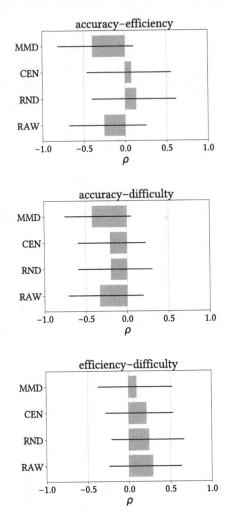

Fig. 7. Spearman's ρ between the measures accuracy, efficiency, and difficulty

Referring to (2) accuracy–difficulty, there is only negative correlation present. This means for all methods that if the task seemed easy, there were more accurate results. However, there are subtle differences in the effect sizes. The MMD-critic method ($\rho = -0.419$) has the largest negative correlation coefficient. This is a moderate monotonic decreasing relationship and the best result.

Finally, we observe only positive correlations for (3) efficiency–difficulty. The centroid-based clusterings method ($\rho = 0.207$), the random method ($\rho = 0.237$), and the raw method ($\rho = 0.282$) have similar correlation coefficients. These values indicate a weak monotonic increasing correlation. This means that when the time needed for finishing the task increased, which is equivalent to a decreasing efficiency, the perceived difficulty of assigning the reference contribution to the

Table 6. Paired mean difference effect sizes of the difficulties, bias-corrected and accelerated 95% bootstrap confidence intervals (CI, 1000 resamples), and associated p-values computed with the Wilcoxon matched-pairs signed rank test ($\alpha = 0.05$). The largest absolute effect size is in bold. The absolute value of the effect size equals the effect size of m_1 minus m_2.

Method m_1	Method m_2	Effect size (m_2 minus m_1) [CI]	p
MMD	CEN	-0.206 [-0.651, 0.111]	0.408
MMD	RND	-0.064 [$-0.38q$, 0.683]	0.917
MMD	RAW	-0.174 [-0.730, 0.492]	0.350
CEN	RND	**0.270** [-0.111, 0.968]	0.601
CEN	RAW	-0.032 [-0.508, 0.619]	0.747
RND	RAW	-0.238 [-0.730, 0.286]	0.275

Table 7. Spearman's ρ and the bias-corrected and accelerated 95% bootstrap confidence intervals (CI, 1000 resamples)

Method	accuracy—efficiency [CI]	accuracy—difficulty [CI]	efficiency—difficulty [CI]
MMD	-0.394 [-0.81, 0.10]	-0.419 [-0.75, 0.05]	0.090 [-0.37, 0.52]
CEN	0.074 [-0.46, 0.55]	-0.203 [-0.59, 0.22]	0.207 [-0.28, 0.53]
RND	0.133 [-0.40, 0.61]	-0.196 [-0.59, 0.30]	0.237 [-0.21, 0.66]
RAW	-0.253 [-0.68, 0.25]	-0.332 [-0.71, 0.19]	0.282 [-0.24, 0.63]

associated cluster increased. However, this does not apply to the MMD-critic method ($\rho = 0.090$), because there is almost no effect present.

5.5 Exploration

The user study system recorded the participants' mouse positions (in screen coordinates) every two seconds. We use this data as an indicator of exploration activity, with the assumption that the participants moved the mouse while reading the contributions. To assess whether the participants explored all groups during the task, we analyze the kernel density estimation of the mouse positions for each method. Figure 8 depicts these kernel density estimations. The results show similar results across all methods. The participants explored the first and second groups the most (first and second columns), with the third group being explored less. The raw method best reflects the three-column layout of the user study system's graphical user interface. This suggests that some participants explored the contributions towards the end of the groups. Based on these results, we can conclude that there were no abnormalities.

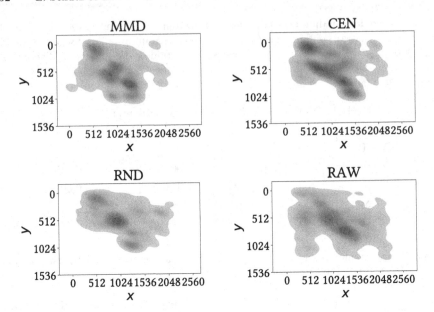

Fig. 8. Kernel density estimation plots of the computer mouse positions (in screen coordinates) per method

6 Conclusions and Future Work

We motivated for the need for clustering explanations in general and especially in digital public participation processes so that citizens and public administration can explore the contributions more easily. We introduced a novel centroid-based clusterings method for finding prototypes and criticisms in clusterings. They are directly inferred from the clustering. The user study results show that both the centroid-based clusterings method and the MMD-critic method are suitable for explaining clustered contributions. However, the centroid-based clusterings method outperforms the MMD-critic algorithm regarding accuracy, efficiency, and perceived difficulty.

Nonetheless, there is potential for future work. We only considered the raw representations of the contributions, i.e., we displayed the raw content to the participants of the user study. We plan to investigate other representations for the prototypes and criticisms. For example, content summarization or the inclusion of topic tags could enhance the comprehensibility of the prototypes and criticisms.

Acknowledgements. We would like to thank the German Research Foundation (Deutsche Forschungsgemeinschaft (DFG)) - project number 491460386 - and the Open Access Publication Fund of the Anhalt University of Applied Sciences for financial support.

References

1. Abdul, A., Vermeulen, J., Wang, D., Lim, B.Y., Kankanhalli, M.: Trends and trajectories for explainable, accountable and intelligible systems: an HCI research agenda. In: Proceedings of the 2018 CHI Conference on Human Factors in Computing Systems, pp. 582:1–582:18. ACM, New York (2018)
2. Allahyari, M., et al.: A brief survey of text mining: classification, clustering and extraction techniques (2017)
3. Arana-Catania, M., et al.: Citizen participation and machine learning for a better democracy. Digit. Gov. Res. Pract. **2**(3), 1–22 (2021). https://doi.org/10.1145/3452118
4. Bezdek, J.C.: Pattern Recognition with Fuzzy Objective Function Algorithms. Advanced Applications in Pattern Recognition, Springer, New York (1981). https://doi.org/10.1007/978-1-4757-0450-1
5. Bobbio, L.: Designing effective public participation. Policy Soc. **38**(1), 41–57 (2018). https://doi.org/10.1080/14494035.2018.1511193
6. Djenouri, Y., Belhadi, A., Fournier-Viger, P., Lin, J.C.W.: Fast and effective cluster-based information retrieval using frequent closed itemsets. Inf. Sci. **453**, 154–167 (2018). https://doi.org/10.1016/j.ins.2018.04.008
7. Doshi-Velez, F., Kim, B.: Towards a rigorous science of interpretable machine learning (2017)
8. Estivill-Castro, V.: Why so many clustering algorithms: a position paper. SIGKDD Explor. Newsl. **4**(1), 65–75 (2002). https://doi.org/10.1145/568574.568575
9. Ezugwu, A.E., et al.: A comprehensive survey of clustering algorithms: state-of-the-art machine learning applications, taxonomy, challenges, and future research prospects. Eng. Appl. Artif. Intell. **110**, 104743 (2022). https://doi.org/10.1016/j.engappai.2022.104743
10. Grave, E., Bojanowski, P., Gupta, P., Joulin, A., Mikolov, T.: Learning word vectors for 157 languages. In: Proceedings of the Eleventh International Conference on Language Resources and Evaluation (LREC 2018). European Language Resources Association (ELRA), Miyazaki, Japan (2018)
11. Gretton, A., Borgwardt, K.M., Rasch, M.J., Schölkopf, B., Smola, A.: A kernel two-sample test. J. Mach. Learn. Res. **13**(25), 723–773 (2012)
12. Gurumoorthy, K.S., Dhurandhar, A., Cecchi, G., Aggarwal, C.: Efficient data representation by selecting prototypes with importance weights. In: 2019 IEEE International Conference on Data Mining (ICDM), pp. 260–269 (2019). https://doi.org/10.1109/ICDM.2019.00036
13. Hampton, J.A.: Concepts as prototypes. In: Psychology of Learning and Motivation, vol. 46, pp. 79–113. Academic Press (2006). https://doi.org/10.1016/S0079-7421(06)46003-5
14. Karim, M.R., et al.: Deep learning-based clustering approaches for bioinformatics. Brief. Bioinform. **22**(1), 393–415 (2020). https://doi.org/10.1093/bib/bbz170
15. Kaufman, L., Rousseeuw, P.J.: Clustering by means of medoids. In: Statistical Data Analysis Based on the L1-Norm and Related Methods, pp. 405–416. Elsevier Science, Amsterdam, North-Holland, New York (1987)
16. Keim, D., Kohlhammer, J., Ellis, G., Mansmann, F. (eds.): Mastering the Information Age: Solving Problems with Visual Analytics. Eurographics Association, Goslar (2010)
17. Kim, B., Khanna, R., Koyejo, O.O.: Examples are not enough, learn to criticize! Criticism for interpretability. In: Lee, D., Sugiyama, M., Luxburg, U., Guyon, I.,

Garnett, R. (eds.) Advances in Neural Information Processing Systems, vol. 29, pp. 2280–2288. Curran Associates, Inc. (2016)

18. Lipton, Z.C.: The mythos of model interpretability: in machine learning, the concept of interpretability is both important and slippery. Queue **16**(3), 31–57 (2018)

19. Lloyd, S.P.: Least squares quantization in PCM. IEEE Trans. Inf. Theory **28**(2), 129–137 (1982)

20. MacQueen, J.: Some methods for classification and analysis of multivariate observations. In: Proceedings of the Fifth Berkeley Symposium on Mathematical Statistics and Probability, Volume 1: Statistics, Berkeley, CA, USA, pp. 281–297. The Regents of the University of California (1967)

21. Murphy, G.: The Big Book of Concepts. MIT Press, Cambridge (2002)

22. Murphy, G.L.: Is there an exemplar theory of concepts? Psychon. Bull. Rev. **23**(4), 1035–1042 (2015). https://doi.org/10.3758/s13423-015-0834-3

23. Ribeiro, M.T., Singh, S., Guestrin, C.: Model-agnostic interpretability of machine learning (2016). https://doi.org/10.48550/ARXIV.1606.05386

24. Romberg, J., Escher, T.: Automated topic categorisation of citizens' contributions: reducing manual labelling efforts through active learning. In: Janssen, M., et al. (eds.) EGOV 2022. LNCS, vol. 13391, pp. 369–385. Springer, Heidelberg (2022). https://doi.org/10.1007/978-3-031-15086-9_24

25. Rosch, E., Mervis, C.B.: Family resemblances: studies in the internal structure of categories. Cogn. Psychol. **7**(4), 573–605 (1975)

26. Schütz, L., Bade, K.: Assessment user interface: supporting the decision-making process in participatory processes. In: Proceedings of the 21st International Conference on Enterprise Information Systems - Volume 2: ICEIS, pp. 398–409. INSTICC, SciTePress (2019). https://doi.org/10.5220/0007719603980409

27. Schütz, L., Bade, K., Nürnberger, A.: Comprehensive differentiation of partitional clusterings. In: Proceedings of the 25th International Conference on Enterprise Information Systems - Volume 2: ICEIS, pp. 243–255. INSTICC, SciTePress (2023). https://doi.org/10.5220/0011762000003467

28. Schütz, L., Raabe, S., Bade, K., Pietsch, M.: Using visual analytics for decision making. J. Digit. Landscape Archit. **2**, 94–101 (2017). https://doi.org/10.14627/537629010

29. Smith, E., Medin, D.: Categories and Concepts. Harvard University Press (1981)

30. Zou, Q., Lin, G., Jiang, X., Liu, X., Zeng, X.: Sequence clustering in bioinformatics: an empirical study. Brief. Bioinform. **21**(1), 1–10 (2018). https://doi.org/10.1093/bib/bby090

FRLL-Beautified: A Dataset of Fun Selfie Filters with Facial Attributes

Shubham Tiwari, Yash Sethia, Ashwani Tanwar, Ritesh Kumar,
and Rudresh Dwivedi[✉]

Department of Computer Science and Engineering, Netaji Subhas University of Technology
(NSUT), Delhi, India
{shubham.cs19,yash.sethia.cs19,ashwani.cs19,ritesh.cs19,
rudresh.dwivedi}@nsut.ac.in

Abstract. There is a need to assess the impact of filters on the performance of
face recognition systems. For this, a standard dataset should be available with
relevant filters applied. Currently, such datasets are not publicly available. Some
datasets which are available with filters applied, are very low in resolution and
thus not relevant for use. To mitigate these limitations, we aim to create a dataset
that provides a face recognition database with filters applied to them. The proposed dataset provides high-quality images with ten different filters applied to
them. These filters vary from beautification filters and AR-based filters to filters
that modify facial landmarks. The wide range of filters including occlusion and
beautification that has been applied to the selfies allow a more diverse set of faces
to be experimented and analyzed with face recognition and other biometric systems. The dataset will contribute further to the set of facial datasets available
publicly. This will allow researchers to study the impact of filters on facial features with a common public benchmark.

Keywords: Face detection · Face recognition · Fun selfie filters ·
Beautification · Snapchat · B612 · FaceApp · FRLL

1 Introduction

1.1 Background

Selfies are really popular in today's world of social media. These social media platforms
also allow the users to add fun selfie filters such as gender swapping, a fake beard or
even letting them change their eye colour. These filters focus on beautifying these selfies
and making them more interactive and giving them mass appeal. Selfies have now found
immense application in biometric identification systems at airports, government IDs,
and several other institutions. However, the augmenting of these selfies with custom
fun selfie filters is now a new threat to these face recognition systems. Such filters
can destroy or modify biometric features that would allow person recognition or even
detection of the face itself. In this work, we have focused on creating a public dataset
of facial images with various filters applied. This will allow researchers to study the
impact of filters on facial features with a common public benchmark.

D. Conte et al. (Eds.): DeLTA 2023, CCIS 1875, pp. 456–465, 2023.
https://doi.org/10.1007/978-3-031-39059-3_30

1.2 Motivations and Contribution

Some of the face filters destroy or modify biometric features to a large extent and can promote the intent to impersonate or deceive face recognition systems. The work of Botezatu et al. [1] involved the use of fun selfie filters. However, the unavailability of their dataset makes the research in the field limited. The authors of Hedman et al. [2] have tried to do similar work on Labelled Faces in the Wild dataset (LFW). To aid in counteracting the above-mentioned issues and allow fundamental studies with a common public benchmark, we contribute with an extension of the Face Research Lab London (FRLL) database of facial images that includes the application of several fun-selfie filters from platforms such as Snapchat [3], FaceApp [4] and B612 [5]. The database includes image enhancement filters (which mostly modify contrast and lightning) as well as augmented reality filters such as "Gender Swap", "Hipster Beard", "Puppy", and "Hair Color- Blonde" to the face image.

We start from the popular FRLL database, which contains 1 baseline image and 9 images of the same person in different poses including Neutral Left, Neutral Right, Smiling Left, and Smiling Right. To the neutral front-facing image we apply the different modifications, generating 10 more selfies, each one containing one particular modification. Initially, we had 102 people with 10 poses resulting in 1020 unfiltered images. The application of 10 fun-selfie filters to all 102 images resulted in 1020 more such images. Hence, the total dataset consists of 1020 unfiltered images + 1020 filtered images which results in the generation of 2040 images. The use of a public and widely employed face dataset allows for replication and comparison. We also include the original (unfiltered) images to allow other researchers to incorporate new filters of their choice, or to train different image reconstruction methods (Table 1).

Table 1. Overview of dataset statistics

Total number of images	2040
Image resolution	1350 × 1350 pixels
Number of face classes	102
Number of gender classes	2
Number of ethnicity classes	4
Total number of profiles for each class	10
Total number of filters applied on each class	10

The major contributions of this paper are described below:

- We contribute an extension of the FRLL database [6] of facial images that include the application of several fun-selfie filters from platforms such as Snapchat [3], FaceApp [4] and B612 [5].
- The presented database includes images with image enhancement filters (which mostly modify contrast and lightning) as well as Augmented reality filters (which add computer-generated effects and superimpose them on faces) such as "Gender Swap", "Hipster Beard", "Puppy", and "Hair Color- Blonde" to the face image.

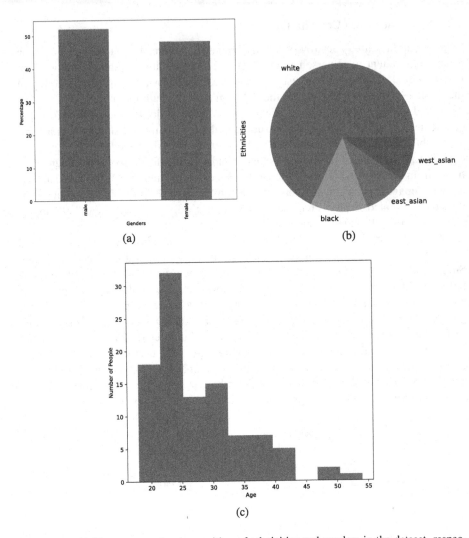

Fig. 1. (a) and (b) represents the composition of ethnicities and genders in the dataset, respectively. (c) denotes the age distribution in the dataset

- The performance of state-of-the-art face recognition methods is also evaluated for our presented database. The evaluation shows how even the best methods are not very effective on filter-applied images (Fig. 1).

2 Significance of Data Construction

1) The dataset can be used to develop new algorithms for face detection or face recognition when the images are post-processed with beautification, occlusion based, or augmented reality filters.

Table 2. Dataset specifications

Broad Area	Computer Vision and Pattern Recognition
Specific subject area	The dataset can be used for research on face detection or face recognition when typical beautification or Fun Selfie filters are used. It contains visible images of faces with different profiles along with their images with these filters.
Type of data	Image (.jpg)
How data were acquired	The data is built from the popular face biometric dataset called the FRLL dataset. The dataset already has images of different 102 different individuals from 10 different angles - neutral front, Smiling front, Neutral left profile, Smiling left profile, Neutral left profile, Smiling left profile, Neutral right profile, Smiling right profile, Neutral right profile and Smiling right profile.
Data format	Raw
Parameters for data collection	The database consists of face photographs of 102 people, containing a large range of variations in pose, expression (smiling and natural), age, gender, and race. Their natural front profile photos are used for the application of selfie filters which include beautification filters, occlusion-based filters, and face modification filters.
Description of data collection	After going through a lot of other publicly available datasets, we decided to use the FRLL Dataset which has images of 102 people along 10 different profiles. Now the next challenge at our hand was to identify and select a good mix of fun selfie filters that include all the different types of filters available these days including beautification, occlusion-based, and AR-based filters. After going through multiple social media and face editing applications popular nowadays and reviewing the different filters they offer for free, we decided to use 3 different FSF apps to apply the beautification and fun selfie filters to the images- FaceApp, Snapchat, and B612. We selected 5 filters for our use case from FaceApp (Haircut, Child, Gender Reverse, Hipster Beard Style, Hair Color - Blonde), 2 filters from B612 (Puppy and So Sad), and 3 filters from Snapchat (Hipster Look, Sparkling Cartoon and Body Mellow Glow). For creating the FSF filtered section of the dataset, we manually transferred these images to our phone, applied the above-mentioned filters to these photos, and transferred them back to our systems.
Data source location	Face Research Lab London Dataset [6]
Data accessibility	GitHub Repository Link: FRLL-Beautified

2) The provided data may aid in developing systems that are resilient to the presence of fun selfie filters. Application examples include the normal operation of face detection/recognition systems for crime investigation on social media or automatic pre-filtering images in personal devices or cloud repositories.

3) The provided data may be used to recreate new social media filters, not considered and benchmark the results against studies made by us or by other users of the

database since we also release the employed images in their original form (unfiltered).

4) The intended purpose of the dataset is to be used in the training and evaluation of face detection or recognition systems when the images are post-processed with Fun Selfie Filters. Application examples include the normal operation of face detection/recognition systems for crime investigation on social media or automatic prefiltering of images in personal devices or cloud repositories. Using a public and widely employed face dataset allows for replication and comparison, an important aspect of academic research (Table 2).

3   Related Work

There has been significant work around facial image recognition. However, only a few datasets have worked on images with filters. Similarly, not a lot of datasets are publicly available with age, gender and ethnicity labels. The dataset of Pilots Parliament Benchmark [7] has around 1000 images with age and gender labels as well. However, it marks ethnicities based on skin color only. The FairFace dataset [8] is a collection of more than 100,000 images in the wild, with age, gender, and ethnicity annotations as well. However, it does not use filtered images. The Labelled Faces in the Wild Beautified dataset [2] includes images with filters applied but the filters used are very basic occlusion and beautification filters. The dataset has a total of 34,000+ images but the image resolution is very low (64×64). Riccio et al. [9] provided a framework for applying filters to images. They used it to beautify the dataset of LFW and FairFace. Although they have mentioned that Augmented Reality-based filters which distort the facial key points can be applied, they have applied only AR beauty filters to these datasets. These datasets also do not have age, gender, or ethnicity labeling. To address the problems with the above-mentioned datasets, we have created this dataset of more than 2000 images, with beautification filters (with or without AR) i.e. Face Mellow Glow, Occlusion-based filters i.e. Puppy, and AR-based key-point distorting filters i.e. Gender-Swap. We also present these images with filters applied with the age, gender, and ethnicity labels. The image quality in our case is also very high (1350×1350). A comparative analysis of all the related works can be shown in Table 3.

Table 3. Summary Statistics of Various Public Face Datasets

Datasets	Number of Images	Age	Gender	Ethnicity	Beauty Filters	Occlusion Filters	Distortion Filters
PPB [7]	1000	✓	✓	Skin Color Only	×	×	×
FairFace [8]	108000	✓	✓	✓	×	×	×
LFW-Beautified [2]	34592	×	×	×	Non-AR	Non-AR	×
OpenFilter [9]	142592	×	×	×	AR	AR	×
FRLL-Beautified (Ours)	**2040**	✓	✓	✓	**All**	**All**	**All**

4   Data Description

To create the dataset, we have deployed the baseline images to apply different fun selfie filters. For this purpose, we have used the FRLL dataset [6]. The FRLL dataset has images of 102 different individuals in 10 different positions as shown in Fig. 2.

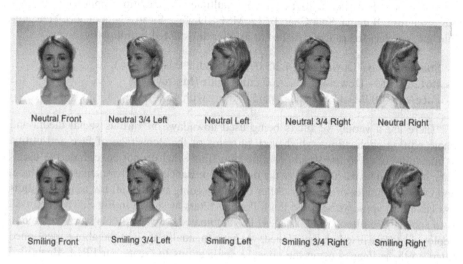

Fig. 2. All the different unfiltered facial profiles available for each person

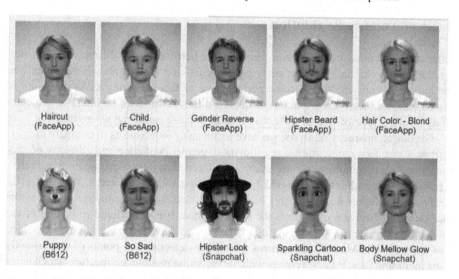

Fig. 3. All the different filters applied to the 'Neutral Front' profile of each person

Next, the task in front of us was to decide the applications to which we want to apply the FSF filters. There were many options that we could have used for our purpose. Our main criteria for selecting the app were the following:-

- Must be popular so that the outcome of this project benefits a larger audience
- Should provide a variety of filters that are very popularly used
- Should be easy to use
- Should allow the application of filters to images in the gallery

Based on these criteria, after trying out multiple image editing applications including Instagram, Banuba, SnapChat, B612, MSQRD, Face Swap, Camera 360, YouCam-Fun, etc. **we decided to use 3 different applications listed in following:**

- **Snapchat** (1B+ Downloads, 4.2/5.0 rating with 29M+ reviews)
- **B612** (500M+ Downloads, 4.2/5.0 rating with 7M+ reviews)
- **FaceApp** (100M+ Downloads, 4.5/5.0 rating with 4M+ reviews)

There are a variety of filters being used nowadays on various social media and photo editing applications which vary in terms of their coverage of face and utility. While some of them are basic beautification filters that do not cover/hide the user's face, others can be used to change the overall appearance (hide/cover a part of the face) and even the gender of the user. After carefully going through all the available filters available in 3 apps, we selected the following filters as shown in Fig. 3. The next task is to apply these filters to the images. The automated softwares/tools were not able to deploy all filters listed above. Hence, we have manually applied the above-mentioned filters to these images using the above 3 applications to create our FRLL-Beautified dataset.

5   Performance Evaluation

After acquiring the dataset, we have compared the performance of different state-of-the-art methods on our dataset. This study is first of its kind in this area where we analyze the impact of applying filters to the performance of face recognition. As discussed in aforementioned sections, we have considered different varieties of filters to critically study how the state-of-the-art methods perform in those cases and where they fail.

For our comparative analysis, we have to consider four aspects to compare the performance, i.e. face recognition, age estimation, gender prediction and ethnicity prediction. For face recognition, we are comparing three state-of-the-art methods namely - **ArcFace + RetinaFace, FaceNet512 + MediaPipe** and **VGGFace + MTCNN** [10]. ArcFace and RetinaFace are deep CNN based models where RetinaFace works as the face detector, ArcFace recognizes the detected face. It is considered to be one of the most effective face recognition system. FaceNet and MediaPipe are libraries provided by Google. While MediaPipe performs robustly in detecting key points of various body parts including face, FaceNet comes with the inception module architecture which helps it reduce the depth and number of trainable parameters without compromising on the accuracy. MTCNN is also one of the best face detector systems which has surpassed even human level accuracy. Combined with VGGFace as the face recognizer, it performs significantly well. Thus these models can be considered to be one of the best systems which we have.

We run the above mentioned systems for all the filter applied images in our self created dataset and calculated a performance metric i.e. **Percentage of Correct Predictions**. For the ith image, let y_i denote the correct label and $\hat{y}_i$ denote the predicted label. If N denotes the total number of images, then the Percentage of Correct Predictions i.e. $\%_{CorrectPredictions}$ is defined as the fraction of correctly predicted images per 100 images tested. The results for the above mentioned systems are shown in Table 4.

$$\%_{CorrectPredictions} = \frac{1}{N} \sum_{i=0}^{N-1} 1(\hat{y}_i = y_i) \times 100 \tag{1}$$

Table 4. Performance analysis of the existing state-of-the-art methods on our dataset

State-of-the-art	$\%_{CorrectPredictions}$	Filters where it failed
ArcFace + RetinaFace	90 %	Hipster Look
		Sparkling Cartoon
FaceNet512 + MediaPipe	83.49 %	Hipster Look
		Sparkling Cartoon
		Child
VGGFace + MTCNN	70.21 %	Hipster Look
		Sparkling Cartoon
		Gender Swap

It is clear from Table 4 that even the best-performing image recognition systems fail at the filters **Hipster Look** and **Sparkling Cartoon** provided by Snapchat. Other filters failing these systems are the **Gender Swap** and **Child** filters provided by FaceApp. Similarly, for the estimation of age, gender, and ethnicities from these filtered images, we used the HyperExtended LightFace [11] which is a facial attribute analysis framework. They claim to have surpassed human-level accuracy with their deep CNN-based models. Age estimation is a regression problem while Gender and Ethnicity estimation are classification problems. Thus we have used different metrics for these two different categories of problems. For age estimation, we have used the metrics, i.e. **R2 Score, Mean Absolute Error, and Mean Squared Error**. Similarly, for gender and ethnicity estimation, we have used the metrics - **Accuracy, Precision, Recall, and F1 Score**. The experimental results of the models in this framework are given in Table 5 and 6.

Table 5. Performance of Age estimation of [11] on our dataset

Performance Metric	Age Model
R2 Score	-0.35
Mean Absolute Error	6.81
Mean Squared Error	67.61

It is evident from the analysis that even the most robust systems claiming to surpass human-level accuracy on unfiltered images do not perform perfectly on filtered images. Thus, the performance of a more robust system should ideally be unaffected by the application of filters. The dataset created clearly demonstrates the gap between the two systems we currently have and the systems which we want to achieve.

Table 6. Performance of Gender and Ethnicity estimation of [11] on our dataset

Performance Metric	Gender Model	Ethnicity Model
Accuracy	0.733	0.754
Precision	0.740	0.437
Recall	0.729	0.303
F1 Score	0.728	0.305

Conclusion and Future Scope

We have created a dataset with FRLL face images by applying fun selfie filters. However, we aim to use this dataset and perform an assessment of the existing state-of-the-art methods in face recognition. We can also move forward and try to develop a filter-resistant face recognition model by training on this dataset. Releasing the subset of images we have employed in its original form (unfiltered) would also allow other researchers to incorporate new filters of their choice. The data contained in the database can be used as-is without filtering or enhancement. Apart from this, the dataset size can be increased by applying ten filters to all the facial profiles of the individuals. Hence, the size of the dataset will increase substantially and can lead to better research and results.

References

1. Botezatu, C., Ibsen, M., Rathgeb, C., Busch, C.: Fun selfie filters in face recognition: impact assessment and removal. IEEE Trans. Biometrics Behav. Identity Sci. **5**, 91–104 (2022)
2. Hedman, P., Skepetzis, V., Hernandez-Diaz, K., Bigun, J., Alonso-Fernandez, F.: LFW-beautified: a dataset of face images with beautification and augmented reality filters (2022)
3. Inc., S.: Snapchat
4. Limited, F.T.: Faceapp
5. Corp., S.: B612
6. DeBruine, L., Jones, B.: Face Research Lab London Set (2021)
7. Buolamwini, J., Gebru, T.: Gender shades: intersectional accuracy disparities in commercial gender classification. In: Friedler, S.A., Wilson, C., (eds.) Proceedings of the 1st Conference on Fairness, Accountability and Transparency, vol. 81, pp. 77–91. Proceedings of Machine Learning Research PMLR (2018)
8. Karkkainen, K., Joo, J.: FairFace: face attribute dataset for balanced race, gender, and age for bias measurement and mitigation. In: Proceedings of the IEEE/CVF Winter Conference on Applications of Computer Vision, pp. 1548–1558 (2021)

9. Riccio, P., Psomas, B., Galati, F., Escolano, F., Hofmann, T., Oliver, N.: OpenFilter: a framework to democratize research access to social media AR filters (2022)
10. Serengil, S.I., Ozpinar, A.:LightFace: a hybrid deep face recognition framework. In: 2020 Innovations in Intelligent Systems and Applications Conference (ASYU), IEEE, pp. 23–27 (2020)
11. Serengil, S.I., Ozpinar, A.: Hyperextended lightface: a facial attribute analysis framework. In: 2021 International Conference on Engineering and Emerging Technologies (ICEET), IEEE, pp. 1–4 (2021)

CSR & Sentiment Analysis: A New Customized Dictionary

Emma Zavarrone and Alessia Forciniti[✉]

IULM University, 20143 Milan, Italy
{emma.zavarrone,alessia.forciniti}@iulm.it

Abstract. Communication concerning the CSR pillars is key to sustainable corporate development. Sentiment analysis (SA) is a sub-area of natural language processing for studying communication through the classification of negative or positive opinions. Measuring sentiment is characterized by pitfalls related to: a) the context, where the polarity classification depends on the domain; b) the methods, if lexicon-based, machine learning, or their combination; c) the language, where the lack of resources (different from English) in literature was observed. Strategic communication based on CSR has no domain resources for investigating sentiment, neither in English nor in other languages. Our contribution is placed within the methodological setting of SA for the sustainability framework. We combined lexicon-based methods with machine-learning ones to build a customized lexicon for analyzing the CSR. The innovation concerns: 1) a domain corpus-based approach for improving a general pre-constructed dictionary; 2) the application for Italian; and 3) the performance assessment through machine learning. We developed an algorithm characterized by a multi-stage model that combines text analysis with network analysis and captures semantic concordances through an index of keyword content in the text. To validate our model from a machine learning perspective, we divided our data collection into five random samples: one sample was utilized as a train set or baseline for the lexicon's implementation, and four were used as test sets. The study showed a notable increase in performance metrics across all samples, demonstrating the effectiveness of our proposal in building a customized lexicon for analyzing CSR in the Italian context.

Keywords: CSR · Sentiment analysis · Lexicon-based and machine learning methods

1 Introduction

The awareness of the paradigm of sustainability [13] has characterized many fields of study, including corporate organizational culture [1], by means of a model of strategic management based on corporate social responsibility (CSR). The changes triggered by the dynamics of globalization and by the influence of transnational economic actors have accelerated the systematic and transversal knowledge of the social, economic, ecological, and political effects connected to corporate activities. Therefore, to focus efforts on the integration of sustainability issues in the corporate organizational culture represents an ethical standard that incorporates the role of business in activities based on social responsibility and on reducing the impact of the negative effects of business [4]. The idea of corporate sustainability by means of the CSR has emerged in the literature concerning business ethics during the mid-1990s to indicate a process of corporate activities oriented to the development of three dimensions: environment, economy, and social context [39]. Successively, corporate sustainable development acquired the shape of a model of strategic management where the main competitive advantage lies in a system of "governance" of transactions and relations with stakeholders and maximizing long-term profits [6]. Many businesses are now focused on explaining their policies and meeting social responsibility standards. The dissemination of CSR information through marketing, product labeling, media relations, CSR reports, and websites has grown in importance as a means of establishing and maintaining a company's credibility in the eyes of stakeholders. While CSR reports and websites are seen as subtle communication, CSR advertising and public relations are thought to be open communication [29], where the subtle communication is highly important, but also more challenging to study.

The widest literature investigated the relationships between CSR and corporate performance [26], while less developed is the integrated view of corporate sustainability management [1] and language used to communicate the CSR to their own stakeholders.

One of the methods used to study communication and obtain an inclusive perspective on a specific domain of interest is sentiment analysis (SA) or opinion mining (OM). This sub-area of natural language processing *NLP* is pivotal [21] for classifying opinions, emotions, and evaluations on a topic or a series of sources.

The analysis of communication and more properly of sentiment represents one of central topic for scholars of communication. In fact, SA finds application in several contexts, such as businesses, politics, social media and governments, to know, for example, opinion on stock markets [2], economic systems [3], policies [19], media effects of phenomena [37], people's opinion (reviews) to make decisions [22], gender-based issues [36], news of biotechnology [25], security issues; despite the sentiment hidden in a written text is generally expressed with ambiguous language [33], and some fields of research can lack agreed-upon conceptualization. In particular, the approaches based on sentiment characterized the last years of scientific contributions aimed at analyzing the CSR [7,38]. The attention on CSR is increasing in different worldwide countries because of how its strategic communication affects stakeholders' choices, and also in Italian context [9–20].

Nevertheless, measuring sentiment in terms of emotionality, negativity, positivity, and subjectivity is a task characterized by pitfalls [15–24], often referred to the context of analysis, the methods, or the language of the texts to be studied. Lexicon-based and machine learning methods are the two major methods for assessing a sentiment, the first based on a dictionary of words and polarity labels, the second focused on training a classifier in a labeled dataset and predicting sentiments using the model it creates, but it has been demonstrated that, in the absence of "good-quality" labeled data for training in a specific domain, a lexicon-based system can be as good as a machine learning system [27].

The lexicon-based approaches are less expensive in terms of time and resources than others, since they are based on pre-constructed lists of polarized terms, where some lexicons are built on general knowledge and others oriented to cover specific domains. However, in the majority of cases, the field of SA application suffers from limits and unavailability of resources. A limitation is often connected to the lack of lexicons able to cover some fields of research, because the polarity classification may depend on the domain, seeing as the meanings of the words are related to the contexts. The general knowledge lexicons cover wide phenomena (politics, sport, social media, etc.), but they are less specific than domain-based ones, and sometimes less adaptable in studying specialized contexts. Another issue is the language, since the majority of resources, both general and domain-based lexicons are developed for English. Consequently, most of the lexicons of other languages are based on translation processes. In addition, the use of general-knowledge dictionaries is predominant for low-resource languages. With reference to the strategic communication of CSR, we currently have no resources for investigating the sentiment of this specific domain, neither in English nor in other languages.

Therefore, the contribution aims to study an ever-evolving context framework such as that of corporate strategic communication on sustainability, whose domain is specific but, at the same time, transversal to different cultural and business aspects, and in a methodological setting characterized by limits and challenges above mentioned.

Our paper falls in the operative limits of *NLP* and aims to propose a successful strategy for correctly investigating the subtle communication of CSR in Italian, by using lexicon-based approach and machine learning.
We explored two main research aims (RQs):

RQ_1: Is it possible to approach a study based on SA to investigate CSR in terms of official communication to stakeholders in Italian?

RQ_2: Can we practically evaluate the performance of lexicon-based methods to analyze the topic of sustainability?

2   State of Art

SA is performed through computational approaches that allow the scalability and replicability of manual coding [42]. Its application may be by means of machine learning approaches, lexicon-based methods, or the combination of these two in a hybrid approach.

With machine learning (ML), a model is learned from data with sentiment labels and then used to the classification of new documents. This technique needs labeled data, which is normally produced through labor-intensive human annotation. Are ML approaches, the naïve Bayes classifiers [18]; Bayesian networks [32]; linear classifiers, such as for instance the support vector machine (SVM) [8]; or decision trees [17].

The lexicon-based (LB) approach is based on polarity lexicons, insofar as a negative, neutral or, positive valence is attributed to a given text based on a list of polarized terms manually created [40] or by automatic or semi-automatic approach [41]. The LB techniques are considered more scalable and resilient about classification across domains than supervised classification methods, which are preferable in specific contexts of domain [43].

In literature several polarities of sentiment have been proposed, some based on three categories: $-1, 0, 1$, to indicate negative, neutral, or positive classifications, while others are based on different intensity scores, going from mostly negative to more positive values. These methods allow the interpretation of a single word based on its context-dependency. In SA lies also the definition of emotional categories (such as joy, anger, sadness, etc.), of which one of the most popular that we can mention is NRC Emotion Lexicon [28], which proposes the positive or negative classification of texts and eight main categories of emotions developed by Plutchik [35]. Some contexts of study instead rely on a binary polarity classification (only positive or negative polarity) [16].

The words classification depends directly on the domain in which each word fits in the context of discourse, and the lexicons of the sense of reference cannot cover all the meanings of the words, which may be specific to a domain. In fact, the SA field of application offers a variety of lexicons, some of which are focused on general sentiment knowledge by including various areas such as sports, politics, and social media [16–31], while others are domain-based and present sector specificities such as the economic-financial environment [14–23]. Although general knowledge lexicons have the advantage of covering a wider range of phenomena, they perform less well than classifications produced by domain-based lexicons, because some terms may not be present in a general knowledge lexicon or may have a different contextual interpretation.

Nevertheless, using pre-constituted dictionaries represents the most popular and faster solution for sentiment classification of texts, and is useful in contexts where the available data are limited or missing. As presented in the section of research problems, the topic of the unavailability of sentiment resources is fundamental, and may principally depends on two reasons: a) unexplored contexts of study, such as the topic of sustainability in business, and b) linguistic issues. This last is one of the biggest limitations for general knowledge or domain-based lexicons, because the largest number of resources developed in literature have been realized in the English language. Attempts to adapt lexicons to other languages often result from English resource translation processes. Also for the Italian language, we find a limited availability of lexicons (often, results of translations) both for generalist and domain-based dictionaries. Two of the best-known Italian resources based on general knowledge are SentiWordNet [10] and National Research Council (NRC) Emotion lexicon [28]. In SentiWordNet, scores are measured on a continuous scale with different intensities from 0 to 1, while the NRC Emotion Lexicon presents a classification based on binary polarity, negative or positive (-1 or 1), and eight categories of emotions (anger, anticipation, disgust, fear, joy, sadness, surprise, and trust). Both resources are not specifically designed to capture the phenomenon of sustainability and may be subject to translation errors, or non-correspondence in the context that the paper plans to analyze.

3   Methodology

The method under development goes in the direction of proposing an innovative model to improve the lexicon's performance and implement the lexical items related to CSR in Italian corporate communication.

The innovative feature lies in three aspects: 1) the investigation of an unmapped domain by means of a domain corpus-based approach and the building of a customized lexicon from a general pre-constructed dictionary; 2) the application for the Italian language; 3) the performance assessment of improvements through machine learning perspective.

Our model is based on a multi-stage approach as designed in Fig. 1 and proposed to follow:

- STEP 1: Selection of two general lexicons for Italian language (SentiWordNet [10]; National Research Council Emotion Lexicon (NRC) [28]).
- STEP 2: Textual data collection on Italian strategic communication of CSR.
- STEP3: Development of an algorithm for building a sentiment-customized lexicon on corporate sustainability by integrating a linguistic approach and network properties [44] based on data collection of STEP 2.
- STEP 4: Comparison between performance measurements [5] before and after the lexicon implementation.

STEP 1: General Lexicons for Italian Language. In our study, we selected two of the most popular general lexicons for Italian language.

The first lexicon is SentiWordNet developed by Esuli & Sebastiani (2006) [10], and implemented in the version 3.0 by Baccianella *et al.*, in 2010 [3], from a corpus-based semantic approach to detect positive, negative and objective (neutral) polarity. It is based on MultiWordNet (MWN) [34], a multilingual lexical database which groups the words with similar meanings attributing them the same polarity, which for Italian

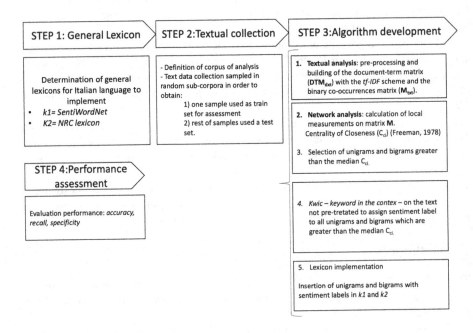

Fig. 1. Flow chart of model.

involves 20,093 lexical items. The scores are measured on a continuous scale from 0 to 1, which - in this context - we converted into a binary sentiment category labelled as negative or positive, by using only the sign of the score and removing the polarity equal to 0 marked as objective or neutral.

The second lexicon is NRC Lexicon developed by Mohammad and Turney in 2010 from an human annotations at sense level rather than at word level using Amazon service called Mechanical Turk to detect simultaneously the semantic orientation (positive or negative) and eight categories of emotions. It consists in 5,468 lexical items. The classification of NRC is based on binary polarity (positive or negative). In this work, we do not consider the emotional classification but only positive and negative polarities.

STEP 2: Textual Data Collection. Textual data collection consists in Sustainability Reports drafted by Italian companies listed which closed the financial year at 31st December 2021, following the Legislative Decree No. 254 of 30th December 2016 [12].

The complete list records 196 companies but, we selected only 130 reports. The missing documents depend on the unavailability of the resource or for the language drafting different from Italian that represents our focus.

We realized a sampling procedure on the 130 reports collected, so obtaining five sub-corpora, each one composed by a random sample of 26 reports. The first sample has been used as a baseline for the development of the next step. Specifically, the corpus of this first sample represents the corpus-based source for realizing the algorithm aimed at the building of a customized lexicon on CSR that implements the Italian general lexicons of STEP 1. In addition, the first sample of reports will represent the training set or baseline of the model to validate the fine-tuning of the other samples that have been used as test sets for the assessment step.

STEP 3: Algorithm Development. The development of the algorithm is aimed at detecting in a coherent way the lexical units (whether they are unigrams or bigrams) useful to implement the generalist lexicons of the Italian language SentiWorkNet and NRC. This algorithmic procedure is used to test whether fine-tuning has improved the lexicon's ability to capture more words within CSR strategic communication, both for positive and negative labels.

1. *Textual Analysis*

The textual analysis procedure concerned the pre-treatment of the baseline sample' corpus through: normalization, lemmatization, removal of punctuations, numbers, company's name, and Italian stop words proposed by ISO 639-1.

Feature extraction was performed by building the lexical table or document-term matrix DTM_{dxt} where the weights system was based on the specificity of a term in inverse proportion to the number of documents in which it occurs expressed by means of the coding *tf-IDF (term frequency-inverse document frequency)*. Each row of lexical table was represented by a CSR report and each column indicated a term. The lexical matrix was subsequently dicotomized to record only the presence or absence (1 or 0) of the words in the 26 CSR documents belonging to the baseline sample. The dichotomized matrix is represented by **DDTM**. In order to keep track of the structure of each text without breaking the links with semantic aspects, we computed the co-occurrences between words by transforming the **DDTM** into an adjacency matrix M_{txt}, in which both rows and columns show the terms and their cells indicate the ties between words in terms of dichotomy of relationship (1 or 0).

2. *Network Analysis*

The adjacent matrix **M** recalls the structural properties of network analysis, in which words can be represented by nodes and co-occurrences as links between them, formally represented with a graph G (V, E), where V are the terms and E are the ties. As with graphic display, the M matrix can be analyzed for global and local network centrality. In this case, we focused on the centrality of nodes to understand which words are more important or have a focal role in CSR strategic communication. For this purpose, we selected unigrams and bigrams based on the measure of local centrality denoted by Freeman's closeness C_{ci} [11], which represents the shortest path from one to the other. After calculating the C_{ci} for all word nodes, we found the median value of C_{ci}, and filtered only unigrams and bigrams with a C_{ci} value higher than its median.

3. *Kwic for Sentiment Vector Labelling*

After detecting unigrams and bigrams with C_{ci} greater than its median value, a positive or negative polarity label should be assigned to each of them. For this task, we used an index known as *kwic*, an acronym for *keyword in context* that makes it easy to find a keyword within the context in which it is inserted. The structure of this index, especially used to write catalogs of libraries, allows you to view the corpus part before and after the unigram or bigram we are looking for, so that you can reconstruct the semantic dimension or of context.

4. *Lexicon Implementation*

The unigrams and bigrams of CSR domain manually labeled as negative or positive have been incorporated within the general lexicons SentiWordNet (k1) and NRC (k2). This procedure represents their implementation aimed to increase their performance to capture the sentiment.

The following code summarizes the above-mentioned steps for the development of our algorithm:

```
Begin
  From DTM(dxt) to M(txt)
    where tij = 1 if
    there is and edge from term i to term j,
    0 = otherwise;
  Compute Closeness centrality (C);
  Let X = C > Median C;
    if X = Dict (k) stop;
    if X NOT =  Dict (k) go to KWIC (X);
  Compute X1 = sentiment vector
  (X-X(Dict(k));
    Add X1 to Dict (k);
End.
```

STEP 4: Performance Assessment. The performance of the implementation of k1 and k2 will be tested through a machine learning approach, where the first sample is used as a baseline or train set to train the fine-tuned sentiment labels, and the remaining four samples are used as domain documents to test the effectiveness of the training, such as test sets.

In particular, we applied performance measures for the binary classification: positive and negative sentiment. We estimated: (a) the overall efficiency of the classifier in terms of accuracy; (b) the effectiveness of the classifier in identifying positive labels as sensitivity or reminder; and (c) the efficacy of the classifier in determining negative labels in relation to specificity.

4   Results and Discussions

4.1   Exploratory Textual Data Analysis

We also analyzed the vocabulary of each sample through an exploratory textual data analysis (EDTA) that replicated the same pre-processing and feature extraction used for the baseline corpus.

EDTA findings on the five sub-corpora show that each sample is characterized by a generalist vocabulary with few specialism or technicalities of the domain concerning the corporate sustainability (Table 1).

The overall lexical diversity of the five corporations based on CSR strategic communication is highlighted through one of the most widely used measures in literature, the type-token ratio (TTR), with values ranging from 30% to 41.3%. Therefore, there is substantial lexical diversity. This could be a favorable element to test the implementation of lexicons on the corpora test sets, as the domain and the lexical forms differ little. The training set could therefore effectively capture the sentiment of the documents.

4.2   Lexicons Performance

The overall sentiment of 130 reports showed a predominance of positive sentiment both by using both general lexicons SentiWordNet and NRC. Specifically, we detected 257,207 negative words and 508,598 positive ones by using SentiWordNet; and 120,905 negative words and 397,615 positive ones with NRC. Both positive and negative sentiments are almost similarly distributed among sub-corpora: $19\% \leq$ positive polarity $\leq 22\%$, while $18\% \leq$ negative polarity $\leq 23\%$.

Our model allowed us to detect 65 unigrams and 82 bigrams greater to median of closeness centrality.

By using the manually annotation procedure to attribute positive or negative valence to each of these item, we determined 57 positive unigrams and 8 negative ones, and 61 positive bigrams and 21 negative ones. The implementation of the two lexicons through the new 147 lexical polarized items has returned a customized-lexicon of SentiWordNet composed of 20,240 items, and of NRC consisting of 5,615 items.

Table 1. EDTA on five sup-corpora.

Sample	N.lemmas	Unigrams	Bigrams
First (baseline)	21,754	gestione (management), gri (global reporting initiative), rischio (risk), emissione (emission),...	consumo energetico (energy consumption), capitale umano (human capital), codice etico (ethics code), lotta corruzione (anticorruption),...
Second	21,867	salute (health), consumo (consumption), impianto (facility), sicurezza (safety),...	aspetto materiale (material aspect), gestione rischio (risk management), pari opportunità (equal opportunity), corporate governance,...
Third	22,496	modello (model), capitale (capital), formazione (training), obiettivo (objective),...	controllo rischio (risk control), collegio sindacale (supervisory board), risorse umana (human resource), corporate governance,...
Fourth	20,915	mercato (model), impatto (impact), gri (global reporting initiative), salute (health),...	gestione responsabile (responsible management), ricerca sviluppo (research development), fonte rinnovabile (renewable source), donna uomo (woman man),...
Fifth	20,644	ambientale (environmental), sviluppo (development), processo (process), sistema (system),...	salute sicurezza (health safety), controllo interno (internal control), materia prima (raw material), etica integrità (ethics integrity),...

4.3 SentiWordNet

The implementation improved SentNordNet in all the measures of performance, with an average improvement of 5.13% (Table 2).

The accuracy presents +7% on average; recall +3.2% on average; and specificity +5.2% on average, demonstrating the effectiveness of domain-based customization on CSR. In particular, the best improvement performance is shown with reference to specificity, such as the effectiveness of the classifier to determine the negative labels.

Table 2. SentiWordNet performance before and after implementation.

Sample	Before Implementation			After implementation		
	Accuracy	Recall	Specificity	Accuracy	Recall	Specificity
First (baseline)	0.53	0.45	0.69	0.59	0.52	0.76
Second	0.53	0.38	0.69	0.60	0.39	0.74
Third	0.51	0.37	0.68	0.57	0.39	0.69
Fourth	0.55	0.36	0.70	0.68	0.39	0.77
Fifth	0.65	0.36	0.70	0.68	0.39	0.75

4.4 NRC Lexicon

Its implementation returned a performance improvements of +4.73% on average of overall performances (Table 3).

Accuracy increased on average by 8.8%; recall improved of 0.8% on average, and specificity increased of 4.6% on average. Also for the NRC lexicon, as for SentiWord-Net, the implementation has further improved specificity and therefore, the classification of polarized unigrams and bigrams as negative.

Table 3. NRC Lexicon performance before and after implementation.

Sample	Before Implementation			After implementation		
	Accuracy	Recall	Specificity	Accuracy	Recall	Specificity
First (baseline)	0.68	0.30.	0.78	0.69	0.32	0.84
Second	0.64	0.28	0.77	0.71	0.28	0.83
Third	0.53	0.28	0.80	0.70	0.28	0.85
Fourth	0.66	0.27	0.79	0.68	0.28	0.81
Fifth	0.52	0.26	0.79	0.69	0.27	0.83

4.5 Comparison of Performances Between Lexicons

The assessment measures computed on both lexicons before the implementation show that NRC performed better in sentiment classification than SentiWordNet. We detected 60.6% on average of accuracy, giving us a better performance of +5.2% compared to SentiWordNet. The same result comes from specificity, with an average performance of 78.6%, denoting a good ability in detecting words with negative polarity and a 9.6% higher performance than SentiWordNet. On the contrary, SentiWordNet performed well on the positive side, showing a greater average percentage of recall equal to 38.4%, with a difference of 10.6% compared to NRC.

The performance measured on implemented lexicons showed an overall improvement both in NRC and SentiWordNet, respectively +4.73% and +5.13%, and this last appears to be the dictionary that on average earned the most in performance (+0.4%).

This shows that SentiWordNet is the lexicon that has most benefited from the implementation procedure. In fact, it increased on average by 0.6% specificity and 2.4% recall compared to the NRC. However, accuracy is steadily higher in the NRC, with an average difference of +1.8% compared to the other lexicon. Along with the average accuracy of 69.4%, the NRC also shows higher percentage averages for specificity, which are 83.2%. Compared to the other lexicon, however, NRC had a smaller increase while SentiWordNet gained 0.6% more improvement. The effectiveness in ranking positive labels, on the other hand, remains higher in SentiWordNet, at 41.6% on average and also denoting a greater increase of 2.4% compared to NRC. Therefore, despite the greater effectiveness of the NRC classifier in identifying negative labels, the implementation of the lexicon has mostly improved SentiwordNet's measures over the NRC. Nevertheless, both general lexicons of the Italian language have shown that the customization of the domain makes the positive and negative classification of the company's strategic communication on sustainability more effective.

5   Conclusion an Further Developments

In this contribution, we study the paradigm of sustainability in Italian corporate culture. In particular, we analyzed the sentiment of corporate strategic communication addressed to their own stakeholders by means of the language adopted in reports drafted by listed companies. In a methodological framework imbued with challenges and limitations for SA and in the specific domain of sustainable development, we proposed an innovative model for the development of a customized lexical-based tool to investigate the phenomenon in Italian. Three areas make up our work's innovative aspect: 1) the analysis of an unmapped domain using a domain corpus-based strategy and the creation of a unique lexicon from a generic pre-constructed dictionary; 2) the application for the Italian language; and 3) the performance evaluation of advancements made through machine learning.

The methodological approach presented in this contribution has proven effective in meeting the two research objectives pursued and outlined in the second paragraph. In fact, with reference to RQ_1, we developed a model that allowed us to study sentiment within the framework of CSR in Italian. We determined the lexical unities (both unigrams and bigrams) most frequently used in official documents on sustainable management that are drawn up by listed companies. We conducted an automatic context analysis of each lexical unit by attributing a semantic-based positive or negative valence. This demonstrated that we can approach a domain-based study for sustainability. To respond effectively to RQ_2 and therefore practically evaluate the possibility of conducting an analysis of CSR based on a lexicon-based method for SA, we implemented two generalist lexicons used for the sentiment of the Italian language and compared performance measures, highlighting the ability of our model to better capture the context of CSR and sustainability in business documents.

More specifically, we employed a multi-stage model that combines text analysis with network analysis to develop an algorithm aimed at the creation of a customized lexicon on CSR that essentially implements two of the most common Italian general lexicons through the use of a textual corpus coming from the social reports of Italian listed companies at December 31st, 2021. Under the cross-validation approach, we separated our data collection into five randomly selected samples: Four were used as test sets, while one served as a training set for the implementation.

The procedure of implementation tested by the machine learning approach and by means of the performance measures of accuracy, recall, and specificity highlighted a significant improvement in performance. This improvement was observed in all samples and to a greater extent with regard to the effectiveness of classifying labels with negative polarity and accuracy. Less efficient is the positive label classifier (recall). Finally, SentiWordNet is the vocabulary that has earned the most in terms of performance improvement after domain implementation. Although both lexicons present specific features of improvement in post-implementation performance, they demonstrate that the items found in the first sample used by training are effective in studying the topic of sustainability in other CSR documents.

Future developments are oriented toward replicating the model proposed for other sources of sustainability in order to improve the process of data training. Furthermore, the next efforts involve the creation of a package that can automate the procedure for detection of unigrams and bigrams proposed in our algorithm and the tagging procedure for the activity of polarity attribution by using keywords in the context. A final perspective is to capture the performance by applying other approaches, such as neural networks. In this direction, we intend to adopt models which train a three-layered neural network from Part of speech findings to learn word representations and build a lexicon for implementing the general dictionaries. The future perspective is to use Continuous Bag of Words Model (CBOW) and Skip-gram. In contrast to Skip-gram, which learns representations by making predictions about each context word based on the target word, CBOW learns representations by making predictions about the target word based on its context words. We can test the results by using the performance metrics, and compare the neural approach with our model to verify the effectiveness.

References

1. Aguinis, H., Glavas, A.: What we know and don't know about corporate social responsibility: a review and research agenda. J. Manage. **38**(4), 932–968 (2012)
2. Arnold, I., Vrugt, E.: Fundamental uncertainty and stock market volatility. Appl. Financ. Econ. **18**(17), 1425–1440 (2008)
3. Baccianella, S., Esuli, A., Sebastiani, F.: SentiWordNet 3.0: an enhanced lexical resource for sentiment analysis and opinion mining. In: Calzolari, N., Choukri, K., Maegaard, B., et al. (eds.) Proceedings of the International Conference on Language Resources and Evaluation, LREC 2010. European Language Resources Association (ELRA), Valletta, Malta (2010)
4. Carpenter, G., White, P.: Sustainable development: finding the real business case. Corp. Environ. Strat.: Int. J. Sustain. Bus. **11** (2), 51–56 (2004)
5. Catelli, R., Pelosi, S., Esposito, M.: Lexicon-based vs. BERT-based sentiment analysis: a comparative study in Italian. Electronics **11**(3), 374 (2022)
6. Chandler, D., Werther, W.B.: Strategic Corporate Social Responsibility. Stakeholders, Globalization, and Sustainable Value Creation, 3ł ed. SAGE Publications Inc., Thousand Oaks, CA(2014)
7. Che, S., Li, X.: RETRACTED ARTICLE: HCI with DEEP learning for sentiment analysis of corporate social responsibility report. Curr. Psychol., 1–1 (2020). https://doi.org/10.1007/s12144-020-00789-y

8. Chin, C.C., Tseng, Y.-D.: Quality evaluation of product reviews using an information quality framework. Decis. Support Syst. **50**(4), 755–768 (2011). https://doi.org/10.1016/j.dss.2010. 08.023

9. Esposito, B., Sessa, M.R., Sica, D., Malandrino, O.: Exploring corporate social responsibility in the Italian wine sector through websites. TQM J. **33**(7), 222–252 (2021)

10. Esuli, A., Sebastiani, F.: SENTIWORDNET: a publicly available lexical resource for opinion mining. In: Proceedings of the Fifth International Conference on Language Resources and Evaluation (LREC 2006). European Language Resources Association (ELRA), Genoa, Italy (2006)

11. Freeman, L.C.: Centrality in social networks: conceptual clarification. Soc. Netw. **1**(3), 215–239 (1978)

12. Gazzetta Ufficiale della Repubblica Italiana: Legislative Decree n. 254 (2016). https://www.gazzettaufficiale.it/eli/id/2017/01/10/17G00002/sg. Accessed Apr 2023

13. Glavic, P., Lukman, R.: Review of sustainability terms and their definitions. J. Clean. Prod. **15**(18), 1875–1885 (2007). https://doi.org/10.1016/j.jclepro.2006.12.006

14. Henry, E.: Are investors influenced by how earnings press releases are written? J. Bus. Commun. **45**(4), 363–407 (2008). https://doi.org/10.1177/0021943608319388

15. Hilbert, M., Barnett, G., Blumenstock, J., Contractor, N., Diesner, J., Frey, S., et al.: Computational communication science: a methodological catalyzer for a maturing discipline. Int. J. Commun. **13**, 3912–3934 (2019)

16. Hu, M., Liu, B.: Mining opinion features in customer reviews. In: Proceedings of the AAAI Conference on Artificial Intelligence, $19^t h$, pp. 755–760 (2004)

17. Hu, Y., Li, W.: Document sentiment classification by exploring description model of topical. Comput. Speech Lang. **25**(2), 386–403 (2011). https://doi.org/10.1016/j.csl.2010.07.004

18. Kang, H., Yoo, S.J., Han, D.: Senti-lexicon and improved Naive Bayes algorithms for sentiment analysis of restaurant reviews. Expert Syst. Appl. **39**(5), 6000–6010 (2012). https://doi.org/10.1016/j.eswa.2011.11.107

19. Kleinnijenhuis, J., Van Hoof, A.M., Van Atteveldt, W.: The combined effects of mass media and social media on political perceptions and preferences. J. Commun. **69**(6), 650–673 (2019). https://doi.org/10.1093/joc/jqz038

20. Lagasio, V., Cucari, N., Åberg, C.: How corporate social responsibility initiatives affect the choice of a bank: empirical evidence of Italian context. Corp. Soc. Responsib. Environ. Manage. **28**(4), 1348–1359 (2021). https://doi.org/10.1002/csr.2162

21. Lengauer, G., Esser, F., Berganza, R.: Negativity in political news: a review of concepts, operationalizations and key findings. Journalism: Theory Pract. Criticism **69**(6), 179–202 (2012) https://doi.org/10.1177/1464884911427800

22. Liu, B.: Sentiment analysis and subjectivity. In: Indurkhya, N. and Damerau, J. (eds.) Handbook of Natural Language Processing (Second Edition), pp. 627–666 (2010)

23. Loughran, T., Mcdonald, B.: When is a liability not a liability? textual analysis, dictionaries, and 10-Ks. J. Finan. **66**(6), 35–65 (2011). https://doi.org/10.1111/j.1540-6261.2010.01625. x

24. Margolin, D.B.: Computational contributions: a symbiotic approach to integrating big, observational data studies into the communication field. Commun. Methods Meas. **13**(1), 229–247 (2019). https://doi.org/10.1080/19312458.2019.1639144

25. Matthes, J., Kohring, M.: The content analysis of media frames: toward improving reliability and validity. J. Commun. **58**(2), 258–279 (2008). https://doi.org/10.1111/j.1460-2466.2008. 00384.x

26. McWilliams, A., Siegel, D.S.: Creating and capturing value: strategic corporate social responsibility, resource-based theory, and sustainable competitive advantage. J. Manage. **37**(5), 1480–1495 (2011)

27. Meriton, X.: Domain independence of Machine Learning and lexicon based methods in sentiment analysis (2020). http://essay.utwente.nl/81995/. Accessed 4 Apr 2023
28. Mohammad, S., Turney, P.: Emotions evoked by common words and phrases: using mechanical Turk to create an emotion lexicon. In: Proceedings of the NAACL-HLT 2010 Workshop on Computational Approaches to Analysis and Generation of Emotion in Text, pp. 26–34. Association for Computational Linguistics, Los Angeles, CA (2010)
29. Morsing, M., Schultz, M.: Corporate social responsibility communication: stakeholder information, response and involvement strategies. Bus. Ethics: Eur. Rev. **15**(42), 323–338 (2006). https://doi.org/10.1111/j.1467-8608.2006.00460.x
30. Mućko, P.: Sentiment analysis of CSR disclosures in annual reports of EU companies. Procedia Comput. Sci. **192**, 3351–3359 (2021). https://doi.org/10.1016/j.procs.2021.09.108
31. Nielsen, F.A.: A new ANEW: evaluation of a word list for sentiment analysis in microblogs. In: Proceedings of the ESWC2011, Workshop on Making Sense of Microposts: Big Things Come in Small Packages, pp. 93-98 (2011) https://doi.org/10.48550/arXiv.1103.2903
32. Ortigosa-Hernandez, J., Rodriguez, J.D., Alzate, L., et al.: Approaching sentiment analysis by using semi-supervised learning of multi-dimensional classifiers. Neurocomputing **92**(90), 98–115 (2012). https://doi.org/10.1016/j.neucom.2012.01.030
33. Pang, B., Lee, L.: Opinion mining and sentiment analysis. Found. Trends Inf. Retrieval **192**, 1–135 (2008)
34. Pianta, E., Bentivogli, L. Girardi, C.: MultiWordNet: developing an aligned multilingual database. In: Proceedings of the $1^{s}t$ International WordNet Conference, Mysore, India, pp. 293–302 (2002)
35. Plutchik, R.: A general psychoevolutionary theory of emotion. In: R. Plutchik & H. Kellerman (eds.) Emotion: Theory, Research and Experience. Theories of Emotion, pp. 3–33 (1980)
36. Rodgers, S., Thorson, E.: A socialization perspective on male and female reporting. J. Commun. **53**(4), 658–675 (2003). https://doi.org/10.1111/j.1460-2466.2003.tb02916.x
37. Shin, J., Thorson, K.: Partisan selective sharing: the biased diffusion of fact-checking messages on social media. J. Commun. **67**(24), 233–255 (2017). https://doi.org/10.1111/jcom.12284
38. Song, Y., Wang, H., Zhu, M.: Sustainable strategy for corporate governance based on the sentiment analysis of financial reports with CSR. Financ. Innov. **4**(1), 1–14 (2018). https://doi.org/10.1186/s40854-018-0086-0
39. Steurer, R., Langer, M.E., Konrad, A., Martinuzzi, A.: Corporations, stakeholders and sustainable development: a theoretical exploration of business-society relations. J. Bus. Ethics **61**, 263–281 (2005)
40. Tong, R.: An operational system for detecting and tracking opinions in on-line discussions. In: Working Notes of the ACM SIGIR 2001 Workshop on Operational Text Classification, pp. 1–6 (2001)
41. Turney, P.D., Littman, M.L.: Measuring praise and criticism: inference of semantic orientation from association. ACM Trans. Inf. Syst. **21**(4), 315–346 (2003). https://doi.org/10.1145/944012.944013
42. Van Atteveldt, W., Welbers, K., Van der Velden, M.A.C.G.: Studying political decision-making with automatic text analysis. In: Oxford Research Encyclopedia of Politics (2019). https://doi.org/10.1093/acrefore/9780190228637.013.957
43. Vassallo, M., et al.: Polarity Imbalance in Lexicon-based sentiment analysis. In: Proceedings of the Seventh Italian Conference on Computational Linguistics CLiC-it 2020, Bologna (2020)
44. Wasserman, S., Faust, K.: Social Network Analysis: Methods and Applications. Cambridge University Press, Cambridge (1994)

Author Index

D. Conte et al. (Eds.): DeLTA 2023, CCIS 1875, pp. 481–482, 2023.
https://doi.org/10.1007/978-3-031-39059-3

Printed in the United States
by Baker & Taylor Publisher Services